Is the Style Readable?

☐ Each sentence understandable on *first* reading (251)
☐ The most information expressed in the fewest words (263)
☐ Sentences varied in construction and length (274)
☐ Each word chosen for exactness (279)
☐ All definitions double-checked (382)
☐ Abstractions and generalizations replaced by concrete, specific, and exact language (286)
☐ No triteness, overstatements, euphemisms or inappropriate jargon (283)
☐ Tone unbiased and appropriate (291)

Are Form, Format, Visuals and Mechanics Appropriate?

☐ The best document form (letter, memo, report) for the stated purpose and audience (461)
☐ An inviting and accessible format: white space, typeface, etc. (346)
☐ Adequate, clear, and informative headings (359)
☐ Adequate visuals, to clarify meaning and create interest (302)
☐ All visuals properly introduced, integrated, and discussed (338)
☐ All pages numbered and in order (352)
☐ All needed supplements: title page, abstract, etc. (367)
☐ Correct spelling, punctuation, and grammar (592)

Technical Writing

Sixth Edition

John M. Lannon

University of Massachusetts, Dartmouth

HarperCollins*CollegePublishers*

Senior Acquisitions Editor: Jane Kinney
Developmental Editor: Marisa L. L'Heureux
Production Coordination, Text, and Cover Design: York Production Services
Compositor: Compset Inc.
Printer and Binder: Malloy Lithographing, Inc.
Cover Printer: The Lehigh Press, Inc.

Technical Writing, Sixth Edition

Copyright © 1994 by John Michael Lannon

Library of Congress Cataloging-in-Publication Data

Lannon, John M.
    Technical writing / John M. Lannon.—6th ed.
      p.   cm.
    Includes bibliographical references (p.) and index.
    ISBN 0-673-52294-6
    1. Technical writing.   I. Title.
  T11.L24 1993   93-36987
  808'.0666—dc20   CIP

96 9 8 7 6 5

Brief Contents

Contents

PART IV   GRAPHIC AND DESIGN ELEMENTS   301

Chapter 14   Designing Visuals   302

Chapter 18 Descriptions and Specifications 404

Chapter 19 Procedures and Processes 427

Preface

Technical Writing, sixth edition, is a comprehensive and flexible introduction to technical and professional communication. Designed for classes in which students from a variety of majors are enrolled, the book addresses a wide range of interests. Rhetorical principles are explained, illustrated, and applied to an array of assignments, from brief memos and summaries to formal reports and proposals. To help students develop awareness of audience and accountability, exercises embody the writing demands that are typical throughout college and on the job.

ORGANIZATION

Following a brief overview of technical writing in Chapter 1, the remaining text has five major sections:

Part I: Writing for Readers in the Workplace treats job-related writing as a problem-solving process. Students learn to think critically about the informative, persuasive, and ethical dimensions of their communications. Also, they learn about adapting to rapidly changing communication technologies, to interpersonal challenges of collaborative writing, and to the various needs and expectations of global audiences.

Part II: Information Retrieval, Analysis, and Synthesis treats research as a deliberate inquiry process. Students learn to formulate significant research questions; to explore primary and secondary sources; to record, evaluate, interpret, and document their findings; and to summarize for economy, accuracy, and emphasis.

Part III: Sequence, Shape, and Style in a Document demonstrates strategies for organizing and expressing messages that readers can follow and understand. Students learn to control their material and to develop a style that connects with readers.

Part IV: Graphic and Design Elements treats the rhetorical implications of graphics, page design, and document supplements. Students learn to enhance a document's access, appeal, visual impact, and usability.

Part V: Specific Documents and Applications applies earlier concepts and strategies to the composing of technical documents. Various letters, memos, reports, and proposals offer a balance of examples from the workplace and from student writing. Each sample document has been chosen so that students can emulate it easily.

Finally, the **appendixes** offer a brief handbook of grammar, usage, and mechanics; interviews with four writers on the job; and a sample proposal, progress report, and final report for an actual workplace project.

THE FOUNDATIONS OF *TECHNICAL WRITING*

- More than a value-neutral exercise in "information transfer," workplace writing typically is a complex social transaction. Each rhetorical situation places specific interpersonal, ethical, legal, and cultural demands on the writer.

- Writers with no rhetorical awareness overlook the decisions that are crucial for effective writing. Only by defining their rhetorical problem and asking the important questions can writers formulate an effective response.

- As well as being *communicators*, today's workplace professionals increasingly are *consumers* of information, who need to be skilled in the methods of inquiry, retrieval, evaluation, and interpretation that comprise the research process.

- Although it follows no single, predictable sequence, the writing process is not a collection of random activities; rather, it is a set of deliberate decisions in problem solving. Beyond emulating this or that model document, students need to understand that effective writing requires critical thinking.

- A technical writing classroom typically contains an assortment of students with varied backgrounds. The textbook, then, should offer explanations that are thorough, examples and models that are broadly intelligible, and goals that are rigorous yet collectively achievable. And the book should be flexible enough to allow for various course plans.

- As an alternative to reiterating the textbook material, classroom workshops apply textbook principles by focusing on the students' writing. These workshops call for an accessible, readable, and engaging book to serve as a comprehensive reference.

NEW TO THIS EDITION

- New material (in Chapter 5) on critical thinking about ethical issues: avoiding fallacies, applying reasonable criteria for ethical judgment, confronting ethical dilemmas, avoiding communication abuse, deciding when and how to take an ethical stand, and anticipating the consequences.

- A new chapter (6) on adapting to communication technology, collaborative relationships, and global audiences. Coverage includes telecommuting, electronic mail, global networks, hypermedia, paperless documents, online documentation, hypertext applications in writing and research; the social nature of collaboration, conflict in collaborative groups, guidelines for managing and evaluating collaborative projects; cultural influences on audience expectations and interpretations, guidelines for analyzing multicultural audiences, a checklist for intercultural documents.

- Two fully revised chapters (8 and 9) on research methods for the information age. New coverage includes thinking critically about the research process; designing a focused and balanced inquiry; evaluating and interpreting findings; exploring automated resources; designing surveys; assessing validity, reliability, and certainty; recognizing the influence of bias (in database sources, direct observation, and interpretation); avoiding causal and statistical fallacies; paraphrasing and integrating quoted material; choosing a system of parenthetical documentation (MLA, APA, or numerical); reassessing one's research process (a checklist).

- A new section (in Chapter 10) on distinctions among forms of summarized information: closing summary, informative abstract, executive summary, and descriptive abstract.

- A new section (in Chapter 13) on the rhetorical implications of sexist usage.

- A new section (in Chapter 19) on assessing a document's usability according to "human factors."

- A new section (in Chapter 23) on the role of critical thinking in the formulation, evaluation, and refinement of recommendations.

- A new analytical report (in Chapter 23) on hazards posed by electromagnetic radiation.

- New material (in Chapter 24) on peer evaluation of oral reports.

- More on rhetorical, legal, and ethical considerations in word choice, definitions, product descriptions, instructions, and other forms of communication.

- A new art program and greater emphasis on visual communication.

- More annotated writing samples, to highlight rhetorical features.

- More applications suitable for collaborative work.

- A comprehensive educational package including an instructor's manual with test bank, chapter quizzes, and master sheets for overhead or opaque projection, and a packet of acetate transparencies.

ACKNOWLEDGEMENTS

Many of the refinements in this edition were inspired by generous and insightful suggestions from the following reviewers: Debby Andrews, University of Delaware; Nancy A. Engemann; Pamela Gardner, Salt Lake Community College; Nancy MacKenzie, Mankato State University; Mary Massirer, Baylor University; Fiore Pugliano, University of Pittsburgh; Mark Rollins, Ohio University; Lorraine Saar, Macomb Community College; Tess Scogan, Auburn University; Edith K. Weinstein, Community and Technical College of the University of Akron. Thank you all.

At the University of Massachusetts, Raymond Dumont was a constant source of help and ideas. For much of the new material on research, I am indebted to librarians Shaleen Barnes, Ross LaBaugh, and Charles McNeil. Many other colleagues, graduate students, teaching assistants, and alumni offered countless suggestions. And my students gave me feedback and inspiration.

Thanks to editors Jane Kinney and Marisa L'Heureux of HarperCollins Publishers for their expertise and commitment to this project. Thanks also go to Andrew Roney of HarperCollins Publishers and Susan Bogle of York Production Services for their thoughtful attention to every detail, and especially to Angela Gladfelter of York Production Services, whose humor, efficiency, and skill made a difficult production job much easier for all of us.

A special thank-you to those I love: Chega, Daniel, Sarah, Patrick, and Max.

John M. Lannon

Introduction to Technical Writing

Technical Writing Serves Practical Needs

Writing Is Part of Most Careers

Writers Today Need Better Skills Than Ever

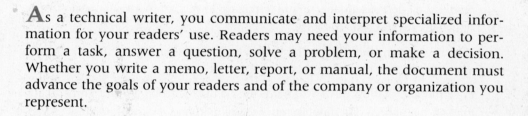

As a technical writer, you communicate and interpret specialized information for your readers' use. Readers may need your information to perform a task, answer a question, solve a problem, or make a decision. Whether you write a memo, letter, report, or manual, the document must advance the goals of your readers and of the company or organization you represent.

TECHNICAL WRITING SERVES PRACTICAL NEEDS

Unlike poetry or fiction, which appeal mainly to our *imagination*, technical documents appeal to our *understanding*. Technical writing therefore rarely seeks to entertain, create suspense, or invite differing interpretations. Those of you who have written any type of lab or research report already know that technical writing has little room for ambiguity.

To serve practical needs in the workplace, technical documents must be reader oriented and efficient.

Technical Documents Are Reader Oriented

Instead of focusing on the writer's desire for self-expression, a technical document addresses the reader's desire for information. This doesn't mean your writing should sound like something produced by a robot, without any personality (or *voice*) at all. Your document may in fact reveal a lot about you (your competence, knowledge, integrity, and so on), but it rarely focuses on you personally. Readers are interested in *what you have done, in what you recommend,* or in *how you speak for your company;* they have only a professional interest in *who you are* (your feelings, hopes, dreams, visions). A personal essay, then, would not be technical writing. Consider this essay fragment:

Focuses on the feelings

> Computers are not a particularly forgiving breed. The wrong key struck or the wrong command typed is almost sure to avenge itself on the inattentive user by banishing the document to some electronic trash can.

This personal view communicates a good deal about the writer's resentment and anxiety but very little about computers themselves.

The following example can be called technical writing because it focuses (see italics) on the subject, on what the writer has done, and on what the reader should do:

Focuses on the subject

> On VR 320 terminals, *the BREAK key* is adjacent to keys used for text editing and special functions. Too often, users inadvertently strike the BREAK key, causing the program to quit prematurely. To prevent the problem, *we have* modified all database management terminals: to quit a program, *you must* now strike BREAK twice successively.

This next example also can be called technical writing because it focuses on what the writer recommends:

Focuses on the recommendation

> I recommend that our VAX 780 hardware be upgraded by a maximum addition to main memory, a disk input/output control unit, and an additional disk storage unit. This expansion will (1) increase the number of simultaneous terminal users from 60 to 80, (2) increase the system's responsiveness, and (3) provide sorely needed disk storage for word processing and company databases.

And so your document never makes you "disappear," but it does focus on that which is most important to the readers.

Technical Documents Strive for Efficiency

Professors read to *test* our knowledge; colleagues, customers, and supervisors read to *use* our knowledge. Workplace readers hate waste and demand efficiency; they might not read a document from beginning to end. They want only as much as they need: "When it comes to memos, letters, proposals, and reports, there's no extra credit for extra words. And no praise for elegant prose. Bosses want employees to get to the point—quickly, clearly, and concisely" (Spruell 32). Efficient documents save time and energy in the workplace.

In any system, efficiency is the ratio of useful output to input. For the product that comes out, how much energy goes in?

ENERGY
(input) ⟶ SYSTEM ⟶ PRODUCT
(output)

When a system is efficient, the output nearly equals the input.

Similarly, a document's efficiency can be measured by how hard readers work to understand the message. Is the product worth the reader's effort?

READER SPENDS
ENERGY ⟶ DOCUMENT ⟶ READER GETS
THE MESSAGE
(input) (output)

No reader should have to spend ten minutes deciphering a message worth only five minutes. Consider, for example, this wordy message:

> At this point in time, we are presently awaiting an on-site inspection by vendor representatives relative to electrical utilization adaptations necessary for the new computer installation. Meanwhile, all staff are asked to respect the off-limits designation of said location, as requested, due to liability insurance provisions requiring the on-line status of the computer.

An inefficient message

Inefficient documents drain the readers' energy; they are too easily misinterpreted; they waste time and money. Notice how hard we had to work with the message above to extract information that could be expressed this efficiently:

> Hardware consultants soon will inspect our new computer room to recommend appropriate wiring. Because our insurance covers only an *operational* computer, this room must remain off limits until the computer is fully installed.

A more efficient message

When readers sense they are working too hard, they tune out the message—or they stop reading altogether.

Inefficient documents have varied origins. Even when the information is accurate, errors like the following make readers work too hard:

- more (or less) information than readers need
- irrelevant or uninterpreted information
- confusing organization
- jargon or vague technical expressions readers cannot understand
- more words than readers need

- uninviting appearance or confusing layout
- no visual aids when readers need them

Far more than a *list* of information, an efficient document sorts, organizes, and interprets that information to suit the audience's needs, abilities, and interests.

Instead of merely "happening," an efficient document is carefully designed to include these elements:

- *content* that makes the document worth reading
- *organization* that guides readers and emphasizes important material
- *style* that is economical and easy to read
- *visuals* (graphs, diagrams) that clarify concepts and relationships
- *format* (layout, typeface) that is accessible and appealing
- *supplements* (abstracts, appendices) that enable readers with different needs to read only those sections required for their work

Reader orientation and efficiency are more than abstract "rules": Writers are accountable for their documents. In questions of liability, faulty writing is no different from any other faulty product. If your inaccurate or unclear or incomplete information leads to injury or damage or loss, *you* can be held legally responsible.

WRITING IS PART OF MOST CAREERS

Although you might not anticipate a career as a "writer," your writing skills will be tested routinely in situations like these:

- proposing various projects to management or to clients
- writing progress reports to the boss
- contributing articles to the employee newsletter
- describing a product to employees or customers
- writing procedures and instructions for employees or customers
- justifying to management a request for funding or personnel
- editing and reviewing documents written by colleagues
- designing material that will be read on a computer screen

You might write alone or as part of a collaborative team.

Your value to any organization will depend on how clearly and persuasively you communicate. Many working professionals spend at least 40 percent of their time writing or dealing with someone else's writing. The higher their position, the more they write (Barnum and Fisher 9–11). Here is a corporate executive's description of some audiences you can expect:

> The technical graduate entering industry today will, in all probability, spend a portion of his or her career explaining technology to lawyers—some friendly and some not—to consumers, to legislators or judges, to bureaucrats, to environmentalists and to representatives of the press.   (Florman 23)

Audiences differ in needs and expectations

And these audiences, among countless others, expect to read efficient documents.

Here is what two top managers for an auto maker say about the effect a document can have on the organization *and* on the writer:

> A written report is often the only record which is made of results that have come out of years of thought and effort. It is used to judge the value of the *person's* work and serves as the foundation for all future action on the project. If it is written clearly and precisely, it is accepted as the result of sound reasoning and careful observation. If it is poorly written, the results presented in it are placed in a bad light and are often dismissed as the work of a careless or incompetent worker.   (Richards and Richards 6)

Writing is an indicator of job performance

Good writing gives you and your ideas *visibility* and *authority* within your organization. Bad writing, on the other hand, is not only useless to readers and politically damaging to the writer, but also expensive: The estimated cost of communication in American business and industry is more than $75 billion yearly. And roughly 60 percent of the writing is inefficient: unclear, misleading, irrelevant, deceptive, or otherwise wasteful of time and money (Max 5–6).

Whatever your career plans, you can expect to be a part-time technical writer. Employers first judge your writing by your application letter and résumé. In a large organization, your future may be decided by executives you've never met. One concrete measure of your job performance will be your letters, memos, and reports. As you advance, communication skill becomes more important than technical background. The higher your goals, the better you need to communicate.

WRITERS TODAY NEED BETTER SKILLS THAN EVER

Computers have brought us the Information Age. Information has become our ultimate product—a product whose volume more than doubles in each decade.

In education, industry, business, and government, people create, gather, analyze, and distribute information electronically around the world. But whether the information itself finally appears on a printed page or a computer screen, it needs to be *written*. A computer can transmit data (or facts), but it cannot give *meaning* to the information—only the writer and reader can do so.

Automation has increased the speed and volume of communication, but excessive information actually can *impede* or *prevent* communication. Despite our advances in communication technology, information still needs to be "processed" by the human brain, as depicted in Figure 1.1. Unfortunately, it is easy for people to process information erroneously (to interpret the wrong *meaning*).

Accidents that make headlines often result from human error in processing information—situations in which the complexity of information overwhelms the person receiving it (Wickens 2). The 1979 release of radiation and near-meltdown at Three-Mile Island; the 1984 chemical explosion in Bhopal, India; the 1989 runway collision at Los Angeles airport—these disasters occurred because vital information had been misunderstood. Similar but less publicized disasters occur routinely in the workplace. Today, more than ever, writers need to sort, organize, and interpret their material so readers can understand it.

Which information is relevant to this situation?
Which information is most important?
What does it mean?
How does it relate to earlier information?
What should be done?

FIGURE 1.1 The Brain Processes Information, to Find Meaning

Today's readers often lack technical expertise, but they do *use* the technology. Some "nontechnical" workers who rely on technology include the manager using a teleconferencing network, the data-entry clerk using a computer terminal, or the bank teller using a check-verification system. Writers tell laypeople how to use software, or how to operate hardware, medical equipment, telephone systems, precision instruments, and all kinds of automated products, from computer games to microwave ovens. With so much information required, and so much available, no writer can afford to "let the facts speak for themselves."

EXERCISES

1. Locate a brief example of a technical document (or a section of one). Make a photocopy, bring it to class, and explain why your selection can be called technical writing.

2. Research the kinds of writing you will do in your future career. (Begin with the *Dictionary of Occupational Titles* in your library.) You might interview a member of your chosen profession. Why will you write on the job? For whom will you write? Explain in a memo to your instructor. (See pages 498–499 for memo elements and format.)

3. In a memo to your instructor, describe the skills you seek to develop in your technical writing course. How exactly will you apply these skills to your career?

4. Assume a friend in your major thinks that writing skills are obsolete, and that secretaries or word processors can "fix" any writing. Write your friend a letter, explaining why you think these assumptions are mistaken. Use examples to support your position. (See pages 461–472 for letter elements and format.)

5. Write a memo to your boss, justifying reimbursement for this course. Explain how the course will help you become more effective on the job.

COLLABORATIVE PROJECT

Introducing a Classmate

Class members often will work together this semester. So that everyone can become acquainted, your task is to introduce to the class the person seated next to you. (That person, in turn, will introduce you.) To prepare your introduction, follow this procedure:

   a. Exchange with your neighbor whatever personal information you think the class needs: background, major, career plans, course goals, and so on. Each person gets five minutes to tell her or his story.

    b. Take careful notes; ask questions if you need to.

    c. Take your notes home and select only what you think will be useful to the class.

    d. Prepare a one-page memo telling your classmates *who this person is.* (See pages 498–499 for memo elements and format.)

    e. Ask your neighbor to review the memo for accuracy; revise as needed.

    f. Present the class with a 5-minute oral paraphrase of your memo, and submit a copy of the memo to your instructor.

Writing for Readers in the Workplace

Problem Solving in Workplace Writing

Technical Writers Face Four Related Problems

Problem Solving Requires Critical Thinking

■　　　■　　　■

Virtually all professionals specialize in solving problems (how to build this bridge, how to repair that equipment, how to improve this product, how to diagnose that ailment). And central to all specialized activity is the problem of *communicating* effectively—a complex problem we might examine more easily by considering several subordinate problems.

TECHNICAL WRITERS FACE FOUR RELATED PROBLEMS

No matter how sophisticated our communication technology, computers cannot *think* for us. More specifically, computers cannot offer solutions to these four problems:

- the **information problem**, because different readers have different information needs in different situations
- the **persuasion problem**, because people often disagree about what the information means and about what should be done
- the **ethics problem**, because the interests of your company may conflict with the interests of your readers
- the **adaptation problem**, because communication technology requires writers to cope with new tools, working relations, and audiences

Chapter 1 explained that information has to have meaning for its audience. But people differ in their interpretations of the facts, and so they may need persuading that one viewpoint is preferable to another. Persuasion, however, is a powerful and often unethical strategy. Even the most "useful" and "efficient" document could be deceptive or harmful to certain people. Therefore, solving the persuasion problem doesn't mean manipulating readers by using "whatever works," but rather building a case from honest and reasonable interpretation of the facts.

Beyond these traditional problems *(information, persuasion, ethics)*, is that of adapting to profound changes in communication technology. We are expected to master electronic *tools* that revolutionize the way information is obtained, written, distributed, and the way much of it is read (on computer screens instead of printed pages). Often we are expected to write in *collaboration* with a team, linked companywide or worldwide via computer network. Finally, in this age of multinational corporations and global markets, we are expected to address *audiences* from other cultures. Figure 2.1 offers one way of visualizing how these four problems relate.

The following scenario illustrates how a typical professional confronts the fourfold problem of communicating in the workplace.

The Information Problem

Sarah Burnes has worked two months as a chemical engineer for Millisun, a leading maker of cameras, multipurpose film, and photographic equipment. Sarah's first major assignment is to evaluate the plant's incoming and outgoing water. (Contaminants in water can taint film during production, and the production process itself can pollute outgoing water.) Management wants an answer to this question: How often should we change water filters? The filters are very expensive and difficult to change, halting the production process for up to a whole day at a time. The company wants as much "mileage" as possible from these filters, without incurring government fines or tainting the film production.

Sarah will study endless printouts of chemical analysis, review current research, and do some testing of her own. When she finally decides on what all the data means, Sarah will prepare a recommendation report for her bosses.

Later, she will write a manual, instructing employees how to check and change the filters. Trying to cut costs, the company has asked Sarah to design this manual using its new desktop publishing system.

"Can I provide accurate and useful information?"

Sarah's report, above all, needs to be accurate; otherwise, the company gets fined or lowers production. Once she has "processed" all the information, she has to solve the problem of giving readers what they need: *How much explaining should I do? How will I organize? Do I need visuals?* And so on.

In other situations, Sarah will face a persuasion problem as well: for example, when decisions must be made or actions taken on the basis of

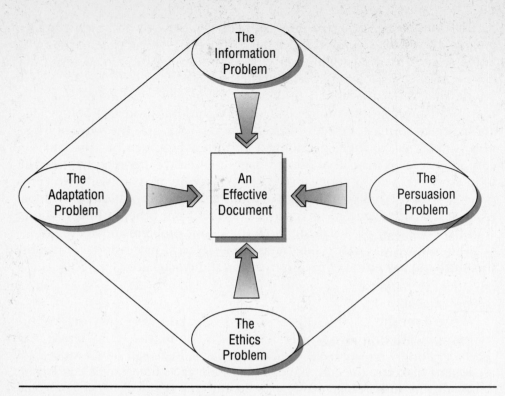

FIGURE 2.1 Writers Face Four Related Problems

incomplete or inconclusive facts or conflicting interpretations (Hauser 72). In these instances, Sarah will need to win reader acceptance for *her* view. Her writing will have to be persuasive as well as informative.

The Persuasion Problem

Millisun and other electronics producers are located on the shores of a small harbor, the port for a major fishing fleet. For twenty years prior to 1980, these companies discharged directly into the harbor effluents containing metal compounds, PCB's, and other toxins. Sarah is on a team from these companies, assigned to work with the Environmental Protection Agency to clean up the harbor. Much of the team's collaboration occurs via electronic mail.

Enraged citizens are demanding immediate action, and the companies themselves are anxious to put this public-relations nightmare behind them. But in its analysis, Sarah's team discovers that any type of cleanup would stir up harbor sediment, possibly dispersing the pollution into surrounding waters and the atmosphere. (Many of the contaminants are airborne.) Therefore, premature action actually might *increase* danger. But team members disagree on the degree of risk and on how to proceed.

Sarah's communication here takes on a persuasive dimension: She and her team members first have to resolve their own conflicts and produce an environmental-impact report that reflects the team's consensus. If the report recommends further study, Sarah will have to justify the delays to her bosses and the public-relations office. She will have to make readers understand the dangers as well as she understands them.

In the above situation, the facts are neither complete nor conclusive, and views conflict about what these facts mean. Sarah will have to balance the various "political" pressures and make a case for *her* interpretation. Moreover, as company *spokesperson*, Sarah will be expected to take a position that protects her company's interests. Some elements of Sarah's persuasion problem: *Are other interpretations possible? Is there a better way? Can I expect political fallout?*

Whenever she writes, Sarah will have to reckon with the ethical implications of her writing, with the question of "doing the right thing." And some situations will present hard choices. For instance, Sarah might feel pressured to overlook or sugarcoat or suppress facts that would be costly or embarrassing to her company. Or sometimes the best technical solution to a problem might be a poor solution in human terms (as when a heavy industry decreases local pollution by building a smokestack to disperse the emissions over hundreds of miles).

The Ethics Problem

To ensure compliance with OSHA[1] standards for worker safety, Sarah is assigned to test the air purification system in Millisun's chemical division. After finding the filters hopelessly clogged, she decides to test the air quality and discovers dangerous levels of benzene (a potent carcinogen). She reports these findings in a memo to the production manager, with an urgent recommendation that all employees be tested for benzene poisoning. The manager phones and tells Sarah to "have the filters replaced, and forget about it." Now Sarah has to decide what to do next: bury the memo in some file cabinet or defy her boss and send copies to other readers who might take action.

"Can I be honest and still keep my job?"

---

[1]Occupational Safety and Health Administration.

Situations that jeopardize truth and fairness present the hardest choices of all: remain silent and look the other way or speak out and risk being fired. Some elements of Sarah's ethics problem: *Is this fair? Who might benefit or suffer? What other consequences could this have?*

In addition to solving these various problems, Sarah has to reckon with the implications of communication technology: Much of her writing, done at the computer, will be produced in collaboration with others (editors, managers, graphic artists); and her audience will extend beyond her own culture.

_____

The Adaptation Problem

Millisun, because of recent mergers, has become a multinational corporation, with branches in eleven countries, all connected by computer network. Sarah can expect to collaborate with co-workers from diverse cultures on research and development, with government agencies of the host countries on safety issues, patents and licensing rights, product liability laws, and environmental concerns.

In order to standardize the sensitive management of the toxic, volatile, and even explosive chemicals used in film production, Millisun is developing automated procedures for quality-control, troubleshooting, and emergency response to chemical leakage. Sarah has been assigned to work with a group of colleagues to prepare computer-based instructional packages for all personnel involved in Millisun's chemical management worldwide.

_____

Besides learning to use the latest communication technology, Sarah will have to develop working relationships with people she has never met, people she knows only via a machine. Some elements of her adaptation problem: *How do I keep up with the technology? How can I get along with all the people on this project? How do I connect with audiences with whom I share no common language or culture?*

For Sarah Burnes or any of us, writing is a process of *discovering* what we want to say, "a way to end up thinking something [we] couldn't have started out thinking" (Elbow 15). Throughout this process in the workplace, we rarely work alone, but instead rely on others for information, help in writing, and feedback (Grice 29–30). We must satisfy not only our audience, but also our employer, whose goals and values ultimately shape the document (Selzer 46–47). And so almost any writing for readers outside our organization will be *reviewed* before it is finally approved.

PROBLEM SOLVING REQUIRES CRITICAL THINKING

Because writing is nothing less than *thinking* on paper, it is never merely a "by-the-numbers" exercise. But even though we cannot boil writing down to a formula, we can improve our chances for success through the strategy of *critical thinking*.

Critical thinking is a process of testing the worth of information or the strength of our ideas. Instead of accepting the idea at face value, we examine, evaluate, verify, analyze, weigh alternatives, and consider consequences—at every stage of that idea's development. We employ critical thinking to examine our evidence and our reasoning, to solve problems, and to test the effectiveness of our solutions.

To solve writing problems as critical thinkers, we carry out the four stages of the *writing process*:

1. We work with the information.
2. We plan the document.

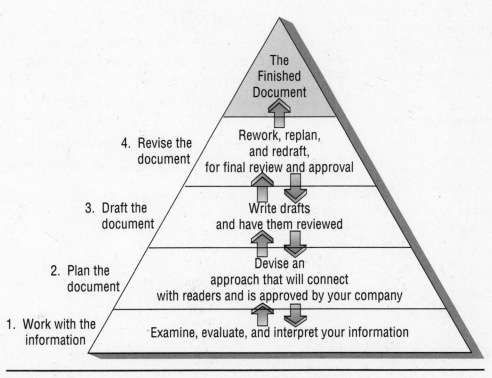

FIGURE 2.2 A Critical-Thinking Model for Technical Writing

1. Work with the information:

- Have I defined the problem accurately?
- Is the information complete, accurate, reliable, and unbiased?
- Can it be verified?
- How much of it is useful?
- Do I need more information?
- What do these facts mean?
- Do the facts conflict?
- Are other interpretations or conclusions possible?
- Is a balance of viewpoints represented?
- What, if anything, should be done?
- Is it honest and fair?
- Is there a better way?
- Who might benefit or suffer?
- What other consequences might this have?
- Should I reconsider?

2. Plan your document:

- When is it due?
- What do I want it to do?
- Who is my audience, and why will they read it?
- What do they need to know?
- What are the "political realities" (feelings, egos, cultural differences, etc.)?
- How will I organize?
- What format and visuals should I use?
- Whose help will I need?

3. Draft your document:

- How do I begin, and what comes next?
- How much is enough?
- What can I leave out?
- Am I forgetting anything?
- How will I end?
- Who needs to review my drafts?

4. Revise your document:

- How does this draft measure up?
- Does it do what I want it to do?
- Is the content worthwhile?
- Is the organization sensible?
- Is the style readable?
- Is everything easy to find?
- Does everything look good?
- Is everything accurate, complete, appropriate, and correct?
- Who needs to review and approve the final version?
- Does it advance my organization's goals?
- Does it advance my audience's goals?

FIGURE 2.3 Critical Thinking in the Writing Process

3. We draft the document.

4. We revise the document.

Each stage calls for deliberate decisions. And during this process our best ideas often occur (Dumont 14).

One engineering professional describes how critical thinking enriches every stage of the writing process:

> Good writing is a process of thinking, writing, revising, thinking, and revising, until the idea is fully developed. An engineer can develop better perspectives and even new technical concepts when writing a report of a project. Many an engineer, at the completion of a laboratory project, senses a new interpretation or sees a defect in the results and goes back to the laboratory for additional data, a more thorough analysis, or a modified design.   (Franke 13)

Writing sharpens thinking

Figure 2.2 depicts critical thinking during the writing process. As the arrows indicate, no one stage of the process is complete until *all* stages are complete. Like the exposed tip of an iceberg, the finished document provides the only visible evidence of our labor. On the job, we often need to complete these stages under pressure of deadline.

Figure 2.3 lists the kinds of questions we answer in moving through the writing process. Chapter 7 will illustrate one writer's critical thinking as he prepares a sensitive report for his bosses.

EXERCISE (Individual or collaborative)

Appendix C (pages 648–667) shows a sequence of documents prepared in response to a workplace writing situation. Read this appendix and evaluate how effectively writer Mike Cabral considers and addresses each of the four related problems discussed in this chapter. As guidelines, use the following questions:

- What information problems does Mike face, and how does he solve them?
- What persuasion problems are involved?
- What are the ethical issues, and how does Mike address them?
- What adaptation problems does Mike confront?

In a memo (pages 498–499) to your instructor, explain your evaluation.

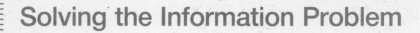

Solving the Information Problem

Assess Readers' Information Needs

Identify Levels of Technicality

Develop an Audience-and-Use Profile

Brainstorm for a Useful Message

All technical writing is for readers who will use and react to your information. You might write to *define* something—as to insurance customers who want to know what *variable annuity* means; to *describe* something—as to an architectural client who wants to know what a new addition to her home will look like; to *explain* something—as to a stereo technician who wants to know how to eliminate bass flutter in your company's new line of speakers. As depicted in Figure 3.1, you solve your information problem by enabling readers to understand exactly what you mean. And so, an essential part of your critical thinking involves **audience analysis**, to learn all you can about readers and how they will use your document.

Most readers would prefer not to have to read your document at all. They are not interested in how smart or eloquent you are, but they *do* want to find what they need, quickly and easily.

ASSESS READERS' INFORMATION NEEDS

Good writing connects with readers by recognizing their different backgrounds, needs, and preferences. A single message may appear in several versions for several audiences. For instance, an article about a new cancer treatment might appear in a medical journal read by doctors and nurses. A less technical version might be published as part of a medical textbook read by medical and nursing students. Or an even more nontechnical version might appear in *Good Housekeeping*. All three versions treat the same topic, but each meets the needs of its specific audience.

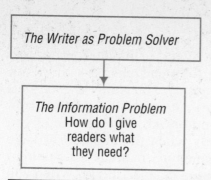

FIGURE 3.1 One Problem Confronted by Writers

Technical writing is intended to be *used*. You become the teacher and the reader becomes the student. Because your readers know less than you,[1] they will have questions:

- *What is the purpose of this document?*
- *Who should read the document?*
- *What is being described or explained?*
- *What does it look like?*
- *How do I do it?*
- *How did you do it?*
- *Why did it happen?*
- *When will it happen?*
- *Why should we do it?*
- *How much will it cost?*
- *What are the risks?*

Each specific question is a subset of one overriding question: *What, exactly, do you mean?* You always write to show a specific audience what you mean; and during the process of making your meaning clear, you often discover new meanings for yourself.

Your readers' information needs resemble your own. If Chapter 1 of your introductory math textbook covered differential equations (advanced calculus), the author would be ignoring your needs. For your background and purposes, the chapter would be useless. All useful messages connect with the reader's level of understanding.

---

[1]Sometimes, of course, you will have to deal with readers who *think* they know something, but who really do not know it well enough (see Chapter 4).

IDENTIFY LEVELS OF TECHNICALITY

When you write for a close acquaintance (friend, computer crony, psychology classmate, co-worker, engineering colleague, chemistry professor who reads your lab reports, or your supervisor), you know a good deal about your reader's background. And so you automatically adapt your report to that reader's knowledge, interests, and needs. But sometimes you write for less defined audiences, particularly when the audience is large (when you are writing a journal article, a computer manual, a set of first-aid procedures, or a report of your investigation of an accident). When you have only a general notion about your audience's background, you must decide whether your message should be *highly technical, semitechnical,* or *nontechnical.*

The Highly Technical Message

Readers at your specialized level expect the technical facts and figures they need, without long explanations. The following report of treatment given to a heart attack victim is highly technical. The writer, an emergency room physician, is reporting to the patient's doctor. This reader needs an exact record of the patient's symptoms and treatment.

A Technical Version

Mr. X was brought to the emergency room by ambulance at 1:00 A.M., September 27, 1993. The patient complained of severe chest pains, dyspnea, and vertigo. Auscultation and EKG revealed a massive cardiac infarction and pulmonary edema marked by pronounced cyanosis. Vital signs: blood pressure, 80/40; pulse 140/min; respiration, 35/min. Lab: wbc, 20,000; elevated serum transaminase; urea nitrogen, 60 mg%. Urinalysis showed 4+ protein and 4+ granular casts/field, indicating acute renal failure secondary to the hypotension.

The patient received 10 mg of morphine stat, subcutaneously, followed by nasal oxygen and 5% D & W intravenously. At 1:25 A.M. the cardiac monitor recorded an irregular sinus rhythm, indicating left ventricular fibrillation. The patient was defibrillated stat and given a 50 mg bolus of Xylocaine intravenously. A Xylocaine drip was started, and sodium bicarbonate administered until a normal heartbeat was established. By 3:00 A.M., the oscilloscope was recording a normal sinus rhythm.

As the heartbeat stabilized and cyanosis diminished, the patient received 5 cc of Heparin intravenously, to be repeated every six hours. By 5:00 A.M. the BUN had fallen to 20 mg% and vital signs had stabilized: blood pressure, 110/

60; pulse, 105/min; respiration, 22/min. The patient was now conscious and responsive.

This highly technical report is clear only to the medical expert. Because her reader has the background to understand the message, this writer defines no technical terms (pulmonary edema, sinus rhythm). Nor does she interpret lab findings (4 + protein, elevated serum transaminase). She uses abbreviations her reader understands (wbc, BUN, 5% D & W). Because her reader knows the reasons for specific treatments and medications (defibrillation, Xylocaine drip), she includes no theoretical background. Her report answers concisely the main questions she can anticipate from her reader: What happened? What treatment was given? What were the results?

The Semitechnical Message

One broad class of readers may have some technical background, but less than the experts. For instance, first-year medical students have specialized knowledge, but not as much as second-, third-, and fourth-year students. Yet students in all four groups could be considered semitechnical readers. When you write for a semitechnical audience, identify the *lowest* level of understanding in the group, and write to that level. Too much explanation is better than too little.

Here is a partial version of the earlier medical report. Written at a semitechnical level, it might appear in a textbook for first-year medical or nursing students, in a report for a medical social worker, in a patient's history for the medical technology department, or in a monthly report for the hospital administration.

A Semitechnical Version

Examination by stethoscope and electrocardiogram revealed a massive failure of the heart muscle along with fluid buildup in the lungs, which produced a cyanotic **discoloration of the lips and fingertips from lack of oxygen.**

The patient's blood pressure at 80 mm Hg (systolic)/40 mm Hg (diastolic) was **dangerously below its normal measure of 130/70.** A pulse rate of 140/minute was **almost twice the normal rate of 60—80.** Respiration at 35/ minute was more than **twice the normal rate of 12—16.**

Laboratory blood tests yielded a white blood cell count of 20,000/cu mm (normal value: 5,000—10,000), **indicating a severe inflammatory response by the heart muscle.** The elevated serum transaminase enzymes (produced in quantity only when the heart muscle fails) confirmed the earlier diagnosis. A blood urea nitrogen level of 60 mg% (normal value: 12—16 mg%) **indicated**

Informed but nonexpert readers need enough explanations to understand what the facts mean

that the kidneys had ceased to filter out metabolic waste products.
The 4+ protein and casts reported from the urinalysis (normal value: 0) **revealed that the kidney tubules were degenerating as a result of the lowered blood pressure.**

The patient immediately received **morphine to ease the chest pain,** followed by **oxygen to relieve strain on the cardiopulmonary system,** and an intravenous solution of **dextrose and water to prevent shock.**

The version explains (in boldface) the raw data. Exact dosages are not mentioned because the readers are not treating the patient. Normal values of lab tests and vital signs, however, make interpretation easier. (Expert readers would know these values.) Knowing what medications the patient received would be especially important to the lab technician, because some medications affect test results. For a nontechnical audience, however, the message needs further translation.

The Nontechnical Message

Readers with no specialized training expect technical data to be translated into terms they understand. Nontechnical readers are impatient with abstract theories but want enough background to help them make the right decision or take the right action. They are bored by long explanations but frustrated by bare facts not explained or interpreted. They expect a report that is clear on first reading, not one that requires review or study.

Following is a nontechnical version of the earlier medical report. The physician might write this version for the patient's spouse who is overseas on business, or as part of a script for a documentary film about emergency room treatment.

A Nontechnical Version

General readers
need everything
translated into
terms they
understand

Heart sounds and electrical impulses both were abnormal, **indicating a massive heart attack caused by failure of a large part of the heart muscle.** The lungs were swollen with fluid and the lips and fingertips showed **a bluish discoloration from lack of oxygen.**

Blood pressure was **dangerously low, creating the risk of shock.** Pulse and respiration were **almost twice the normal rate, indicating that the heart and lungs were being overworked** in keeping oxygenated blood circulating freely.

Blood tests confirmed the heart attack diagnosis and **indicated that waste products usually filtered out by the kidneys were building up in**

the bloodstream. Urine tests showed that the kidneys were failing as a result of the lowered blood pressure.

The patient was given **medication to ease his chest pain, oxygen to ease the strain on his heart and lungs, and intravenous solution to prevent his blood vessels from collapsing and causing irreversible shock.**

Nearly all interpretation (in boldface), this nontechnical version omits all mention of medications, lab tests, or normal values, because these have no meaning for the reader. The writer merely summarizes events and explains the causes of the crisis and the reasons for the particular treatment.

In some other situation, however (say, in a jury trial for malpractice), the nontechnical audience might need information about specific medication and treatment. Such a report would, of course, be much longer—a short course in emergency coronary treatment.

Each version of the medical report is useful *only* to readers at a specific level. Doctors and nurses have no need for the explanations in the two latter versions, but they do need the specialized data in the first. Beginning medical students and paramedics might be confused by the first version and bored by the third. Nontechnical readers would find both the first and second versions meaningless.

Primary Versus Secondary Readers

Whenever you prepare a single document for readers with diverse backgrounds, classify your readers as *primary* or *secondary*. Primary readers usually are those who requested the document and who will use it as a basis for decisions or actions. Secondary readers are those who will carry out the project, who will advise the primary readers about their decision, or who will somehow be affected by this decision. They will read your document (or perhaps only part of it) for information that will help them get the job done, or provide educated advice, or stay abreast of new developments.

Often these two audiences differ in technical background. Primary readers may require highly technical messages, and secondary readers may need semitechnical or nontechnical messages—or vice versa. When you must write for audiences at different levels, follow these guidelines:

1. If the document is short (a letter, memo, or anything less than two pages), rewrite it at various levels for various readers.

2. If the document exceeds two pages, address the primary readers. Then provide appendixes for secondary readers (technical appendixes when secondary readers are technical, or vice versa). Letters

of transmittal, informative abstracts, and glossaries are other supplements that help nonspecialized audiences understand a highly technical report. (See pages 367–379 for how to use and prepare appendixes and other supplements.)

The next scenario shows how some documents must be tailored for both primary and secondary readers.

Tailoring a Document for Different Readers

Different readers have different information needs

You are a metallurgical engineer in a Detroit consulting firm. Your supervisor has asked that you test the fractured rear axle of a 1990 Delphi pickup truck recently involved in a fatal accident. Your job is to determine whether the fractured axle *caused* or *resulted from* the accident.

After testing the hardness and chemical composition of the metal and examining microscopic photographs of the fractured surfaces (fractographs), you conclude that the fracture resulted from stress that developed *during* the accident. Now you must report your procedure and your findings to a variety of readers.

"What do these findings mean?"

Because your report may serve as evidence in court, you must explain your findings in meticulous detail. But your primary readers (the decision makers) will be nonspecialists (the attorneys who have requested the report, insurance representatives, possibly a judge and a jury), and so you will have to translate your report, explaining the principles behind the various tests, defining specialized terms such as "chevron marks," "shrinkage cavities," and "dimpled core," and showing the significance of these features as evidence.

"How did you arrive at these conclusions?"

Secondary readers will include your supervisor and outside consulting engineers who will be evaluating your test procedures and assessing the validity of your findings. Consultants will be focusing on various parts of your report, to verify that your procedure has been exact and faultless. For these readers, you will have to include appendixes spelling out the technical details of your analysis: *how* hardness testing of the axle's case and core indicated that the axle had been properly carburized; *how* chemical analysis ruled out the possibility that the manufacturer had used inferior alloys; *how* light-microscope fractographs revealed that the origin of the fracture, its direction of propagation, and the point of final rupture indicated a ductile fast fracture, not one caused by torsional fatigue.

In this situation, the primary readers need to know *what your findings mean*, whereas the secondary readers need to know *how you arrived at your conclusions*. Unless you serve the needs of each group independently, your information will be worthless.

DEVELOP AN AUDIENCE-AND-USE PROFILE

When you write for a particular reader or a small group of readers, you can focus sharply on your audience by asking specific questions:

1. Who wants the document? Who else will read it?

2. Why do they want the document? How will they use it? What purpose do I want to achieve?

3. What is the technical background of the primary audience? Of the secondary audience?

4. How much does the audience already know about the subject? What material will have informative value?

5. What exactly does the audience need to know, and in what format? How much is enough?

6. When is the document due?

To answer these questions consider the suggestions that follow, and use some version of the Audience-and-Use Profile Sheet on page 32 for all your writing.

Reader Characteristics

Identify the primary readers by name, job title, and speciality (Martha Jones, Director of Quality Control, B.S. and M.S. in mechanical engineering). Are they superiors, colleagues, or subordinates? Are they inside or outside your organization? What is their attitude toward this topic likely to be? Are they apt to accept or reject your conclusions and recommendations? Will your report be good or bad news? For readers from other countries, how might cultural differences affect their expectations and interpretations? (See Chapter 6.)

Identify also those secondary readers who might be interested in or affected by your document, or who will affect the primary reader's perception or use of your document.

Purpose of the Document

Find out why readers want the document and how they will use it by asking: Do they merely want a record of activities or progress? Do they expect only raw data, or conclusions and recommendations as well? Will readers act immediately on the information? Do they need step-by-step instructions? Will the document be read and discarded, filed, or published? In your audience's view, *what* is most important? What purpose should this document achieve?

Readers' Technical Background

Colleagues who speak your technical language will understand raw data. Supervisors responsible for several technical areas may want interpretations and recommendations. Managers who have limited technical knowledge expect definitions and explanations. Clients with no technical background expect versions that spell out what the facts mean to *them* (to their health, pocketbook, business prospects). However, none of these generalizations might apply to *your* situation.

By assessing your readers' technical background, you can avoid insulting their intelligence or writing over their heads. For a mixed audience, aim for the lowest level of technicality.

Readers' Knowledge of the Subject

Do not waste time rehashing information readers already have. Readers expect something *new* and *significant*. Writing has informative value[2] when it (1) conveys knowledge that is new *and* worthwhile to the intended audience; (2) reminds the audience of something they know but ignore; or (3) offers fresh insight about something familiar, a new understanding.

The informative value of any message is measured by its relevancy to the writer's purpose and the audience's needs. Assume for instance, that you have no computer background (only general knowledge), but are trying to decide whether to take a computer course. In this situation, which of these bits of information would you find useful?

1. Interest in computers has grown immensely in the last decade.
2. Roughly 80 percent of businesses are computer-dependent.
3. The first digital computer was built by Howard Aiken.
4. Information can be transmitted rapidly by computer.

Number 2 is new to many novices, and relevant to your needs and my purpose (to help you decide). Because 1 repeats common knowledge, it has no informative value here. Although 3 offers an unfamiliar fact, it is not relevant to my purpose and your needs. (But 3 could be relevant in another situation: if you were already in a computer course.) Everyone would agree with 4, and so it offers you nothing worthwhile.

[2]Adapted from James L. Kinneavy's assertion that discourse ought to be unpredictable, in *A Theory of Discourse* (Englewood Cliffs: Prentice, 1971).

A message also has informative value if it offers fresh insight on familiar facts. As this book's audience, for instance, you expect to learn about technical writing, and my purpose is to help you do so. In this situation, which of these statements would you find useful?

1. Technical writing is hard work.
2. Technical writing is a process of making deliberate decisions in response to a specific situation. In this process, you discover important meanings in your topic, and give your readers the information they need to understand your meanings.[3]

Statement 1 is no news to anyone who has ever picked up a pencil, and so it has no informative value for you. But 2 offers a new way of thinking about something familiar. Even if you've done much writing, and struggled through decisions about punctuation, organization, and so on, you probably have not viewed writing as entailing the critical thinking discussed in this book (and illustrated in Chapter 7). Because 2 provides new insight into a familiar activity, you can say it has informative value.

The more nonessential information readers receive, the more likely they will overlook or misinterpret the important material. Take the time to determine what your readers need, and try to give them just that.

Appropriate Details and Format

The details in your report *(How much is enough?)* will depend on what you have learned about your readers and their needs. Were you asked to "keep it short" or to "be comprehensive"? Can you summarize some material, or does it all need spelling out? How deeply do readers need to know this material? What length will they tolerate? Are the primary readers most interested in conclusions and recommendations, or do they want all the details? Have they requested a letter, a memo, a short report, or a long, formal report with supplements (title page, table of contents, appendixes, and so on)? What kinds of visuals (charts, graphs, drawings, photographs) make your material more accessible? What level of technicality will connect with primary readers?

High technicality	The diesel engine generates 10 BTUs per gallon of fuel, as opposed to the conventional gas engine's 8 BTUs.
Low technicality	The diesel engine yields 25 percent better fuel mileage than its gas-burning counterpart.

[3]My thanks to Robert M. Hogge, U.S. Air Force Academy, for this definition.

Every professional in the Information Age has to keep abreast of technology, in order to use it effectively. What one has to know changes quickly and often. For instance, a senior engineer (and supervisor) might know less about VLSI circuits (very-large-scale integrated circuits—the heart of microprocessors) than a recent graduate.

Due Date

Does your report have a deadline? Find out. Allow plenty of time to collect data, to write, and to revise. If possible, ask primary readers to review an early draft and to suggest improvements.

BRAINSTORM FOR A USEFUL MESSAGE

When you begin working with an idea, content is raw material: ideas, insights, statistics, facts, or examples—anything that advances your meaning, that helps you answer this question: *How can I find something worthwhile to say, something that will convey my meaning?*

Outlines for specific writing tasks appear in later chapters. But the technique of *brainstorming* is one good way to find worthwhile content. The aim is to get all possible raw material *on paper.* Business and technical people use group brainstorming to develop ideas for new projects, marketing campaigns, and problem solving.

Some people brainstorm as a first writing step, before deciding about purpose and audience needs, just to get started. Others brainstorm only after writing a rough draft. Regardless of the sequence, good writers usually brainstorm at some stage in the process, to ensure they discover *all* the material readers might find useful.

The brainstorming procedure is simple: concentrate on your writing situation, and jot down *every thought.* Don't stop to judge relevance or worth, and don't worry about complete sentences or spelling. Just write everything that comes to mind. The more, the better. Trust your imagination; even the wildest idea might lead to a valuable insight.

Brainstorming invariably produces more information than you can use. And chances are you will discover more while writing your letter, memo, or report. From this broad inventory, you select only the useful material—namely, worthwhile content. (Chapter 7 shows how a working professional brainstorms while preparing an important memo to his superiors.)

EXERCISES

1. Locate a short article from your field. (Or select part of a long article or a section from one of your textbooks for an advanced course.) Choose a piece written at the highest level of technicality you understand and then translate the piece for a layperson, as in the example on page 22. Exchange translations with a classmate from a different major. Read your neighbor's translation and write a paragraph evaluating its level of technicality. Submit to your instructor a copy of the original, your translated version, and your evaluation of your neighbor's translation.

2. Assume that a new employee is taking over your job (part time or full time) because you have been promoted. Identify a specific problem in your old job that could cause difficulty for the new employee. Write for the employee instructions for avoiding or dealing with the problem. Before writing, perform an audience analysis by answering (on paper) the questions on page 25. Then brainstorm for details. Submit to your instructor your audience analysis, your brainstorming list, and your instructions.

3. Assume you live in the Northeast, and citizens of your state are voting on a solar energy referendum that would channel millions of tax dollars toward solar technology. These two paragraphs are versions of a message designed to help you, as a voter, make an educated decision. Do both messages have informative value, or does one merely repeat common information? Explain.

Solar power offers a realistic solution to the Northeast's energy problems. In recent years, the cost of fossil fuels (oil, coal, and natural gas) has risen sharply while supplies continue to decline. High prices and short supply will prolong a worsening energy crisis. Because solar energy comes directly from the sun, it is an inexhaustible resource. Using this energy to heat and air-condition our buildings, as well as to provide electricity, we could decrease substantially our consumption of fossil fuels. In turn, we would be less dependent on the unstable Middle East for our oil supplies. Clearly, solar power is a good alternative to conventional sources of energy.

Solar power offers a realistic solution to the Northeast's energy problems. To begin with, solar power is efficient. Solar collectors installed on fewer than 30 percent of roofs in the Northeast would provide more than 70 percent of the area's heating and air-conditioning needs. Moreover, solar heat collectors are economical, operating for up to twenty years with little or no maintenance. These savings recoup the initial cost of installation within only ten years. Most important, solar power is safe. It can be transformed into electricity through photovoltaic cells (a type of storage battery) in a noiseless process that produces no air pollution—unlike coal, oil, and wood combustion. In sharp contrast to its nuclear counterpart, solar power produces no toxic waste and poses no catastrophic danger of meltdown. Thus, massive conversion to solar power would ensure abundant energy and a safe, clean environment for future generations.

COLLABORATIVE PROJECT

Form teams according to major (electrical engineering, biology, etc.), and respond to the following situation.

Assume your team has received the following assignment from your major department's chairperson: An increasing number of first-year students are dropping out of the major because of low grades or stress or inability to keep up with the work. Your task is to prepare a "Survival Guide," for distribution to incoming students. This one- or two-page memo should focus on the challenges and the pitfalls and should include a brief motivational section, along with whatever else your team feels readers need.

Analyze Your Audience

Use the following questions as a guide for developing your audience-and-use profile. (One set of possible responses to these questions is shown.)

- *Who is my audience?* Incoming students in the major (and faculty).
- *How will readers use the information?* To develop a sense of what to expect and how to proceed (say, in managing workloads or meeting deadlines).
- *How much is the audience likely to know about this topic?* Very little. They need everything spelled out.
- *What else does the audience need to know?* They need answers to questions like these: How big is the problem? How can it affect me? What are the department's expectations? How much homework will I need to do? How should I budget my time? Can I squeeze in a part-time job? Are there any skills I should try to acquire beforehand (say, word processing or graphics and basic design skills)? Where do most first-year students make their big mistakes? (Based on your own experience, can your team anticipate any other questions?)
- *What attitudes or misconceptions about this topic is the audience likely to have?*
    Any who are overly optimistic ("No problem!") will need to visualize the real challenges ahead.
    Any who are overly pessimistic ("I'm dead for sure!") will need encouragement, along with the facts.
    Any who are indifferent ("Who cares?") will need some motivation, along with the facts.
    Whatever combination of attitudes the audience holds, we have to address each attitude—as well as we can identify it.
- *What probable attitude does the audience have toward the writers?* (Are we seen as trustworthy, sincere, threatening, arrogant, or what?) Since we are all stu-

dents, readers will likely identify with and trust us to an extent. They'll probably realize we're on their side.

- *Who will be affected by this document?* Mostly the incoming students (primary audience), and possibly the department.
- *In this situation, how can we characterize the audience's temperament and probable reaction?* Most readers should be eager for this information and should take it seriously.
- *Do we risk alienating anyone?* Gifted students who don't know the meaning of failure might feel patronized or offended. Some faculty might resent any suggestions that courses are too demanding, and so we don't want to editorialize. The purpose of this piece is informative and advisory—not evaluative.
- *How did this document originate, and how long should it be?* Because it was requested by the department and not by the primary audience, we can't expect students to tolerate more than a page or two.
- *What material will be most important to this audience?* They will want clear advice about what and what not to do.
- *What arrangement would be most effective for this audience and purpose?* We should provide brief background on the dropout problem, discuss its causes, suggest ways to survive, and end on a positive note of encouragement and motivation.
- *What tone would this audience expect?* We are all students; a friendly, relaxed and positive (to avoid panic), but serious tone seems best.
- *What is the document's intended effect on its audience?* If it manages to connect it will, we hope, cut down the dropout rate.

Follow the model in Figure 3.2 for designing a profile sheet to record your audience-and-use analysis. (Feel free to improve on the design and content of this model.)

Devise a Plan for Achieving Your Goal

From the audience traits you have identified, develop a plan for communicating your information. Express your goal and plan in a statement of purpose.

> The purpose of this document is to explain the challenges and pitfalls of the first year in our major. We will show how dropouts have increased, discuss what seems to go wrong, give advice on avoiding some common mistakes, and emphasize the benefits of remaining in the program.

Plan, Draft, and Revise Your Document

Brainstorm for worthwhile content (for this exercise, make up some reasons for the dropout rate if you need to), do any research that may be needed, write a workable draft, and revise until it represents your team's best work.

Audience Identity and Needs

Primary reader(s): _____ *(name, title)*

Secondary reader(s): _____

Relationship: _____ *(client, employer, other)*

Intended use of document: _____ *(perform a task, solve a problem, other)*

Prior knowledge about this topic: _____ *(knows nothing, a few details, other)*

Additional information needed: _____ *(background, only bare facts, other)*

Probable questions: _____ ?

_____ ?

_____ ?

_____ ?

_____ ?

Audience's Probable Attitude and Personality

Attitude toward topic: _____ *(indifferent, skeptical, other)*

Probable objections: _____ *(cost, time, none, other)*

Probable attitude toward this writer: _____ *(intimidated, hostile, receptive, other)*

Persons most affected by this document: _____

Temperament: _____ *(cautious, impatient, other)*

Probable reaction to document: _____ *(resistance, approval, anger, guilt, other)*

Risk of alienating anyone: _____

Audience Expectations About the Document

Reason document originated: _____ *(audience request, my idea, other)*

Acceptable length: _____ *(comprehensive, concise, other)*

Material important to this audience: _____ *(interpretations, costs,*

conclusions, other)

Most useful arrangement: _____ *(problem-causes-solutions, other)*

Tone: _____ *(businesslike, apologetic, enthusiastic, other)*

Intended effect on this audience: _____ *(win support, change behavior, other)*

Due date: _____

FIGURE 3.2 Audience-And-Use Profile Sheet

Appoint a team member to present the finished document (along with a complete audience-and-use analysis) for class evaluation, comparison, and response.

Alternative Projects

a. In a one- or two-page memo to *all* incoming students develop a "First-Year Survival Guide." Spell out the least information anyone should know in order to get through the first year.

b. In one or two pages, describe the job outlook in your field (prospects for the coming decade, salaries, subspecialties, promotional opportunities, etc.). Write for high school seniors interested in your major. Your team's description will be included in the career handbook published by your college.

c. Identify an area or situation on campus that is dangerous or inconvenient or in need of improvement (endless cafeteria lines, poorly lit intersections or parking lots, noisy library, speeding drivers, inadequate dorm security, etc.). Observe the situation as a group during a peak-use period. Spell out the problem in a letter to a specified decision maker (dean, campus police chief, head of food service) who presumably will use your information as a basis for action.

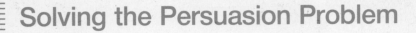

Solving the Persuasion Problem

Assess the Political Realities

Expect Reader Resistance

Know How to Connect with Readers

Ask for a Specific Decision

Never Ask for Too Much

Recognize All Constraints

Support Your Claims Convincingly

Observe Persuasion Guidelines

■   ■   ■

Chapter 3 explained how writers face the *information problem: How can I make readers understand exactly what I mean?* But workplace writers face a *persuasion problem* as well, outlined in Figure 4.1. Persuasion means trying to influence people's thinking or win their cooperation. And the size of your persuasion problem depends on who your readers are, how you want them to respond, and how strongly they are committed to their position.

You face a persuasion problem whenever you express a viewpoint readers might dispute. Viewpoints ordinarily are expressed in a thesis or a *claim* (a statement of the point you are trying to prove). For instance, you might want readers to *recognize* facts they've ignored:

A claim about what the facts are	The O-rings in the space shuttle's booster rockets have a serious defect that could have disastrous consequences.

Or you might want to influence how they *evaluate* the facts:

A claim about what the facts mean	Taking time to redesign and test the O-rings is better than taking unacceptable risks to keep the shuttle program on schedule.

Or you might want readers to *take immediate action:*

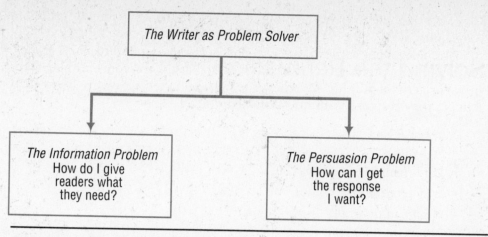

FIGURE 4.1 Two Problems Confronted by Writers

We should call for a delay of tomorrow's shuttle launch because the risks simply are too great.

A claim about what should be done

Whenever an audience disagrees about what things mean or what is better or worse or what should be done, you face a persuasion problem.

Your own letters, memos, and reports will be asking readers to accept and act on claims like these:[1]

- We cannot meet this production deadline without sacrificing quality.
- We're doing all we can to correct your software problem.
- This hiring policy is discriminatory.
- Our software is superior to the competing brand.
- We should begin this project immediately.
- I deserve a raise.

Your goal might be to convert readers to a different way of thinking, to reinforce one particular way of thinking, or to create a new way of thinking. In any event, you need to make the best case for seeing things *your* way.

ASSESS THE POLITICAL REALITIES

Besides their varied backgrounds, different readers have different attitudes. On one level, your readers are consumers of information; on another they are *human beings*, who react on the basis of their personality and feelings,

[1]This list of sample claims was inspired by Jeanette W. Gilsdorf ("Executives' and Academics' Perception"). The full citation appears in Works Cited, page 670.

and who create their own meanings for what they have just read (Littlejohn and Jabusch 5).

Any document can evoke different reactions—depending on a reader's temperament, preferences, interests, fears, biases, misconceptions, ambitions, or general attitude. Whenever readers feel their views are being challenged, they respond with questions like these:

- *Says who?*
- *So what?*
- *Why should I?*
- *Why rock the boat?*
- *What's in it for me?*
- *What's in it for you?*
- *What does this really mean?*
- *Will it mean more work for me?*
- *Will it make me look bad?*

Some readers might be impressed and pleased by your suggestions for increasing productivity; some might feel offended or threatened; others might think you are trying to make yourself "look good" or make them look bad. People can read much more between the lines than what is actually on the page. Such are the political realities of writing in any organization.

If you have worked with others, you already know something about office politics: how some people seek favor, status, or power; how some resent, envy, or intimidate others; how some are easily threatened. Author, teacher, and writing consultant Robert Hays sums up the writer's political situation this way:[2] "A writer must labor under political pressures from boss, peers, and subordinates. Any conclusion affecting other people can arouse resistance" (19). Some readers might resist your suggestion for shortening lunch breaks, cutting expenses, or automating the assembly line. Or your document might be seen as an attempt to undermine your boss.

No one wants bad news; some people prefer to ignore it (as the O-ring flaw on the shuttle *Challenger*'s booster rockets made all too clear). If you know something is wrong, that a project or product is unsafe, inefficient, or worthless, you have to decide whether "to try to change company plans; to keep silent; to 'blow the whistle'; or to quit" (Hays 19). Does your organization encourage or discourage outspokenness and constructive criticism? Find out—preferably before you accept the job. Ignoring political

---

[2] I am indebted to Professor Hays's article for excellent suggestions about analyzing and addressing political realities faced by writers. The full citation appears in Works Cited, page 670.

realities, you might write something that violates expectations and hurts your career.

EXPECT READER RESISTANCE

When people haven't made up their minds about what to do or think, they are more likely to be receptive to persuasive influence:

> We are all consumers as well as providers of persuasion. Daily, we open OUR-SELVES to the persuasion of others. We need others' arguments and evidence. We're busy. We can't and don't want to discover and reason out everything for ourselves. We look for help, for short cuts, in making up our minds.   (Gilsdorf, "Write Me" 12)

We rely on persuasion to help us make up our minds

In a world overwhelmed by information, persuasion can help people "proc-ess" the information and decide on its meaning.

People who already have decided what to do or think, however, don't like to change their minds without good reason. And sometimes, even for the best reasons, people refuse to budge. The O-ring claims on page 34 were made—and convincingly supported—by engineers before the shuttle *Challenger* exploded on January 28, 1986. That such claims were ignored by decision makers illustrates how audiences can resist the most compelling arguments.

Whenever you question someone's stand on an issue, or try to change their behavior, expect resistance. One researcher explains why persuasion is so difficult:

> By its nature, informing "works" more often than persuading does. While most people do not mind taking in some new facts, many people do resist efforts to change their opinions, attitudes, or behaviors.   (Gilsdorf, "Executives' and Academics' Perception" 61)

Once our minds are made up, we tend to hold stubbornly to our views

The bigger the readers' stake in the issue, the more personal their involve-ment will be, and the more resistance you can expect.

When people do "yield" to persuasion, they yield either grudgingly, willingly, or enthusiastically (as in Figure 4.2). Researchers categorize these responses as *compliance, identification,* or *internalization* (Kelman 51–60):

- *Compliance:* "I'm yielding to your demand in order to get a reward or to avoid punishment. I really don't accept it, but I feel pressured and so I'll go along to get along."

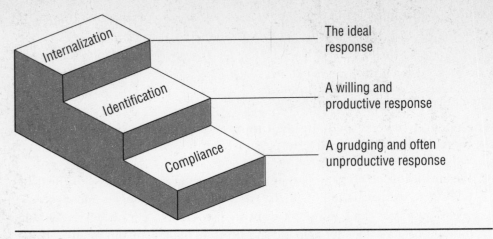

FIGURE 4.2 The Levels of Response to Persuasion

- *Identification:* "I'm yielding to your appeal because I like and believe you, I want you to like me, and I feel we have something in common."
- *Internalization:* "I'm yielding because what you're saying makes good sense and it fits my goals and values."

Although compliance is sometimes a necessary response (as in military orders or workplace safety regulations) nobody likes being "pushed." Effective persuasion relies on identification or internalization. If readers merely comply or feel coerced, then you probably have lost their loyalty and goodwill—and as soon as the threat or reward disappears, you will lose their compliance as well.

KNOW HOW TO CONNECT WITH READERS

Persuasive people know when to merely declare what they want, when to reach out and create a relationship, when to appeal to reason—or when to employ some combination of these strategies. These three strategies for connecting have been categorized as *hard, soft,* and *rational* (Kipnis and Schmidt, 40–46). Let's call them the *power connection,* the *relationship connection,* and the *rational connection,* as shown in Figure 4.3.

For an illustration of these different connections, picture the following situation: Your company, XYZ Engineering, has just instituted a fitness program, based on findings that healthy employees work better, take fewer sick days, and cost less to insure. This program offers clinics for smoking, stress reduction, and weight loss, along with group exercise. In your second month on the job you receive this notice through Electronic-mail:

Power Connection

To:	All Employees	5/20/93
From:	G. Maximus, Human Resources Director	
Subject:	<u>Physical Fitness</u>	

On Monday, June 10, all employees will report to the company gymnasium at 8:00 A.M. for the purpose of choosing a walking or jogging group. Each group will meet 30 minutes three times weekly during lunch hour.

Orders readers to show up

How would you react to this memo—and to the person who wrote it? Here the writer seeks nothing more than compliance. Although the writer speaks of "choosing," you are given no real choice but simply ordered to show up. This kind of *power connection* is typically used by bosses and others in power. And while it may or may not achieve its goal, the power connection almost surely will alienate its audience.

Now assume instead that you received the following version of our memo (page 40). How would you react to this message and its writer?

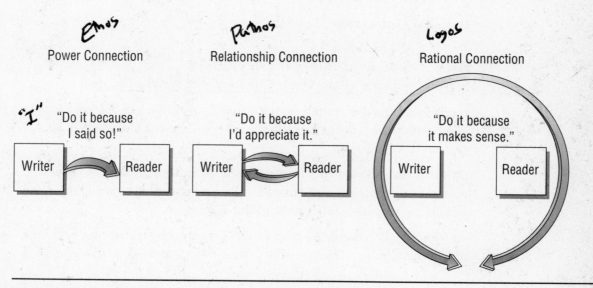

FIGURE 4.3   Three Strategies for Connecting with an Audience

Relationship Connection

To: All Employees   5/20/93
From: G. Maximus, Human Resources Director
Subject: An Invitation to Physical Fitness

Invites readers to participate

I realize most of you spend lunch hour playing cards, reading, or just enjoying a bit of well-earned relaxation in the middle of a hectic day. But I'd like to invite you to join our lunchtime walking/jogging club.

Leaves the choice to the reader

We're starting this club in hopes that it will be a great way for us all to feel better. Why not give it a try?

This version evokes a sense of identification, of shared feelings and goals. Instead of being commanded, readers are invited—they are given a real choice. This *relationship connection* establishes goodwill.

Often, the biggest factor in persuasion is an audience's perception of the writer. Audiences are more receptive to people they like, trust, and respect.[3] Of course, you would be unethical in appealing to the relationship or in faking the relationship merely to hide the fact that you had no evidence to support your claim (Ross 28). And despite its personal appeal, the relationship connection might strike some readers as too "chummy" to carry any real authority—and so the request might be ignored.

Here is a third version of our memo. As you read, think about the ways it makes a persuasive case.

Rational Connection

To: All Employees   5/20/93
From: G. Maximus, Human Resources Director
Subject: Invitation to Join One of Our Jogging or Walking Groups

Presents authoritative evidence

I want to share a recent study from the *New England Journal of Medicine*, which reports that adults who walk two miles a day could increase their life expectancy by three years.

Other research shows that 30 minutes of moderate aerobic exercise, at least three times weekly, has a significant and long-term effect in reducing stress, lowering blood pressure, and improving job performance.

Offers alternatives

As a first step in our exercise program, XYZ Engineering is offering a variety of daily jogging groups: The One-Milers, Three-Milers, and Five-Milers. All groups

---

[3]No matter how good the relationship, audiences also need to (a) find the claim believable ("Exercise will help me feel better") and (b) feel the claim is relevant ("I personally need this kind of exercise").

will meet at designated times on our brand new, quarter-mile, rubberized clay track.

For beginners or skeptics, we're offering daily two-mile walking groups. And for the truly resistant, we offer the option of a Monday-Wednesday-Friday two-mile walk.

Coffee and lunch breaks can be rearranged to accommodate whichever group you select.

Why not take advantage of our hot new track? As small incentives, XYZ will reimburse anyone who signs up as much as $100 for running or walking shoes, and will even throw in an extra fifteen minutes for lunch breaks. And with a consistent turnout of 90 percent or better, our company insurer may be able to eliminate everyone's $200 yearly deductible in medical costs.

Offers a compromise

Leaves the choice to the reader

Offers incentives

Here the writer shows willingness to compromise ("If you do this, I'll do that"). This *rational connection* communicates respect for the reader's intelligence *and* for the relationship by presenting good reasons, a variety of alternatives, and attractive incentives—all framed as an invitation. Whenever an audience is willing to listen to reason, the rational connection stands the best chance of succeeding.

Keep in mind that each kind of connection (or some combination) can work in particular situations. But no cookbook formula exists. Each situation calls for careful decisions on the writer's part.

ASK FOR A SPECIFIC DECISION

Unless you are giving an order, diplomacy is essential in connecting with readers. But don't be afraid to ask for the specific decision you want, preferably at the end of the message:

Studies show that the moment of decision is made easier for people when we show them what the desired action is, rather than leaving it up to them. . . . Without this directive, people may misunderstand or lose interest in the entire message. No one likes to make decisions: there is always a risk involved. But if the writer asks for the action, and makes it look easy and urgent, the decision itself looks less risky and the entire persuasive effort has a better chance of succeeding.   (Cross 3)

Let people know exactly what you want

Let readers know what you want them to do or think.

NEVER ASK FOR TOO MUCH

No amount of persuasion will move people to accept something they consider unreasonable. And the definition of *reasonable* depends on the individual. Employees at XYZ, for example, will differ as to which walking/jogging option they might accept.

Options offered by XYZ

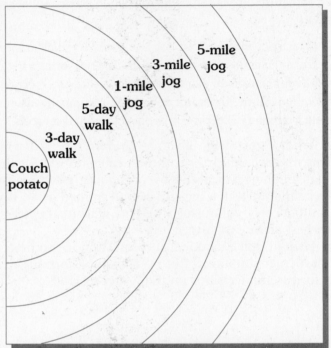

To the jock writing the memo, a daily 5-mile jog might seem perfectly reasonable. But some employees would think it outrageous. XYZ's program therefore has to offer something most of its audience (except, say, couch potatoes and those in poor health) accept as reasonable. Any request that exceeds its audience's range of acceptance (Sherif 39–59) is doomed.

RECOGNIZE ALL CONSTRAINTS

Persuasive communicators observe certain limits or restrictions imposed by their situation. These are the *constraints*, and they govern what should or should not be said, who should say it and to whom, when and how it should be said, and through which medium (printed document, computer screen, telephone, face to face, and so on). In the workplace you need to account for constraints like those in Figure 4.4.

Organizational Constraints

Organizations often have their own "official" constraints: for instance, schedules, deadlines, budget limitations, writing style, the way a document is organized and formatted, and its chain of distribution throughout the organization. But writers also face unofficial constraints like these:

Most organizations have clear rules for interpreting and acting on (or responding to) statements made by colleagues. Even if the rules are unstated, we know who can initiate interaction, who can be approached, who can propose a delay, what topics can or cannot be discussed, who can interrupt or be interrupted, who can order or be ordered, who can terminate interaction, and how long interaction should last. (Littlejohn and Jabusch 143)

Decide carefully when to say what to whom

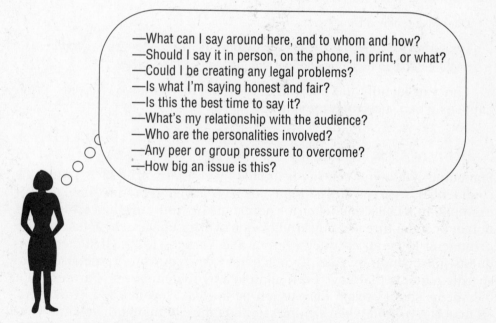

—What can I say around here, and to whom and how?
—Should I say it in person, on the phone, in print, or what?
—Could I be creating any legal problems?
—Is what I'm saying honest and fair?
—Is this the best time to say it?
—What's my relationship with the audience?
—Who are the personalities involved?
—Any peer or group pressure to overcome?
—How big an issue is this?

FIGURE 4.4 Every Writing Situation Poses Its Own Constraints

The exact rules vary among organizations, depending on whether communication channels are open and flexible or closed and rigid, on whether employee participation in decision making is encouraged or discouraged.

Although the rules of the game mostly are unspoken, anyone who ignores them (say, in going over a supervisor's head with a complaint or suggestion) invites disaster.

Airing even the most legitimate gripe in the wrong way through the wrong medium to the wrong person can be fatal to your work relationships and your career. The following memo, for instance, is likely to be interpreted by the executive officer as petty and whining behavior, and by the maintenance director as a public attack.

> Dear Chief Executive Officer:
>
> Please ask the Maintenance Director to get his people to do their job for a change. I realize we're all short-staffed, but I've gotten fifty complaints this week about the filthy restrooms and overflowing wastebaskets in my department. If he wants us to empty our own wastebaskets, why doesn't he let us know?
>
> cc: Maintenance Director

Instead, why not address the memo directly to the key person—or better yet—phone the person instead?

> Dear Maintenance Director:
>
> I wonder if we could meet to exchange some ideas about how our departments might be able to help one another during these staff shortages.

Can you identify the unspoken rules where you have worked? What happens when such rules are ignored?

Legal Constraints

Sometimes what you can say is limited by contract or by laws protecting confidentiality or customers' rights, or laws affecting product liability. For example, in a collection letter for nonpayment, you can threaten legal action but cannot threaten any kind of violence or publicize the refusal to pay or pretend to be an attorney (Varner and Varner 31–40). If someone requests information on one of your employees, you can "respond only to specific requests that have been approved by the employee. Further, your comments should relate only to job performance which is documented" (Harcourt 64). And when writing sales literature or manuals, you and your company are liable for faulty information that leads to injury or damage.

Know how the law applies to any document you prepare. Suppose, for instance, an employee drops dead while participating in the new jogging program you've marketed so persuasively. Could you and your company be liable? Perhaps you should require physical exams and stress tests (at company expense) for participants.

Ethical Constraints

While legal constraints are defined by federal and state laws, ethical constraints are defined by good conscience, honesty, and fair play. For example, it may be perfectly legal to promote a new pesticide by emphasizing its effectiveness while downplaying its carcinogenic effects; whether such action is *ethical*, however, is another issue entirely. To earn people's trust, you will find that "saying the right thing" involves more than legal considerations.

Persuasive skills carry tremendous potential for abuse. There is a difference between honestly presenting your best case to influence the reader and using deception to exploit the reader. (Chapter 5 is devoted to various ethics problems in communication.)

Time Constraints

Persuasion often is a matter of good timing. Should you delay saying your piece, say it immediately, or what? Let's assume you're trying to "bring out the vote" among members of your professional society on some hotly debated issue: say, whether to refuse work on any project related to biological warfare. You might want to wait until you have all the information you need or until you've analyzed the situation and planned a strategy. But you don't want to delay so long that rumors, misinformation, or paranoia cause people to harden their position *and* their resistance to your appeals. If delay might place the situation beyond your control, you might have to speak out sooner than you would like.

Social and Psychological Constraints

Too often, what we say can be misunderstood or misinterpreted. Here are just a few of the "human" constraints routinely encountered by communicators.

- *Relationship between communicator and audience:* Are you writing to a superior, a subordinate, or an equal? (Try not to appear dictatorial to subordinates or not to shield superiors from bad news.) How well do you and your audience know each other? Can you joke around or should you be dead serious? Do you get along or have a history of conflict? Do you trust and like one another? What you say and how you say it—and how it is interpreted—will be influenced by the relationship.

- *Audience's personality:* Researchers claim that "some people are easier to persuade than others, regardless of the topic or situation" (Littlejohn 136). Any reader's ability to be persuaded might depend on such personality traits as confidence, optimism, self-esteem, willingness to be different, desire to conform, open- or closed-mindedness, or regard for power (Stonecipher 188–89). The less your audience is open to persuasion, the harder you have to work. If you sense that your audience is totally resistant, you may want to back off—or give up altogether.

- *Audience's sense of identity and affiliation as a group:* How close-knit is the group? Does it have a strong sense of identity (as, say, union members or conservationists or engineering majors)? Will group loyalty or pressure to conform prevent certain appeals from working? Address the group's collective concerns.

- *Perceived size and urgency of the problem or issue:* In the audience's view, how big is this issue or problem? Has it been understated or overstated? Big problems are more likely to cause people to exaggerate their fears, anxieties, loyalties, and resistance to change—or to desperately seek some quick and easy solution. Assess the problem realistically: You don't want to downplay it, but you don't want to cause panic, either.

Writers who can assess a situation's constraints avoid serious blunders and can develop their message for greatest effectiveness.

SUPPORT YOUR CLAIMS CONVINCINGLY

The persuasive argument is the one that makes the best case in the audience's view. The strength of your case depends on the *reasons* you offer to support your claims.

Persuasive claims are backed up by reasons that have meaning for the reader

When we seek a project extension, argue for a raise, interview for a job, justify our actions, advise a friend, speak out on issues of the day . . . we are involved in acts that require good reasons. Good reasons allow our audience and ourselves to find a shared basis for cooperating. . . . In speaking and writing, you can use marvelous language, tell great stories, provide exciting metaphors, speak in enthralling tones, and even use your reputation to advantage, but what it comes down to is that you must speak to your audience with reasons they understand.   (Hauser 71)

To see how reasons serve in persuasion, imagine yourself in the following situation: As documentation manager for Bemis Software, a rapidly growing company, you supervise preparation and production of all user manuals. The present system for producing manuals is inefficient because three respective departments are involved in (1) assembling the required material, (2) word processing and designing, and (3) publishing the manuals. As a result, much time and energy are wasted as a manual goes back and forth among software specialists, communication specialists, and the art and printing department. After studying the problem and calling in a consultant, you decide that greater efficiency could be achieved if desktop publishing software were installed in all computer terminals. This way, all employees involved could contribute to all three phases of the process. To sell this plan to bosses and co-workers you will need good reasons, in the form of *evidence* and *appeals to readers' needs and values* (Rottenberg 104–06).

Offer Convincing Evidence

Evidence is any information that supports your claim. Common types of evidence are statistics, examples, and expert testimony.

Statistics.   Numbers can be convincing. Before reading about other details, many workplace readers are most interested in the "bottom line" (Goodall and Waagen 57).

> After a cost/benefit analysis, our accounting office estimates that an integrated desktop publishing network will save Bemis 30 percent in manual production costs and 25 percent in production time—savings that will enable the system to pay for itself within one year.

Give the numbers

Any statistics you present have to be accurate, trustworthy, and easy for readers to understand and verify. (See pages 185–187 for ways to avoid faulty statistical reasoning.) Always cite your source.

Examples.   By showing specific instances of your point, examples help audiences *visualize* the idea or the concept. The best way to explain what you mean by "inefficiency" in your company is to show one or more instances of it occurring:

The figure illustrates the inefficiency of Bemis's present system for producing manuals:

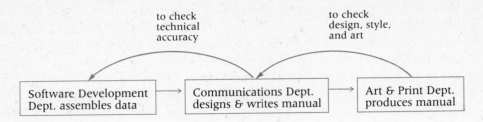

A manual typically goes back and forth through this cycle three or four times, wasting time and effort in all three departments.

Good examples have persuasive force; they give readers something solid, a way of understanding even the most surprising or unlikely claim. Use examples that the audience can identify with, and that fit the point they are designed to illustrate.

Expert Testimony.   A claim gains authority when it can be supported by views of recognized experts.

Ron Catabia, nationally recognized networking consultant, has studied our needs and strongly recommends we move ahead with the integrated network.

To be credible, however, an expert has to be unbiased and considered reliable by the audience.

While solid evidence can be persuasive, evidence alone isn't always enough. At Bemis, for example, the "bottom-line" might be very persuasive for company executives, but could mean little to some managers and employees who will be asking: Does this threaten my authority? Will I have to work harder? Will I fall behind? Is my job in danger? These readers will have to perceive some benefit beyond company profit.

Appeal to Common Goals and Values

Audiences are less likely to resist someone who shares their goals and values. If you hope to connect, you have to identify some goals you and the audience have in common: "What do we all want most?"

Bemis employees, like most people, share these goals: job security, a sense of belonging, control over their jobs and destinies, a growing and fulfilling career. Any persuasive recommendation will have to take these goals into account. For example,

> I'd like to show how desktop publishing skills, instead of threatening anyone's job, would only increase career mobility for all of us.

Our goals are shaped by our values (qualities we believe in, things we stand for): friendship, loyalty, ambition, honesty, self-discipline, equality, fairness, achievement, among others (Rokeach 57–58). Beyond appealing to common goals, you can appeal to shared values.

At Bemis, for example, you might appeal to the commitment to quality and achievement shared by the company and by individual employees:

> None of us needs reminding of the fierce competition in the software industry. The improved collaboration among networking departments will result in better manuals, keeping us on the front line of quality and achievement.

Give your audience reasons that have real meaning for *them* personally.

Persuasion, at best, is risky business because no single collection of strategies always works. Your approach will depend on the people involved, the setting, the relationships, and the variables discussed in this chapter. And with a truly resistant audience, even the best arguments may fail.

But even though you have no way of knowing whether your efforts will succeed, careful planning greatly improves your chances.

OBSERVE PERSUASION GUIDELINES

Later chapters offer specific guidelines for various persuasive documents. But beyond attending to the particular requirements of this or that document, remember this principle:

No matter how brilliant, any argument rejected by its audience is a failed argument.

If readers dislike you or conclude that your argument has no meaning for them personally, they usually reject *anything* you say. Connecting with an audience means being able to see things from their perspective. The following guidelines can help you make that connection.

1. *Assess the political climate.* Can you be outspoken? Who will be affected by your document? How will they react? How will your motives be interpreted? The better you assess readers' political feelings, the less likely your document will backfire. Do what you can to earn confidence and goodwill:

   - Be diplomatic; try not to make anyone look bad.

- Remember your place in the organization, and don't overstep.
- Don't expect anyone to be perfect—including yourself.
- Ask your intended reader(s) to review early drafts.

When reporting company negligence, dishonesty, stupidity, or incompetence, expect political fallout. Decide beforehand whether you want to keep your job or your dignity (more in Chapter 5).

2. *Learn the unspoken rules.* Know the constraints on what you can say, to whom you can say it, and how and when you can say it.

3. *Be clear about what you want.* Diplomacy is important, but people won't like having to guess about your purpose.

4. *Never make a claim or ask for something you know readers will reject outright.* Be sure readers can live with whatever you're requesting or proposing. Offer a genuine choice.

5. *Anticipate your audience's reaction.* Will they be defensive, surprised, annoyed, angry, or what? Try to address their biggest objections beforehand.

6. *Decide on a connection (or combination of connections).* Does the situation call for you to merely declare your position, appeal to the relationship, or appeal to common sense and reason?

7. *Avoid an extreme persona.* **Persona** is the image or impression of the writer's personality suggested by the tone of a document. Resist the urge to "sound off," no matter how strongly you feel, because audiences tune out aggressive people regardless of how sensible the argument is. Try to be likable and reasonable. Admit the imperfections in your case—a little humility never hurts.

8. *Find points of agreement with your audience.* Focusing early on a shared value or goal or concern can reduce conflict and help win agreement on later points.

9. *Never distort the opponent's position.* A sure way to alienate people is to cast the opponent as more of a villain or simpleton than the facts warrant.

10. *Try to concede* something *to the opponent.* Surely the opposing case is based on at least one good reason. Acknowledge the merits of that case before arguing for your own. Instead of seeming like a know-it-all, show some empathy and willingness to compromise.

11. *Use only your best material.* Not all your reasons or appeals will have equal strength or significance. Decide which material—from your *audience's* view—best advances your case.

12. *Make no claim or assertion unless you can support it with good reasons.* "Just because" does not constitute adequate support!

13. *Use your skills responsibly.* Persuasive skills are easily abused. People who feel bullied or manipulated or deceived most likely will become your enemies.

Figures 4.5 and 4.6 illustrate how our guidelines are employed in actual persuasive situations. The letter in Figure 4.5 is from a company that distributes systems for generating electrical power from recycled steam (cogeneration). President Tom Ewing writes a persuasive answer to a potential customer's question: "Why should I invest in the cogeneration system you are proposing for my plant?" As you read the letter, notice the kinds of evidence and appeals that support the opening claim. Notice also how the writer focuses on reasons important to the reader.

Figure 4.6 shows a school-related example of effective persuasion. Our writer, Christopher Biddle, has applied earlier to a state university for midyear acceptance. But, facing severe budget cuts, the university decided not to review any applications for midyear transfer. And so Brook appeals directly to the university president.

As you read, think about the way this letter makes a human connection and how it appeals to reason. Befitting the relationship with this reader, the tone is forceful, but respectful and *reasonable*. Brook is careful to ask for nothing the reader would consider extreme or outrageous: instead of special consideration ("Make an exception for me") Brook asks that *all* transfer applications be reviewed ("Give us all a fair chance"). How else does Brook make a persuasive case here? (Each of these letters, by the way, achieved its purpose: Tom Ewing made the sale and Christopher Biddle was accepted.)

EXERCISES

1. Assume you work for a technical marketing firm proud of its reputation for honesty and fair dealing. A handbook being prepared for new personnel includes a section titled "How to Avoid Abusing Your Persuasive Skills." All employees have been asked to contribute to this section by preparing a written response to this question:

> Share a personal experience in which you or a friend were the victim of persuasive abuse in a business transaction. In a one- or two-page memo, describe the situation and explain exactly how the intimidation or manipulation or deception occurred.

Write the memo and be prepared to discuss it in class.

EWING
POWER SYSTEMS

July 20, 1993

Mr. Richard White, President
Southern Wood Products
Box 84
Memphis, TN 37162

Dear Dick,

The writer states his claim

In our meeting last week, you asked me to explain why we have such confidence in the project we are proposing. Let me outline what I think are excellent reasons.

Offers first reason

Gives example

First, you and Don Smith have given us a clear idea of your needs, and our recent discussions confirm that we fully understand these needs. For instance, our proposal specifies an air-cooled condenser rather than a water-cooled condenser for your project because water in Memphis is expensive. And besides saving money, an air-cooled condenser will be easier to operate and maintain.

Offers second reason

Appeals to shared value (quality)

Gives example

Gives statistic
Further examples

Appeals to reader's goal (security)

Second, we have confidence in our component suppliers and they have confidence in this project. We don't manufacture the equipment; instead, we integrate and package cogeneration systems by selecting for each application the best components from leading manufacturers. For example, Alias Engineering, the turbine manufacturer, is the world's leading producer of single-stage turbines, having built more than 40,000 turbines in 70 years. Likewise, each component manufacturer leads the field and has a proven track record. We have reviewed your project with each major component supplier, and each guarantees the equipment. This guarantee is of course transferable to you and is supplemented by our own performance guarantee.

Offers third reason

Third, we have confidence in the system design. We developed the CX Series specifically for applications like yours, in which there is a need for both a condensing and a backpressure turbine. In our last meeting, I pointed out the cost, maintenance, and performance benefits of the CX Series. And although the CX Series is an innovative design, all components are fully proven in many other applications, and our suppliers fully endorse this design.

Cites experts

FIGURE 4.5 Supporting a Claim with Good Reasons

Richard White, July 20, 1993, p.2

Finally, and perhaps most important, you should have confidence in this project because we will stand behind it. As you know, we are eager to establish ourselves in Memphis-area industries. If we plan to succeed, this project must succeed. We have a tremendous amount at stake in keeping you happy.

If I can answer any questions, please phone me. We look forward to working with you.

Sincerely,

EWING POWER SYSTEMS, INC.

Tom Ewing

Thomas S. Ewing
President

FIGURE 4.5 Supporting a Claim with Good Reasons *Continued*

Dear President Mason:

Introduces himself and the problem

As fall semester ends, I face bleak prospects: After three semesters of excellent work as a special student, I had planned to begin the new year fully enrolled at State U. Enter the budget cuts.

Having reached the credit limit for special students, I'm now looking at the end of my academic career for at least a semester, if not forever.

States the claim (as a question)

I realize that the budget cuts mandated drastic measures on your part. However, the decision not to review mid-year transfer applications may have already served its purpose. If so, might there be time to review those applications and admit the very best applicants on a space-available basis?

Establishes early agreement

Acknowledges reader's dilemma

Appeals to reader's sense of fairness

Of all the tough decisions you've had to make in this financial crisis, the decision not to review mid-year applications must have been one of the toughest. On one hand, you have a responsibility not to "water down" the education of those students already admitted. On the other hand, you have a responsibility toward students who have been working hard at junior colleges or as special students, all in hopes of being admitted this winter. It's not fair for the presently enrolled students to have their classes overcrowded by mid-year transfers; but then again it's not fair for transfer students to be denied a chance.

Concedes a risk in what is requested

The legislature understandably needs to receive the message loud and clear that State U is struggling. And I see the danger of legislators interpreting transfer admissions as a sign of "business as usual"—a sign that State U has weathered these cuts and thus might be able to handle further cuts.

Tries to represent the reader's position fairly

I don't envy your role in the decision process, and I'm in no position to fault your handling of the issue. Maybe this is just the kind of "blood" our legislature had to see.

Concedes the merit of the reader's position

Through state house demonstrations (of which I was a part), countless protest letters and phone calls, and political SOS's from higher education officials, we have implored the legislators to ease up. And they have been forced to listen. Almost as soon as they had announced the latest round of cuts, the legislature felt compelled to soften the blow by almost one-third. Maybe the university's political hardball helped.

Restates the claim

But the point has now been made. State U has profited from its decision not to review mid-year applications, and there still may be time to counteract the losses.

FIGURE 4.6 A Persuasive Request

Besides benefiting the students themselves, mid-year transfers would benefit the university. A transfer student occupying an otherwise empty seat costs State U no money; tuition dollars in fact bring money in. And assuming that only the best students are admitted, they enhance the entire student body.

Of course, some majors already are overcrowded, and so transfers would have to be admitted on a space-available basis. The Registrar assures me that he has worked with the admissions office in earlier years to fill last-minute spaces with transfers as late as mid-January. And the Admissions Director claims that his office's role in this process is "the easy part."

I realize that even the "easy part" might not be so easy at this late stage, but I would love to see the university give it a try.

Picture State U after the budget cuts and what do you see? The football, basketball, and hockey teams are still playing. The radio station is still broadcasting.

What's missing from this picture? Students. A couple of dozen highly motivated transfer students. So what's the big deal? The university can get by without these students.

The big deal is that the university exists to educate. Without a football team you can still have a university. Without a track team, without a swimming pool, even without a radio station or newspaper you can still have a university. As long as teachers are teaching students, there is still a university. But take away the students and what's the point? Students should come first, and they should go last.

Respectfully,

Christopher B. Biddle

Christopher B. Biddle

Marginal annotations:

- Supports the claim with appeals to shared goals and values
- Anticipates and addresses a probable objection
- Cites expert opinion
- Concedes difficulties and restates the claim
- Supports the claim with examples
- Appeals to shared values
- Closes with the best reason

FIGURE 4.6 A Persuasive Request   *Continued*

2. Find an example of an effective persuasive letter. In a memo to your instructor, explain why and how the message succeeds. Base your evaluation on the persuasion guidelines, pages 49–51. Attach a copy of the letter to your evaluation memo. Be prepared to discuss your evaluation in class.

Now, evaluate a poorly written document, explaining how and why it fails.

3. Think about some change you would like to see on your campus or at your part-time job. Perhaps you would like to make something happen such as a campus-wide policy on plagiarism, changes in course offerings or requirements, more access to computers, a policy on sexist language, or a day-care center. Or perhaps you would like to improve something such as the grading system, campus lighting, the system for student evaluation of teachers, or the promotion system at work. Or perhaps you would like to stop something from happening such as noise in the library or sexual harassment at work. (Christopher Biddle's letter on pages 54–55 offers a useful model.)

Decide who, exactly, you want to persuade, and write a memo to that audience. Anticipate carefully your audience's implied questions, such as:

- *Do we really have a problem or need?*
- *If so, should we care enough about it to do anything?*
- *Can the problem be solved?*
- *What are some possible solutions?*
- *What benefits can we anticipate? What liabilities?*

Can you envision additional audience questions? Do an audience-and-use analysis based on the profile sheet, page 60.

Don't think of this memo as the final word, but as a consciousness-raising introduction that gets the reader to acknowledge that the issue deserves attention. At this early stage, highly specific recommendations would be premature and inappropriate.

4. Challenge an attitude or viewpoint that is widely held by your audience. Maybe you want to persuade your classmates that the time required to earn a Bachelor's degree should be extended to five years or that grade inflation is watering down your school's education. Maybe you want to claim that the campus police should (or should not) wear guns. Or maybe you want to ask students to support a 10-percent tuition increase in order to make more computers and software available.

Do an audience-and-use analysis based on the profile sheet, page 60. Write specific answers to the following questions: What are the political realities? What kind of resistance could you anticipate? How would you connect with readers? What about their range of acceptance? Any other constraints? What reasons could you offer to support your claim?

In a memo to your instructor, submit your plan for presenting your case. Be prepared to discuss your plan in class.

COLLABORATIVE PROJECT

Often, workplace readers need to be *persuaded* to accept recommendations that are controversial or unpopular. This project offers practice in dealing with the persuasion problems of communicating within organizations.

Divide into teams. Assume that your team agrees strongly about one of these recommendations and is seeking support from classmates and instructor (and administrators, as potential readers) for implementing the recommendations.

Choose One Goal

a. Your campus Writing Center always needs qualified tutors to help first-year composition students with writing problems. On the other hand, students of professional writing need to sharpen their own skill in editing, writing, motivation, and diplomacy. All students in your class, therefore, should be assigned to the Writing Center during the semester's final half, to serve as tutors for twenty hours (beyond normal course time).

b. To prepare students for communicating in an automated work environment, at least one course assignment (preferably the long report) should be composed, critiqued, and revised on disk. Students not yet skilled on a word processor will be required to develop the skill by midsemester.

c. This course should help individuals improve at their own level, instead of forcing them to compete with stronger or weaker writers. All grades, therefore, should be Pass/Fail.

d. To prepare for the world of work, students need practice in peer evaluation as well as self-evaluation. Because this textbook provides definite criteria and checklists for evaluating various documents, students should be allowed to grade each other and to grade themselves. These grades should count as heavily as the instructor's grades.

e. In preparation for writing in the workplace, no one should be allowed to limp along, just getting by with minimal performance. This course, therefore, should carry only three possible grades: A, B, or F. Those whose work would otherwise merit a C or D would instead receive an Incomplete, and be allowed to repeat the course as often as needed to achieve a B grade.

f. To ensure that all graduates have adequate communication skills for survival in a world in which information is the ultimate product, each student in the college should pass a writing proficiency examination as a graduation requirement.

Analyze Your Audience

Your audience here consists of classmates and instructor (and possibly administrators). From your recent observation of this audience, what reader characteristics can you deduce? Use these questions as a guide for developing your audience-and-use profile. (One possible set of responses is given for Goal *e*.)

- *Who is my audience?* Classmates and instructor (and possibly some college administrators).
- *How will readers use my information?* Readers will decide whether to support our recommendation for limiting possible grades in this course to three: A, B, or F.
- *How much is the audience likely to know already about this topic?* Everyone here is already a grade expert, and will need no explanation of the present grading system.
- *What else does the audience need to know?* The instructor should need no persuading; he or she knows all about the quality of writing expected in the workplace. But some of our classmates probably will have questions like these: Why should we have to meet such high expectations? How can this grading be fair to the marginal writers? How will I benefit from these tougher requirements? Don't we already have enough work here?

  We will have to answer questions by explaining how the issue boils down to "suffering now" or "suffering later," and that one's skill in communication will determine one's career advancement.
- *What attitude or misconceptions about the topic is our audience likely to have?*

  If they are indifferent ("Who cares?"), we have to encourage them to care by helping them understand the eventual career benefits of our recommended grading system.

  If they are interested ("Tell me more!"), we have to hold their interest and gain their support.

  If they are skeptical ("How could this plan ever work?"), we have to show concretely how the plan could succeed.

  If they are biased ("I hate writing."), we have to emphasize how greatly writing matters in the workplace.

  If they are misinformed ("I'll have secretaries or word processors to fix up my writing."), we have to provide the correct information, backed by solid evidence.

  If they are defensive ("Why pick on me?"), we have to persuade them that our idea is constructive, that we are genuinely supportive, and that we have their best interests in mind.

  If they have realistic objections ("Weaker writers will be unfairly penalized," or "Some students can't afford more than one semester for this course."), we need to offer a plan addressing these objections, and to show how the benefits can outweigh the objections.

  Whichever combination of attitudes our audience holds, we have to do our best to satisfy each reader's objections—as well as we can identify them. And if some readers are dead set against the idea, we can't expect to convert them, but we can encourage them at least to consider our position.

- *What is the audience's attitude toward the writer(s), before anyone has read the document? (Is the writer seen as trustworthy, sincere, threatening, arrogant, meek, underhanded, or what?)* We need to gain our readers' confidence, to make sure they interpret our motives not as totally self-serving or elitist, but as sincere and caring about the welfare of the whole group.
- *What is the organizational climate? (Is it competitive, repressive, cooperative, creative, resistant to change, or what?)* In this class, we can all feel comfortable about speaking out.
- *Who will be most affected by this document?* The primary audience, our classmates.
- *In this situation, how can we characterize the audience's temperament and probable reaction?* Many classmates are likely to feel threatened, and probably will react initially with resentment and resistance.
- *Do we risk alienating anyone?* Yes, especially students whose writing never has earned a grade of B or better.
- *How did this document originate, and how long should it be?* Because we writers initiated the document, we can't expect readers to tolerate a long, involved presentation. To be persuasive, however, we do have to make our case concrete.
- *What material will be most important to this audience?* They will want a clear picture of the benefits in such an apparently radical plan.
- *What arrangement would be most effective for this audience and purpose?* We should propose the new grading system, offer our reasons, point out the benefits, show how the system could operate, and close with a request for support.
- *What tone would this audience expect?* We are all at least acquainted; a friendly, conversational tone seems best.
- *What is this document's intended effect on its audience?* If it manages to connect with its audience, this document will win their support for our recommendation.

Follow the model in Figure 4.7 for designing a profile sheet to record your audience-and-use analysis, and to duplicate for use throughout the semester. (Feel free to improve on the design and content of our model.)

Devise a Plan for Achieving Your Goal

From the audience traits you have identified, develop a plan for justifying your recommendation. Express your goal and plan in a statement of purpose.

The purpose of this document is to convince classmates that our recommendation for an A/B/F grading system deserves their support. We will explain how skill in workplace writing affects career advancement, how higher standards for grading would help motivate students, and how our recommendation could be implemented realistically and fairly.

Audience Identity and Needs

Primary reader(s): _____ *(name, title)*

Secondary reader(s): _____

Relationship: _____ *(client, employer, other)*

Intended use of document: _____ *(perform a task, solve a problem, other)*

Prior knowledge about this topic: _____ *(knows nothing, a few details, other)*

Additional information needed: _____ *(background, only bare facts, other)*

Probable questions: _____ ?

_____ ?

_____ ?

_____ ?

_____ ?

Audience's Probable Attitude and Personality

Attitude toward topic: _____ *(indifferent, skeptical, other)*

Probable objections: _____ *(cost, time, none, other)*

Probable attitude toward this writer: _____ *(intimidated, hostile, receptive, other)*

Organizational climate: _____ *(receptive, repressive, creative, other)*

Persons most affected by this document: _____

Temperament: _____ *(cautious, impatient, other)*

Probable reaction to document: _____ *(resistance, approval, anger, guilt, other)*

Risk of alienating anyone: _____

Audience Expectations About the Document

Reason document originated: _____ *(audience request, my idea, other)*

Acceptable length: _____ *(comprehensive, concise, other)*

Material important to this audience: _____ *(interpretations, costs, conclusions, other)*

Most useful arrangement: _____ *(problem-causes-solutions, other)*

Tone: _____ *(businesslike, apologetic, enthusiastic, other)*

Intended effect on this audience: _____ *(win support, change behavior, other)*

Due date: _____

FIGURE 4.7 Audience-And-Use Profile Sheet

Plan, Draft, and Revise Your Document

Brainstorm for worthwhile content, do any research that may be needed, write a draft, and revise as often as needed to produce a document that stands the best chance of connecting with your audience.

Appoint a member of your team to present the finished document (along with a complete audience-and-use analysis) for class evaluation and response.

Solving the Ethics Problem

Recognize Unethical Communication

Expect Social Pressure to Produce Unethical Communication

Never Confuse Team Play with *Groupthink*

Rely on Critical Thinking for Ethical Decisions

Anticipate Some Hard Choices

Never Depend Only on Legal Guidelines

Understand the Potential for Communication Abuse

Know Your Communication Guidelines

Decide Where and How to Draw the Line

■   ■   ■

Chapters 3 and 4 explain how audience analysis helps us tailor informative and persuasive communication (so we can complete the project, win the contract, or the like). But an *effective* message (one that achieves its purpose) isn't necessarily an *ethical* message. Think of examples from advertising: "Our artificial sweetener is composed of proteins that occur naturally in the human body (amino acids)" or "Our potato chips contain no cholesterol." Such claims are technically accurate but misleading: amino acids in certain sweeteners can alter body chemistry to cause headaches, seizures, and possibly brain tumors; potato chips are loaded with saturated fat—which produces cholesterol. While the advertisers' facts may be accurate, they often are incomplete and they imply misleading conclusions.

Whether the miscommunication occurs deliberately or through neglect, a message is unethical when it leaves readers at a disadvantage or prevents readers from making their best decision. Therefore, writers ultimately face the threefold problem outlined in Figure 5.1. Ethical communication is measured by standards of honesty, fairness, and concern for everyone involved (Johannesen 1).

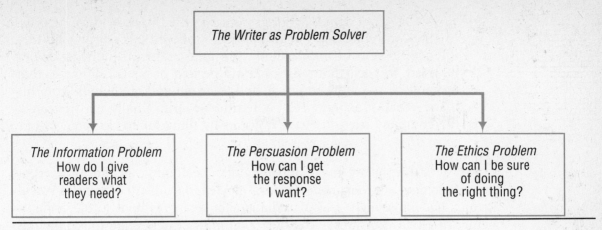

FIGURE 5.1   Three Problems Confronted by Writers

RECOGNIZE UNETHICAL COMMUNICATION

Thousands of people are injured or killed yearly in avoidable accidents—the result of communication that prevented intelligent decision making. Following are descriptions of tragedies caused ultimately by unethical communication.

- On March 3, 1974, as a Turkish Airlines DC-10 leaving Paris reached 12,000 feet, a cargo door burst open, causing a crash that killed all 346 people. Immediate cause: a poorly designed locking mechanism gave way under high pressure. Ultimate cause: the faulty design had been recognized and documented since 1969, and cargo doors had burst open during a 1970 test and a 1972 flight (in which the pilot managed to land safely). On January 27, 1972, a head engineer wrote a memo to his superiors warning that "in the twenty years ahead of us, DC-10 cargo doors will come open, and I expect this to usually result in the loss of the airplane." This flaw easily could have been corrected. But, unwilling to admit design errors and under pressure from competitors to get the DC-10 flying as early as possible, manufacturers suppressed these data (Burghardt 4–5).

Unethical communication has consequences

- In the early 1980s, thousands of workers (or their survivors) filed personal-injury suits against a leading manufacturer of asbestos products. The plaintiffs held the company responsible for respiratory afflictions ranging from lung cancer to emphysema. Immediate cause: exposure to asbestos fibers. Ultimate cause: for roughly 50 years the company hid from its workers the deadly facts about asbestos exposure—and even denied workers access to their own medical records. In 1963, the company's medical director offered this reason to justify the suppression of information: "As long as [the employee] is not disabled, it is felt that he should not be told of his condition so that he can live and work in peace, and the company can benefit from his many years of experience." When the coverup finally was revealed, the company tried to evade lawsuits by filing for bankruptcy (Mokhiber 15).

These catastrophes make for dramatic headlines, as did the recent revelation that a government-operated nuclear facility in Hanford, Washington, had knowingly leaked radiation for years without local residents being informed. But more "routine" examples of deliberate miscommunication rarely are publicized. Messages like the following succeed by *saying whatever works*.

- A person lands a great job by exaggerating his credentials, experience, or expertise.
- A marketing specialist for a chemical company negotiates a huge bulk sale of its powerful new pesticide by downplaying its carcinogenic hazards.
- To meet the production deadline on a new auto model, the test engineer suppresses, in her final report, data indicating that the fuel tank could explode upon impact.
- A manager writes a strong recommendation to get a friend promoted, while overlooking someone more deserving.

Can you recall some instances of miscommunication from news reports or perhaps from personal experiences on the job or in school? What were some of the effects of this miscommunication? How might the problems have been avoided?

To save face or escape blame or get ahead, anyone might be tempted to say what people want to hear, or to suppress bad news or make it seem "rosier." Some of these decisions are not simply black and white. Here is one engineer's description of the "gray area" in which issues of product safety and quality often are decided:

Ethical decisions are not always "black and white"

The company must be able to produce its products at a cost low enough to be competitive. . . . To design a product that is of the highest quality and consequently has a high and uncompetitive price may mean that the company will not be able to remain profitable, and be forced out of business. (Burghardt 92).

Do you emphasize to a customer the need for extra careful maintenance of this highly sensitive computer—and risk losing the sale? Or do you downplay maintenance requirements, focusing instead on the computer's positive features? Do you tell a white lie so as not to hurt a colleague's feelings, or do you "tell it like it is" because you're convinced that lying is wrong in any circumstance? The decisions we make in these ethical gray areas often are influenced by the pressures we feel.

EXPECT SOCIAL PRESSURE TO PRODUCE UNETHICAL COMMUNICATION

Pressure to get the job done can cause normally honest people to break the rules. At some point in your career you might have to choose between doing what your employer wants ("just follow orders" or "look the other way") and doing what you know is right. Maybe you will be pressured to ignore a safety hazard in order to meet a project deadline:

> Just as your automobile company is about to unveil its hot, new pickup truck, your safety engineering team discovers that the reserve gas tanks (installed beneath the truck but *outside* the frame) can explode on impact in a side collision. The company has spent a small fortune developing and producing this new model, and doesn't want to hear about this problem.

Pressure to "look the other way"

Companies often face the contradictory goals of *production* (which means *making* money on the product) and *safety* (which means *spending* money to avoid accidents that may or may not happen). And when productivity receives exclusive priority, safety concerns may suffer (Wickens 434–36). Thus it seems no surprise that well over 50 percent of managers studied nationwide feel "pressure to compromise personal ethics for company goals" (Golen et al. 75). And these pressures come in varied forms (Lewis and Reinsch 31):

- the drive for profit
- the need to beat the competition (other organizations or co-workers)
- the need to succeed at any cost, as when superiors demand more productivity or savings without questioning the methods
- an appeal to loyalty—to the organization and to its way of doing things

Figure 5.2 depicts how such pressures can add up.

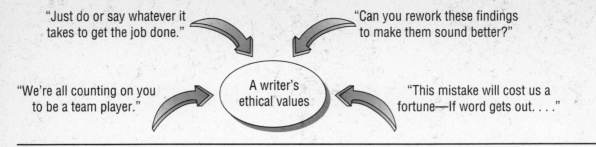

"Just do or say whatever it takes to get the job done."

"Can you rework these findings to make them sound better?"

"We're all counting on you to be a team player."

A writer's ethical values

"This mistake will cost us a fortune—If word gets out. . . ."

FIGURE 5.2 How Workplace Pressures Can Influence Ethical Values

Here is a vivid reminder of how organizational pressure can result in disastrous communication: On January 28, 1986, the space shuttle *Challenger* exploded 43 seconds after launch, killing all seven crew members. Immediate cause: two rubber O-ring seals in a booster rocket permitted hot exhaust gases to escape, igniting the adjacent fuel tank. (See Figure 5.3.) Ultimate cause: the O-ring hazard had been recognized since 1977 and documented by engineers, but largely ignored by management. (Managers had claimed that the O-ring system was safe because it was "redundant": each primary O-ring was backed up by a secondary O-ring.)

Moreover, in the final hours, engineers argued against launching because that day's low temperature would drastically increase the danger of both primary and secondary O-rings failing. But, under pressure to meet schedules and deadlines, managers chose to relay only a highly downplayed version of these warnings to the NASA decision makers who were to make *Challenger*'s fatal launch decision.[1]

The following analysis of key events and documents illustrates the role of miscommunication in a tragic but avoidable accident.

1. More than six months before the explosion, officials at Morton Thiokol, Inc., manufacturer of the booster rockets, received a memo from engineer R. M. Boisjoly (Presidential Commission 49). Boisjoly described how exhaust gas leakage on an earlier flight had eroded O-rings in certain noncritical joints. The memo emphatically warned of possible "catastrophe" in some future shuttle flight if O-rings should fail to seal a critical joint. Boisjoly called for renewed attention to the O-ring problem. Marked COMPANY PRIVATE, Boisjoly's urgent message was never passed on to top-level decision makers at NASA.

[1]For detailed analysis of miscommunication in the *Challenger* disaster, see D. A. Winsor, Roger C. Pace, Robert C. Rowland, and Dennis S. Gouran et al. Full citations appear in Works Cited, pages 671–672.

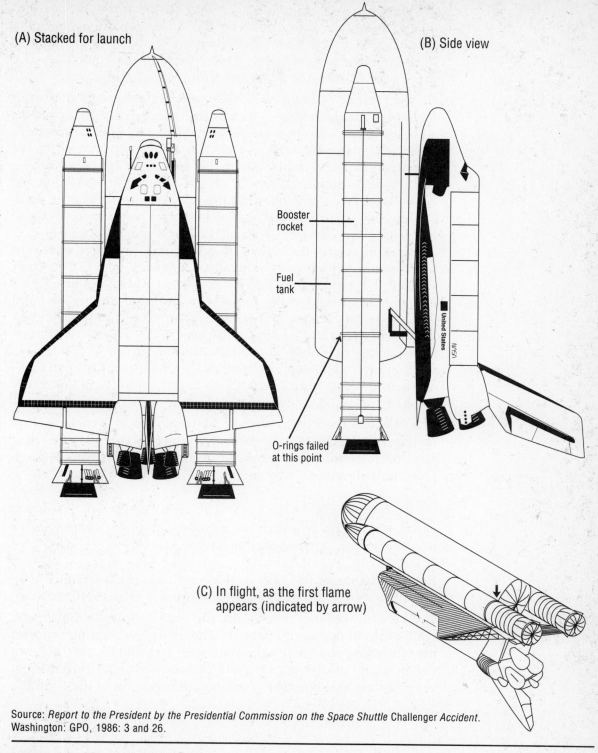

(A) Stacked for launch

(B) Side view

Booster
rocket

Fuel
tank

United States

NASA

O-rings failed
at this point

(C) In flight, as the first flame
appears (indicated by arrow)

Source: *Report to the President by the Presidential Commission on the Space Shuttle* Challenger *Accident.*
Washington: GPO, 1986: 3 and 26.

FIGURE 5.3 Views of *Challenger*

2. During the months preceding *Challenger*'s fatal launch, Boisjoly and other Morton Thiokol engineers complained, in writing, about the lack of management attention or support on the O-ring issue. But these complaints were largely ignored within the company—and they never reached the customer, NASA decision makers.

3. On the evening of January 27, managers and engineers at Morton Thiokol debated whether to recommend the January 28 launch. In addition to the yet-unsolved problem of O-ring erosion from exhaust gases, engineers were especially concerned that predicted low temperatures could harden the rubber O-rings, preventing them from sealing the joint at all. No prior flight had launched below 53 degrees Fahrenheit. At this temperature—and even up to 75 degrees—some blow-by (exhaust leakage) had occurred. O-ring temperature on January 28 would be barely above 30 degrees.

   Arguing against the launch, Roger Boisjoly presented a chart to his Thiokol colleagues (Presidential Commission 89). The chart showed that the O-rings would take longer to seal because lower temperatures would make the rubber hard and less pliable, and that the worst exhaust leakage had occurred on a January 1985 flight, when the O-ring temperature was 53 degrees. At 30 degrees, the O-rings might not seal at all, Boisjoly warned.

   Boisjoly and his supporters made their point, and Thiokol management decided to recommend no launch until the temperature reached at least 53 degrees.

4. In a teleconference with NASA's Marshall Space Center, Thiokol managers relayed their no-launch recommendation to the next level of decision makers. But the recommendation was rejected. One Thiokol manager's testimony:

   . . . Mr. Mulloy [a NASA official] said he did not accept that recommendation, and Mr. Hardy said he was appalled that we would make such a recommendation. (Presidential Commission 94)

   Refusing to accept the facts or the engineers' interpretations (i.e., the truth), NASA asked Thiokol to reconsider its recommendation.

5. The Thiokol staff met once again, the engineers and one manager continuing to oppose the launch. The managers then met *without* the engineers, and the reluctant manager was told to "take off your engineering hat and put on your management hat" (Presidential Commission 93). Despite the engineering evidence to the contrary,

Thiokol managers finally concluded that the O-ring system had an acceptable margin of safety (Presidential Commission 108).

6. And so, despite continued engineering objections, Thiokol management reversed its recommendation (Presidential Commission 97) to Marshall and Kennedy Space Centers. Top NASA decision makers never were told about Thiokol engineers' objections to the low-temperature launch or about the level of concern about O-ring erosion in prior shuttle flights. The rest is history.

Here are some conclusions of the Presidential Commission investigating the fatal launch decision (104):

- The Commission was troubled by what appears to be a propensity of management at Marshall to contain potentially serious problems and to attempt to resolve them internally rather than communicate them forward.

- The Commission concluded that the Thiokol management reversed its position and recommended the launch . . . at the urging of Marshall and contrary to the views of its engineers in order to accommodate a major customer.

Unethical communication played a key role in the *Challenger* disaster.

NEVER CONFUSE TEAM PLAY WITH *GROUPTHINK*

Any successful organization relies on team spirit and *collaboration* (page 96), everyone cooperating to get the job done. But team spirit is not the same as blindly following orders. While teamwork is productive, *groupthink* is destructive. Psychologist Irving L. Janis defines groupthink as "a mode of thinking that people engage in when they are deeply involved in a cohesive in-group, when members' strivings for unanimity override their motivation to realistically appraise alternative courses of action" (9).

Groupthink occurs when group pressure prevents individuals from questioning, criticizing, or "making a wave." Group members feel a greater need for acceptance and a sense of belonging than for critically examining the issues. In a conformist climate, critical thinking is impossible. Anyone who has lived through adolescent peer pressure has already experienced a version of groupthink.

Yielding to pressure can be especially tempting in a large company or on a complex project, where individual responsibility is easy to camouflage in the crowd:

Lack of accountability is deeply embedded in the concept of the corporation. Shareholders' liability is limited to the amount of money they invest. Managers' liability is limited to what they choose to know about the operation of the company. And the corporation's liability is limited by Congress (the Price-Anderson Act, for example, caps the liability of nuclear power companies in the aftermath of a nuclear disaster), by insurance, and by laws allowing corporations to duck liability by altering their . . . structure. (Mokhiber 16)

All kinds of people work at all levels on a major project (for instance, the production of a new passenger airplane). Countless decisions at any level have far-reaching effects on the whole project (as in the decision to ignore the DC-10's faulty locking mechanism). But with so many people collaborating, identifying those responsible for an error often is impossible—especially when the error is one of omission, that is, of *not* doing something that should have been done (Unger 137).

People commit unethical acts inside corporations that they never would commit as individuals representing only themselves. (Bryan 86)

> My job is to put together a persuasive ad for this brand of diet pill. It is someone else's job to make sure the claims are accurate and cause the customer no harm. It was someone else's job to make sure the stuff was safe. My job is only to promote the product. If someone does get hurt, it won't be my problem!

FIGURE 5.4 Groupthink Can Be a Handy Hiding Place

After completing their assigned task, employees too often assume their job is done. Figure 5.4 depicts the kind of thinking that enables people to deny personal responsibility for the consequences of their communication.[2]

Groupthink was exemplified by the decision to relay to top NASA officials only downplayed warnings about *Challenger*'s launch. In such cases, the group suffers an "illusion of invulnerability," which creates "excessive optimism and encourages taking extreme risks" (Janis 197).

Groupthink encourages the distortion of messages from subordinates to superiors. Research shows that subordinates tend to downplay or suppress bad news in their reports to superiors, instead "stressing what they think the superior wants to hear" (Littlejohn and Jabusch 159). Morton Thiokol's final launch recommendation to NASA serves as a memorable example.

RELY ON CRITICAL THINKING FOR ETHICAL DECISIONS

Because of their impact on people and on your career, ethical decisions challenge your critical thinking skills:

- *How can I know the "right thing" in this situation?*
- *What do I owe to whom in this situation?*
- *What values do I want to stand for in this situation?*
- *What is likely to happen if I do X, or Y?*

Can you rely on more than intuition or conscience in navigating the "gray areas" of ethical decisions? How will you make a convincing case against danger or folly to a roomful of people caught up in groupthink, people who ask "How do we know your way is right? Says who?" or who urge you to "take off your engineering hat and put on your management hat"?

Although ethical issues resist simple formulas, you can avoid two major fallacies that obstruct good judgment.

The Fallacy of "Doing One's Thing"

One way to oversimplify a complex ethical issue is through a misguided notion known as **ethical relativism:** "Since we have no way to agree on right or wrong, it all depends on personal preference. *Right*, then, is what I think is right!" Such denial of any reasonable criteria of course legitimizes

---

[2]My thanks to Judith Kaufman for this idea.

any action, including the atrocities of Hitler, Charles Manson, or other fanatics who claim the "right" intentions.

The Fallacy of "One Rule Fits All"

The opposite extreme of ethical relativism is **absolutism,** the inflexible notion that one set of criteria governs all ethical decisions. If, for example, you abide absolutely by the rule "Thou shall not lie," you would violate this rule under no circumstance (even, say, to save your family from criminal harm).

Reasonable Criteria for Ethical Judgment

Somewhere between the extremes of relativism and absolutism are **reasonable criteria** (standards of measurement that most people would consider acceptable). These criteria for ethical judgment take the form of **obligations, ideals,** and **consequences** (Ruggiero 55–56; Christians et al. 17–18).

Obligations are the responsibilities we have to everyone involved:

- *Obligation to ourselves,* to act in our own self-interest and according to good conscience.
- *Obligation to clients and customers,* to stand by the people to whom we are bound by contract—and who pay the bills.
- *Obligation to our company,* to advance its goals, respect its policies, protect confidential information, and expose misconduct that would harm the organization.
- *Obligation to co-workers,* to promote their safety and well-being.
- *Obligation to the community,* to preserve the local economy, welfare, and quality of life.
- *Obligation to society,* to consider the national and global impact of our actions.

When the interests of these parties conflict—as they often do—we have to decide very carefully where our primary obligations lie. Can we honor all these obligations all the time?

Ideals are "notions of excellence" (Ruggiero 55), the positive values that we believe in or stand for: loyalty, friendship, courage, compassion, dignity, fairness, and whatever qualities that make us who we are.

Consequences are the beneficial or harmful results of our actions. Consequences may be immediate or delayed, intentional or unintentional, obvious or subtle (Ruggiero 56). Some consequences are easy to predict; some aren't so easy; some are impossible.

Figure 5.5 depicts the relationship among these three criteria.

The above criteria help us understand why even good intentions can produce bad judgments, as in the following situation:

> Someone observes . . . that waste from the local mill is seeping into the water table and polluting the water supply. This is a serious situation and requires a remedy. But before one can be found, extremists condemn the mill for lack of conscience and for exploiting the community. People get upset and clamor for the mill to be shut down and its management tried on criminal charges. The next thing you know, the plant does close, 500 workers are without jobs, and no solution has been found for the pollution problem. (Hauser 96)

Because of their zealous dedication to the *ideal* of a pollution-free environment, the above group failed to anticipate the *consequences* of their protest or to respect their *obligation* to the community's economic welfare.

To analyze the miscommunication surrounding the *Challenger* disaster in light of these criteria: The group's sense of *obligation* to its client (i.e.,

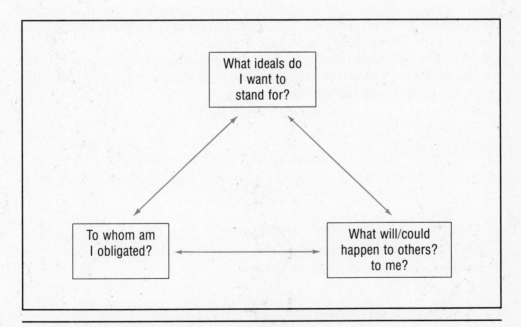

FIGURE 5.5 Reasonable Criteria for Ethical Judgment

NASA) caused it to ignore the *ideal* of honesty and the probable *consequences* of its recommendation to launch.

Ethical Dilemmas

Ethics decisions are especially frustrating when no single answer seems acceptable:

> [An ethical] dilemma exists whenever the conflicting obligations, ideals, and consequences are so very nearly equal in their importance that we feel we cannot choose among them, even though we must. (Ruggiero 91)

In private and public ways, such dilemmas are inescapable. For example, political candidates speak of drastic plans to eliminate the federal deficit in five years. One could argue that such a dedication to the *consequences* (or results) would violate our *obligations* (to the poor, the sick, etc.) and our *ideals* (of compassion, fairness, etc.). On the basis of our three criteria, how else might the deficit issue be considered?

Ethical dilemmas confront the medical community. For instance, in late 1992, a brain-dead, pregnant woman in Western Europe was kept alive for several weeks until her child could be delivered. Also in 1992, a California court debated the ethics of using medical findings of Nazi doctors who had experimented on prisoners in concentration camps. In terms of our three criteria, how might these dilemmas be considered?

As impossible as such dilemmas may seem, critical thinking directs our decision; without oversimplifying the issue or reaching a hasty judgment, we must consider all the criteria involved.

ANTICIPATE SOME HARD CHOICES

Communicators' ethical choices basically are concerned with honesty in choosing to reveal or conceal information:

- *What, exactly, do I report, and to whom?*
- *To whom do I owe loyalty?*
- *How much do I reveal or conceal?*
- *How do I say what I have to say?*
- *Could misplaced obligation to one party be causing me to deceive others?*

For illustration of a hard choice that working professionals might face when they communicate, consider the following scenario:

A Hard Choice

You are an assistant structural engineer working on the construction of a nuclear power plant near a far northern city. After years of construction delays and cost overruns, the plant has finally received its limited operating license from the Nuclear Regulatory Commission.

During your final inspection of the nuclear core containment unit on February 15, you discover a 10-foot-long, hairline crack in a section of the reinforced concrete floor, within 20 feet of the area where the cooling pipes enter the containment unit. (The especially cold and snowless winter likely has caused a frost heave under a small part of the foundation.) The crack has either just appeared or was overlooked by NRC inspectors on February 10.

The crack could be perfectly harmless, caused by normal settling of the structure; and this is, after all, a "redundant" containment system (a shell within a shell). But then again, the crack could signal some kind of serious stress on the entire containment unit, which furthermore could damage the entry and exit cooling pipes or other vital structures.

You phone your boss, who is just about to leave on a ski vacation, and who tells you, "Forget it; no problem," and hangs up.

You know that if the crack is reported, the whole start-up process scheduled for February 16 will be delayed indefinitely. More money will be lost; excavation, reinforcement, and further testing will be required—and many people with a stake in this project (from company executives to construction officials to shareholders) will be furious—especially if your report turns out to be a false alarm. All segments of plant management are geared up for the final big moment. Media coverage will be widespread. As the bearer of bad news— and bad publicity—you suspect that, even if you turn out to be right, your own career could be hurt by what some people will see as your overreaction that has made them look bad.

On the other hand, ignoring the crack could compromise the system's safety, with unforeseeable consequences. Of course, no one would ever be able to implicate you: the NRC has already inspected and approved the containment unit, leaving you and your boss and your company in the clear. You have very little time to decide: start-up is scheduled for tomorrow, at which time the containment system will become intensely radioactive.

What would you do? Come to class prepared to justify your decision on the basis of the obligations, ideals, and consequences involved.

> You may have to choose between the goals of your organization and what you know is right

Working professionals commonly face similar choices, the product of conflicting goals and expectations, the pressure to meet deadlines and achieve results, to be a "loyal" employee, a "team player," to consider "the bottom line." Often, these choices have to be made alone or on the spur of the moment, without the luxury of meditation or consultation.

NEVER DEPEND ONLY ON LEGAL GUIDELINES

Can the law tell you how to communicate ethically? Sometimes. If you stay within the law, are you being ethical? Not always. Legal standards "sometimes do no more than delineate minimally acceptable behavior." In contrast, ethical standards "often attempt to describe ideal behavior, to define the best possible practices for corporations" (Porter 183). Some perfectly legal actions:

Deception often is legal

- The investigative TV show "20/20" exposed the common and legal practice for trucks to haul garbage or toxic substances (such as formaldehyde—an embalming chemical) one way and then haul food products (such as juice concentrates) on the return trip—without ever informing the customer.

- It is perfectly legal to advertise a cereal made with oat bran (which allegedly lowers cholesterol) without mentioning that another ingredient in the cereal is coconut oil (loaded with cholesterol).

Lying is rarely illegal, except in cases of lying under oath or breaking a contractual promise (Wicclair and Farkas 16). But putting aside these and other illegal lies, such as defamation of character or lying about a product so as to cause injury, we see plenty of room for the kinds of "legal" lies depicted in Figure 5.6. Later chapters cover other kinds of legal lying, such as page design that distorts the real emphasis or words that are deliberately unclear or misleading or ambiguous.

What then are a communicator's legal guidelines? Besides obscenity laws (not especially relevant here), workplace writing is regulated by the five types of laws described on page 77.

"Count on us to meet your deadline!"

Promises you know you can't keep

"This product will last you for years!"

Assurances you haven't verified

"Trust our experts to solve your problem!"

Credentials you don't have

"You're our #1 priority!"

Inflated claims about your commitment

FIGURE 5.6 Some Legal Lies in the Workplace

- *Laws against libel* prohibit any false written statement that maliciously attacks or ridicules anyone. A statement is considered libelous when it damages someone's reputation, character, career, or livelihood or when it causes humiliation or mental suffering. Material that is damaging but *truthful* would not be considered libelous unless it were used intentionally to cause harm. In the event of a libel suit, a writer's ignorance is no defense: even when the damaging material has been obtained from a source presumed reliable, the writer (and publisher) are legally accountable.[3]

- *Copyright laws* protect the ownership rights of authors—or of their employers, in cases where the writing was done as part of one's employment (Girill 48).

- *Law protecting software* makes illegal duplication of copyrighted software a felony. Conviction for a first offense carries up to five years in prison and fines up to $250,000. The Software Publisher's Association claims that software piracy cost the industry $2.4 billion in 1990 alone ("On Line" 29).

- *Laws against deceptive or fraudulent advertising* make it illegal, for example, to falsely claim or imply that a product or treatment will cure cancer, or to represent and sell a used product as new. Fraud can be defined as "lying that causes another person monetary damage" (Harcourt 64).

- *Liability laws* define the responsibilities of authors, editors, and publishers for damages resulting from the use of incomplete, unclear, misleading, or otherwise defective information. The misinformation might be about a product (say, failure to warn about the toxic fumes from a spray-on oven cleaner) or a procedure (misleading instructions in an owner's manual for using a tire jack). And even if misinformation is given out of ignorance, the writer is liable (Walter and Marsteller 164–65). Later chapters deal with specific liabilities for particular documents, especially product and mechanism descriptions and instructions.

Laws regulating communication practices are few because such laws traditionally have been seen as threats to our freedom of speech (Johannesen 86).

Many companies have legal departments you can turn to if you have doubts about the legality of a document.

[3]Thanks to my colleague Peter Owens for the material on libel.

Recognizing that ethical behavior is a matter of commitment, and not of legislation, most professions have developed their own ethics guidelines. If your field has its own formal code, obtain a copy.

UNDERSTAND THE POTENTIAL FOR COMMUNICATION ABUSE

On the job, you write in the service of your employer. Your effectiveness is judged by how well your documents speak for the company and advance its interests and agendas (Ornatowski 100–01). You walk the proverbial line between telling the truth and doing what your employer expects (Dumbrowski 97).

Workplace writing influences the thinking, actions, and welfare of different people: customers, investors, co-workers, the public, policy makers—to name a few. And these people are victims of communication abuse whenever we give them information that is less than the truth as we know it. Following are some examples of such abuses.

Suppressing Knowledge the Public Deserves

Except for disasters that make big news (Bhopal, *Challenger*, Chernobyl) people hear plenty about the *wonders* of technology: how fluoride eradicates tooth decay, how smart bombs never miss, how nuclear power will solve our energy problems—but they rarely hear about the failures or the dangers (Staudenmaier 67).

In fact, the pressure to downplay failure sometimes results in censorship. For instance, some prestigious science journals have refused to publish studies linking chlorine and fluoride in drinking water with cancer risk, and fluorescent lights with childhood leukemia. The papers allegedly were rejected as part of widespread suppression of news about dangers of technological products (Begley 63).

Exaggerating Claims about Technology

Organizations that have a stake in a particular technology are especially tempted to exaggerate its benefits, potential or safety.

Unwarranted
claims help
technology sell

An entrepreneur needs financiers. Scientists in a large corporation need advocates high enough in the hierarchy to allocate funds. And government-supported researchers at universities and national labs have an obvious incentive to overstate their progress and understate the problems that lie ahead: the better the chances for success, the more money an agency is willing to shell out. (Brody 40)

If your organization depends on outside funding, you might find yourself pressured to make unrealistic promises.

Divulging or Stealing Proprietary Information

Proprietary information is considered the exclusive property of the organization that had it produced. Proprietary documents include company records, trade secrets, market research, minutes of meetings, plans and specifications (Lavin 5). Even though such information has some legal protection, it is still vulnerable to sabotage or theft. How responsibly you handle proprietary information can spell prosperity or disaster for your organization.

Mismanaging Electronic Information

With so much information stored in databases (by schools, employers, government agencies, mail-order retailers, credit bureaus, banks, credit card companies, insurance companies, pharmacies) questions of how we combine, use, and disseminate the information become increasingly important (Finkelstein 471). Moreover, a database is easier to alter than its printed equivalent; one simple command can wipe out or transform the "facts."

Withholding Information People Need to Do Their Jobs

Nowhere is the adage that "information power" more true than among coworkers. One sure way to sabotage a colleague is to withhold vital information about the task at hand.

Beyond these deliberate communication abuses is this reality: *all* information is a matter of personal or social interpretation (Dumbrowski 97); therefore, "objective reporting," practically speaking, is impossible. What we say on the job and how we say it are influenced by the expectations of our employer and by our own self-interest.

KNOW YOUR COMMUNICATION GUIDELINES

How do we balance self-interest with the interests of others—the organization, the public, our customers? How can we be "practical" and "responsible" at the same time? Here are two basic guidelines for ethical communication (Clark 194):

1. *Give the audience everything it needs to know.* To see things as clearly as you do, people need more than just a partial view. Don't bury readers in needless details, but do make sure they get all the facts and get them straight.

2. *Give the audience a clear understanding of what the information means.* Even when all the facts are known, they can be misinterpreted. Do all you can to ensure that your readers understand the real meaning, as you know it.

We have seen how the *Challenger* tragedy resulted from decision makers knowing too little or misunderstanding what they did know (Clark 194): "Low temperatures will harden the O-rings and will compromise their ability to seal the rocket joint." The *meaning* of this data depended on whether it was interpreted as a technical fact ("Exhaust leakage means an explosion") or a social fact ("Another launch delay means our company looks bad") (Ornatowski 98).

- To the engineers—the technical experts, but not the decision makers—the technical fact meant that the risk was serious enough to warrant a launch delay.

- To lower-level decision makers, the social fact meant that the risk was acceptable because of the social pressure to launch on schedule.

- To the ultimate decision makers at NASA, the data meant little, because lower-level decision makers provided information that was neither sufficient nor clear enough for the risk to be understood, or even recognized.

Because of social pressure to keep an employer happy, vital facts and truthful interpretations were suppressed.

The Ethics Checklist on page 82 incorporates additional guidelines from various chapters. Use the checklist for any document you prepare or for which you are responsible.

DECIDE WHERE AND HOW TO DRAW THE LINE

Suppose your employer asks you to do something unethical—say, altering data to cover up a violation of federal pollution standards. If you decide to resist, your choices seem limited: resign or go public (i.e., blow the whistle).

But these alternatives aren't realistic (Rubens 330). Walking away from a job isn't easy. And whistle-blowing, although courageous, can mean career disaster. Many organizations refuse to hire anyone blacklisted as a whistle-blower (Wicclair and Farkas 19). And even if you aren't fired, expect your job to become hellish. Here is one communicator's gloomy assessment, based on personal experience:

> Most of the ethical infractions [you] witness will be so small that blowing the whistle will seem fruitless and self-destructive. And leaving one company for another may prove equally fruitless, given the pervasiveness of the problem. (Bryan 86)

Know what to expect

Moreover, laws offer little protection for whistle-blowers. In general, employers are immune to lawsuits by employees who have been dismissed unfairly but who have no contract or union agreement specifying length of employment. The *employment-at-will rule* stipulates that employees can be dismissed at any time and for any cause, or even for no cause, at the employer's discretion (Unger 94).

Some laws do supersede the employment-at-will rule, and additional laws to protect whistle-blowers are in the works. But anyone who takes on the company without the backing of a union or other powerful group can expect disruption of life and career. Where you draw the line (on saving your integrity instead of your job) will be strictly your own decision.

If you do decide to take a stand, be reasonable and cautious, and follow these suggestions (Unger 127–130):

- *Get your facts straight, and get them on paper.* Don't blow matters out of proportion, but do keep a "paper trail" in case of legal proceedings.
- *Appeal your case in terms of the company's interests.* Instead of being pious and judgmental ("This is a racist and sexist policy, and you'd better get your act together"), focus on what the company stands to gain or lose ("Promoting too few women and minorities makes us vulnerable to legal action").
- *Aim your appeal toward the right person.* If you have to "go to the top," find someone who knows enough to appreciate the problem and who has enough clout to make something happen.
- *Get professional advice.* Contact an attorney and your professional society for advice about your legal rights.

Before accepting a job offer, do some discrete research about the company's ethical reputation. (Of course you can learn only so much about a company before actually working there.) Some companies have ombudspersons, who help employees lodge complaints. Others offer hotlines for advice on ethics problems or for reporting violations. Also, realizing that good ethics are good business, companies increasingly are developing codes for personal and organizational behavior. Without such supports, don't expect to last long as an ethical employee in an unethical organization.

And remember that very few employers tolerate any public statement, no matter how truthful, that makes the company look bad.

A Final Note: Sometimes, the "right choice" is obvious; sometimes, not so obvious. No one has any sure way of always knowing what to do. This chapter is only an introduction to the inevitable hard choices that, throughout your career, will be yours to make and to live with.

AN ETHICS CHECKLIST FOR COMMUNICATORS[4]

Use this checklist to help your documents reflect reasonable, ethical judgment. (Numbers in parentheses refer to the first page of discussion.)

☐ Do I avoid exaggeration, understatement, sugarcoating, or any distortion or omission that leaves readers at a disadvantage? (62)

☐ Do I know what I'm talking about, instead of "faking" certainty? (76)

☐ Am I being honest and fair? (62)

☐ Have I explored all sides of the issue? (180)

☐ Are my information sources valid, reliable, and unbiased? (180)

☐ Do I actually believe what I'm saying, instead of being a mouthpiece for groupthink or advancing some hidden agenda? (69)

☐ Would I still advocate this position if I were held publicly accountable for it? (70)

☐ Do I provide enough information and interpretation for readers to understand the truth as I know it? (80)

☐ Am I reasonably sure this document harms no innocent persons or damages their reputation? (77)

☐ Am I respecting all legitimate rights to privacy and confidentiality? (79)

[4]This list is largely adapted from Brownell and Fitzgerald, page 18; Bryan, page 87; Johannesen, pages 21–22; Unger, pages 39–46; Yoos, pages 50–55. Full citations appear in Works Cited, pages 671–672.

☐ Do I inform readers of the consequences or risks (as I am able to predict) of what I'm advocating? (71)

☐ Do I state the case clearly, instead of hiding behind jargon and generalities? (279)

☐ Do I give candid feedback or criticism, if it is warranted? (71)

☐ Am I providing copies of this document to everyone who has the right to know about it? (467)

☐ Do I credit all contributors and sources of ideas and information? (188)

EXERCISES

1. Prepare a memo (one or two pages) for distribution to first-year students in which you introduce the ethical dilemmas they will face in college. For instance:

- If you receive a final grade of A by mistake, would you inform your professor?
- If the library loses the record of books you've signed out, would you return them anyway?
- Would you plagiarize—and would that change in your professional life?
- Would you support the lowering of standards for underprivileged students if you felt your school's status would be compromised?
- Would you allow a friend to submit a paper you've written for some other course?

What other ethical dilemmas can you envision? Tell your audience what to expect, and give them some *realistic* advice for coping. No sermons, please.

2. Only about two dozen companies cause one-third of toxic-waste pollution in the U.S.—and it's all perfectly legal. Who are the biggest polluters, and why do they get away with it?

3. In your workplace communications, you may end up facing hard choices concerning what to say, how much to say, how to say it, and to whom. And whatever your choice, it will have definite consequences. Be prepared to discuss the following cases in terms of the obligations, ideals, and consequences involved. Can you think of similar choices you or someone you know has already faced? What happened?

- While traveling on assignment that is being paid for by your employer, you visit an area in which you would really like to live and work, an area in which you have lots of contacts but never can find time to visit on your own. You have five days to complete your assignment, and then you must report on your activities. You complete the assignment in three days. Should you spend the remaining two days checking out other job possibilities, without reporting this activity?

- As a marketing specialist, you are offered a lucrative account from a cigarette manufacturer; you are expected to promote the product. Should you accept the account? Suppose instead the account were for beer, junk food, suntanning parlors, or ice cream. Would your choice be different? Why, or why not?

- You have been authorized to hire a technical assistant, and so you are about to prepare an advertisement. This is a time of threatened cutbacks for your company. People hired as "temporary," however, have never seemed to work out well. Should your ad include the warning that this position could be only temporary?

- You are one of three employees being considered for a yearly production bonus, which will be awarded in six weeks. You've just accepted a better job, at which you can start anytime in the next two months. Should you wait until the bonus decision is made before announcing your plans to leave?

- You are marketing director for a major importer of coffee beans. Your testing labs report that certain African beans contain roughly twice the caffeine of South American varieties. Many of these African varieties are big sellers, from countries whose coffee bean production helps prop otherwise desperate economies. Should your advertising of these varieties inform the public about the high caffeine content? If so, how much emphasis should this fact be given?

- You are research director for a biotechnology company working on an AIDS vaccine. At a national conference, a researcher from a competing company secretly offers to sell your company crucial data that could speed discovery of an effective vaccine. Should you accept the offer?

4. Study the following scenario and complete the assignment: You belong to the Forestry Management Division in a state whose year-round economy depends almost totally on forest products (lumber, paper, etc.) but whose summer economy is greatly enriched by tourism, especially from fishing, canoeing, and other outdoor activities. The state's poorest area is also its most scenic, largely because of the virgin stands of hardwoods. Your division has been facing growing political pressure from this area to allow logging companies to harvest the trees. Logging here would have good and bad consequences: for the foreseeable future, the area's economy would benefit greatly from the jobs created; but traditional logging practices would erode the soil, pollute waterways, and decimate wildlife, including several endangered species—besides posing a serious threat to the area's tourist industry. Logging, in short, would give a desperately needed boost to the area's standard of living, but would put an end to many tourist-oriented businesses and would change the landscape forever.

You've been assigned to weigh the economic and environmental impacts of logging, and prepare recommendations (to log or not to log) for your bosses, who will use your report in making their final decision. To whom do you owe the most loyalty here: the unemployed or underemployed residents, the tourist businesses (mostly owned by residents), the wildlife, the land, future generations? The choices are by no means simple. In cases like this, it isn't enough to say that we should "do the right thing," because we are sometimes unable to predict the consequences of a particular action—even when it seems the best thing to do. In a memo to your instructor, tell what action you would recommend and explain why. Be prepared to defend your ethical choice in class on the basis of the obligations, ideals, and consequences involved.

COLLABORATIVE PROJECT

Workplace decisions with ethical implications often are made *collaboratively* by members of a project or management team. This project offers practice in collaborative decision making on a growing ethical problem spawned by new technology.

Divide into groups. Assume you are a junior manager at Killawatt, a major producer of consumer electronics, from digital watches to cordless phones to computer games to the newest gadgetry. Your team is composed of managers from various departments such as research and development, computer services, human resources, accounting, production, and corporate relations. You have been brought together to tackle the following problem.[5]

Analyze the Problem

Killawatt is facing a threefold management problem:

1. A two-million dollar loss in the recent quarter, caused by an economic downturn, necessitates severe austerity measures. Upper management wants to eliminate internal losses from such sources as

- excessive and unauthorized use of photocopy and fax machines, digitizers, laser printers, and other costly equipment
- personal phone calls on the WATS line
- personal or excessive use of database retrieval services such as Compuserve and Dialog (costing as much as $5.00 per minute), for stock market quotations, investment information, and the like
- extra-long lunch hours and coffee breaks
- personal use of company cars and trucks
- unlimited purchases of company products at a 40 percent employee discount

In rosier economic periods all these "perks" were tolerated by Killawatt. But today's hard times make such benefits unaffordable.

2. A more crucial and long-term problem for the company is to find ways of identifying its most and least productive employees. Presumably, this step could

[5]Adapted from "Monitoring on the Job," by Gary T. Marx and Sanford Sherizen (Nov./Dec. 1986): 63–72. Reprinted with permission from *Technology Review,* copyright 1986.

not only increase productivity, but might also provide a more selective basis for salary increases, bonuses, promotions, and even any layoffs that might later be unavoidable.

3. A third and somewhat related problem is for Killawatt to plug all security leaks. In the past two years, major company breakthroughs in product technology have been leaked to competitors. And details of the company's recent fiscal problems have been leaked to the press, hurting stock value and scaring investors.

In light of these problems with perks, productivity, and security, Killawatt is exploring ways to monitor its employees. (Monitoring equipment, although expensive, could be capitalized, depreciated, or written off as a business expense; resulting tax benefits would largely offset equipment costs.) At this preliminary stage, you have been asked to assess the feasibility of such a plan in terms of its impact on employee relations.

Like many co-workers, you have heard your share of horror stories about perfectly "legal" monitoring in some other companies. For example, security officers for one government contractor rummage through employees' desk drawers after work hours. Abuses at other companies include

- Random checking of employees' computers for unauthorized files (such as personal correspondence, coursework, home designs, and the like).
- Devices that monitor computer workstations by keeping track of the number of keystrokes, errors, corrections, words typed per minute, customers processed per hour—even the number and length of visits to the bathroom.
- Companywide and random polygraph (lie detector) testing to identify dishonest employees.
- Cameras or microphones to keep track of productivity, behavior, and even conversations.
- Telephone devices that keep track of each call: the extension from which it was made, the number dialed, time and length of call, and even a recording of the conversation. Such devices presumably help expose security leaks by keeping track of who is talking to whom about what.
- Programs that tell workers how their performance is measuring up against co-workers', or some with "relaxation" sounds or other subliminal messages designed to increase concentration or speed or morale.

While advocates of monitoring at Killawatt acknowledge the potential for abuse, they argue that a conscientious program could benefit the employees as well as the company. Besides keeping everyone "honest," monitoring could provide an objective measure of productivity, without the interpersonal and sometimes discriminating clashes between supervisor and employee. The technology, in effect, would ensure equal treatment across the board. Annual performance appraisals could then be based on reliable criteria that would enhance a supervisor's "subjective" impressions of the employee's contribution.

But these arguments have not persuaded you or your team members. In your view the plan raises troubling ethical questions:

- What are the employees' rights in this issue?
- Is the plan fair?
- Could the monitoring plan backfire? Why?
- How would the plan affect employee's perceptions of the company?
- How would the plan affect employee morale and loyalty and productivity?
- What should be done to avoid alienating employees?
- Are there acceptable alternatives to monitoring?

Your team struggles with these and other questions you identify when you meet to plan your response. How will your team respond? What position will you advocate? What are the human costs and benefits of the proposed plan? What obligations, ideals, and consequences are involved? What plan can your team offer instead? How can you be persuasive here?

Plan Your Response

As your team plans its response to management, think of key points you want to emphasize.

Analyze Your Audience(s)

Your primary audience consists of executives and managers, many who see nothing wrong with monitoring. If your team's memo (followed by a meeting) manages to persuade your bosses, then you will be expected to win employee acceptance of your plan—and so your secondary audience will consist of employees who mostly are strongly opposed to any type of monitoring. Prepare audience-and-use profiles (pages 32, 60) of each audience.

Prepare and Present Your Documents

In a memo to your bosses, argue against their plan and make a *persuasive* case for your alternative. Be explicit about your objections and about the *ethical* problems you envision. Brainstorm for worthwhile content, do any research that may be needed, write a workable draft, and revise using the Ethics Checklist (page 82).

Assume that your argument succeeds with your bosses, and prepare a memo to employees that is reassuring and persuasive in eliciting their cooperation with your plan.

Appoint a team member to present the finished documents (along with complete audience profiles) for class evaluation and response.

Solving the Adaptation Problem

Computers Diversify the Ways We Communicate

Computers Diversify Our Work Relationships

Computers Diversify Our Audiences

■   ■   ■

In addition to information, persuasion, and ethics problems, we confront the problem of coping with the diverse challenges created by communication technology, as described in Figure 6.1.

The Adaptation Problem

How do I adjust to diversity in the writer's tools, work relationships, and audiences?

FIGURE 6.1 An Additional Problem Confronted by Writers

Following are some implications of technology for today's communication in the workplace:

- Computer networks connect people across town or across the globe. Though much of this communication occurs via voice or video, much continues to be *written*. (Even "paperless writing" on a computer screen must address the problems outlined in earlier chapters.)

- Instant communication enables multiple writers to work on a single document. Rapidly disappearing is our notion of a solitary writer, working alone in some quiet corner.

- Communication technology has spawned a global economy, a global marketplace, and a global audience as well.

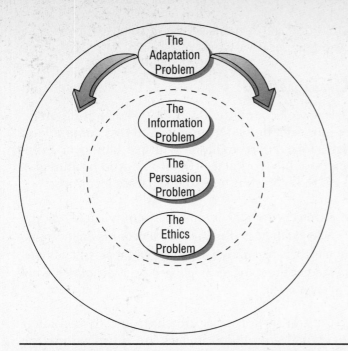

FIGURE 6.2 Problem Solving Has Assumed a New Dimension

In short, more people than ever are likely to *work* on a document, and more are likely to *read* it.

Figure 6.2 depicts the problem-solving environment in which writers must function.

COMPUTERS DIVERSIFY THE WAYS WE COMMUNICATE

In an age in which information (or knowledge) is the product and computers are the vehicle, people from any location can work together:

> The Information Age has allowed companies that used to be housed in one location to fan out across the globe. Scientists are almost as likely to collaborate with colleagues on the next continent as across campus. And industrial projects are so large that those engaged in them couldn't possibly work in close proximity. (Brittan 44)

Computer networks are redefining the "workplace"

Roughly 4 million researchers, students, and businesspeople, for example, use *Internet*, a worldwide network for academic information (Wilson 17).

Fiber optic networks connect consultants and colleagues from everywhere, enabling all types of collaborative work.

Technology Facilitates Telecommuting

More than 5 million Americans telecommute, (Weiss 17) researching, writing, and performing other jobs on a home terminal linked to a company or worldwide network. By the year 2000, millions will use laptop or even palmtop computers as their portable workplace (Halal 11). And multimedia soon will enable co-workers to see each other from separate locations.

As more information "travels," more people can "stay put"

For example, a team of architects from all around the country could "meet" via picturephone. They could examine and even revise blueprints online, look at sketches various team members had drawn, read and edit reports, and note from facial expressions whether the source was pleased or disappointed with the way work was going. (Weiss 17)

Of course, the prospect of transactions that are electronically mediated (instead of face-to-face) is worrisome. Video images are no substitute for human presence (Brittan 50). And studies indicate that people resist the idea of communicating exclusively via machine (Halal 14). As we search for new ways of relating to coworkers and audiences, we fret over the specter of a dehumanized workplace and the computer's ultimate impact on human relations.

"Good grief, Bradbury! How long have you been working at home?"

Source: Drawing by C. Barsotti; Copyright © 1992 The New Yorker Magazine, Inc.

Despite these anxieties, however, we can look forward to greater *collaboration* in which people share their own knowledge and benefit from the knowledge of others.

Writers Today Use Multipurpose Tools

Electronic writing tools streamline our research, planning, drafting, and revising. Figure 6.3 depicts the multipurpose tools that are discussed on the following pages.

Research and Reference.   Workplace writing typically is based on research, increasingly being done via computer. An *online data base* accessed by personal computer will provide the latest facts and figures on the stock market, global energy consumption, toxic-waste sites—virtually any topic.

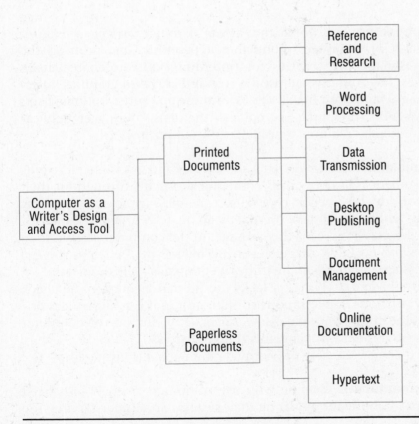

FIGURE 6.3 The Writer's Electronic Tools

Compact disks (CDs) offer a growing array of information, from electronic encyclopedias (twenty volumes on a compact disk) to the latest edition of *Books in Print*. Moreover, *hypertext* systems (page 93) allow users to explore a topic from countless angles and to any depth.

Word Processing and Document Management. Besides eliminating much of the mechanical drudgery of writing and revising, *word processing* provides design and visual options. "Groupware" (group authoring systems) enables writers in different locations to edit, proofread, compare drafts, add graphics, and comment on each other's work (Brittan 44; Easton et al. 35–36). Through such a network colleagues can work jointly, preparing, editing, or revising different parts of a document (say, a proposal or an annual report).

Finished documents can be *managed* electronically, filed and catalogued for easy retrieval. Documents or parts of documents used repeatedly ("boilerplate") can be retrieved, modified as needed, and inserted in some other document.

Desktop Publishing. Beyond the actual writing, *desktop publishing* includes page design, graphics production, typesetting, and printing—all done on a personal computer. Desktop publishing software enables users to divide pages into columns, adjust the size and shape of graphics, insert predrawn images, and print headlines. With these and other options, users produce newsletters, brochures, pamphlets, manuals, and other publications that traditionally required graphic artists and print shops.

Data Transmission. *Electronic mail* allows users to transmit electronic copies of documents via networked computer terminals throughout their company or worldwide. Specific codes direct the message to any electronic "mailbox" the sender designates. Alerted by an on-screen signal, the recipient "opens" the mailbox, reads the message on the computer screen, then responds, files the message, or deletes it. Because E-mail enables the exchange of messages anyplace at any time, it eliminates "telephone tag."

E-mail does have drawbacks: it lacks the human element of feedback that we enjoy in phone conversations (Brittan 44); and E-mail reduces privacy, in that the message could show up on any number of computer screens besides the intended recipient's (Finkelstein 422).

For printed copy, *FAX* networks provide immediate transmission.

Online Documentation. People who use computers in their jobs need instructions and training for operating their systems and understanding the many features.

Although computers come with printed manuals, the computer itself is becoming the preferred training medium. In computer-based training, *online documentation* on the screen itself explains how the system works and how to use it. Types of online documentation include error messages, reference guides, tutorial lessons, interactive exercises with immediate feedback, and help and review options to accommodate different learning styles. Instead of leafing through a printed manual, users find what they need by typing a simple command.

While online documentation offers an exciting new communications medium, the preparation of electronic instruction presents unique challenges (discussed in the next section).

Hypertext. One form of online documentation increasingly used in tutorial and training software is *hypertext,* information in electronic form designed for nonlinear reading. Unlike printed text, designed to be read front-to-back, a hypertext document offers various informational paths through the material, the particular path (or paths) determined by *what* and *how much* a reader wants to know. *Hypermedia* expands the applications of hypertext by adding graphics, sound, and animation.

Besides its instructional value, hypertext is a research and reference tool for navigating complex cross-references and retrieving information electronically.

In a hypertext system, a topic can be explored from any angle, at any level of detail. Assume, for instance, you are researching the AIDS epidemic with the use of a hypertext database. The database contains chunks of related topics, organized in a network (or web) of files linked electronically (Horton 22), as in Figure 6.4.

After accessing the initial file ("The AIDS Epidemic"), you navigate the network in any direction you wish, choosing which file to open and where to go next. (The files themselves might be printed words, graphics, sound, or animation.) Freed from the fixed page sequence of a printed document, users "customize" their searches.

Hypertext also accommodates different levels of audience needs by offering multiple "layers" of information, from general to specialized, as depicted in Figure 6.5.

Communication specialist William Horton explains the instructional power achieved through hypertext "layering":

Paper . . . lacks depth. All information must be on the same level or layer. With hypertext, however, the screen can have deeper reserves of information. These deeper layers do not clutter the screen, but are available if needed.

Hypertext can increase a reader's *depth* of knowledge

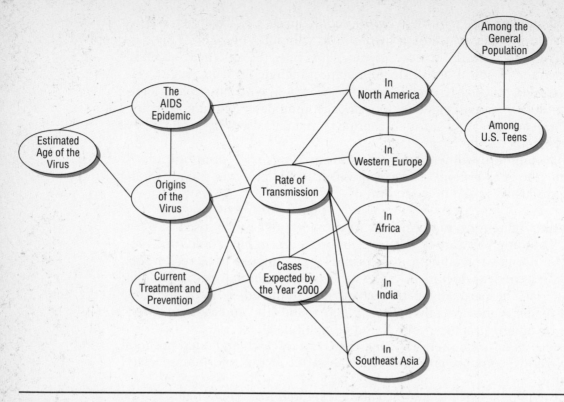

FIGURE 6.4 Topics (or Files) in a Hypertext Network Are Linked Electronically

[In Figure 6.5] Topic A is the top layer. It is written at the level of detail and generality appropriate for mainstream users [seeking basic information].

If this surface explanation is sufficient, the reader need look no deeper. However, Topics B and C lie just below the surface and may be brought forth by the selection of their references within the text of Topic A. . . . Even deeper layers are in Topics D, E, and F. (25)

Persons navigating a hypertext document "invent" their own text, and so can discover combinations, relationships, and chains of knowledge that could not possibly be expressed in the sequential pages of a printed document (Bernstein 42).

Following is an application of hypertext instruction in the auto industry:

A mechanic can zoom from a picture of a car engine to a video of a specific malfunctioning part, and then move directly to the text that tells how to fix it. (Morse 7)

For all their potential as research and instructional tools hypertext documents also have limitations:

- Confronting countless possible paths through the material, readers can get lost in "hyperspace." *(Where do I go next? How do I get out? How do I organize?)* Ease of navigation depends on how effectively individual chunks (or "nodes") of information are segmented and linked.

- Readers can resist a document that leaves them responsible for organizing their own learning. They generally prefer to rely on the writer for an orienting framework (Horton 26).

- Hypertext systems do not always enhance learning. They can in fact interfere with understanding and impede performance (Barfield, Hasel-

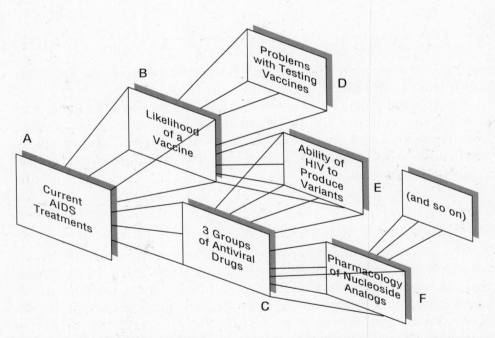

Source: Visual adapted from Horton, William. "Is Hypertext the Best Way to Document Your Product?" *Technical Communication* 39.1 (1991): 25.

FIGURE 6.5 Hypertext Topics Can Be Layered, to Serve Different Audiences korn, and Weatbrook 22, 27). Users in one study took longer to read a hypertext document than the equivalent paper document (Rubens 36).

Despite the limitations of its early stages, hypertext remains a promising medium that challenges communicators to make decisions like these:

- How much information should be included on one screen?
- How can each chunk be written so it can be read in any order and still make sense (Horton 27)?
- How can the material best be linked for easiest navigation of various possible paths (Nickels-Shirk 191)?
- Which combination of media should be employed (printed words, speech, sound effects, animation, music) (Horton 27)?

Clearly, creating hypertext documents is no one-person job: it requires writers, graphic artists, computer specialists, animators, and the like. This is one example of how communication technology makes collaboration both possible *and* necessary.

COMPUTERS DIVERSIFY OUR WORK RELATIONSHIPS

The growing complexity of science and technical projects requires that specialists work cooperatively.

Whether the product is a complex mechanism or a complex document, people work on it as a group. Shorter documents (letters, memos, short reports) usually are written alone, but, manuals, proposals, annual reports, newsletters, policy statements, and long reports usually are written collaboratively (Easton et al. 34).

Much Workplace Writing Is Collaborative

Today's documents, printed or paperless, face high standards for worthwhile content, sensible organization, readable style, and inviting format. Thus the organization producing, say, a software manual, relies on a team of specialists: writers, programmers, software engineers, graphic artists, editors, reviewers, marketing representatives, and lawyers (Debs, "Recent Research" 477). More than merely coauthoring, collaboration involves various specialists working to achieve a consensus (Debs, "Collaborative Writing" 37). Even a document with a "single" author can be a collaborative product. Figure 6.6, for instance, traces the process of writing, editing, revising, and reviewing to produce a document collaboratively.

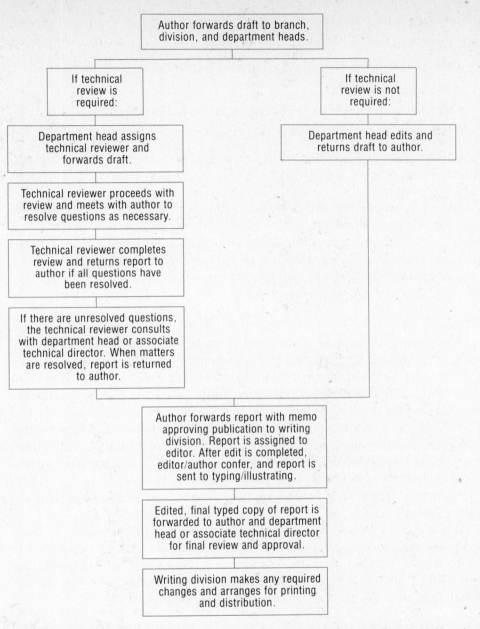

Author forwards draft to branch, division, and department heads.

If technical review is required:

If technical review is not required:

Department head assigns technical reviewer and forwards draft.

Department head edits and returns draft to author.

Technical reviewer proceeds with review and meets with author to resolve questions as necessary.

Technical reviewer completes review and returns report to author if all questions have been resolved.

If there are unresolved questions, the technical reviewer consults with department head or associate technical director. When matters are resolved, report is returned to author.

Author forwards report with memo approving publication to writing division. Report is assigned to editor. After edit is completed, editor/author confer, and report is sent to typing/illustrating.

Edited, final typed copy of report is forwarded to author and department head or associate technical director for final review and approval.

Writing division makes any required changes and arranges for printing and distribution.

Source: *NUSC Technical Publication Guidelines*. Naval Underwater Systems Center, Technical Information Dept., 1977.

FIGURE 6.6 How One Organization Collaborates to Produce a Report

Collaborative Writing Is a Social Enterprise

Teamwork has obvious benefits (Alred, Oliu, and Brusaw 37):

- We benefit from various perspectives and strengths.
- We receive feedback.
- In explaining and defending our ideas, we sharpen them.
- We avoid the stress and the sole responsibility of writing alone.

Instead of the actual writing, some team members might do the research, editing, reviewing, or proofreading. But each person expects a say about the final product.

The more important the document, the more it gets reviewed and revised for accuracy, appropriateness, usability, and legality (Kleimann 521). This process of review, revision, and verification makes workplace writing a "social enterprise" (Beard and Rymer 1), in which people have to put aside personal differences in order to get the job done.

Group Effort Can Lead to Interpersonal Conflict

As in any group enterprise, collaborating writers sometimes have difficulty working as a team.[1] For instance, some members don't get along because of differences in personality or working style or commitment or ability to take criticism. Some members disagree about what the group should accomplish, when it should meet, how much labor is required, who should be in charge, who should do what. Others disagree about ethical questions, about graphics or page design, about style, about document quality, about who should have final say. (And these interpersonal problems actually can worsen when the group transacts via electronic network.)

Figure 6.7 depicts the types of pressures that lead to conflict in collaborative groups.

While group conflict may be inevitable, it can also be beneficial: To transcend the limitations of *groupthink* (page 69), any group has to synthesize the *best* from each member. And the best ideas are often a synthesis of conflicting viewpoints. Conflict, well-managed, can generate dialogue that enhances understanding and critical thinking (Alred, Oliu, and Brusaw 36).

Collaboration Requires Definite Guidelines

Collaboration requires planning, self-assessment, and cooperation. The following guidelines focus on classroom projects in which people meet face-

[1] Adapted from Bogert and Butt, page 51; Burnett, pages 533–34; Hill-Duin, pages 45–46; Debs, "Collaborative Writing," page 38; Nelson and Smith, page 61. Full citations appear in Works Cited, pages 673–674.

FIGURE 6.7 How Pressures of a Group Effort Can Add Up

to-face, but these guidelines apply also to the workplace and to collaboration via electronic network.[2]

1. *Appoint a group manager.* This person will assign tasks, enforce deadlines, set the agenda, serve as chair/moderator for meetings, consult with supervisors (or instructor), and "run the show."

2. *Define the overall task.* Compose a purpose statement (as on pages 31, 59), so each member understands what needs to be done, and why, and for whom.

3. *Decide how the group will be organized.* Some possibilities:

   a. The group researches and plans together, but each person writes a different part of the document.

   b. Some members plan and research; one person incorporates this material into a complete draft; others review, revise, and produce the final version.

   c. Some other arrangement.

[2]Adapted from Debs, "Collaborative Writing," pages 38, 41; Hill-Duin, pages 45–46; Morgan, pages 540–41. Full citations appear in Works Cited, pages 673–674.

4. *Divide the task.* Who will be responsible for which parts of the document or which phases of the project (research, planning, drafting, reviewing, revising, typing, and document design)? Would the whole group be more effective with certain tasks (say, planning and editing)? Should only one person be responsible for the final revision? Which jobs are hardest?

5. *Evaluate each person's capabilities.* Who could best contribute what, exactly? Can anyone use page-design or graphics software? Are some stronger writers, or editors? If the report is to be presented orally, who is best suited?

6. *Decide on a meeting schedule and format.* How often will the group meet, and for how long? Who will take notes (or minutes, page 509), or will people take turns? Will the instructor attend, or participate?

7. *Establish a procedure for responding to others' work.* Will reviewing and editing be done in writing, face-to-face, as a group, one-on-one, or via computer? Will this process be overseen by the project manager, or the instructor?

8. *Establish procedures for dealing with group problems.* How will gripes and disagreements be aired (to the manager, the whole group, the "offending" individual)? How will disputes be resolved (by vote, the manager, the instructor)? How will irrelevant discussion be avoided or curtailed? Expect conflict, but try to use it positively.

9. *Establish a timetable.* With specific completion dates for each phase, everyone will know what is due, and when.

10. *Establish procedures and criteria for evaluating everyone's contribution, and design the evaluation forms.* Will the manager assess each member's performance and, in turn, be evaluated by members? Will members evaluate each other? What are the criteria (dependability, cooperation, effort, attitude)? Figure 6.8 depicts one possible form for a manager's evaluation of members. Equivalent criteria for evaluating the project manager could include "ability to organize the team," "fairness in assigning tasks," "ability to resolve conflicts," "ability to motivate," or other essential traits.

11. *Prepare a project management plan.* Record, *on paper,* all project elements: task assignments, completion dates for various drafts and phases, meeting schedule, and so on (Figure 6.9). Distribute copies of the plan to members and the instructor.

12. *Submit progress reports regularly.* Progress reports (pages 503, 639) enable everyone to track activities, problems, and rate of progress.

Performance Appraisal for _____

(After each item place an X in the column that applies.)

	Superior	Acceptable	Unacceptable
Dependability			
Cooperation			
Effort			
Quality of work			
Ability to meet deadlines			

Project Manager's signature

FIGURE 6.8 Sample Form for Evaluating Team Members

Each member might keep a journal of personal observations, not only for progress reports, but for overall evaluation of the project.

COMPUTERS DIVERSIFY OUR AUDIENCES

Our world is becoming a global community linked electronically, sharing social, political, and financial interests that demand cooperation as well as competition. As political barriers fall, products often have parts manufactured in one country (from imported materials), assembled elsewhere, and marketed in other countries (say, cars assembled in the United States for a "foreign" automaker, or farm equipment manufactured in the Far East for a "U.S." company). Even medical and environmental research crosses national boundaries, and professionals in all fields transact with colleagues from other cultures (Figure 6.10).

What documents are involved in global transactions? Page 103 offers a sampling of the almost limitless kinds of writing addressed to various levels of users (Weymouth 143).

Management Plan Sheet

Project title:

Audience:

Project manager:

Team members:

Statement of purpose:

Specific assignments

Research:

Planning:

Drafting:

Revising:

Preparing final document:

Presenting oral briefing:

Due Dates

Research due:

Plan and outline due:

First draft due:

Reviews due:

Revision due:

Final document due:

Progress report(s) due:

Work Schedule

Group meetings:	date	place	time	note-taker
#1				
#2				
#3				
etc.				

Mtgs. w/instructor:
   #1
   #2
   etc.

Miscellaneous

How will disputes and grievances be resolved?

How will performances be evaluated?

Other issues?

FIGURE 6.9 Sample Plan Sheet for Managing a Collaborative Project

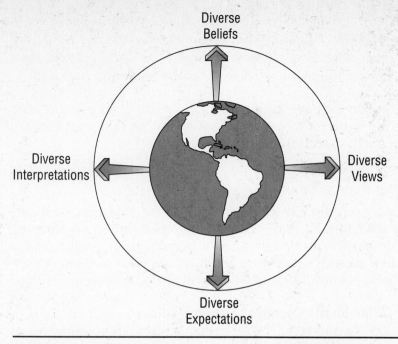

FIGURE 6.10 Global Transactions Enrich Audience Diversity

- scientific reports and articles on AIDS and other communicable diseases
- studies of global pollution and industrial emissions
- specifications for hydroelectric dams and other engineering projects
- operating instructions for video games, appliances, and other electronic equipment
- catalogs, promotional literature, and repair manuals

To communicate across national boundaries, these documents must address the expectations of their intended audiences. Besides language differences, communications must respect *cultural* differences. One writer offers this helpful definition of *culture*.

> Our accumulated knowledge and experiences, beliefs and values, attitudes and roles—in other words, our cultures—shape us as individuals and differentiate us as a people. Our cultures, inbred through family life, religious training, and educational and work experiences. . . . manifest themselves . . . in our thoughts and feelings, our actions and reactions, and our views of the world.

Culture shapes a reader's expectations and interpretations

Most important for communicators, our cultures manifest themselves in our information needs and our styles of communication. In other words, our cultures define our expectations as to how information should be organized, what should be included in its content, and how it should be expressed. (Hein 125)

Proficiency in the language of a particular culture is an asset for any communicator, but is no substitute for sensitivity to the values and frames of reference unique to that culture.

Cultures Differ in Expectations

All readers expect worthwhile content, sensible organization, readable style, and appealing format, but cultures define these criteria differently. Whether the document is to be translated or to remain in English, intercultural communication demands new levels of audience awareness. Following is a sampling of how cultural differences affect reader expectations.

Expectations Differ about Content.   Cultures differ in their notions of worthwhile *content*. German readers, for example, value thoroughness and complexity; they expect details, facts, background, and explanations, all in businesslike tone. But Japanese readers expect multiple perspectives on the material, with liberal use of graphics and a friendly, encouraging tone (Hein 125–26).

U.S. business culture expects messages that spell out the meaning directly, but some Asian cultures prefer indirect and somewhat ambiguous messages, which leave explanations and interpretation for readers to decipher (Leki 151; Martin and Chaney 276–77).

Expectations Differ about Organization.   A document considered well organized by one culture might confuse or offend another. For instance, English paragraphs typically proceed from an orienting statement to specific support (as shown in Chapter 12); any digressions seem to violate *unity*. But some cultures consider digression a sign of intelligence. To English readers, the long introductions and digressions in, say, Spanish or Russian documents might seem tedious and confusing, but those cultures might view the more direct organization of English as simple-minded (Leki 151).

Expectations differ even among same-language cultures. British letters, for instance, typically express the bad news directly up front, instead of the indirect approach preferred in the United States. Therefore, a bad-news letter appropriate for U.S. readers could be interpreted by British readers as evasive (Scott and Green 19).

Expectations Differ about Style.   Cultures differ in *style* expectations. As opposed to "plain English," for example, Germans prefer long sentences and complex language, to convey an idea's full complexity. Japanese style can seem abstract and elusive, tempered with feeling and respect (Hein 125–26) and relying on the liberal use of "please" and other terms of politeness and praise (Mackin 349–50). Documents in Spanish tend to be more formal than in English, and German documents rely more heavily on passive voice (Weymouth 144).

A translated document might lack equivalent words to convey the exact meaning (Mackin 349). Also, a language may have multiple versions of the same expression, one version usually considered more formal and polite (Caswell-Coward 266). For instance, the French "How are you?" can be the formal *"Comment allez-vous?"* or the informal *"Comment ça va?"* The wrong version can seem pretentious or presumptuous, so be aware of your audience: offensive communication could alienate them, toward your company *and* your culture (Sturges 32).

Expectations Differ about Correctness.   One culture's *grammar and usage* may be incorrect for another. For instance, Arabic and Persian sentences typically begin with a lowercase letter, and "I" appears in lowercase (Leki 149). Even hyphens and other punctuation differ among languages.

Expectations Differ about Format.   A *format* appropriate in one culture may be inappropriate in another. For instance, Arabic and Persian text is written through the line instead of on the line, and from right to left instead of left to right (Leki 149). In other cultures, readers move up and down, instead of across the page. Be aware that a particular culture might be offended by icons, or by a typeface that seems too plain or fancy (Weymouth 144).

Expectations Differ about Legal Standards.   Does the document meet the country's *legal standards* for safety, health, and other regulations? Instructions for any product requiring assembly or operation have to carry appropriate warnings. Each country has standards with which product literature must comply. Inadequate documentation, as judged by that country's standards, can result in a lawsuit (Caswell-Coward 264–66; Weymouth 145).

Audience analysis, at best, is an imprecise art. But we communicate more deliberately when we are sensitive to the cultural background that influences a reader's expectations.

Global Communication Requires Cultural Fluency

Cultural fluency means understanding not only a culture's language, but also its behaviors, values, and attitudes, as in Figure 6.11 (Beamer 293–94).

More specific questions appear in Figure 6.12. (But each situation raises different questions.) And although the answers help reveal an audience's frame of reference, they provide only a partial picture.

Stereotypes Do Not Equate with Cultural Fluency

Awareness of a culture's characteristics can help us understand its frame of reference. For example, Korean society is male-dominated, and so women face inequity in the workplace. Japanese prefer group decision-making, while Koreans have a top-down, authoritarian structure (Lee and Jablin 203, 205). One researcher offers additional stereotypes and cautions on page 108.

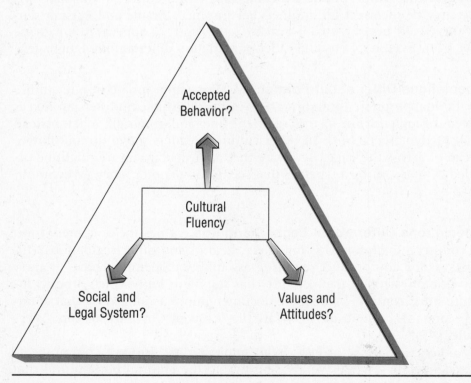

FIGURE 6.11 Analyzing Behaviors, Values, and Attitudes of Multicultural Audiences

Questions for Analyzing Multicultural Audiences

ABOUT ACCEPTED BEHAVIOR
- Does the culture observe certain formalities for making requests, expressing disagreement, offering criticism or praise?

- Is there an accepted form for greetings and introductions (first names, last names, titles)?

- Do people enjoy haggling during negotiation, or is this seen as bad taste?

- Do people interact on a more casual or formal level?

ABOUT THE SOCIAL AND LEGAL SYSTEM
- Is the social/political system inflexible or open? Does it have a history of fiercely democratic and egalitarian ideals or rank-conscious, authoritarian rule?

- Does the legal system enforce contractual arrangements? How are bribes looked upon? How important to this culture is the law in general? Who makes the contract decisions? How formal is the contract process—a simple handshake or an extensive legal document? Is gift-giving seen as bribery or as a display of respect?

ABOUT VALUES AND ATTITUDES
- What is the culture's attitude toward technology and change, competition, youth versus age, gender equality, risk taking, status?

- Does the culture value immediate reward or delayed gratification, progress or stability?

- Do people see themselves as rugged individuals or more group oriented?

- Do people emphasize results or relationships?

- Do people believe strongly in luck or destiny?

- How do people view our culture (with admiration, contempt, trust, fear)?

Source: Questions adapted from Beamer, Linda. "Learning Intercultural Communication Competence." *The Journal of Business Communication* 29.3 (1992): 293–95; Martin, Jeanette S. and Lillian H. Chaney. "Determination of Content for a Collegiate Course in Intercultural Business Communication by Three Delphi Panels." *The Journal Of Business Communication* 29.3 (1992): 271–77.

FIGURE 6.12 Questions for Analyzing Multicultural Audiences

Arabs like to stand very near a listener when speaking; Chinese always initially refuse hospitality when offered; Latin Americans only like to do business with people who show consideration for their family affairs. These categories enable us to familiarize an unfamiliar culture, and to make it comprehensible. . . . [But] these examples are stereotypes. . . . They may be accurate to some degree, but they are limited insights, revealing only a part of the whole culture. Two or three stereotypes of one culture, or even a dozen or a score, do not equal an understanding of that culture. (Beamer 293–94)

No amount of stereotyped features can replace our intuitive sensitivity toward those with whom we transact.

Getting to "know" a different culture can take a lifetime, but awareness and thoughtfulness go a long way to help us avoid misunderstanding or misinterpretation.

Global Communication Requires Ethical Standards

Beyond the inadvertant types of misunderstanding described earlier, intercultural documents carry great potential for the deliberate communication abuses described in Chapter 5. For instance, based on its level of business experience, technological development, or financial need, a particular culture might be especially vulnerable to deception, manipulation, or other exploitation. Communicators are responsible for ensuring that all documents in all cultural contexts embody universal standards of honesty and fairness.

A CHECKLIST FOR INTERCULTURAL DOCUMENTS[3]

Use this checklist to verify that your documents respect audience diversity. (Page numbers in parentheses refer to the first page of discussion.)

☐ Does the document avoid offending anyone? (105)

☐ Is the document sensitive to the culture's social and political values? (107)

☐ Does the document conform to the safety and regulatory standards of the country? (105)

☐ Does the document provide the expected level of detail? (104)

[3]This list was largely adapted from Caswell-Coward, page 265; Weymouth, page 144; Beamer, pages 293–95; Martin, pages 271–77. Full citations appear in Works Cited, pages 673–674.

☐ Does the document avoid possible misinterpretation? (104)

☐ Is the document organized in a way that readers will consider appropriate? (104)

☐ Does the document observe interpersonal conventions important to the culture (accepted forms of greeting or introduction, politeness requirements, first names, last names, and so on)? (107)

☐ Does the document's tone reflect the appropriate level of formality or casualness one would expect? (107)

☐ Is the document's style appropriately direct or indirect? (105)

☐ Is the document's format consistent with the culture's expectations? (105)

☐ Does the document embody universal standards for ethical communication? (108)

EXERCISE

Keep a journal during a collaborative project, noting carefully what succeeded and what didn't, what interpersonal conflicts developed and how they were resolved, what other issues contributed to progress or delay, the role and effectiveness of electronic tools, and so on. In a memo report to your classmates and instructor, summarize the achievements and setbacks in your project and prepare recommendations for improving collaboration on future projects.

NOTE: In this evaluation/recommendation report, avoid attacking or blaming or offending anyone. Offer constructive suggestions for improving collaborative work *in general*.

COLLABORATIVE PROJECT (GROUP OR INDIVIDUAL)

Assume that you work for an environmental consulting firm that is under contract with various countries for a range of projects, including these:

- A plan for rain forest regeneration in Latin America and Sub-Saharan Africa
- A plan to decrease industrial pollution in Eastern and Western Europe
- A plan for "clean" industries in developing countries
- A plan for organic agricultural development in Africa and India
- A joint American/Canadian plan to decrease acid rain
- A plan for developing alternative energy sources in the Near East, Far East, and Southeast Asia

Each project will require environmental impact statements, feasibility studies, grant proposals, and a legion of other documents, often prepared in collaboration with members of the host country, and, in some cases, prepared by your company for audiences in the host country: from political, social, and industrial leaders to technical experts and so on.

For such projects to succeed, people from different cultures have to communicate effectively and sensitively, creating goodwill and cooperation.

Before your company begins work in earnest with a particular country, your co-workers will need to develop some degree of cultural fluency. Your assignment is to select a country and to research that culture's behaviors, attitudes, values, and social system in terms of how these variables influence the culture's communication preferences and expectations. What should your colleagues know about this culture in order to communicate effectively and diplomatically?

Prepare a recommendation report in memo form (as on pages 499–501), and be prepared to present your findings in class.

The Problem-Solving Process Illustrated

Critical Thinking in the Writing Process

A Sample Writing Situation

Your Own Writing Situation

■   ■   ■

Every writing situation requires deliberate decisions for *working with the information* and for *planning, drafting,* and *revising* the document. Some of these decisions are illustrated in Figure 7.1. Each writer approaches the process through a sequence of decisions that works best for *that* person. And no group of decisions is complete until *all* groups are complete. This looping structure of the writing process is illustrated in Figure 7.2.

CRITICAL THINKING IN THE WRITING PROCESS

The writing process is a *critical thinking* process: the writer makes a series of deliberate decisions in response to a situation. And the actual "writing" (putting words on the page) is only a small part of the overall process— probably the least significant part.

In this chapter, we will follow one working writer through an important writing situation; we will see how he solves his unique information, persuasion, ethics, and adaptation problems and how he designs a useful and efficient document.

A SAMPLE WRITING SITUATION

The company is Microbyte, maker of dedicated word processors and portable microcomputers. The writer is Glenn Tarullo (B.S., Management; Minor: Computer Science). Glenn has been on the job two months as Assis-

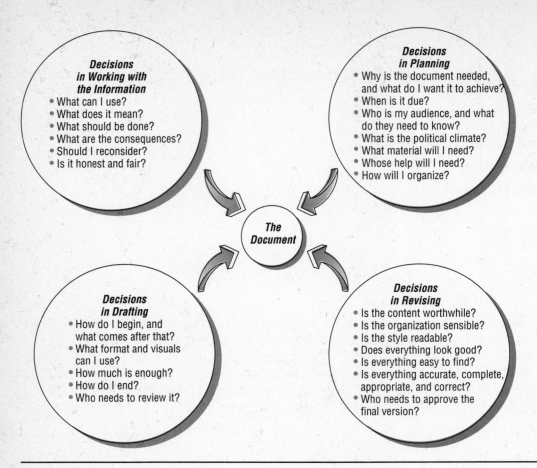

Decisions in Working with the Information
- What can I use?
- What does it mean?
- What should be done?
- What are the consequences?
- Should I reconsider?
- Is it honest and fair?

Decisions in Planning
- Why is the document needed, and what do I want it to achieve?
- When is it due?
- Who is my audience, and what do they need to know?
- What is the political climate?
- What material will I need?
- Whose help will I need?
- How will I organize?

The Document

Decisions in Drafting
- How do I begin, and what comes after that?
- What format and visuals can I use?
- How much is enough?
- How do I end?
- Who needs to review it?

Decisions in Revising
- Is the content worthwhile?
- Is the organization sensible?
- Is the style readable?
- Does everything look good?
- Is everything easy to find?
- Is everything accurate, complete, appropriate, and correct?
- Who needs to approve the final version?

FIGURE 7.1 Typical Decisions During the Writing Process

tant Training Manager for Microbyte's Marketing and Customer Service Division.

For three years, Glenn's boss, Marvin Long, periodically has offered a training program for new managers. Long's program combines an introduction to the company with instruction in management skills (time management, motivation, communication). Long seems satisfied with his two-week program, but has asked Glenn to evaluate it and write a report as part of a company move to upgrade personnel procedures.

Glenn knows his report will be read by Long's boss, George Hopkins (Assistant Vice President, Personnel), and Charlotte Black (Vice President, Marketing, the person who devised the upgrading plan). Copies will go to other division heads, and to the division's chief executive.

Decisions in the writing process are recursive: No one stage of decisions is complete until *all* stages are complete.

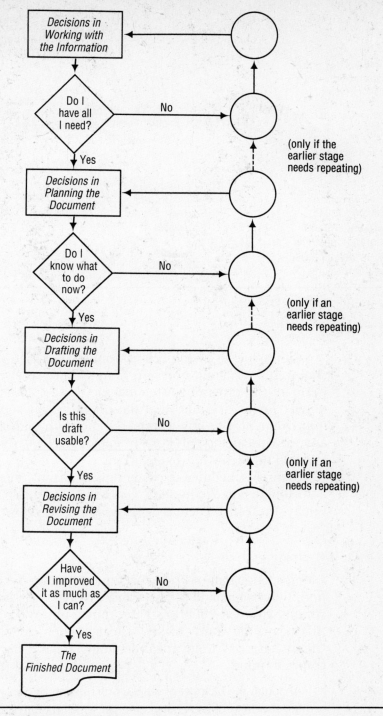

FIGURE 7.2 A Flowchart for the Writing Process

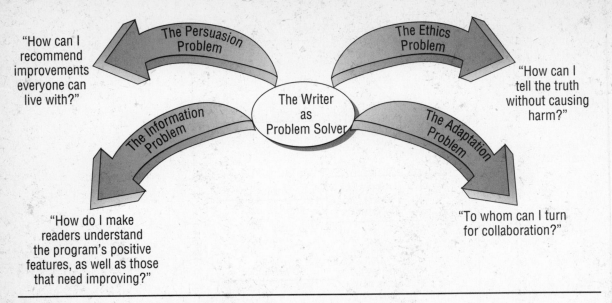

"How can I recommend improvements everyone can live with?"

The Persuasion Problem

The Ethics Problem

"How can I tell the truth without causing harm?"

The Writer as Problem Solver

The Information Problem

The Adaptation Problem

"How do I make readers understand the program's positive features, as well as those that need improving?"

"To whom can I turn for collaboration?"

FIGURE 7.3 Glenn's Fourfold Problem

Glenn spends two weeks (Monday, October 3 to Friday, October 14) sitting in and taking notes on Long's classes. On October 14, the trainees evaluate the program. After reading these evaluations and reviewing his notes, Glenn concludes that the program was successful, but could stand improvement. But how can he tell the truth without harming or offending anyone (instructors, his boss, or guest speakers)? Figure 7.3 depicts Glenn's fourfold problem.

Glenn is scheduled to present his report in conference with Long, Black, and Hopkins on Wednesday, October 19. Right after the final class (1 P.M., Friday, the 14th), Glenn begins work on his report.

Working with the Information

Glenn spends half of Friday afternoon fretting over the details of his situation, the readers and other people involved, the political realities and constraints and consequences. (He knows no love is lost between Long and Black, and he wants to steer clear of their ongoing conflict.) By 3 P.M., Glenn hasn't written a word. Desperate, he decides to write whatever comes to mind:

Glenn's first draft

> Although the October Management Training Session was deemed quite successful, several problems have emerged which require our immediate attention.

- Too many of the instructors had poor presentation skills. A few never arrived on time. One didn't stick to the topic but rambled incessantly. Jones and Wells seemed poorly prepared. Instructors in general seemed to lack any clear objectives. Also, because too few visual aids were used, many presentations seemed colorless and apparently bored the trainees.

- The trainees (all new people) were not at all cognizant of how the company was organized or functioned. So the majority of them often couldn't relate to what the speakers were talking about.

- It is my impression that this was a weak session due to the fact that there were insufficient members (only five trainees). Such a small class makes the session a waste of time and money. For instance, Lester Beck, Senior Vice President of Personnel, came down to spend over one hour addressing only a handful of trainees. Another factor is that with fewer trainees in a class, less dialogue occurs, with people tending to just sit and get "talked at."

- Last but not least, executive speakers generally skirted the real issues, saying nothing about what it was really like to work here. They never really explained how to survive politically (e.g., never criticize your superior; never complain about the hard work or long hours; never tell anyone what you *really* think; never observe how few women are in executive or managerial positions, or how disorganized things seem to be). New employees shouldn't have to learn these things the hard way.

In the final analysis, if these problems can be addressed immediately, it is my opinion we can look forward in the future to effectuating management training sessions of even higher quality than those we now have.

Glenn completes this draft of his written reaction to the meeting at 5:10 P.M. Displeased with the results, but not sure how to improve the piece, he asks an experienced colleague for advice and feedback. Blair Cordasco, a senior project manager, has already helped Glenn several times. Cordasco agrees to read Glenn's draft over the weekend.

At 8:05 Monday morning, Cordasco reviews the document with Glenn. First, she points out obvious style problems: wordiness ("due to the fact that"), jargon ("effectuating"), triteness ("in the final analysis"), implied bias ("weak presentation," "skirted"), among others. Can you identify other style problems in Glenn's draft?

Cordasco points out other problems: The piece is disorganized. And even though Glenn is being honest, he isn't being particularly fair: the emphasis is too critical (making Glenn's boss look bad to his superiors); and the views are too subjective (no one is interested in hearing Glenn gripe about the company's political problems). The report lacks persuasive force because it contains little useful advice for solving the problems he identifies. In this form, the report will only alienate people, and harm Glenn's career.

The tone is bossy and judgmental. Glenn is in no position to make this kind of *power connection* (p. 38); he needs to be more fair, diplomatic, and reasonable.

Planning the Document

Glenn realizes he needs to begin by *focusing* on his writing situation. His audience-and-use analysis goes like this:[1]

I'd better decide *exactly* what my primary reader wants.

Long requested the report, but only because Black developed the scheme for division-wide improvements. And so I really have two primary readers: my boss and the big boss.

My major question here: Am I including enough detail for all the bosses? And the answer to this question will require answers to more specific questions:

<table>
<tr><td rowspan="4">Anticipated
readers'
questions</td><td>What are we doing right, and how can we do it better?</td></tr>
<tr><td>What are we doing wrong, and does it cost us money?</td></tr>
<tr><td>Have we left anything out, and does it matter?</td></tr>
<tr><td>How, specifically, can we improve the program, and how will those improvements help the company?</td></tr>
</table>

Because all readers have participated in these sessions (as trainees, instructors, or guest speakers), they don't need background explanations.

I should begin with the *positive* features of the last session. Then I can discuss the problems and make recommendations. Maybe I can eliminate the bossy and judgmental tone by *suggesting improvements* instead of *criticizing weaknesses*. Also, I could be more persuasive by describing the *benefits* of my suggestions.

Glenn realizes that if he wants successful future programs, he can't afford to alienate anyone. After all, he wants to be seen as a loyal member of the

---

[1]Throughout this chapter, Glenn's analysis will address *all* the areas illustrated on the audience-and-use profile sheet (page 60).

company, yet preserve his self-esteem and demonstrate he is capable of making objective recommendations.

Now, I have a clear enough sense of what to do.

Statement of purpose The purpose of my document is to provide my supervisor and interested executives with an evaluation of the workshop by describing its strengths, suggesting improvements, and explaining the benefits of these changes.

From this plan, I should be able to revise my first draft. But . . . that first draft lacks important details. I should brainstorm to get *all* the details (including the *positive* ones) I want to include.

Glenn's Brainstorming List.   Glenn's first draft touched on several topics; incorporating them into his brainstorming, he came up with the following list.

1. better-prepared instructors and more visuals
2. on-the-job orientation *before* the training session
3. more members in training sessions
4. executive speakers should spell out qualities needed for success
5. beneficial emphasis on interpersonal communication
6. need follow-up evaluation (in six months?)
7. four types of training evaluations:
   a. trainees' reactions
   b. testing of classroom learning
   c. transference of skills to the job
   d. impact of training on the organization (high sales, more promotions, better-written reports)
8. videotaping and critiquing of trainee speeches worked well
9. acknowledge the positive features of the session
10. ongoing improvement ensures quality training
11. division of class topics into two areas was a good idea
12. additional trainees would increase classroom dialogue
13. the more trainees in a session, the less time and money wasted
14. instructors shouldn't drift from the topic
15. on-the-job training to give a broad view of the division

16. clear course objectives, to increase audience interest and to measure the program's success
17. Marvin Long has done a great job with these sessions over the years

By 9:05 A.M., the office is hectic. Glenn puts his list aside to spend the day on work piling up. Not until 4 P.M. does he return to his report.

Now what? I should delete whatever my audience already knows or doesn't need, or whatever seems unfair or insincere: 7 can go (this audience needs no lecture in training theory); 14 is too negative and critical—besides, the same idea is stated more positively in 4; 17 is obvious brown-nosing, and I'm in no position to make such grand judgments.

Maybe I can unscramble this list by arranging items within categories (strengths, suggested changes, and benefits) from my statement of purpose.

Glenn's Brainstorming List Rearranged.   Notice here how Glenn discovers additional *content* (see italic type) while he's deciding about *organization.*

Strengths of the Workshop

- division of class topics into two areas was useful
- emphasis on interpersonal communication
- videotaping of trainees' oral reports, followed by critiques

Well, that's one category done. Maybe I should combine *suggested changes* with *benefits,* since I'll want to cover them together in the report.

Suggested Changes/Benefits

- more members per session would increase dialogue and use resources more efficiently
- varied on-the-job experiences before the training sessions would give each member a broad view of the marketing division
- executive speakers should spell out qualities required for success *and future sessions should cover professional behavior, to provide trainees with a clear guide*
- follow-up evaluation in six months *by both supervisors and trainees would reveal the effectiveness of this training and suggest future improvements*
- clear course objectives and more visual aids to increase *instructor efficiency* and audience interest

Now that he has a fairly sensible arrangement, Glenn can get this list into report form, even though he probably will think of more material to add as he works. Since this is *internal* correspondence, he employs a memo format.

Drafting the Document

Glenn then produces a usable draft—one containing just about everything he wants to cover. (Sentences are numbered for our later reference.)

[1]In my opinion, the Management Training Session for the month of October was somewhat successful. [2]This success was evidenced when most participants rated their training as "very good." [3]But improvements still are needed.

[4]First and foremost, a number of innovative aspects in this October session proved especially useful. [5]Class topics were divided into two distinct areas. [6]These topics created a general-to-specific focus. [7]An emphasis on interpersonal communication skills was the most dramatic innovation. [8]This innovation helped class members develop a better attitude toward things in general. [9]Videotaping of trainees' oral reports, followed by critiques, helped clarify strengths and weaknesses.

[10]There is a detailed summary of the trainees' evaluations attached. [11]Based on these and on my past observations, I have several suggestions.

- [12]All management training sessions should have a minimum of ten to fifteen members. [13]This would better utilize the larger number of managers involved and the time expended in the implementation of the training. [14]The quality of class interaction with the speakers would also be improved with a larger group.
- [15]There should be several brief on-the-job training experiences in different sales and service areas. [16]These should be developed *prior* to the training session. [17]This would provide each member with a broad view of the duties and responsibilities in all areas of the marketing division.
- [18]Executive speakers should take a few minutes to spell out the personal and professional qualities essential for success with our company. [19]This would provide trainees a concrete guide to both general company and individual supervisors' expectations. [20]Additionally, by the next training session we should develop a presentation dealing with the attitudes, manners, and behavior appropriate in the business environment.
- [21]Do a six-month follow-up. [22]Get feedback from supervisors as well as trainees. [23]Ask for any new recommendations. [24]This would provide a clear assessment of the long-range impact of this training on an individual's job performance.

A later draft

- [25]We need to demand clearer course objectives. [26]Instructors should be required to use more visual aids and improve their course structure based on these objectives. [27]This would increase instructor quality and audience interest.
- [28]These changes are bound to help. [29]Please contact me if you have further questions.

Although now developed and organized, this version still is some way from the finished document on pages 123–24. Glenn has to make further decisions about his style, content, arrangement, audience, and purpose.

Revising the Document

At 8:15 Tuesday morning, Glenn decides on a sentence-by-sentence revision for worthwhile content, sensible organization, and readable style.

Sentence 1 begins with a needless qualifier, has a redundant phrase, and sounds insulting ("somewhat successful"). Sentence 2 should be in the passive voice, to emphasize the training—not the participants. Also, 1 and 2 are choppy and repetitious, and so I'll combine them.

Original	In my opinion, the Management Training Session for the month of October was somewhat successful. This success was evidenced when most participants rated their training as "very good." (28 words)
Revised	The October Management Training Session was successful, with training rated "very good" by most participants. (15 words)[2]

Sentence 3 is too blunt. I need an orienting sentence here that forecasts content diplomatically. I can be honest without being so negative.

Original	But improvements still are needed.
Revised	A few changes—beyond the recent innovations—should result in even greater training efficiency.

In sentence 4, "First and foremost" is trite, "aspects" is a clutter word, and word order needs changing to improve the emphasis (on innovations) and to lead into the examples.

---

[2]Notice throughout how careful revision sharpens the writer's meaning while cutting needless words.

Original	First and foremost, a number of innovative aspects in this October session proved especially useful.
Revised	Especially useful in this session were several program innovations.

Sentences 5 and 6 need combining, and content in 5 needs beefing up: *name* the two areas. If 7 is labeled the "most dramatic innovation," it ought to come last, for emphasis. In 8, "things in general" is too indefinite. And 7 and 8 need combining. Sentence 9 seems okay. All three examples would be more readable in a *list*, and with parallel phrasing (maybe starting each with an "ing" phrase to signal the "action" verb, as in 9).

Original	Class topics were divided into two distinct areas. This created a general-to-specific focus. An emphasis on interpersonal communication was the most dramatic innovation. This helped class members develop an enthusiastic and relaxed attitude toward things in general. Videotaping of trainees' oral reports, followed by critiques, helped clarify strengths and weaknesses.
Revised	• Dividing class topics into two areas created a general-to-specific focus: *the first week's coverage of company structure and functions created a context for the second week's coverage of management skills.*[3] • Videotaping and critiquing trainees' oral reports clarified strengths and weaknesses. • Emphasizing interpersonal communication skills *(listening, showing empathy, and reading nonverbal feedback) generated enthusiasm and a sense of ease about the group, their training, and the company.*

And while I'm thinking about format, a couple of clear headings ("Workshop Strengths," "Suggested Changes/Benefits") would segment the text, improving readability and appearance.

Glenn follows this editing and revising process throughout. (Exercise 3 asks you to identify the remaining changes.) Wednesday morning, after

---

[3]Notice that this revision has more words, but also much more concrete and specific detail (in italic type). Completeness of information *always* takes priority over word count.

much revising and proofreading, Glenn prints out the *final* draft shown in Figure 7.4.

Glenn's final report is both informative and persuasive. But this document did not appear magically. Glenn made deliberate decisions about purpose, audience, content, organization, and style. Most important, he *spent time revising.*[4]

YOUR OWN WRITING SITUATIONS

Writers work in different ways. Some begin by brainstorming. Some begin with an outline. Others hate outlining and simply write and rewrite. Some write a quick draft before thinking through their writing situation. Introductions and titles often are written last. No one "step" in the process is complete until *the whole* is complete. (Notice, for instance, how Glenn sharpens his content *and* style while he organizes.) Every document you write will require *all* these decisions but rarely will you make them in the same sequence from day to day.

No matter what the sequence, *revision* is a fact of life. It is the one *constant* in the writing process. When you've finished a draft, you have, in a sense, only begun. Sometimes you will have more time to compose than Glenn did, sometimes much less. Whenever your deadline allows, leave time to revise.

EXERCISES

1. Think of a course you've taken that had both definite strengths and weaknesses. Assume that the instructor has asked you to evaluate the course and recommend improvements. Your secondary audience will be your instructor's chairperson and the Dean of Faculty. The instructor is up for tenure and is someone you like very much. You need to be candid in your report, but you don't want to make the instructor look bad. Write the report, being as specific as possible. (*Note:* Use a fictional name for the instructor and the course.)

2. Assume that you are a training manager for XYZ Corporation. After completing this first section of the text and the course, what advice about the writing process would you have for a beginning writer who will frequently need to write reports on the job? In a one- or two-page (single-spaced) memo to new employees, explain the writing process briefly, and give a list of guidelines these beginning writers can follow.

---

[4]A special thanks to Glenn Tarullo for his perseverance. I made his task doubly difficult by having him explain each of his decisions during this writing process.

MICRO*BYTE*

October 19, 1993

To: Marvin Long
From: Glenn Tarullo
Subject: October Management Training Program: Evaluation
         and Recommendations

The October Management Training Session was successful, with training rated as "very good" by most participants. A few changes, beyond the recent innovations, should result in even greater training efficiency.

Workshop Strengths
Especially useful in this session were several program innovations:

—Dividing class topics into two areas created a general-to-specific focus: The first week's coverage of company structure and functions created a context for the second week's coverage of management skills.

—Videotaping and critiquing trainees' oral reports clarified their speaking strengths and weaknesses.

—Emphasizing interpersonal communication skills (listening, showing empathy, and reading nonverbal feedback) created a sense of ease about the group, the training, and the company.

Innovations like these ensure high-quality training. And future sessions could provide other innovative ideas.

Suggested Changes/Benefits
From the trainees' evaluation of the October session (summary attached) and my observations, I recommend these additional changes:

—We should develop several brief (one-day) on-the-job rotations in different sales and service areas before the training session. These rotations would give each member a real-life view of duties and responsibilities throughout the company.

—All training sessions should have at least 10 to 15 members. Larger classes would make more efficient use of resources and improve class-speaker interaction.

FIGURE 7.4 Glenn's Final Draft

Long, Oct. 19, 1993, page 2

Supports each
recommendation
with convincing
reasons

—We should ask instructors to follow a standard format (based on definite course objectives) for their presentation, and to use visuals liberally. These enhancements would ensure the greatest possible instructor efficiency and audience interest.

—Executive speakers should spell out personal and professional traits essential in our company. Such advice would give trainees a concrete guide to both general company and individual supervisor expectations. Also, by the next training session, we should assemble a presentation dealing with the attitudes, manners, and behavior appropriate in business.

—We should do a six-month follow-up of trainees (with feedback from supervisors as well as ex-trainees) to gain long-term insights, to measure the influence of this training on job performance, and to help design advanced training.

Closes by appealing to
shared goals (efficiency
and profit)

Inexpensive and easy to implement, these changes should produce more efficient training.

Copies: B. Hull, C. Black, G. Hopkins, J. Capilona, P. Maxwell,
            R. Sanders, L. Hunter

FIGURE 7.4 Glenn's Final Draft *Continued*

3. Compare Glenn's second draft (page 119) with his final draft (pages 123–24). Identify all improvements in content, arrangement, and style besides those discussed on pages 120–22.

COLLABORATIVE PROJECT

Revising a Document

After several weeks in Technical Writing class, you have a good sense of *what* material is covered and of *how* the material is taught. Now, imagine that at a recent meeting your school's Advanced Writing Committee passed this motion:

> Because of the popularity of Technical Writing, and the 200 percent enrollment increase within two years, many new sections of the course have been added. To ensure a unified program, we suggest that all sections follow a standard syllabus and similar teaching approaches.

For help in developing a standard model, the committee has decided to survey students about to complete the Technical Writing course. Each student has been asked to submit a memo evaluating the section, with suggestions for improvement. The responses will form a databank for the committee's decisions about a course model that meets students' needs.

As a guide for evaluation, the committee has provided this question:

> How well do the content and teaching approach in your Technical Writing course fulfill your needs and expectations? How can this course best be taught, and what material should be covered? Be specific in your evaluation of strengths and weaknesses. Along with suggestions for improvement, explain how a specific change or improvement would benefit you.

Assume that Frank White, a student in some other section, has responded with this memo:

> Date: April 25, 19XX
> To: The Advanced Writing Committee
> From: Frank White, Technical Writing Student
> Subject: Section 1499: Evaluation and Recommendation
>
> This course is providing me with a great deal of useful information. Also, the teacher usually manages to hold the attention of the class very well. Learning about writing can be a pretty boring experience, but this class hardly ever is boring because the teacher does such a good job of making the material interesting. Also, the teacher is a very nice person. I'm happy to say that I've learned a number of approaches that have helped me improve my writing.

Writing skills are important in just about anyone's career, and so I'm glad we have the chance to take a Technical Writing course before graduation. Having the course as a requirement is a good idea, because many students (myself included) probably would avoid any course that requires this much writing. It isn't until we get here that we realize how worthwhile (and difficult!) this course is for everyone. I'd like to see every section taught the way this one is: by a teacher who knows how to get a tough job done.

The only complaint I have against this teacher is the fact that he talks too much about computers. I know that computers are important, but I'd like to learn to use one for my writing instead of just being exposed to computers in general. And too many writing assignments have been saved for the latter part of the course. Everything is just stacking up, leaving me buried and confused.

Also, the class is much too large, and the layout is awful. With so many students, the teacher has no way of providing individual attention, and so too many students "get lost in the crowd." We need a classroom layout that would make group editing easier.

I really enjoyed the audiovisual presentations in class, and would like to see more of them. All in all, the concrete stuff was always the most useful.

In general, the material has been covered well, except for the material on oral communication.

With a few minor changes, this course would be excellent.

Before Frank can submit this memo, it will need heavy revision for content that is informative and persuasive and ethical, organization that is easy to follow, and style that is readable. Working in teams, revise Frank's memo.

a. As a first step, complete an audience-and-use profile sheet (page 60) based on the following data, as well as on the details given earlier.

Primary audience: The writing committee and the department chair. Several committee members also are on the tenure committee, and they are likely to use this information for an additional purpose: to evaluate Frank's instructor for tenure. These are the decision makers. Above all, Frank wants to convey a *positive* impression of his instructor.

Secondary audience: Frank's instructor (whom Frank likes, but who deserves an honest and detailed evaluation). Frank wants to be fair to his instructor, but he also wants to make some realistic suggestions for improving a course so important to everyone's career.

b. Assume you helped Frank prepare the brainstorming list he should have prepared *before* writing the previous draft. To visualize what it was that specifically pleased or displeased Frank, imagine that these items would have appeared on his list—if Frank had taken the time to brainstorm:

- instructor is always willing to help students individually
- instructor always takes time in class to answer all questions thoroughly
- emphasis on planning, drafting, and revising for a specific audience is helpful
- instructor spends a lot of time encouraging us
- instructor spends too much time talking about computers and automated offices—what I need is to develop strong writing skills by *using* the computer
- now and then we should have a guest lecturer from business and industry
- we should spend more time discussing the documents *we* have prepared
- instructor spends too much time emphasizing mistakes—not enough time on the positive
- instructor should spend more time on writing, and less than the present four weeks on oral communication
- when they are used (which is not often enough), the opaque and overhead and slide projectors make things more vivid and interesting
- instructor gives a lot of feedback on our papers
- the class should have an IBM PC and a Macintosh computer (since these are the kinds we have in our campus micro labs), to illustrate how automation can affect the writing and revising process
- we should have a class with several round tables that seat 4–6 students for editing groups
- class size should be reduced from 30 to no more than 20 students
- tutoring should be available on a regularly scheduled basis for students with any type of writing problem
- we should spend at least one full week on job applications
- too few editing assignments early in the course; thus, too many at the end
- instructor should begin discussing the long report (term project) as early as the first or second week

From Frank's original draft and from his brainstorming list, select only that which is useful and appropriate for *this* audience and purpose. Compose a final draft of Frank's memo. Appoint one team member to present the document in class.

Information Retrieval, Analysis, and Synthesis

Acquiring Information

Thinking Critically About the Research Process

Searching the Literature

Using Electronic Information Services

Exploring Primary Sources

■   ■   ■

Significant research serves some purpose: to answer a question, to make an evaluation, to establish a principle. Through research you can find your own answers, check your opinions against the facts, reach a conclusion that has the best chance of being valid. In the workplace few major decisions are made without research, with the findings recorded in a written report.

Depending on its sources, research is classified as *primary* or *secondary*. Primary research is an original, first-hand study of your topic or problem; primary sources include observation, interviews, questionnaires, inquiry letters, personal experiments, analysis of samples, fieldwork, or company records. Secondary research is information published by other researchers, which you then apply to your own topic or problem; secondary sources are *about* primary sources, and include journal articles, encyclopedias, textbooks, and handbooks. Most workplace research draws from both primary and secondary sources.

Research strategies and resources differ widely among disciplines. This chapter focuses on research for preparing a formal report (covered in Chapter 23).

THINKING CRITICALLY ABOUT THE RESEARCH PROCESS[1]

Research is a process of *problem solving*, in which certain procedures follow the sequence in Figure 8.1.

[1]My thanks to UMD librarian Shaleen Barnes for inspiring this entire section.

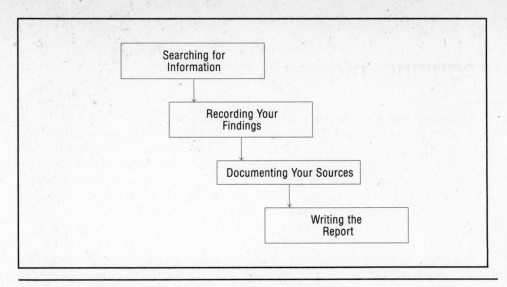

FIGURE 8.1 Procedural Parts of the Research Process

Later parts of the chapter treat these more "procedural" parts of the process, and Chapter 23 deals with preparing the actual research report.

But research is not merely a "by-the-numbers" set of procedures ("First, do this; then, do that"). Intertwined with procedural parts are the "thinking" parts in Figure 8.2, the many decisions that accompany any legitimate inquiry. Let's consider each "thinking" part of the research process.

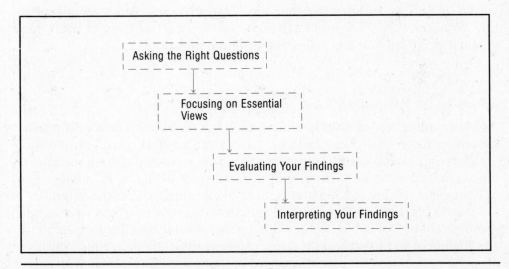

FIGURE 8.2 The "Thinking" Parts of the Research Process

Asking the Right Questions

The answers you uncover will depend on the questions you ask. Assume, for instance, that you are faced with the following scenario:

Defining and Redefining a Research Question

You are the public-health manager for a small, New England town in which high-tension power lines run within 100 feet of the elementary school. Parents are concerned about danger from electromagnetic radiation (EMR) emitted by these power lines, in energy waves known as electromagnetic fields (EMFs). Town officials ask you to research the issue and prepare a report to be distributed at the next town meeting in six weeks.

Your first task is to identify the exact question or questions you want to answer. Initially, it looks as if the major question might be this one: *Do the power lines pose any real danger to our children?* But after some phone calls around town and discussions at the coffee shop, you figure out that townspeople actually have three major questions about electromagnetic fields: *What are they? Do they endanger our children? What can be done?*

To answer these major questions, you need to consider a range of subordinate questions, like those in the Figure 8.3 tree chart.

As research progresses, this chart will grow. For instance, after some preliminary reading, one of the first things you learn is that electromagnetic fields radiate not only from power lines but from *all* electrical equipment, and even from the earth itself. And so you face this additional question: *Do power lines present the greatest hazard as a source of EMFs?*

You begin to wonder whether the greater hazard comes from power lines or from other sources of EMF exposure. Critical thinking, in short, has enabled you to define and refine the essential questions.

Focusing on Essential Views

Assume you've settled on this central question: *Do electromagnetic fields from various sources endanger our children?* Now you can start considering information sources to consult (journals, interviews, reports, database searches). For a fair and accurate picture, you need all sides of the story, as depicted in Figure 8.4. Figure 8.5 illustrates some likely sources of information.

We do research to discover the right answer—or the answer that stands the best chance of being right. Rather than settling for the first or most comforting or most convenient answer, we are ethically obliged to consider all perspectives. Even "expert" testimony may not be the final word, be-

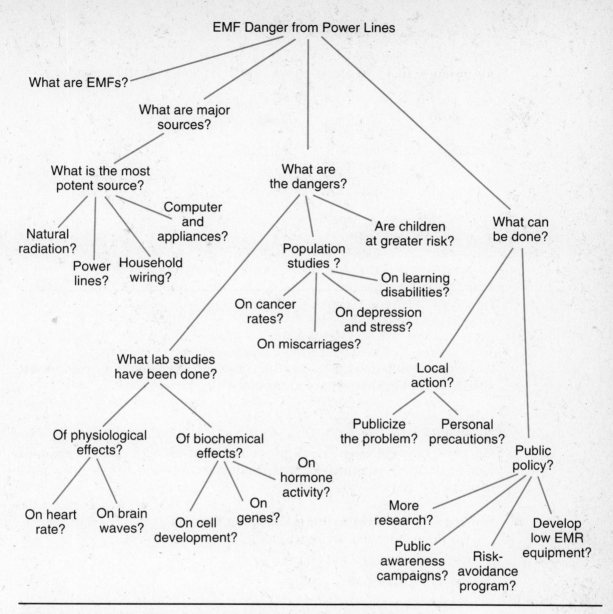

FIGURE 8.3 How the Right Questions Help Define a Research Problem

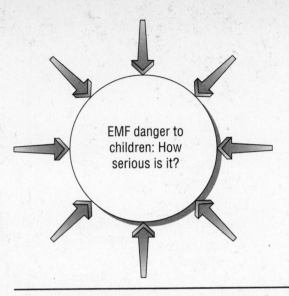

FIGURE 8.4 Effective Research Considers Multiple Perspectives

cause experts often disagree. To reach an objective conclusion, you need to survey the entire spectrum of viewpoints.

Evaluating Your Findings

Once you have collected everything you need, work with your information, first to decide how much of it is legitimate and then to decide what it means.

Questions for Evaluating a Particular Finding

- Is this information, accurate, reliable, and unbiased?
- Can the claim be verified by the facts?
- How much (if any) is useful?
- Is this the whole story?
- Does something seem missing?
- Do I need more information?

Not all findings have equal value. Some might be distorted or incomplete or misleading or irrelevant. Information might be tainted by the bias of the source. With such an emotional issue involving children, a source might understate or overstate certain facts, depending on whose interests that source represents (power company, government agency, parent organization, and so on).

Ethical researchers rely on legitimate evidence that fairly represents *all* views. They don't merely emphasize findings that support their own biases or assumptions.

Interpreting Your Findings

Once you have decided which findings seem legitimate, decide what they all mean.

> **Questions for Interpreting Your Findings**
> - What do all these data mean?
> - Do any findings conflict?
> - Are other interpretations possible?
> - Should I reconsider the evidence?
> - What are my conclusions?
> - What, if anything, should be done?

The interpretation should fit the evidence and lead to an accurate conclusion—an overall judgment about what the findings mean. Perhaps you will reach a definite conclusion. (For example, "The evidence about EMF dangers seems persuasive enough for us to be concerned and to take the following actions.") Perhaps you will not.

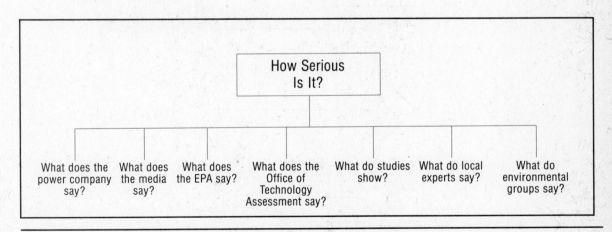

FIGURE 8.5 A Range of Essential Viewpoints

No single stage is
complete until *all*
stages are complete

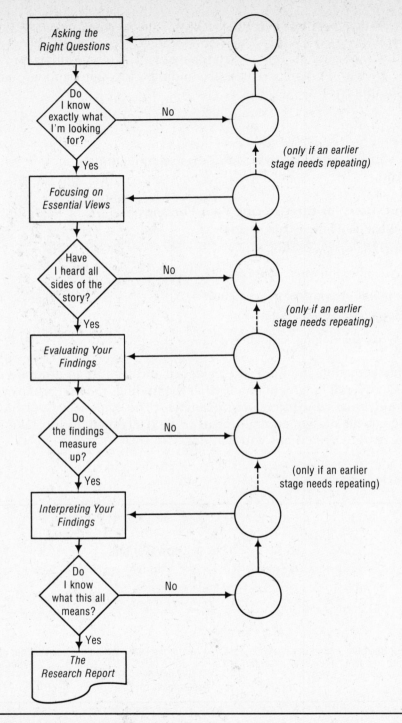

FIGURE 8.6 Critical Thinking in the Research Process

Even the best research can produce contradictory or indefinite conclusions. For instance, some scientists question the studies linking electromagnetic radiation to health hazards. (They claim these studies cannot be replicated or are flawed by statistical or procedural errors.) They point out that, while some EMF studies indicate increased cancer risk, others indicate beneficial health effects. Other scientists claim that stronger EMFs are emitted by natural sources, such as earth's magnetic field, than by electrical sources (McDonald 5). You would have to consider all views before deciding that one outweighs the others—or that only time will tell.

Never force a simplistic conclusion on a complex issue. Sometimes the best solution you can come up with is an indefinite conclusion: "Although controversy continues over the extent of EMF hazards, we can all take simple precautions to reduce our exposure." Keep in mind that a wrong conclusion is far worse than no definite conclusion at all.

Figure 8.6 shows the critical-thinking decisions crucial to worthwhile research: asking the right questions about your topic, your sources, your findings, and your conclusions. Like the writing process, the research process is recursive: no single stage is complete until all stages are complete. The quality of your entire research project will be determined by the quality of your *thinking* at each stage.

SEARCHING THE LITERATURE

No matter how much you already know, you can learn something from your library. Inexperienced researchers, however, are sometimes confused about where to begin searching the literature. Options appear in Figure 8.7.

Where you begin a library search depends on whether you seek background and basic facts or the latest information. If you are an expert in the field, you might simply do a computerized database search or browse through specialized journals. But if you have only limited knowledge or you need to focus your topic, you probably want to begin with books and other general reference works.[2] These sources can be located through the card catalog.

The Card Catalog

Printed Catalog Entries. All books, reference works, indexes, periodicals and other materials held by a library usually are listed in its card catalog

[2]UMass Dartmouth librarian Ross LaBaugh advises students to begin with the popular, general literature, and then to work toward journals and other specialized sources: "The more accessible the source, the less valuable it is likely to be."

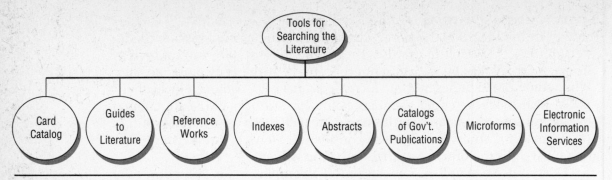

FIGURE 8.7 Ways You Can Search the Literature

under three headings: *author, title,* and *subject.* You thus have three access points for retrieving an item, as shown in Figures 8.8, 8.9, and 8.10.

First decide whether you seek a specific title, an author, or material about a given subject. Then look in the card catalog under one of those three access points (title, author, or subject).

Library of Congress Guide to Subject Headings. If you know neither authors nor titles of works on your subject, use the *subject* listing. To identify related subject headings under which you might find material on your topic, consult the *Library of Congress Subject Headings* (large, red books in the card catalog area). For material on electromagnetic radiation, for instance, you might scan the listings under *Electromagnetic energy;* among other entries, you would see these:

> **Electromagnetic waves**
>> Radiation
>> Waves
>> Electric waves
>> Electromagnetic fields
>> Atmospheric radiation
>> Microwaves
>> Solar radiation
>
> **Electromagnetism in medicine**
>> Medicine, electromagnetism in
>> Electrotherapeutics

You now have an array of subjects under which to search for useful material in the card catalog.

Call number ——————— RA 569.3
                     B75
Author ————————————— **Brodeur, Paul**
Title ——————————————— Currents of death: power lines, computer
                     terminals, and the attempt to cover up
                     their threat to your health.

Publisher, date ——————— New York: Simon & Schuster, 1989.
Physical features ——————— 333 p., 24 cm.
ISBN number ——————————— ISBN 0–671–67845–0.

Other headings ——————— 1. Electromagnetic fields—Health aspects.
under which this   2. ELF magnetic fields—Health aspects. 3.
work is cataloged   Electric lines—Health aspects. 4. Computer
                       terminals—Health aspects.

FIGURE 8.8   Catalog Card Classified by Author

RA 569.3
B75
Currents of death.
Brodeur, Paul

FIGURE 8.9   Partial Catalog Card Classified by Title

RA 569.3
B75
Electromagnetic fields—Health aspects
Brodeur, Paul
Currents of death: power lines. . . .

FIGURE 8.10   Partial Catalog Card Classified by Subject

Electronic Catalog Entries.   In place of printed entries, libraries increasingly are automating their card catalogs. Fast and easy to use, electronic
card catalogs offer additional access points (beyond *author, title,* and *subject*)
including:

- *Descriptor:* for retrieving works on the basis of a key word or phrase in the subject heading ("electromagnetic" or "power lines and health").

- *Document type:* for retrieving works in a specific format (videotape, audiotape, compact disk, motion picture).
- *Organizations and Conferences:* for retrieving works produced under the name of an institution or professional association (Brookings Institution or American Heart Association).
- *Publisher:* for retrieving works produced by a particular publisher (Little, Brown and Co.).
- *Combination:* for retrieving works by combining any available access points (a book about a particular subject by a particular author or institution).

Figure 8.11 shows the first three screens you might encounter in an automated search using the descriptor *ELECTROMAGNETIC*. You could also narrow your search, say, by combining the key words *ELECTROMAGNETIC and HEALTH HAZARDS*. If the computer responds to your descriptors with a "no record" screen, consult the *Library of Congress Subject Headings* (p. 138) for other possible key terms.

Through a computer network such as *Internet,* an electronic catalog can be searched from home or office or anywhere in the world. (The search illustrated in Figure 8.11 was done from this author's university office via campus network.)

Caution: Any misspelling or typographical error in entering key terms can result in a false indication of "no record."

Guides to Literature

If you simply don't know which books, journals, indexes, and reference works are available for your topic, consult a guide to literature. For a general list of books in various disciplines, see Walford's *Guide to Reference Material* or *Sheehy's Guide to Reference Books.*

To see listings for sources in scientific and technical literature, consult Malinowsky and Richardson's *Science and Engineering Literature: A Guide to Reference Sources* or White's *Sources of Information in the Social Sciences.* For sources in specific disciplines, consult specialized guides such as *Using the Chemical Literature: A Practical Guide* or the *Encyclopedia of Business Information Sources.* Ask your librarian about literature guides for your discipline.

Reference Works

Reference works include the various resources shown in Figure 8.12. These can be a good starting point because they provide background and can lead

You begin by pressing any key, and the computer responds with this screen:

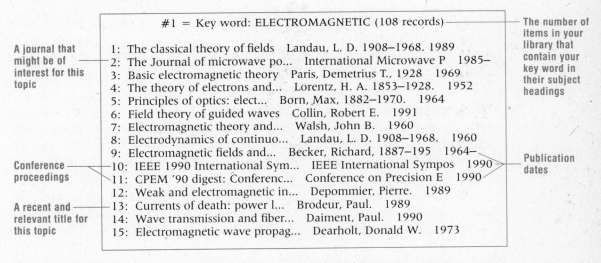

This first screen lists your options for getting help or for searching the catalog from various access points

```
Type of searches:              Press Help key for HELP

 1 AU  = Author                 8 PU  = Publisher
 2 OC  = Organization or conference   9 SH  = Subject heading
 3 TI  = Title                 10 DT  = Document type
 4 UT  = Uniform or collective title  11       Combination
 5 DE  = Descriptor            12       ISBN
 6 CN  = Call number           13       ISBN
 7 SE  = Series                14       Numeric

Enter the NUMBER of your search request and press RETURN:
```

After selecting the DE search mode, you type in your key word (ELECTRO-MAGNETIC), and then press RETURN. This next screen appears (the first of several with all 108 entries):

```
              #1 = Key word: ELECTROMAGNETIC (108 records)

 1: The classical theory of fields   Landau, L. D. 1908–1968. 1989
 2: The Journal of microwave po...   International Microwave P   1985–
 3: Basic electromagnetic theory   Paris, Demetrius T., 1928   1969
 4: The theory of electrons and...   Lorentz, H. A. 1853–1928.   1952
 5: Principles of optics: elect...   Born, Max, 1882–1970.   1964
 6: Field theory of guided waves   Collin, Robert E.   1991
 7: Electromagnetic theory and...   Walsh, John B.   1960
 8: Electrodynamics of continuo...   Landau, L. D. 1908–1968.   1960
 9: Electromagnetic fields and...   Becker, Richard, 1887–195   1964–
10: IEEE 1990 International Sym...   IEEE International Sympos   1990
11: CPEM '90 digest: Conferenc...   Conference on Precision E   1990
12: Weak and electromagnetic in...   Depommier, Pierre.   1989
13: Currents of death: power l...   Brodeur, Paul.   1989
14: Wave transmission and fiber...   Daiment, Paul.   1990
15: Electromagnetic wave propag...   Dearholt, Donald W.   1973
```

A journal that might be of interest for this topic

Conference proceedings

A recent and relevant title for this topic

The number of items in your library that contain your key word in their subject headings

Publication dates

You select entry #13 and then press return. The computer responds with detailed bibliographic information on your selected item.

This screen shows the electronic equivalent of the printed catalog entry.

```
Selection:
01-0211132

AUTHOR      Brodeur, Paul.
TITLE       Currents of death: power lines, computer terminals, and
            the attempt to cover up their threat to your health
PUBLISHER   New York: Simon & Schuster
DATE        c1989
PHYS. FEAT. 333 p.; 24 cm.
SUBJECTS    Electromagnetic fields—Health aspects.
            ELF electromagnetic fields—Health aspects.
            Electric lines—Health aspects.
            Computer terminals—Health aspects.
```

FIGURE 8.11   Searching an Electronic Card Catalog

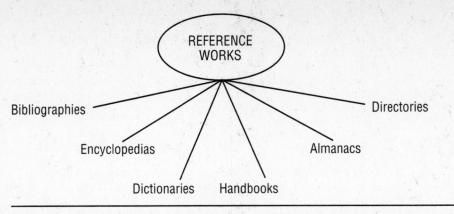

FIGURE 8.12 Common Reference Works for Technical Disciplines

to more specific information. One drawback to reference books is that some may be outdated.

All reference works will be indexed in the *Subject* card catalog, with a "Ref." designation above the call number.

Bibliographies. Bibliographies are lists of publications about a subject, within specified dates. Although they provide a comprehensive view of major sources, bibliographies quickly become dated. Ask a librarian about recent bibliographies on your subject. (Some bibliographies are issued yearly, or even weekly.)

Annotated bibliographies (which include an abstract for each entry) are most helpful because they can help you identify the most useful sources. A sample listing of bibliographies (shown with annotations):

Bibliographic Index. Updated three times yearly, a listing (by subject) of bibliographies that contain at least fifty citations; to see which bibliographies are published in your field, begin here.

A Guide to U.S. Government Scientific and Technical Resources. A list of everything published in these broad fields by the government.

Bibliographic Guide to Business and Economics. A list of all major business and economic publications.

Health Hazards of Video Display Terminals: An Annotated Bibliography. One of many bibliographies focused on a highly specific subject.

Shorter, more specific bibliographies appear as parts of books and journal articles. To locate bibliographies that are whole volumes in themselves, look in the card catalog under "Bibliography" as a subject or title heading. For recent sources on electromagnetic radiation, for example, you might begin with the *Bibliographic Index*.

Encyclopedias. Use encyclopedias to quickly find basic information (which might be outdated). Sample listings:

Encyclopaedia Britannica

Encyclopedia of Building and Construction Terms

Encyclopedia of Banking and Finance

Encyclopedia of Food Technology

Journals, newsletters, and other publications from professional organizations (such as the American Medical Association or the Institute of Electrical and Electronics Engineers) are a valuable source of specialized information. The *Encyclopedia of Associations* offers a yearly listing of over 30,000 societies and organizations worldwide that range from agricultural to scientific and technical. For information on electromagnetic radiation, you might want to contact environmental organizations such as the Audubon Society or the Sierra Club.

Dictionaries. Besides carrying general definitions, dictionaries can focus on specific disciplines or they can give biographical information. Sample listings:

Webster's Third New International Dictionary of the English Language. Considered the best general dictionary.

Dictionary of Engineering and Technology

Dictionary of Telecommunications

Dictionary of Scientific Biography

Handbooks. Handbooks amass key facts (including formulas, tables, advice, and examples) about a field in condensed form. Often aimed at users experienced in the field, some handbooks may not be useful to newcomers. Sample listings:

Business Writer's Handbook

Civil Engineering Handbook

The McGraw-Hill Computer Handbook

Almanacs.   Ranging from general to specific, almanacs have factual and statistical data. Sample listings:

World Almanac and Book of Facts

Almanac for Computers

Almanac of Business and Industrial Financial Ratios

Directories.   Directories offer information about organizations, companies, people, products, services, statistics, or careers, often including addresses and phone numbers. This material usually is updated annually. Sample listings:

The Career Guide: Dun's Employment Opportunities Directory

Directory of Computer Software

Standard & Poor's Register of Corporations, Directors, and Executives

Directory of New England Manufacturers

Directory of American Firms Operating in Foreign Countries

Directory of International Statistics

International Directory of Computer Viruses

A growing number of directories, such as the *Construction Directory*, are accessible by computer.

Reference books can be found for every discipline. In researching electromagnetic radiation, you might start with titles such as the *McGraw-Hill Encyclopedia of Science and Technology* (updated yearly) or the *McGraw-Hill Directory of Scientific and Technical Terms*. Look for your subject in the card catalog, then check for any subheadings such as "Handbooks," "Manuals," or "Dictionaries." These often will be the books that get you started.

Indexes

Indexes are lists of books, newspaper articles, journal articles, or other works, as shown in Figure 8.13. They are excellent sources for current information. Because different indexes list sources in different ways, always read the introductory pages for instructions. Or ask a librarian for help.

Book Indexes.   All books currently being published (up to a set date) are listed in book indexes by author, title, or subject. Sample indexes (shown with annotations):

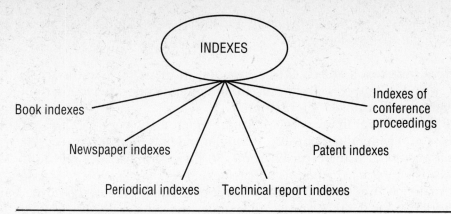

FIGURE 8.13 Useful Indexes for Technical Disciplines

Books in Print. An annual listing of all books published in the United States.

Cumulative Book Index. A monthly worldwide listing of books in English.

Forthcoming Books. A listing every two months of U.S. books to be published.

Scientific and Technical Books and Serials in Print. An annual listing of literature in science and technology.

New Technical Books: A Selective List with Descriptive Annotations. Issued 10 times yearly.

Technical Book Review Index. A monthly listing (with excerpts) of book reviews.

Medical Books and Serials in Print. An annual listing of works from medicine and psychology.

In research on electromagnetic radiation, you might check the current issue of *New Technical Books in Print* or *Scientific and Technical Books and Serials in Print.* But no book is likely to offer the very latest information because of the time required to publish a book manuscript (from several months to one year).

Newspaper Indexes.   Most newspaper indexes list articles by subject. The *New York Times Index* is best known, but other major newspapers have their own indexes. Sample titles:

Boston Globe Index
Christian Science Monitor Index
Wall Street Journal Index

For your research on electromagnetic radiation, you might check recent editions of the *Boston Globe Index*.

Periodical Indexes.   For recent information in magazines and journals, consult periodical indexes. To find useful indexes, first decide whether you seek general or specialized information.

One most general index is the *Magazine Index,* a subject index (on microfilm) of 400 general periodicals. A popular index is the *Readers' Guide to Periodical Literature,* listing articles from 150 general magazines and journals. Because the *Readers' Guide* is updated every few weeks, you can locate current material. In research on electromagnetic radiation, you would find these entries in the 1991 *Readers' Guide* under the subject heading, "Electromagnetic waves":

Physiological effects

See also

Electromagnetic therapy

Computer danger zone [generation of electromagnetic fields from VDTs] G. Rebeck. il *Utne Reader* p18-19 Ja/F '91

Danger in the air [birth defects linked to satellite dishes in Vernon, N.J.] A. M. Cunningham. il *Ladies' Home Journal* 108:60+ Jl '91

Delayed EPA study says evidence "suggests link" between ELF, disease. A. Davidson. *Byte* 16:34 Mr '91

Don't let your monitor be the death of you. W. Taylor. *PC Computing* 4:228-9 Ja '91

Electricity and cancer. T. O. Bakke. il *Home Mechanix* 87:22-3 Jl/Ag '91

Electricity and cancer: how to be on the safe side. M. J. Schnatter. il *McCall's* 118:18 Jl '91

Electricity under study as cancer risk. il *Successful Farming* 89:55 O '91

Electromagnetic fields: in search of the truth. A. M. Cunningham. il *Popular Science* 239:86-90+ D '91

EMFs and cancer [press coverage] L. Slesin. *Columbia Journalism Review* 30:17-18 My/Je '91

Fickle fields: EMFs and epidemiology [children; study by Stephanie J. London] K. F. Schmidt. *Science News* 140:357 N 30 '91

As in many indexes, subjects and their subheads in the *Readers' Guide* are listed alphabetically. Each article is listed in this order: article title, author (if known), periodical title, volume and page numbers, and date. Codes for abbreviated journal titles are at the front of each volume. General indexes such as the *Readers' Guide* list only nonspecialized sources of information.

For specialized information, consult indexes that list journal articles in specific disciplines, such as *Ulrich's International Periodicals Directory*. Another comprehensive source of specialized information, the *Applied Science and Technology Index* carries a monthly listing, by subject, of articles in more than 200 scientific and technical journals. In research on electromagnetic radiation, you would find these entries, under the heading "Electromagnetic fields," in the June 1992 issue of the *AS&T Index:*

Physiological effect

Alleged health effects of electromagnetic fields: misconceptions in the scientific literature. J. Jauchem. bibl (p194-5) *J Microw Power Electromagn Energy* 26 no4:189-95 '91

Are power lines bad for you? A. Coghlan. il *New Sci* 134:22-3 Ap 11 '92

Design of a UHF applicator for rewarming of cryopreserved biomaterials. S. Evans and others. bibl diags *IEEE Trans Biomedic Eng* 39:217-25 Mr '92

Electric & magnetic fields: managing an uncertain risk. G. L. Hester. bibl diags *Environment* 34:6-11+ Ja/F '92

Initial results for automated computational modeling of patient-specific electromagnetic hyperthermia. M. J. Piket-May and others. bibl (p234-6) il diags *IEEE Trans Biomedic Eng* 39:226-37 Mr '92

Power politics: playing with children's lives? A. Philips. *Electron World Wirel World* 97:277-80 Ap '92

For business articles, consult the *Business Periodicals Index*, with its monthly subject listing of articles and book reviews from 270 business periodicals.

Other broad indexes that cover specialized fields in general include the *Business Index* (on microfilm) and the *General Science Index*. The *Statistical Reference Index* lists statistical works not published by the government.

Along with these broad indexes, some disciplines have their own specific indexes. Sample listings:

Agricultural Index

Education Index

Energy Index

F&S Index of Corporations and Industries

Environment Index

Index to Legal Periodicals

International Nursing Index

Ask your librarian about the best indexes for your topic, and about those (such as the *Engineering Index*) that can be searched by computer.

Citation Indexes.   Citation indexes enable researchers to trace, through the literature, the development and refinement of a published idea, concept, or theory. Using a citation index, you can track down the specific publication(s) in which the original material has been cited, quoted, applied, critiqued, verified, or otherwise amplified (Garfield 200). In short, you can answer this question: Who else has said what about this idea?

The *Science Citation Index,* a quarterly publication, provides a system for cross-referencing important articles on science and technology worldwide. Both the *Science Citation Index* and its counterpart, the *Social Science Citation Index,* can be searched by computer.

Technical Report Indexes.   Countless government and private-sector reports written worldwide offer specialized and highly current information. (Proprietary or security restrictions, of course, restrict public access to certain corporate or government documents.) Sample indexes for these reports:

Scientific and Technical Aerospace Reports

Government Reports Announcements and Index

Monthly Catalog of United States Government Publications

U.S. government report indexes are discussed on page 151.

Patent Indexes[3].   Over 75,000 patents yearly are issued in the United States to protect individual and company rights to new inventions, products, or processes. These patents are just a fraction of the roughly one-half million issued worldwide. As information specialists Schenk and Webster explain, patents are an excellent and often overlooked source of current information: "Since it is necessary that complete descriptions of the invention be included in patent applications, one can assume that almost everything that is new and original in technology can be found in patents." Sample indexes:

Index of Patents Issued from the United States Patent and Trademark Office

NASA Patent Abstracts Bibliography

World Patents Index

---

[3]Adapted from Margaret T. Schenk and James K. Webster. Consult their work for detailed treatment of patent information and its sources, and for invaluable discussions of information sources in general. Full citation appears in Works Cited, page 675.

Online information about patents in fiber optics, lasers, or other technologies can be obtained through databases such as Hi Tech Patents, Data Communications, and through WPI (World Patents Index).

Indexes of Conference Proceedings.   Schenk and Webster point out that many of the papers presented at the more than 10,000 yearly professional conferences are collected and then indexed in printed or computerized listings such as these:

Proceedings in Print

Index to Scientific and Technical Proceedings

Engineering Meetings (an *Engineering Index* database)

The very latest ideas or explorations or advances in a field often are presented during such proceedings, before appearing as journal publications.

Abstracts

Beyond indexing various works, abstracts briefly describe each article. The abstract can save you from going all the way to the journal in order to decide whether to read the article or to skip it.

Abstracts usually are titled by discipline. A sample list:

Biological Abstracts

Computer Abstracts

Engineering Index

Environment Abstracts

Excerpta Medica

Forestry Abstracts

International Aerospace Abstracts

Metals Abstracts

In researching electromagnetic radiation, you would see this entry in the December 1989 *Energy Research Abstracts*, under the subject heading "Electromagnetic Fields":

50907   (DOE/BPA-945) **Electrical and biological effects of transmission lines: A review.** Lee, J.M. Jr., et al. USDOE Bonneville Power Administration, Portland, OR (USA). Jun 1989. 106p. Sponsored by U.S. DOE Management & Administration. Order Number DE89017650. Available from NTIS, PC A06/MF A01 - OSTI; GPO Dep.

This review describes the electrical properties of a-c and d-c transmission lines and the resulting effects on plants, animals, and people. Methods used by BPA to mitigate undesirable effects are also discussed. Although much of the information in this review pertains to high-voltage transmission lines, information on distribution lines and electrical appliances is included. The electrical properties discussed are electric and magnetic fields and corona: first for alternating-current (a-c) lines, then for direct current (d-c).

Abstracts (such as *Energy Research Abstracts* and *Pollution Abstracts*) increasingly are searchable by computer. Check with your librarian.

For some current research, you might consult abstracts of doctoral dissertations in *Dissertation Abstracts International*.

Locating the Source.   If your library does not hold the article you need, use the OCLC terminal (pages 153–54) to identify a holding library, and request the article through interlibrary loan.

Access Tools for United States Government Publications

The federal government publishes maps, periodicals, books, pamphlets, manuals, monographs, annual reports, research reports, and a bewildering array of other information. Types of information available to the public include presidential proclamations, congressional bills and reports, judiciary rulings, some reports from the Central Intelligence Agency, and publications from all other government agencies (Departments of Agriculture, Commerce, Transportation, and so on). Even many unpublished government documents can be obtained through the Freedom of Information Act (Lavin 8). A few of the countless titles available in this gold mine of information:

Effects of New York's Fiscal Crisis on Small Business

Economic Report of the President

Major Oil and Gas Fields of the Free World

Decisions of the Federal Trade Commission

Journal of Research of the National Bureau of Standards

Siting Small Wind Turbines

Much of this information can be searched online as well as in printed volumes. Your best bet for tapping this valuable but complex resource is to ask the librarian in charge of government documents for help. If your library does not hold the publication you seek, it can be obtained through an interlibrary loan.

Here are the basic access tools for documents issued or published at government expense as well as for many privately sponsored documents.

- *The Monthly Catalog of the United States Government*, the major access to government publications and reports, is indexed by author, subject, and title. These indexes provide you with the catalog entry number that leads you, in turn, to a complete citation for a work.

- *Government Reports Announcements & Index* is a listing published every two weeks by the National Technical Information Service (NTIS),[4] a federal clearinghouse for scientific and technical information—all stored in a computer database. The collection has summaries of over 900,000 federally sponsored research reports published and patents issued since 1964. About 70,000 new summaries are added annually in 22 subject categories, from aeronautics to medicine and biology. Full copies of reports are available from NTIS.

- *The American Statistics Index*, a yearly guide to statistical publications by the U.S. government, is divided in two sections: *Index* and *Abstracts* (an index with summaries). The *Index* volume lists material by subject and provides geographic (U.S., state, and so on), economic (income, occupation, and so on), and demographic (sex, marital status, race, and so on) breakdowns. The *Index* volume refers you to a number and entry in the *Abstracts* volume.

In addition, the government issues *Selected Government Publications*, a monthly list of 150 titles (with descriptive abstracts). These titles range from highly general (*Questions About the Oceans*) to highly technical (*An Emission-Line Survey of the Milky Way*).

The government also publishes bibliographies on hundreds of subjects, from "Accidents and Accident Prevention" to "Home Gardening of Fruits and Vegetables." Ask your librarian for information about these subject bibliographies.

Microforms

Microform technology enables vast quantities of printed information to be reproduced and stored on rolls of microfilm or packets of microfiche. (This material is read on machines that magnify the reduced image. Ask your librarian for assistance.) Among the growing array of microform products

[4]A branch of the U.S. Department of Commerce.

are government documents, technical reports, newspapers, business directories, and translated documents from worldwide (Lavin 12).

A valuable business resource, for example, is *Business NewsBank,* a microfiche index to articles on business and economic development from over 450 U.S. cities. Also on microfilm are specialized indexes such as the *Business Index,* and more general indexes such as the *Magazine Index.*

USING ELECTRONIC INFORMATION SERVICES

Libraries increasingly offer computerized services. In addition to electronic card catalogs (p. 139) and hypertext systems (p. 93), access to libraries worldwide is facilitated by *Internet* or some similar network. And some libraries expect to store electronic versions of all their material within 25 years (Watkins 19). Figure 8.14 shows common search tools.

Compact Disks and Diskettes

One compact disk (CD) stores an entire encyclopedia and provides instant access to any part. CD technology offers reference resources like these: *Science Citation Index, Population Statistics, Ulrich's International Guide to Periodicals,* and products ranging from corporate directories to government reports. Even weekly job listings from the nation's largest Sunday newspapers are becoming available on CDs ("On Line" 19).

One popular index on compact disk is *InfoTrac*™, a monthly listing of articles from 900 business, technical, and general magazines and journals. As in printed indexes, InfoTrac's entries are arranged by subject headings and subheadings. If you type in "electromagnetic radiation," the system responds with a list of entries you can print out.

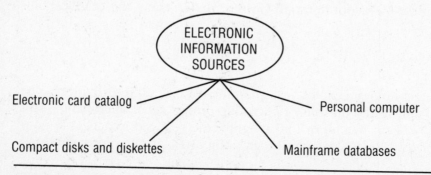

FIGURE 8.14 Options for Searching for Literature Electronically

Other databases on CD include *Government Publications Index*™, a monthly listing of U.S. government literature, and *LegalTrac*™, a monthly listing of entries from 750 legal publications. *Lotus One Source*™ provides corporate profiles and financial summaries for over 15,000 U.S. companies, research reports on stocks, and article summaries from major business sources. Ask at your library about disk-based indexing services.

For smaller bodies of specialized information (census and stock-market data, economic profiles of U.S. cities, and so on), files on diskettes can be purchased (Lavin 19).

Mainframe Databases

Most college libraries subscribe to retrieval services that can access thousands of individual databases stored on centralized computers. From a library terminal (or in some cases, a microcomputer), you can access indexes, journals, books, monographs, dissertations, and reports. Compared with CDs, mainframe databases tend to be more specialized and more current, sometimes updated daily.

Mainframe retrieval services offer three types of databases, or some combination: *bibliographic*, *full-text*, and *factual* (Lavin 14). Bibliographic databases list publications in a field; entries can be searched according to author, title, subject, document type (report, article, dissertation) or other access point. Some bibliographic databases include abstracts of each entry.

Full-text databases display the entire article or document (usually excluding graphics) directly on the computer screen, and then will print the article on command.

Factual databases provide specialized facts of all kinds: global and up-to-the minute stock quotations, weather data, lists of new patents filed, and credit ratings of major companies, to name a few.

Until recently, searching mainframe databases has required the skills of a librarian or other trained professional. But new, "end-user systems" now offer simplified menus for each step of the search process (Lavin 16).

Three popular database services are discussed in the following section. The first service (OCLC) helps you locate titles you have already identified as useful. The next two (Dialog and BRS) help you identify useful titles.

The Online Computer Library Center (OCLC).　You can easily compile a comprehensive list of works on your subject at any library that belongs to the Online Computer Library Center. The OCLC database, in Columbus, Ohio, stores more than 19 million records with the same information found in a printed card catalog. Using the library's computer terminal, you type in author or title. Within seconds, you get a free listing of the publication you seek and information about where to find it. If your library doesn't have the publication, your librarian can activate the Interlibrary Loan System (ILS). The system forwards requests to libraries holding the material. Once a lender indicates (via its terminal) that it will supply the material, the system stops forwarding the request, and notifies your librarian that the request has been filled. Your order will arrive at the library by mail in a week or two. (Pages 441–446 show instructions for using an OCLC terminal.)

OCLC recently has added a variety of bibliographic databases, along with a more user-friendly search system (Lavin 66). Ask if these enhancements are available in your library.

Dialog.   Many libraries subscribe to Dialog, a comprehensive technical database. You retrieve information by typing in key terms that enable the computer to scan bibliography lists for titles containing those terms. Say you need information on possible *health hazards* from *household electrical equipment*. You instruct the computer to search Medline, a medical database (one of 150 Dialog databases in science, technology, medicine, business, and so on), for titles including the words italicized above (or synonymous words, such as *risk, danger, appliances*). The system would provide full bibliographies and abstracts of the most recent medical articles on your topic.

Besides bibliographic information on published works, dissertations, and conference papers, Dialog also provides financial and product information about companies, names and addresses of company officers, statistical data, and patent information. Here are just a few of Dialog's databases:

Career Placement Registry

Claims/U.S. Patents

Conference Papers Index

Electronic Yellow Pages (for Retailers, Services, Manufacturers)

Enviroline

Index Medicus

International Software Database

Oceanic Abstracts

U.S. Exports

Water Resources Abstracts

Despite its expense ($150 hourly or more for many of its databases), a sizable number of college libraries subscribe to Dialog. Companies who have full-time database researchers also subscribe.

BRS. Bibliographic Retrieval Services (BRS) is another popular database providing bibliographies and abstracts from life sciences, physical sciences, business, or social sciences. These are a few from the more than fifty BRS databases:

American Chemical Society Journals

Dissertation Abstracts International

Government Reports Announcements & Index

Harvard Business Review

International Pharmaceutical Abstracts

Military and Federal Specifications and Standards

Monthly Catalog of United States Government Publications

PATDATA (U.S. Patents)

Pollution Abstracts

Robotics Information Database

College libraries increasingly offer BRS service.

A Sample Automated Search. Assume you are continuing research on electromagnetic radiation. You have searched the "manual" indexes, and for a comprehensive view you decide on an automated search, using your library's Dialog service. You ask a librarian for help, and the two of you sit at the terminal and begin.

After logging into the Dialog system, you instruct the computer to search the *Enviroline* database, using the key words *electromagnetic*, *health*, *hazard*. The computer responds with a listing of articles in that database whose titles or abstracts contain a combination of those key words.[5] Here is the partial list of titles.

Partial Listing of Article Titles from an Electronic Search

1. Video Display Terminals and the Risk of Spontaneous Abortion
2. Sharpening the Focus in EMI Research
3. Extremely Low-Frequency Electromagnetic Fields and Cancer: the Epidemiologic Evidence.

The title that appears most relevant to your topic is 3, and so you instruct the computer to print the full bibliographic information on this article.

---

[5]In a "full-text" database, the computer would search for works containing that combination of key words *anywhere* in the text.

The Full Citation for One Article from an Electronic Search

Access number

Title

Author and affiliation

Periodical

Abstract

0224095 Enviroline Number: 92–006348

Extremely Low-Frequency Electromagnetic Fields and Cancer: the Epidemiologic Evidence

Bates, Michael N.
Univ. of California, Berkeley

Env. Health Perspectives, Nov 91, v95, p147 (10)

Journal article. An overview of data is presented from past studies on the relationship between low-frequency electromagnetic fields generated by alternating current and cancer incidences. The electromagnetic fields usually have both electric and magnetic field components. Epidemiologic studies are cited that involve residential exposures of children and adults, and those involving occupational exposure that have examined leukemia only, brain cancer only, and all cancers combined. The evidence for carcinogenic effects of low-frequency field exposure is strongest for brain in both occupationally exposed adults and in children, and not as strong for leukemia. Risk assessment for carcinogenic effects from adult residential exposures is inconclusive, but should be evaluated further. (56 references, 4 tables)

After reading the abstract, you decide to obtain the complete article, and so you make a note to check your library's holdings or to order a copy through interlibrary loan—or via the computer terminal. You turn again to the list of titles to see if others seem promising.

Retrieval Services for Home and Office

On a home or office computer, you can search source databases. For a small fee, you can join *Compuserve Information Service* or *Dow-Jones News Retrieval System,* and gain access to energy news, stockmarket quotes, corporate news releases, new-product news, business reports, medical news, and many other reference sources.

Internet users now can access *ORBIT Search Service* (for scientific, medical, and technical information) and *BRS Search Service* (for medical, pharmaceutical, and life-sciences information) ("On Line" 19). Ask your librarian for information about commercial databases in your field.[6]

---

[6]For a major source of information on databases in various disciplines, see these works: Martha E. Williams et al., eds., *Computer-Readable Data Bases: A Directory and Data Sourcebook.* Urbana: Ill.: American Society for Information Science; *The Database Book,* a directory of the world's databases, published by On-line, Inc.; *The Data Base Directory,* published by Knowledge Industry; *Directory of On-line Databases.* Santa Monica, Calif.: Cuadra Associates, Inc., published quarterly since 1979.

Benefits and Limitations of Automated Searches

Online searches have several advantages over manual searches (that is, flipping pages by hand).

- They are rapid: you can review ten or fifteen years of an index in minutes.
- They are detailed: beyond listing titles and sources, an automated search often provides abstracts.
- They are current: the index usually comes online about six weeks before the printed copies.
- They are thorough: the system can search not only for titles but for key words (or word combinations) found in the title *or* the abstract.
- They are efficient: you can extract from the database only the information you need, without tracking material that turns out useless.

But automated searches have limitations as well. Most computerized bibliographies include no entries before the mid-1960s; earlier information requires a manual search. Also, a manual search provides the whole "database" (the bound index or abstracts). As you browse, you often *randomly* discover something useful. This randomness, of course, is impossible with an automated search, which can give you the illusion of having surveyed all that is known on your topic.[7] Finally, automated searches can be expensive, depending on how many databases you search and how long you spend online. (The average cost of a BRS search is about $30.) Some schools offer students one free search, but if your school doesn't, you pay the cost.

For any automated search, keep in mind that a manual (random) search is almost always needed as well. And a thorough search calls for a preliminary conference with a trained librarian.

EXPLORING PRIMARY SOURCES

Interviews

Although libraries are a good secondary research source, you may also have to consult firsthand, or primary, sources as shown in Figure 8.15.

---

[7]UMass Dartmouth librarian Charles McNeil cautions against assuming that computer access yields the best material: "The material in the computer is what is cheapest to put there." Librarian Ross LaBaugh alerts users to a built-in bias in databases: "The company that assembles the bibliographic or full-text database often includes a disproportionate number of its own publications." Like any collection of information, a database can reflect the biases of its assemblers.

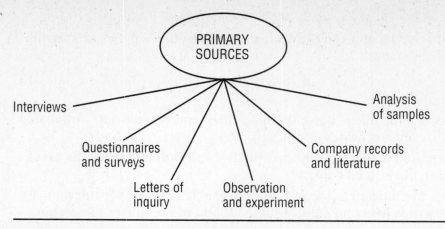

FIGURE 8.15 Sources for Firsthand Research

An excellent primary source for data that cannot be found in any publication is the personal interview. Much of what an expert knows may never be submitted for publication (Pugliano 6).[8]

Identify Your Purpose. Know *exactly* what you are looking for. Suppose, for example, you decide to interview three authorities on electromagnetic radiation. Determine the information you seek from each respondent:

Purpose statement

The purpose of my interview with Anne Hector, Chief Engineer at Northport Electric Company [a town-owned utility], is to discuss a recent company survey of community awareness and attitudes about electromagnetic radiation. Also, I will inquire about the company's plans and procedures for risk avoidance.

As soon as you can write out your purpose clearly, contact your respondent.

Contact Your Respondent. Phone or write to request the interview, at your respondent's convenience. If you plan a phone interview, call or write in advance, asking the respondent to stipulate a convenient time for the interview. Always give respondents ample notice and time to prepare.

---

[8]Of course, an "expert" opinion can be just as mistaken or biased as anyone else's. Like the wise patient who seeks a "second opinion" about a serious medical condition, the wise researcher seeks a balance or range of expert opinions about a complex problem or controversial issue—not only from, say, the company engineer or the environmentalist, but from independent and presumably more objective third parties as well (such as a professor or a journalist who has studied the issue).

Prepare Well.   Plan questions that elicit the specific information you seek. Write out each question on a separate notecard (on which you can summarize the response). Above all, know your subject. Effective interviews usually involve more than asking a set of preplanned questions and recording the responses. A productive interview often resembles a good conversation, in which the interviewer follows up on interesting comments with additional, unplanned questions.

Make your questions clear and specific. Vague, unspecific questions elicit an unfocused answer or leave the respondent asking "What do you mean?":

| How is your company dealing with the EMR problem? — *Vague*

Which problem—public relations, public awareness, potential liability, danger to electrical workers, danger to the community, or what? Here are clear and specific versions:

| What specific procedures is your company developing for risk avoidance? — *More focused*
| Does the company have plans for promoting community awareness of EMR?

Avoid questions that can be answered with a simple "yes" or "no," because such answers provide little information:

| Do you think technology can find a way to decrease EMR hazards? — *Uninformative*

Instead, phrase your questions to elicit detailed information:

| What technological solutions are being proposed or considered? — *More informative*
| How would you rate their potential effectiveness?

Avoid questions that reflect a particular bias or that invite a particular stance on the issue:

| Don't you agree that EMR hazards have been overstated? — *Biased*

Instead, allow the respondent to express her or his own views without your influence:

| Do you think EMR hazards have been understated or overstated? — *More impartial*

Be ready with follow-up questions that will enable you to probe beneath the surface of an issue:

| Why or why not? — *Follow-up questions*
| What more needs to be done?
| What would you recommend?

The responses you obtain will only be as good as the questions you ask.

Maintain Control.   You—not your respondent—are responsible for conducting a competent and productive interview. These suggestions should help:

1. Dress neatly and arrive on time.
2. Begin by thanking your respondent, in advance.
3. Restate the purpose of your interview.
4. Tell your respondent why you feel he or she can be helpful.
5. Discuss your plans for using the information.
6. Ask (preferably before you arrive) if your respondent objects to being quoted or taped. (Although the most accurate means for recording responses, taping can make people uncomfortable.)
7. Avoid small talk.
8. Ask your questions clearly and directly, following the order in which you prepared them.
9. Be assertive but courteous. Ask pointed questions, but remember that the respondent is doing you a favor.
10. Let your respondent do most of the talking. Keep your opinions to yourself.
11. Guide the interview. If your respondent wanders, politely bring the conversation back on track (unless the additional information is useful).
12. Be a good listener. Don't stare out the window, doodle, or ogle office staff while your respondent is speaking.
13. Be prepared to explore new areas of questioning. A respondent's answers often reveal new directions for the interview.
14. Keep note-taking to a minimum. Record all numbers, statistics, dates, names, and other precise data, but don't transcribe responses word for word. Simply record significant words and phrases that later can refresh your memory.
15. If interviewing several people about the same issue, standardize your questions. Ask each respondent identical questions in identical order and phrasing. Avoid random comments that could influence one person's responses.
16. Don't hesitate to ask for clarification or further explanation.
17. When all your questions have been answered, ask for any additional comments. If the interview is to be published, ask respondents to review the final draft before you quote them in print.

18. Offer to provide a copy of the document in which this information will be used.

19. Finally, thank your respondent, and leave promptly.

20. As soon as you leave, write out a summary while the responses are fresh in your memory.

Here is the partial text of an interview on persuasive challenges. Notice how the interviewer probes, seeks clarification, and follows up on offered data.

An Informative Interview

(The respondent: Z. Quark, Assistant to the Vice-president for Corporate Relations in an emerging software firm. The topic: The Role of Persuasion in Workplace Communication.)

Q. *Could you please summarize your communication responsibilities?*
A. The corporate relations office oversees three departments: customer service (which handles claims, adjustments, and queries), public relations, and employee relations. My job is to supervise the production of all documents generated by this office.

Q. *Isn't that a lot of responsibility?*
A. It is, considering we're trying to keep some people happy, getting others to cooperate, and trying to get everyone to change their thinking and see things in a positive light. Just about *every* document we write has to be persuasive.

Probing and following up

Q. *What exactly do you mean by "persuasive"?*
A. The best way to explain is through examples of what we do. The customer service department responds to problems like these: some users are unhappy with our software because it won't work for a particular application, or they find a glitch in one of our programs, or they're confused by the documentation, or someone wants the software modified to meet a specific need. For each of these complaints or requests we have to persuade our audience that we've resolved the problem or that we're making a genuine effort to resolve it quickly.

The public relations department works to keep up our reputation through links outside the company. For instance, we keep in touch with this community, with consumers, the general public, government and educational agencies. . . .

Seeking clarification

Q. *Could you be more specific? "Keeping in touch" doesn't sound much like persuasion.*

Seeking clarification

A. Okay, right now we're developing programs with colleges and universities, in which we offer heavily discounted software, backed up by an extensive support network (regional consultants, an 800 phone hotline, and workshops). We're hoping to persuade them that our software is superior to our well-entrenched competitor's. And locally we're offering the same kind of service and support to business clients.

Following up

Q. *What about employee relations?*
A. Day to day we face the usual kinds of problems: trying to get 100 percent employee contributions to the United Way, or persuading employees to help out in the community, or getting them to abide by new company regulations restricting smoking or to limit personal phone calls. Right now, we're facing a real persuasive challenge. Because of market saturation, software sales have flattened across the board. This means temporary layoffs for roughly 28 percent of our employees. Our only alternative is to persuade *all* employees to accept a 10-percent salary and benefit cut until the market improves.

Probing

Q. *And just how do you persuade employees to accept a cut in pay and benefits?*
A. Basically, we have to make them see that by taking the cut, they're really investing in the company's future—and, of course, in their own.

(The interview continues.)

Surveys and Questionnaires

Surveys enable us to develop profiles (generalized views) about the concerns, preferences, attitudes, or behaviors of a large, identifiable group (a *target group*) by analyzing responses from representatives of that group (a *sample group*). Although surveys can be conducted through personal interviews, the preferred and more efficient method is the questionnaire.

Granted, interviews have certain advantages over questionnaires. Face-to-face, you can clarify your questions and explore unexpected areas. But an interview can also intimidate and inhibit a respondent.

Questionnaires save time and are an inexpensive way to survey a large group. Respondents can answer privately and anonymously—and thus more candidly, with time to think about their answers.

Getting people to respond to a questionnaire, however, is difficult. Expect less than a 30-percent response, perhaps far less, and allow plenty of time for responses to be returned.

Increase your chances for a good response by following these suggestions:

- *Define your purpose*. Why are you doing this survey? What, exactly, are you looking for? How will the information be used?

- *Define your sample group*. Who are the intended respondents? How many? Generally, the larger the sample surveyed, the more valid the results (assuming a well-chosen and representative sample). Will the group be randomly chosen? A cross section?

- *Define your method*. What type of data (opinions, ideas, facts, figures) will you collect? Is timing important? How will you administer the questionnaire—in person, by mail, by phone? (Phone and in-person surveys yield fast results and high response rates, but mail surveys are inexpensive and confidential.) How will you collect, record, analyze, and report your data (Lavin 277)?

- *Decide on the types of questions* (Adams and Schvaneveldt 202–12; Velotta 390). Survey questions should be easy to understand and hard to misinterpret. Decide between questions that are *open-ended* or *closed-ended*. Open-ended questions allow respondents to express exactly what they're thinking or feeling in a word, phrase, sentence, or short essay:

How much do you know about electromagnetic radiation at our school?
What do you think should be done about EMR at our school?

Open-ended questions

Since one never knows what people will say, open-ended questions are a good way to uncover attitudes and obtain unexpected information. One disadvantage is that essay-type questions are hard to answer and tabulate, especially when answers are illegible or ambiguous.

When you know all the possible attitudes in your target group, and you want to measure where various people stand on the issue, choose closed-ended questions:

Are you interested in joining a group of concerned parents?
YES _____ NO _____

Closed-ended questions

Characterize your degree of concern about the EMF issue at our school.
HIGH _____ MODERATE _____ LOW _____ NO CONCERN _____

Circle the number that indicates your view about the town's proposal to spend $20,000 to hire its own EMF consultant.

1 . . . . 2 . . . . 3 . . . . 4 . . . . 5 . . . . 6 7
Strongly   No   Strongly
Approve   Opinion   Disapprove

Depending on their design, closed-ended questions elicit various kinds of data. For instance, respondents may be asked to *rate* one item on a scale (from high to low, best to worse, and so on) or to *rank* two or more items (in order of importance, desirability, and so on). Other questions measure percentages or frequency ("How often do you . . . ?" ALWAYS _____OFTEN _____SOMETIMES _____ RARELY _____NEVER _____). The sample questionnaire on page 167 illustrates various designs for closed-ended questions.

Respondents generally are more willing to check or circle an item rather than to make more personal revelations in essay form. Easy to answer, tabulate, and analyze, closed-ended questions also provide a more "objective" measure because all responses are expressed in identical terms.

- *Ensure the survey's validity and reliability* (Adams and Schvaneveldt 79–97; Burghardt 174–75; Velotta 391). Research is *valid* when (1) it measures what you want to measure and (2) when it measures accurately and precisely. An accurate survey is free of procedural errors (faulty questions, faulty interpretations, or other errors in survey technique); a precise survey is free of random errors (delays in administering the survey, insufficient responses, or other unpredictable variations in the survey conditions). Valid survey questions enable each respondent to interpret each question exactly as you intended.

  Research is *reliable* when its results are consistent, for instance when a respondent gives identical answers to the same survey given twice or to different versions of the same questions. Reliable survey questions enable all respondents to interpret the questions in the same way.

  Much of your technical writing will be based on secondary research findings, and so you will need to assess the validity and reliability of other people's research as well as your own by asking these questions: Was the sample group representative and large enough? Were the right questions asked? Were the responses tabulated and interpreted correctly?

- *Design an engaging introduction and opening questions.* Your introduction should make respondents feel that the survey is relevant to their concerns and that their answers matter. Whenever possible, explain how respondents will benefit from your findings, or offer an incentive (such as a copy of your final report). Keep the tone conversational, like a person talking to people.

A survey introduction

Your answers to these questions about your views on proposed state handgun legislation will be appreciated. Your state representative will tabulate all responses so that she may continue to speak accurately for *your* views in legislative session. Thank you.

Some writers include a cover letter with a questionnaire (as on page 166).

Begin with the easiest questions. Respondents are inclined to answer those questions that are general or more interesting or nonthreatening. Once they commit themselves to these initial questions, they are likely to complete any difficult questions that might follow.

- *Keep the questionnaire short.* Try to limit questions and response space to two sides of a single page. And phrase the questions in sentences that are as short as possible.

- *Make each question unambiguous and unbiased.* If different respondents interpret the same question differently, or if they misinterpret the question, the survey is unreliable or invalid. Consider, for instance, this ambiguous question:

Do you favor foreign aid? YES _____   NO _____

An ambiguous question

"Foreign aid" might mean military, economic, or humanitarian aid, all three, or two out of three. Some respondents might support one or two types, but not all three—and their level of support might depend on the survey's timing (as in a period of military conflicts, widespread famine, or global recession). Consequently, responses to the above question, no matter how well analyzed, would produce a meaningless or misleading statistic, such as "Only 40 percent of Americans favor foreign aid," when the accurate conclusion might be "Over 95 percent of Americans favor some form of foreign aid." (See pages 185–87 for analyzing statistical findings.) Allow for these differences by rephrasing the above question:

Do you favor (check all that apply):
_____ Our current foreign aid program of military, economic, and humanitarian assistance to other countries?
_____ Greater military assistance in our foreign aid program?
_____ Less military assistance?
_____ Greater humanitarian assistance?
_____ Less humanitarian assistance?
_____ Greater economic assistance?
_____ Less economic assistance?
_____ No foreign aid program at all?
_____ Don't know

A clear and incisive question

Avoid influencing respondents with *loaded questions* that invite or advocate a particular viewpoint or bias, as in these examples:

Is foreign aid a waste of taxpayer's money on foreigners?

YES _____   NO _____

Is a wealthy nation morally responsible for helping the less fortunate?

YES _____   NO _____

Emotionally loaded words ("waste," "foreigners," "morally responsible," "radicals," "bureaucracy," "system") in a supposedly impartial survey are unethical because they carry built-in judgments that manipulate people's responses (Hayakawa 40).

- *Structure the questions so that responses can be tabulated easily, validly, and reliably.* Decide on the best form of response: yes-or-no, multiple-choice, true-false, fill-in-the-blank, rating scale, order of ranking, and so on. But be sure that the form of response will enable you to measure accurately what you intended to measure.

  To ensure a full range of possible responses, include options such as "Other _____," "Don't know," "Not Applicable," or an "Additional Comments" section.

- *Add "personal touches" that encourage reader response.* Try to simplify the task and engage the reader. With mailed questionnaires, include a stamped, return-addressed envelope. Research experts Adams and Schvaneveldt point out that response rates improve when personal touches are included: letters and surveys individually typed, respondents addressed by name, the researcher's signature followed by a title—and even when postage stamps are used instead of a postal meter imprint (206).

The questionnaire on pages 167–68, sent to presidents of local companies, is prefaced by the following cover letter explaining the questionnaire's purpose. The clearer the rationale for a questionnaire, the more inclined readers are to respond.

Questionnaire Cover Letter

As part of my course work, I am preparing a report on Southeastern Massachusetts University's professional communication program. This survey will help determine how SMU's communication program can better serve students and the community.

I plan to assess services and needs by studying

1. the communication needs of local companies and industries
2. the feasibility of on-campus and in-house seminars in communication

3. the feasibility of offering a broader variety of professional communication courses at SMU

Please take a few minutes to respond to this survey. Your response is important because it provides the data needed for my study. Survey results, conclusions, and recommendations will be published in the fall issue of *The Business and Industry Newsletter*. (I will gladly send a copy.)

One last favor: Kindly mail the survey to

> Lynne Taylor
> House 10, SMU Dorms
> North Dartmouth, MA 02747

I had hoped to include an addressed, stamped envelope for your convenience, but my budget for this project has been depleted by printing and mailing costs.

Sincerely,

Lynne Taylor

The following questionnaire is designed to elicit specific responses that can then be easily classified and tabulated.

Communication Questionnaire for Executives

1. Describe your type of company (manufacturing, high tech, banking, etc.).

   _____

2. How many people do you employ? (Please check one.)

   _____ 5–25   _____ 100–150
   _____ 25–50   _____ 150–300
   _____ 50–100   _____ 300–450

3. How do you usually communicate with employees?

   _____ by memo
   _____ in person
   _____ through supervisors and managers

4. Which types of writing are done in your company? (Label by frequency: never, rarely, sometimes, often, most often.)

   _____ office memos   _____ catalogs
   _____ manuals   _____ advertisements
   _____ procedures   _____ house organs
   _____ letters   _____ other (specify)
   _____ reports   _____

5. Who does most of the writing in your organization? (Give title.)

   _____

6. Please characterize the writing effectiveness in your organization.

_____ good _____ fair _____ poor

7. Does your company have writing guidelines?

_____ no _____ yes (Please describe briefly below.)

_____

8. Do you ever hire outside writers for specific projects?

_____ no _____ yes

9. Do you have an in-house program for communication training?

_____ no _____ yes

10. Rank the usefulness of the following topics in a writing program (with 1 as most important through 9 as least important).

_____ organizing information _____ audience awareness

_____ summarizing information _____ grammar

_____ precise phrasing _____ sales writing

_____ formatting reports _____ other (specify)

_____ editing for style _____

11. Rank these skills in order of importance (1 being most important).

_____ reading _____ speaking to groups

_____ writing _____ speaking face to face

_____ listening _____ dictating

12. Do you provide tuition reimbursement for employees?

_____ no _____ yes

13. Would you participate in a contract learning program in which SMU writing students would work for you?

_____ no _____ yes

14. Should SMU offer Saturday seminars in Professional Communication?

_____ no _____ yes

15. Which courses, beyond introductory professional writing courses, should SMU offer? (Rank in numerical order of preference.)

_____ Report Writing _____ Documentation

_____ Proposal Writing _____ Sales Writing

_____ Procedure Writing _____ other (specify)

_____ Public Speaking _____

Additional comments: _____

_____

Caution: Interpret statistical results carefully (as explained in Chapter 9). With a political questionnaire, people with extreme opinions respond more often than people with moderate views; consequently, your data may not reflect a cross-section. Supplement questionnaire responses with other sources whenever possible.

Some reports include the full text of the interview or questionnaire (with questions, answers, and tabulations) in an appendix (Chapter 16).

Other Primary Sources

Whenever possible, explore all primary sources by writing letters, checking records, or observing and analyzing directly.

Inquiry Letters or Calls.   Letters or calls are handy for obtaining specific information from government agencies, legislators, private companies, university research centers, trade associations and research foundations such as the Brookings Institution and the Rand Corporation (Lavin 9). Letters are discussed in Chapter 20.

Organizational Records and Publications.   Company records (reports, memos, computer printouts, and so on) are a good primary source for data. Most organizations also publish pamphlets, brochures, annual reports, or prospectuses for consumers, employees, investors, or voters. But be alert for bias in company literature. If you were evaluating the safety measures at a local nuclear power plant, you would want the complete picture. Along with the company's literature, you would want studies and reports from government agencies and publications from environmental groups.

Personal Observation and Experiment.   If possible, amplify and verify your findings with a firsthand look. Observation should be your final step, because you now know what to look for. Have a plan. Know how, where, and when to look, and jot down observations immediately. You might even take photos or make drawings.

Informed observations can pinpoint real problems. Here is an excerpt from a report investigating low morale at an electronics firm. This researcher's observations and interpretation are crucial in defining the problem:

> Our survey revealed that employees were unaware of any major barriers to communication. The 75 percent of employees with positive work attitudes said they felt free to talk to their managers, but the managers, in turn, estimated that only 50 percent of employees felt free to talk to them.
>
> The problem involves misinterpretation. Because managers don't ask for complaints, employees are afraid to make them, and because employees never ask for an evaluation, they never get one. Both sides have inaccurate percep-

Direct observation can be essential

tions of what the other wants, and because of ineffective communications, each side fails to realize that its perceptions are wrong.

Keep in mind that even direct observation can lack validity: for instance, you might be biased about what you see (say, focusing on the wrong events or ignoring something important), or, instead of behaving normally, people who know they are being observed might exhibit behavior they think you expect (Adams and Schvanveldt 244).

An experiment is a controlled form of observation designed to *verify an assumption* (e.g., the role of fish oil in preventing heart disease) or to *test something untried* (the relationship between background music and worker productivity). Each specialty has its own guidelines for experiment design.

Analysis of Samples.   Workplace research can involve collecting and analyzing samples: water or soil or air, for contamination and pollution; foods, for nutritional value; ore, for mineral value; or plants, for medicinal value. Investigators analyze material samples to find the cause of an airline accident. Engineers analyze samples of steel, concrete, or other building materials to determine their load-bearing capacity. Medical specialists analyze tissue samples for disease.

EXERCISES

1. Begin researching for the analytical report (Chapter 23) due at semester's end. Complete these steps. (Your instructor might establish a timetable.)

Phase One: Preliminary Steps

a. Choose a topic of *immediate practical importance,* something that affects you or your community directly. (See page 173 for a list of possible topics.)

b. Identify a specific audience and its intended use of your information. Complete an audience-and-use profile (page 60).

c. Narrow your topic, and check with your instructor for approval.

d. Make a working bibliography to ensure sufficient primary and secondary resources. Don't delay this step!

e. List things you already know about your topic.

f. Write a clear statement of purpose and submit it in a proposal memo (pages 519–521) to your instructor.

g. Develop a tree chart of possible questions (as on page 133).

h. Make a working outline.

Phase Two: Collecting Data (Read Chapter 9 in preparation for this phase.)

a. In your research, move from general to specific; begin with general reference works for an overview.

b. Skim your material, looking for high points.

c. Take selective notes. Don't write everything down! Use notecards.

d. Plan and administer (or distribute) questionnaires, interviews, and letters of inquiry.

e. Whenever possible, conclude your research with direct observation.

f. Evaluate and interpret your findings.

g. Use the page 203 checklist to reassess your research methods and reasoning.

Phase Three: Organizing Your Data and Writing Your Report

a. Revise and adjust your working outline, as needed.

b. Compose an audience-and-use analysis, like the sample on pages 568–570.

c. Fully document all sources of information.

d. Proofread carefully and add all needed supplements (title page, letter of transmittal, abstract, summary, appendix, glossary—Chapter 16).

Due Dates: To Be Assigned by Your Instructor

List of possible topics due:
Final topic due:
Proposal memo due:
Working bibliography and working outline due:
Notecards due:
Copies of questionnaires, interview questions, and inquiry letters due:
Revised outline due:
First draft of report due:
Final draft with supplements and documentation due:

2. Using the card or electronic catalog, locate and list the full bibliographic data (author's name, title of work, place and publisher, date) for five books in your field or on your semester report topic, all published within the past year.

3. Consult the *Library of Congress Subject Headings* for alternative headings under which you might find information in the card catalog for your semester report topic.

4. List five major reference works in your field or on your topic by consulting Sheehy, Walford, or a more specific guide to literature.

5. List the titles of each of these specialized reference works in your field or on your topic: a bibliography, an encyclopedia, a dictionary, a handbook, an almanac (if available), and a directory.

6. Identify the major periodical index in your field or on your topic. Locate a recent article on a specific topic (e.g., use of artificial intelligence in medical diagnosis). Photocopy the article and write an informative abstract.

7. Consult the appropriate librarian and identify two databases you would search for information on the topic in Exercise 3.

8. Identify the major abstract collection in your field or on your topic. Using the abstracts, locate a recent article. Photocopy the abstract and the article.

9. Using technical report indexes, locate and summarize three recent reports on *one* specific topic in your field. Provide complete bibliographic information.

10. Using patent indexes, locate and describe three recently patented inventions in your field, and provide complete bibliographic information.

11. Using indexes of conference proceedings, locate and summarize three recent conference papers on *one* specific topic in your field. Provide complete bibliographic information.

12. Using the *Monthly Catalog* or *Government Reports Announcements and Index*, locate and photocopy a recent government publication in your field or on your topic.

13. Determine whether your library offers OCLC or InfoTrac services. Use whichever service is available to locate the titles of two current books or articles in your field or on your topic.

14. If your library offers students a free search of mainframe databases, ask your librarian for help in preparing an electronic search for your semester report.

15. Revise these questions to make them appropriate for inclusion in a questionnaire:

    a. Would a female president do the job as well as a male?

    b. Don't you think that euthanasia is a crime?

    c. Do you oppose increased government spending?

    d. Do you feel that welfare recipients are too lazy to support themselves?

    e. Are teachers responsible for the decline in literacy among students?

    f. Aren't humanities studies a waste of time?

    g. Do you prefer Rocket Cola to other leading brands?

16. Arrange an interview with a successful person in your field. List general areas for questioning: job opportunities, chances for promotion, salary range, requirements, outlook for the next decade, working conditions, job satisfaction, and so on. Compose interview questions; conduct the interview; and summarize your findings in a memo to your instructor.

COLLABORATIVE PROJECT

Your instructor will divide your class into small groups, each of which will decide on a campus or community issue or some other topic worthy of research. Elect a group manager to assign and coordinate tasks. At project's end, the manager will provide a performance appraisal by summarizing, in writing, the contribution of each team member. Assigned tasks will include planning, information gathering from primary and secondary sources, document preparation (including visuals) and revision, and classroom presentation.

Do the research, write the report, and present your findings to the class. (Your instructor may assign Chapter 23 in conjunction with this project.)

Here are some possible research topics:

a. Survey student, faculty, and administration about some proposed curriculum change or about some other controversial campus issue. Compare your findings with nationwide or statewide statistics about attitudes on this issue.

b. As much as 30 percent of groundwater in some states is contaminated. Find out how the quality of local groundwater measures up to national averages. Has the quality increased or decreased over the last ten years? What are the major elements affecting local water quality? What is the outlook for the next decade? Can local residents feel safe drinking tap water?

c. Identify the main qualities employers seek in job applicants. Have employers' expectations changed over the last ten years? If so, why? What is the chance that your generation will face six or seven career changes in your lifetime? What should people do to prepare?

d. Find out which geographic area of the United States is enjoying the greatest prosperity and population growth (or which area is suffering the greatest hardship and population decrease). What are the major reasons? Trace the recent history of this change.

e. Older homes can present a frightening array of toxic hazards: termite spray, wood preservatives, urea formaldehyde insulation, radon gas, chemical contamination of well water, lead paint, asbestos, and so on. Research the major effects of these hazards for someone who is thinking of buying an older home.

f. How safe is your school (or your dorm) from "sick-building syndrome"? Are there any dangers from insulation, asbestos, art supplies, cleaning fluids and solvents, water pipes, or the like? Find out, and prepare a report for your classmates.

g. Which area of your state has the cleanest air and groundwater, and which has the most polluted? Write for someone looking for the safest place to raise a family.

h. Has acid rain caused any damage in your area? Write for classmates.

i. Can peanut butter, black pepper, potatoes, or toasted bread cause cancer? Which of the most common "pure" foods can be carcinogenic? Find out, and write a report for the school dietitian.

j. Are there any recent inventions that could help decrease our reliance on fossil fuels in ways that are economically feasible and practical? Find out, and prepare a report for your U.S. senator.

k. What is the very latest that scientists are saying about the implications of global warming from rain forest destruction and ozone depletion? Find out, and prepare a report to be published in a national magazine.

l. Computer screens: How dangerous are they? What does the latest research indicate? Your company wants to know if it can take any precautions to avoid risks to employees and future lawsuits.

m. Say you work for a "Think-Tank" researching this issue. Or you work for a U.S. senator who wants to introduce legislation to curb the abuses. What kinds of privacy violations does present law allow in the workplace? What legislation is pending?

n. Does alcohol consumption have any effect on academic performance? What do the latest studies indicate? Find out, and prepare a report for publication in your campus newspaper.

Recording, Reviewing, and Documenting Findings

Recording Findings

Reviewing Findings

Documenting Sources

Reassessing the Entire Process

■　　　　　■　　　　　■

As you discover material during research, you confront questions like these: *How much is worth keeping? How should I record it? Can I trust this information? What, exactly, does it mean? How will I credit the source?* These latter stages of the research process call for the same quality of critical thinking required by the earlier stages in Chapter 8.

RECORDING FINDINGS

Findings should be recorded in ways that enable you to easily locate, organize, shuffle, and control the material as you work with it. Record primary research findings by using notebooks, photographs, drawings, tape recorder, videotape, or whichever medium suits your purpose. Record secondary research findings in the form of notes.

Taking Notes

Notecards are convenient because they are easy to organize and reorganize.[1] Follow these suggestions for using notecards:

[1] In place of notecards, some researchers now prefer electronic file programs or database management software that allows notes to be filed, shuffled, and retrieved by author, title, topic, date, or other access point.

Record each bibliographic citation exactly as it will appear in your final report

> Nickerson, Robert C. *Fundamentals of Programming in Basic.* Boston: Little, 1981.

FIGURE 9.1 Bibliography Card

1. Begin by making separate bibliography cards for each work you plan to consult (Figure 9.1). Record the complete entry, using the identical citation format that will appear in your actual report. (See pages 191–96 for sample entries.)
2. Skim the entire work to locate relevant material.
3. Go back and decide what to record. (Use a separate card for each item.)
4. Decide how to record the item: as a quotation or a paraphrase. When quoting others directly, be sure to record words and punctuation accurately. When restating or adapting material in your own words, be sure to preserve the original meaning and emphasis.

Quoting the Work of Others

When you borrow exact wording, whether the words were written or spoken (as in an interview or presentation), you must place quotation marks around all borrowed material. Even a single borrowed sentence or phrase, or a single word used in a special way, needs quotation marks, with the exact source properly cited.

If your notes fail to identify quoted material accurately, you might forget to credit the source in your report. Even when this omission is unintentional, writers face the charge of *plagiarism* (misrepresenting as one's own the words or ideas of someone else).

In recording a direct quotation, copy the selection word for word (Figure 9.2) and include the page number(s). If your quotation omits parts of a sentence, use an *ellipsis* (three periods: . . .) to indicate each part that you have omitted from the original. If your quotation omits the end of a sentence, the beginning of the subsequent sentence, or whole sentences or paragraphs, show the ellipsis with four periods (. . . .).

> If your quotation omits parts . . . use an ellipsis. . . . If your quotation omits the end. . . .

Ellipsis within and between sentences

Be sure that your elliptical expression is grammatical and that the omitted material in no way distorts the original meaning.

If you insert your own comments within the quotation, place them inside brackets to distinguish your words from those of your source:

> "This profession [aircraft ground controller] requires exhaustive attention."

Brackets setting off personal comments within quoted material

(For more on brackets, see page 620.)

Nickerson, Robert. p. 34

"The first step in the programming process is understanding the problem to be solved. Understanding the problem involves determining the requirements of the problem and how these requirements can be met."

Place quotation marks around all directly quoted material

FIGURE 9.2 Notecard for a Quotation

Introducing quotations with your own expressions requires attention to audience needs and proper grammar. Generally, integrated quotations are introduced by phrases such as "Jones argues that," "Smith suggests that," so that readers will know who said what. But, more importantly, readers must see the relationship between the quoted idea and the sentence that precedes it. You therefore want to use a transitional phrase that emphasizes this relationship by looking back as well as ahead:

An introduction that unifies a quotation with the discussion

> After you decide to develop a program, "the first step in the programming process. . . ."

Besides showing how each quotation helps advance the main idea you are developing, your integrated sentences should be grammatical:

Quoted material integrated grammatically with the writer's words

> "The agricultural crisis," Marx acknowledges, "resulted primarily from unchecked land speculation."

> "She has rejuvenated the industrial economy of our region," Smith writes of Berry's term as regional planner.

(For quoting long passages and for punctuating at the end of a quotation, see pages 618 and 619.)

Use a direct quotation only when precision or clarity or emphasis requires the exact words from the original. Research writing is more a process of independent thinking, in which you work with the ideas of others in order to reach your own conclusions; you should therefore paraphrase, instead of quoting, much of your borrowed material.

Paraphrasing the Work of Others

We paraphrase not only to preserve the original idea, but also to express it in a clearer or simpler or more direct or emphatic way—without distorting the idea. Paraphrasing means more than changing or shuffling a few words; it means restating the original idea in your own words and giving full credit to the source.

To borrow or adapt someone else's ideas or reasoning without properly documenting the source is plagiarism. To offer as a paraphrase an original passage only slightly altered—even when you document the source—also is plagiarism. Equally unethical is to offer a paraphrase, although documented, that distorts the original meaning.

An effective paraphrase generally displays all or most of the following elements (Weinstein 3):

- reference to the author early in the paraphrase, to indicate the beginning of the borrowed passage

- key words retained from the original, to preserve the meaning
- original sentences restructured and combined, for emphasis and fluency
- needless words from the original deleted, for conciseness
- your own words and phrases that help explain the author's ideas, for clarity
- a citation (in parentheses) of the exact source, to mark the end of the borrowed passage and to give full credit
- preservation of the author's original intent

Figure 9.3 shows an entry paraphrased from the following passage:

Finally the programming process is completed by bringing together all the material that describes the program. This is called *documenting* the program, and the result of this activity is the program's *documentation*. Included in the documentation is the program listing and a description of the input and output data. Documentation enables other programmers to understand how the program functions. Often it is necessary to return to the program after a time to make

Nickerson, Robert. *Fundamentals*

According to Nickerson's definition, a programmer's *documentation* describes the program, and explains how it works. Later programmers use the documentation to understand the program if they need to correct or change it (35).

Signal the beginning of the paraphrase by citing the author, and the end by citing the source

FIGURE 9.3 Notecard for a Paraphrase

corrections or changes. With adequate documentation, it is much easier to understand a program's operation.[2]

Paraphrased material takes no quotation marks, but it has to be documented to acknowledge your debt to the source. Failing to acknowledge ideas, findings, judgments, lines of reasoning, opinions, facts, or insights not considered *common knowledge* (page 189) is plagiarism—even when these are expressed in your own words.

REVIEWING FINDINGS

In working with your research findings, you employ critical thinking to evaluate your sources for dependability and to interpret the material accurately.

Evaluating the Sources

Your sources should be current enough to convey the latest information on your topic. Also, each source should be reputable, relatively unbiased, authoritative, and borne out by similar sources. Say you are researching the alleged benefits of low-impact aerobics for reducing stress among employees at a fireworks factory. You could expect claims in a professional journal such as the *New England Journal of Medicine* to have bases in scientific fact. Also, a reputable magazine, such as *Scientific American,* would be a reliable source of evidence or of informed opinion. On the other hand, you might wisely suspect the claims in supermarket scandal sheets or movie magazines. Even claims in monthly "digests," which offer simplified and mostly undocumented "wisdom" to mass audiences, should be verified.

You would need to interview a representative sample of people who have practiced aerobics for a long time: people of both sexes, different ages, different lifestyles before they began aerobics, and so on. Even reports from ten successful practitioners would be a small sample unless those reports were supported by laboratory data. On the other hand, with a hundred reports from people ranging from students to judges and doctors, you might not have "proved" anything, but the persuasiveness of your evidence would increase.

Your own experience often is an inadequate base for generalizing. You cannot tell whether your experience is representative, regardless of how

[2]This passage and those in Figures 9.2 and 9.3 are adapted from Robert C. Nickerson. *Fundamentals of Programming in BASIC.* Boston: Little, 1983.

long you might have practiced aerobics. Interpret your experience only within the broader context of sources.

Some issues (the need for defense spending or causes of inflation) are always controversial, and will never be resolved. Although we can get verifiable data and can reason persuasively on some subjects, no close reasoning by any expert and no supporting statistical analysis will "prove" anything about a controversial subject. For instance, one could only *argue* (more or less effectively) that federal funds will or will not alleviate poverty or unemployment. Some problems simply are more resistant to solution than others, no matter how reliable and valid the sources.

Amid such difficulties, resist the temptation to report hasty but unverified answers. Better to report no answers than misleading ones. Take no claims for granted, and cross-check all data.

Evaluating the Evidence

Hard evidence consists of facts, examples, statistics, expert testimony, or informed opinion. It can stand up under testing because it can be verified (shown to be true). Soft evidence consists of uninformed opinion or data that was obtained unscientifically. It may collapse under testing unless the opinion is expert and unbiased.

Base your conclusions on hard evidence. Early in your research, you might read an article that makes positive claims about low-impact aerobics, without providing data on measurements of pulse, blood pressure, or metabolic rates. Although your own experience and opinion might agree with the author's, you should not hastily conclude that this form of exercise benefits everyone. So far, you have only two opinions—yours and the author's—without scientific support (e.g., tests of a cross-section under controlled conditions). Conclusions now would rest on soft evidence. Only after a full survey of reliable sources can you decide which conclusions are supported by the bulk of your evidence.

Interpreting the Material

Interpreting the material means trying to reach the truth of the matter: an overall judgment about what the data mean and what conclusion or action they suggest. Unfortunately, however, research does not always yield answers that are clear or conclusive or about which we can be certain.

As possible outcomes of our research, we need to be able to recognize three distinct and very different levels of certainty:

1. The ultimate truth: the *conclusive answer:*

A practical definition of "truth"

Truth is *what is so* about something, the reality of the matter, as distinguished from what people wish were so, believe to be so, or assert to be so. From another perspective, in the words of Harvard philosopher Israel Scheffler, truth is the view "which is fated to be ultimately agreed to by all who investigate."[3] The word *ultimately* is important. Investigation may produce a wrong answer for years, even for centuries. . . . Does the truth ever change? No. . . . One easy way to spare yourself any further confusion about truth is to reserve the word *truth* for the final answer to an issue. Get in the habit of using the words, *belief, theory,* and *present understanding* more often. (Ruggiero 21–22)

People are too often mistaken in their certainty about the *truth.* For example, in the second century A.D., Ptolemy's view of the universe concluded that the earth was its center—and though untrue, this judgment was based on the best information available at that time. And Ptolemy's view survived for thirteen centuries, even after new information had discredited this belief. When Galileo proposed a more truthful view in the fifteenth century, he was labeled a heretic.

Conclusive answers, of course, are the research outcome we seek, but we often have to settle for something less certain.

2. The *probable answer:* the answer that stands the best chance of being true or accurate—given the most we can know at this particular time. Probable answers are subject to revision in the light of new information.

3. The *inconclusive answer:* the realization that the truth of the matter is far more elusive or ambiguous or complex than we expected.

To ensure an accurate outcome, we must decide what level of certainty the findings warrant. Otherwise, we might invent an unwarranted conclusion for inconclusive material.

When the issue is controversial our own bias might cause us to overestimate the certainty of our findings.

Personal bias is a fact of life

Expect yourself to be biased, and expect your bias to affect your efforts to construct arguments. Unless you are perfectly neutral about the issue, an unlikely circumstance, at the very outset . . . you will believe one side of the issue to be right, and that belief will incline you to . . . present more and better arguments for the side of the issue you prefer. (Ruggiero 134)

[3]From *Reason and Teaching.* New York: Bobbs-Merrill, 1973.

Because personal bias is hard to transcend, *rationalizing* often becomes a substitute for *reasoning*:

> You are reasoning if your belief follows the evidence—that is, if you examine the evidence first and then make up your mind. You are rationalizing if the evidence follows your belief—if you first decide what you'll believe and then select and interpret evidence to justify it. (Ruggiero 44)

Reasoning versus rationalizing

Personal bias is inescapable but manageable—as long as we recognize it.

Finding the truth, especially in a complex issue or problem, is often a process of elimination, of ruling out or avoiding errors in reasoning. Following are three common errors that can distort our interpretations.

Faulty Causal Reasoning. Causal reasoning tries to explain *why* something happened or *what* will happen, often very complex questions. Faulty causal reasoning oversimplifies or distorts the cause-effect relationship through errors like these:

Investment builds wealth. [*Ignores the role of knowledge, wisdom, timing, and luck in successful investing.*]

Ignoring other causes

Running improves health. [*Ignores the fact that many runners get injured, and that some even drop dead while running.*]

Ignoring other effects

Right after buying a rabbit's foot, Felix won the state lottery. [*Posits an unwarranted causal relationship merely because one event follows another.*]

Inventing a cause

Poverty causes disease. [*Ignores the fact that disease, while highly associated with poverty, has many causes unrelated to poverty.*]

Confusing correlation with causation

My grades were poor because my exams were unfair. [*Denies the real causes of one's failures.*]

Rationalizing

Because of bias or impatience, we can be tempted to settle for a hasty cause or to confuse possible, probable, and definite causes.

Sometimes a definite cause is apparent (e.g., "The engine's overheating is caused by a faulty radiator cap"), but usually much searching and thought are needed to isolate a specific cause. Suppose you want to answer this question: Why does our state college have no children's daycare facilities? Brainstorming yields these possible causes:

- lack of need among students
- lack of interest among students, faculty, and staff
- high cost of liability insurance
- lack of space and facilities on campus

- lack of trained personnel
- prohibition by state law
- lack of legislative funding for such a project

Say you proceed with interviews, questionnaires, and research into state laws, insurance rates, and availability of personnel. You begin to rule out some items, and others appear as probable causes. Specifically, you find a need among students, high campus interest, an abundance of qualified people for staffing, and no state laws prohibiting such a project. Three probable causes remain: lack of funding, high insurance rates, and lack of space. Further inquiry shows that lack of funding and high insurance rates *are* issues. These obstacles, however, could be eliminated through new sources of revenue: charging a fee for each child, soliciting donations, or diverting funds from other campus organizations. Finally, after examining available campus space and speaking with school officials, you arrive at one definite cause: lack of space and facilities.[4]

When you report on your research, be sure readers can draw conclusions identical to your own on the basis of the evidence. The process might be diagrammed like this:

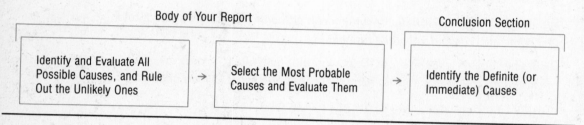

Initially you might have based your conclusions hastily on soft evidence (say, an opinion—buttressed by a newspaper editorial—that the campus was apathetic.) Now you base your conclusions on solid, factual evidence. You have moved from a wide range of possible causes to a narrow range of probable causes, and finally to a definite cause.

Sometimes, finding a single cause is impossible, but this reasoning process can be tailored to most problem-solving analyses. Anything but the simplest effect is likely to have more than one cause. By narrowing the field, you can focus on the real issues.

[4]Of course, one could argue that lack of space and facilities is somehow related to funding. And the college's being unable to find funds or space may be related to student need, which is not sufficiently acute or interest sufficiently high to exert real pressure. Lack of space and facilities, however, appear to be the *immediate* cause.

Faulty Statistical Reasoning. The purpose of statistical analysis is to determine the meaning of a collected set of numbers. In primary research, our surveys and questionnaires (Chapter 8) often lead to some kind of numerical interpretation ("What percentage of respondents prefer *X*?" "How often does *Y* happen?" and so on). In secondary research, we often rely on numbers collected by primary researchers.

Numbers have special appeal because they seem more precise, more objective, and less ambiguous than words. They are easier to summarize, measure, compare, and analyze. But, through error or abuse, numbers can be totally misleading.

> One journalist explains how radio or television "phone-in" surveys produce grossly distorted "data": Although ninety percent of callers, say, express support (or opposition) toward this or that viewpoint, the people who call tend to be those with the greatest anger or the strongest feelings about the issue— usually a mere two or three percent of the total audience (Fineman 24).

Before relying on any set of numbers, we need to know exactly where they come from, how they were collected, and how they were analyzed (Lavin 275–76).

Faulty statistical reasoning produces conclusions that are unwarranted, inaccurate, or downright deceptive. Here are some typical fallacies:

- *The meaningless statistic:* when exact numbers are used to quantify something so inexact or vaguely defined that it should only be approximated (Huff 247; Lavin 278): "Only 38.2 percent of college graduates end up working in their specialty." "Boston has 3,247,561 rats." An exact number looks impressive, but it can hide the fact that certain subjects (child abuse, cheating in college, virginity, drug and alcohol abuse on the job, eating habits) cannot be counted exactly because respondents don't always tell the truth (because of denial or embarrassment or merely guessing).

- *The undefined average:* when the mean, median, and mode are confused in determining an average (Huff 244; Lavin 279). The *mean* is the result of adding up the value of each item, and then dividing by the number of items. The *median* is the result of ranking all the values from high to low, and then choosing the middle value. The *mode* is the value that occurs most often.

Each of these three measurements represents some kind of average. But unless we know which "average" is being presented, we cannot possibly interpret the figures accurately.

Assume, for instance, that we are computing the average salary among female vice presidents at XYZ Corporation (ranked from high to low):

Vice President	Salary
"A"	$90,000
"B"	90,000
"C"	80,000
"D"	65,000
"E"	60,000
"F"	55,000
"G"	50,000

In the above example, the mean salary (total salaries divided by people) equals $70,000; the median salary (middle value) equals $65,000; the mode (most frequent value) equals $90,000. Each is, legitimately, an "average," and each could be used to support or refute a particular assertion (for example, "Women receive too little" or "Women receive too much").

Research expert Michael Lavin sums up the potential for bias in reporting averages:

> Depending on the circumstances, any one of these measurements may describe a group of numbers better than the other two. . . . [But] people typically choose the value which best presents their case, whether or not it is the most appropriate to use. (279)

Unethical use of statistics misleads and manipulates the audience.

- *The distorted percentage figure:* when percentages are reported without explanation of the original numbers used in the calculation (Adams and Schvaneveldt 359; Lavin 280): "Seventy-five percent of respondents prefer our brand over the competing brand"—without mention that only four people were surveyed. Or "Sixty-six percent of employees we hired this year are women and minorities, compared to the national average of forty percent"—without mention that only three people have been hired this year, by a company that employs three hundred (mostly white males). Even the most impressive looking numbers can be misleading.

Another fallacy in reporting percentages is in failing to account for the margin of error. For example, a claim that the majority of people surveyed prefer Brand X might be based on the fact that 51 percent of respondents expressed this preference; but if the survey carried a 2 percent margin of error, the claim could be invalid.

- *The bogus ranking:* when items are compared on the basis of ill-defined criteria (Adams and Schvaneveldt 212; Lavin 284): "Last year, the Batmobile was the number-one selling car in America"—without mention that some competing car makers actually sold *more* cars to private individuals, and that the Batmobile figures were inflated by hefty sales to rental-car companies and corporate fleets. Unless we know how the ranked items were chosen and how they were compared (the criteria), a ranking can produce a scientific-seeming number based on a completely unscientific method.

These are only a few examples of statistics that seem highly persuasive but that in fact cannot be trusted. As producers *and* consumers of information, we have an ethical responsibility to seek out the truthful answers, not merely those that are most comforting or convenient.

Even when the statistics are valid and reliable, we need to interpret them realistically. Consider, for example, the legitimate finding that rates for certain cancers double among people who are exposed for prolonged periods to electromagnetic fields. What this statistic may mean is that the incidence of cancer actually increases from 1 in 10,000 to 2 in 10,000.

Specious Conclusions. Specious conclusions are deceptive because they seem correct at first glance, but prove faulty when scrutinized. Conclusions based on soft evidence or faulty reasoning are specious.

Assume you are an education consultant evaluating the accuracy of IQ testing as a measure of intelligence and as a predictor of academic performance. Reviewing the evidence, you find a correlation between low IQ scores and low achievers. You then verify your statistics by examining a cross-section of reliable sources. Should you feel justified in concluding that IQ tests do measure intelligence and predict performance accurately? This conclusion might be specious unless you could show that

1. Neither parents nor teachers nor the children tested had seen individual test scores and had thus been able to develop biased expectations.
2. Regardless of their IQ scores, all children had been exposed to an identical curriculum at an identical pace, instead of being "tracked" on the basis of individual scores.

Your data could be interpreted only within the context of these two variables. Even hard evidence can support specious conclusions, unless it is interpreted within a context that accounts for all major variables.

The critical-thinking strategies we have just considered are part of the process that distinguishes legitimate research from mere information gathering.

DOCUMENTING SOURCES

Documenting research means acknowledging one's debt to each information source. Proper documentation satisfies professional requirements for
ethics, efficiency, and authority.

Why You Should Document

Documentation is a matter of *ethics* in that the originator of borrowed material deserves full credit and recognition. Moreover, all published material
is protected by copyright law. Failure to credit a source could make you
liable to legal action, even if your omission was unintentional.

Documentation is also a matter of *efficiency*. It provides a network for
organizing and locating the world's printed knowledge. If you cite a particular source correctly, your reference will enable interested readers to locate
that source themselves.

Finally, documentation is a matter of *authority*. In making any claim
(say, "A Mercedes-Benz is more reliable than a Ford Taurus") you invite
challenge: "Says who?" Data on road tests, frequency of repairs, resale
value, workmanship, and owner comments can help validate your claim
by showing its basis in *fact*. A claim's credibility increases in relation to the
expert references supporting it. For a controversial topic, you may need to
cite several authorities who hold various views, as in this next example,
instead of forcing a simplistic conclusion on your material:

Citing a balance of views

> Opinion is mixed as to whether a marketable quantity of oil rests under Georges
> Bank. Cape Cod Geologist John Blocke feels that extensive reserves are im-
> probable ("Geologist Dampens Hopes" 3). Oil geologist Donald Marshall is un-
> certain about the existence of any oil in quantity under Georges Bank ("Off-
> shore Oil Drilling" 2). But the U.S. Interior Department reports that the Atlantic
> continental shelf may contain 5.5 billion barrels of oil (Kemprecos 8).

Readers of your research report expect the *complete* picture.

What You Should Document

Document any insight, assertion, fact, finding, interpretation, judgment or
other "appropriated material that readers might otherwise mistake for your
own" (Gibaldi and Achtert 155). Specifically, you must document

- any source from which you use exact wording, or
- any source from which you adapt material in your own words, or
- any visual illustration: charts, graphs, drawings, or the like (see Chapter 14 for documenting visuals).

You don't need to document anything considered *common knowledge:* material that appears repeatedly in general sources. In medicine, for instance, it is common knowledge that foods high in fat cause some types of cancer. Thus, in a research report on fatty diets and cancer, you probably would not need to document that well-known fact. But you would document information about how the fat/cancer connection was discovered, subsequent studies (say, of the role of saturated versus unsaturated fats), and any information for which some other person could claim specific credit. If the borrowed material can be found in only one specific source, and not in multiple sources, document it. When in doubt, document the source.

How You Should Document

Borrowed material has to be cited twice: at the exact place you use that material, and at the end of your document. Documentation practices vary widely, but all systems work almost identically: a brief reference in the text names the source and refers readers to the complete citation, which enables the source to be retrieved.

Many disciplines, institutions, and organizations publish their own documentation manuals. Here are a few:

> *Style Guide for Chemists*
> *Geographical Research and Writing*
> *Style Manual for Engineering Authors and Editors*
> *IBM Style Manual*
> *NASA Publications Manual*

Style guides from various disciplines

When no specific format is stipulated, consult one of the two general manuals discussed in the next section or the *Chicago Manual of Style,* 13th ed., by the University of Chicago Press, which covers documentation in the humanities, related fields, and natural sciences. The formats in any of these three manuals can be adapted to most research writing.

The next section illustrates citations and entries for MLA documentation style, and then presents alternatives with APA (American Psychological Association) style and, briefly, numerical documentation style.

MLA Documentation Style

Traditional documentation used superscripted numbers (like this:[1]) in the text, followed by full references at page bottom (footnotes) or at document's end (endnotes) and, finally, by a bibliography. But a more current form of documentation appears in the *MLA Handbook for Writers of Research Papers,*

3rd ed. New York: Modern Language Association, 1988. MLA style replaces footnote and endnote numbers with parenthetical references, which briefly identify the source(s) (in parentheses). The full documentation then appears in a "Works Cited" section, at report's end.[5]

A parenthetical reference usually includes the author's surname and the exact page number(s) of the borrowed material:

Parenthetical reference in the text

> Cancer risk increases when magnetic fields measure consistently higher than 2.5 milligauss (Abelson 241).

Readers seeking the complete citation for Abelson can move easily to "Works Cited," listed alphabetically by author:

Full citation at document's end

> Abelson, Philip H. "Effects of Electric and Magnetic Fields." Science 21 July
> 1989: 240–46.

This complete citation includes page numbers for the entire article.

Guidelines for Parenthetical References.   For clear and informative parenthetical references, observe these guidelines:

- If your discussion names the author, do not repeat the name in your parenthetical reference; simply give the page number(s):

Citing page numbers only

> Abelson describes recent evidence indicating that cancer risk increases when magnetic fields measure consistently higher than 2.5 milligauss (241).

- If you cite two or more works in a single parenthetical reference, separate the citations with semicolons:

Three works in a single reference

> (Jones 32; Leduc 41; Gomez 293–94)

- If you cite two or more authors with the same surname, include the first initial in your parenthetical reference to each author:

Two authors with identical surnames

> (R. Jones 32)   (S. Jones 14–15)

- If you cite two or more works by the same author, include the first significant word from each work's title, or a shortened version:

Two works by one author

> (Lamont, Biophysics 100–01)   (Lamont, Diagnostic Tests 81)

- If the work is by an institutional or corporate author or if it is unsigned (that is, author unknown), use only the first few words of the institutional name or the work's title in your parenthetical reference:

---

[5]In the MLA system, footnotes (and endnotes) are now used only to comment or expand on material in the text, or to comment on sources or suggest additional sources. Place these notes at page bottom or in a "Notes" section at document's end.

(American Medical Assn. 2)   ("Distribution Systems" 18)

To avoid distracting your readers, keep each parenthetical reference as brief as possible. (One method is to name the source in your discussion, and to place the page number[s] only in parentheses.)

For a paraphrase, place the parenthetical reference *before* the closing punctuation mark. For a quotation that runs into the text, place the reference *between* the final quotation mark and the closing punctuation mark. For a quotation set off (indented) from the text, place the reference two spaces *after* the closing punctuation mark.

Works Cited Entries for Books.   Any citation for a book should contain the following information (found on the book's title and copyright pages): author, title, editor or translator, edition, volume number, and facts about publication (city, publisher, date).

Type the first line of each entry flush with the left margin. Indent the second and subsequent lines five spaces. Double-space within and between each entry. Skip two horizontal spaces after any closing period in an entry,[6] and one space after any comma or colon.

Following are examples of complete citations as they would appear in the "Works Cited" section of your document. Shown italicized after each citation is its corresponding parenthetical reference as it would appear in the text.

Single Author

Kerzin-Fontana, Jane B.   Technology Management: A Handbook.   3rd ed.

    Delmar, NY: American Management Assn., 1992.

Parenthetical reference: (Kerzin-Fontana 3–4)

Identify the state of publication by U.S. Postal Service abbreviations. If the city of publication is well known (Boston, Chicago, and so on), omit the state.

Two or Three Authors

Aronson, Linda, Roger Katz, and Candide Moustafa.   Toxic Waste Disposal

    Methods.   Englewood Cliffs: Prentice, 1991.

Parenthetical reference: (Aronson, Katz, and Moustafa 9)

---

[6]Only those periods that separate different items in the entry (say, author's name from the work's title) are followed by two spaces. Those periods that end an abbreviation within one particular item (say, "29 Dec." or "Mary H. Gordon") are followed by only one space.

More than Three Authors

Santos, Ruth J., et al.   Environmental Crises in Developing Countries.   New
York: Harper, 1990.

Parenthetical reference: (Santos et al. 111-23)

Author(s) Not Named

Structured Programming.   Boston: Meredith, 1989.

Parenthetical reference: (Structured 67)

Two Books by the Same Author

Chang, John W.   Biophysics.   Boston: Little, 1989.

---.   Diagnostic Techniques.   New York: Radon, 1990.

Parenthetical references: (Chang, Biophysics 123–26); (Chang, Diagnostic 87)

When citing more than one work by the same author, do not repeat the
author's name; simply type three hyphens followed by a period. List the
works alphabetically by title.

One or Two Editors

Morris, A. J., and Louise B. Pardin-Walker, eds.   Handbook of New
Information Technology.   New York: Harper, 1993.

Parenthetical reference: (Morris and Pardin-Walker 34)

For more than three editors, name only the first, followed by "et al."

Quotation of a Quotation

Kline, Thomas.   Automated Office Systems.   New York: Random, 1989. p. 97,
qtd. in Sturtevant, John.   White-Collar Productivity.   Boston: Houghton,
1991.

Parenthetical reference: (Kline, qtd. in Sturtevant 116)

When your source (as in Sturtevant, above) has quoted another, refer first
to the original, followed by "qtd. in" (for "quoted in") and the work you
consulted. In your parenthetical reference, give the page number(s) of the
work in which you found the quotation.

Selection in an Anthology (collected works by various authors)

Anderson, Paul V. "What Survey Research Tells Us about Writing at Work."
   Writing in Nonacademic Settings. Ed. Lee Odell and Dixie Goswami.
   New York: Guilford, 1985. 3–83.

Parenthetical reference: (Anderson 31)

The page numbers in the complete citation are for the selection cited from the anthology.

Works Cited Entries for Periodicals. A citation for an article should give this information (as available): author, article title, periodical title, volume or number (or both), date (day, month, year), and page numbers for the entire article—not just the pages cited. List the information in the order given here, as in the following examples.

Magazine Article

Main, Jeremy. "The Executive Yearning to Learn." Fortune 3 May 1982:
   234–48.

Parenthetical Reference: (Main 235–36)

No punctuation separates the magazine title and date. Nor is the abbreviation "p." or "pp." used to designate page numbers.
   If no author is given, list all other information:

"Distribution Systems for the New Decade." Power Technology Magazine 18
   Oct. 1990: 18+.

Parenthetical Reference: ("Distribution Systems" 18)

This article began on page 18 and then continued on page 21. When an article does not appear on consecutive pages, give only the number of the first page, followed immediately by a plus sign. A three-letter abbreviation denotes any month spelled with five or more letters.

Article in a Journal with New Pagination in Each Issue

Thackman-White, Joan R. "Computer-Assisted Research." American Librarian 51.1 (1992): 3–9.

Parenthetical reference: (Thackman-White 4–5)

Because each issue for that year will have page numbers beginning with "1," readers need the number of this issue. The "51" denotes the volume number; the "1" denotes the issue number. Omit "The" or "A" or any other introductory article from a journal or magazine title.

Article in a Journal with Continuous Pagination

Barnstead, Marion H. "The Writing Crisis." Writing Theory 12 (1989):

415–33.

Parenthetical reference: (Barnstead 418)

When page numbers continue from issue to issue for the full year, readers won't need the issue number, because no other issue in that year repeats these same page numbers. (You may, however, include the issue number if you think it will help readers retrieve the article more easily.) The "12" denotes the volume number.

Newspaper Article

Baranski, Vida H. "Errors in Technology Assessment." Boston Times 15 Jan.

1993, evening ed., sec. 2: 3.

Parenthetical reference: (Baranski 3)

When a daily newspaper has more than one edition, cite the specific edition after the date. Omit any introductory article in the newspaper's name (not The Boston Times). If no author is given, list all other information. If the newspaper's name does not contain the city of publication, insert it, using brackets: "Sippican Sentinel [Marion, MA]."

Works Cited Entries for Other Kinds of Materials. Miscellaneous sources range from unsigned encyclopedia entries to nonprint sources to software packages. A full citation should give this information (as available): author, title, city, publisher, date, and page numbers.

Encyclopedia, Dictionary, or other Alphabetic Reference

"Communication." The Business Reference Book. 1987 ed.

Parenthetical reference: ("Communication")

If the entry is signed, begin with the author's name. For any work arranged alphabetically, omit page numbers in both the complete citation and the parenthetical reference.

Personally Conducted Interview

Nasser, Gamel. Chief Engineer for Northern Electric. Personal Interview.

> Rangeley, ME. 2 Apr. 1992.

Parenthetical reference: (Nasser)

Published Interview

Lescault, James. "The Future of Graphics." Executive Views of Automation.

> Ed. Karen Prell. Boston: Haber, 1992. 216–31.

Parenthetical reference: (Lescault 218)

The interviewee's name is placed in the entry's author slot.

Unpublished Letter

Rogers, Leonard. Letter to the author. 15 May 1993.

Parenthetical reference: (Rogers)

Questionnaire

Taylor, Lynne. Questionnaire sent to 612 Massachusetts business executives.

> 14 Feb. 1992.

Parenthetical reference: (Taylor)

Pamphlet or Brochure

Waters, Joan L. Investment Strategies for the 90's. San Francisco: Blount

> Economics Assn., 1990.

Parenthetical reference: (Waters)

If the work is unsigned, begin with its title.

Lecture

Dumont, R. A. "Managing Natural Gas." Lecture at UMass Dartmouth, 15

> Jan. 1993.

Parenthetical reference: (Dumont)

Database Source

Keyes, Langley Carlton. "Profits in Prose." Harvard Business Review 39

> (1961): 105–12; Latham, New York: Bibliographic Retrieval Service, 1984.

Parenthetical reference: (Keyes 107)

Software

Levy, Michael C., et al. Statmaster: Exploring and Computing Statistics.

Computer software. Boston, Little, 1988. IBM PC-DOS 2.0, 512KB, disk.

Parenthetical reference: (Levy)

Name the appropriate computer, the kilobytes, the software format ("disk"), and include any other useful information.

Corporate Author or Government Publication

Presidential Task Force on Acid Rain. Acid Rain and Corporate Profits.

Washington: GPO, 1992.

Parenthetical reference: (Presidential Task Force 108-12)

The shortened version in the parenthetical reference avoids interrupting the flow of the text.

Other Items (unpublished reports, dissertations, and so on)

Author (if known), title (in quotes), sponsoring organization or publisher, date,

page number(s).

For any work that has group authorship (corporation, committee, task force), cite the name of the group or agency in place of the author's name.

The "Works Cited" List.   In your "works cited" section, arrange entries alphabetically by author's surname. When the author is unknown, list the title alphabetically according to its first word (excluding introductory articles). For a title that begins with a digit ("5," "6," etc.), alphabetize the entry as if the digit were spelled out.

The list of works cited in Figure 9.4 accompanies the report on electromagnetic fields, pages 559–567.[7] In the left margin, colored numbers refer to the elements discussed on the page facing Figure 9.4. Near the right margin, italicized labels in brackets identify different types of sources the first time a particular type is cited.

APA Documentation Style

One popular alternative to MLA style appears in the *Publication Manual of the American Psychological Association,* 3rd ed. Washington: American Psy-

---

[7]Normally, of course, this list would appear at the end of the report.

chological Association, 1983. APA style is useful when writers wish to emphasize the publication dates of their references. A parenthetical reference in the text briefly identifies the source, date, and page number(s):

> Beyond motivation and communication skills, interpersonal skills are the ultimate requirement for success in marketing (Splaver, 1987, p. 14).

Reference cited in the text

The full citation then appears in the alphabetic listing of "References," at report's end:

> Splaver, S. (1987).   Your personality and your career.   New York:
> Simon & Schuster.

Full citation at document's end

Because it emphasizes the date, APA style (or some similar author-date style) is preferred in the sciences and social sciences, where information quickly becomes dated.

Guidelines for Parenthetical References.   APA's parenthetical references resemble MLA's (pages 190–91), but a comma separates each item in the reference, and "p." or "pp." precedes the page number(s). When a subsequent reference to a given work follows closely after the initial reference, the date need not be included. For additional guidelines governing parenthetical references, consult the *APA Manual.*

The List of References.   APA's "References" section is an alphabetic listing equivalent to MLA's "Works Cited" section. Like "Works Cited," the reference list includes only those works actually cited. (A bibliography usually would include background works or works consulted as well.) APA entries, however, differ somewhat from MLA entries, as shown in Figure 9.5.[8]

The list of references in Figure 9.5 accompanies the report on technical marketing, pages 570–578. In the left margin, colored numbers denote elements discussed on the page facing Figure 9.5. Near the right margin, italicized labels in brackets identify different types of sources the first time a particular type is cited.

For comprehensive examples of entries in the reference list, consult the *APA Manual* (usually available at the reserve desk).

---

[8]One notable difference: In APA style, only "recoverable" sources appear in the reference list. Therefore, personal interviews and most other unpublished materials are cited in the text only, as shown on page 572.

WORKS CITED

Abelson, Philip H. "Effects of Electric and Magnetic Fields." Science 21 July 1989: 240–46.   [*magazine article*]

Black, Pamela. "Rising Tension over High-Tension Lines." Business Week 30 Oct. 1989: 42–48.

Brodeur Paul. "ANNALS OF RADIATION: The Cancer at Slater School." The New Yorker 7 Dec. 1992: 86+.

---. Currents of Death. New York: Simon, 1989.   [*book—one author*]

---. Interview. Nightline. ABC Television. 9 Mar. 1990.   [*media interview*]

Castleman, Michael. "Electromagnetic Fields." Sierra Jan./Feb. 1992: 21–22.

Halloran-Barney, Marianne B. Energy Service Advisor for County Electric. Personal Interview. Adams, MA. 3 Apr. 1993.   [*personal interview*]

Hecht, Jeff. "Cell Tests Suggest Link between Cables and Cancer." New Scientist 3 Dec. 1987: 27–30.

Jauchem, J. "Alleged Health Effects of Electromagnetic Fields: Misconceptions in the Scientific Literature." Journal of Microwave Power and Electromagnetic Energy 26.4 (1991): 189–95.   [*journal article*]

Kirkpatrick, David. "Can Power Lines Give You Cancer?" Fortune 31 Dec. 1990: 80–85.

Lee., J. M., Jr., et al. Electrical and Biological Effects of Transmission Lines: A Review. U.S. Dept. of Energy. NTIS no. PC A06/MF A01. Washington: GPO, 1989.   [*report*]

McDonald, Kim A. "Some Physicists Criticize Research Purporting to Show Links between Low-Level Electromagnetic Fields and Cancer." Chronicle of Higher Education 8 May 1991, sec. A: 5+.   [*newspaper article*]

Miltane, John. Chief Engineer for County Electric. Personal Interview. Adams, MA. 5 Apr. 1993.

Noland, David. "Power Play." Discover Dec. 1989: 62–68.

Toufexis, Anastasia. "Panic over Power Lines." Time 17 July 1989: 71.

FIGURE 9.4 A List of Works Cited (MLA Style)

Discussion of Figure 9.4

1. Place "Works Cited" title one inch from page top and two spaces above first entry. Make left margin one inch. Double-space within and between each entry. Order entries alphabetically. For numbering this page, follow numbering of text pages.

2. Indent five spaces for the second and subsequent lines of an entry.

3. Place quotation marks around article titles, and underline or italicize periodical or book titles. Capitalize all key words (and articles, prepositions, or conjunctions only if they come first or last).

4. Do not cite a magazine's volume number, even if it is given.

5. Alphabetize multiple works by the same author according to their titles. Use three hyphens and a period for a second work by the same author. Shorten publisher's names (as in "Simon" for Simon & Schuster; "Knopf" for Alfred A. Knopf, Inc.; "GPO" for Government Printing Office; or "Yale UP" for Yale University Press).

6. Alphabetize hyphenated surnames according to the name that appears first.

7. Use a period and two spaces to separate a citation's three major items (author, title, publication data). Skip one space after a comma or colon. Use no punctuation to separate magazine title and date.

8. Include the issue number for a journal with new pagination in each issue. For page numbers of more than two digits, give only the final two digits in the second number.

9. For government reports, name the sponsoring agency and include all available information for retrieving the document. Use the first person's name and "et al." for works with more than three authors or editors.

10. When an article skips pages in a publication, give only the first-page number, followed by a plus sign. Because this is not a regional newspaper, the city of publication need not be inserted in brackets.

11. Use three-letter abbreviations for months with five or more letters.

REFERENCES

Basta, N. (1988, September).   Take a good look at sales engineering.
Graduating Engineer, pp. 84–87.   [*magazine article*]

Campbell, M.K. (1988).   Wanted: sales reps with EE degrees.   IEEE Potentials,
31 (1), 28–29.   [*journal article*]

College Placement Council. (1992).   CPC annual (36th ed.)   Bethlehem, PA:
Author.   [*book with author as publisher*]

Cornelius, H. & Lewis, W. (1983).   Career guide for sales and marketing.
New York: Monarch.   [*book with two authors*]

Electronics sales positions. (1993).   The national job bank.   Holbrook, MA:
Bob Adams, Inc.   [*directory entry—no author*]

Engineering careers. (1990).   The encyclopedia of careers and vocational
guidance (8th ed.). (Vol. 1).   Chicago: J. G. Ferguson.
                                            [*encyclopedia—no author*]

The job outlook in brief. (1992, Spring).   Washington, DC: U.S. Department of
Labor.   [*govt. publication—no author*]

Schranke, R. W. (1985).   EE and MBA: A winning combination?   IEEE
Potentials, 28 (1). 13–15.

Splaver, S. (1987).   Your personality and your career.   New York: Simon &
Schuster.   [*book with one author*]

Tolland, M. (1993, April).   Alternate careers in marketing.   [Presentation]
Electro 93. Conference of electronics developers and engineers, New York.
                                  [*unpublished conference presentation*]

FIGURE 9.5 A List of References (APA Style)

Discussion of Figure 9.5

1. In the APA reference list, include only "recoverable data" (material that readers could retrieve for themselves[9]); cite personal interviews, unpublished lectures, personal correspondence, or the like parenthetically in the text only (as in the report on pages 570–578). Leave margins of 1½ inches on all sides. Double space within and between each entry. Order entries alphabetically. For numbering this page, follow numbering of text pages. Indent three spaces for second and subsequent lines.

2. Give surname and initials only. Write out names of all months. Do not enclose article titles in quotation marks. Underline or italicize periodical titles. Capitalize only the first word in article titles, and all key words in magazine or journal titles. Include "p." or "pp." before magazine page numbers.

3. Underline a journal article's volume number, and give the issue number in parentheses. Do not include "p." or "pp." before journal page numbers.

4. Underline or italicize book titles, and capitalize only the first word. Identify the edition in parentheses.

5. Use ampersands instead of spelling out "and."

6. For a multivolume work, indicate the specific volume.

7. Alphabetize works with no author by the first key word in the title.

8. Write out the publisher's full name.

9. Use brackets to indicate your insertion of explanatory material.

Numerical Documentation

In the numerical system, each work is assigned a number upon first citation. This same number is then used for any subsequent reference to that work. Citations in the text include the work's number, a comma, the abbreviation "p.," and the page number:

> Seventy-five percent of technicians interviewed expressed a desire for further training (2, p. 83).

In the list of references at report's end, works are numbered in alphabetical order or in the order in which first cited in the text. (Use one arrangement

---

[9]According to APA guidelines, an unpublished conference presentation is considered a "recoverable" source. Thus, the final entry appears in Figure 9.5 instead of merely being cited parenthetically in the text.

or the other, consistently.) Otherwise, the format resembles APA style. Here are entries for a reference list in order of first citation in the text. (These, of course, are not alphabetized.)

REFERENCES

1. Donne, M.   Job prospects for college graduates.   Education Digest 28(2): 86—89; 1993.

2. Albey, J.   Modern career choices.   San Francisco: Hamilton; 1990.

3. Crashaw, H., et al.   Careers in the natural sciences.   Dallas: Bovary; 1992.

4. Marsh, A., and C. Smith, eds.   Advice for the job seeker.   Boston: Arngold; 1993.

5. Crashaw, H., et al.   Technology and careers.   Boston: Little, Brown; 1991.

A numbered list of references

The numerical system often is used in the physical sciences (astronomy, chemistry, geology, physics) and the applied sciences (mathematics, medicine, computer science). The sample report in Appendix C employs the documentation style shown here, based on the *CBE Style Manual*, from the Council of Biology Editors.

For specific formats in other disciplines, consult one of these style guides or one your instructor recommends:

American Institute of Physics, *Style Manual*
American Medical Association, *Style Book*
Manual for Authors of Mathematical Papers

Whichever documentation system you choose, use it consistently throughout your report.

REASSESSING THE ENTIRE PROCESS

Chapters 8 and 9 show that the research process is a minefield of potential errors, in what we do and how we reason: we might ask the wrong question(s); we might rely on the wrong sources; we might collect or record data incorrectly; we might analyze or document data incorrectly.

We therefore need to critically examine our methods and our reasoning before reporting findings and conclusions. The following research checklist helps guide our assessment.

CHECKLIST FOR THE RESEARCH PROCESS

Use this checklist to assess your research process. (Numbers in parentheses refer to the first page of discussion.)

Method

☐ Did I ask the right question(s)? (132)

☐ Are the sources current enough to convey the latest that is known? (180)

☐ Is each source reliable, relatively unbiased, and borne out by other, similar sources? (180)

☐ Is the evidence verifiable? (134)

☐ Have I obtained a balance of viewpoints? (132)

☐ Do I have enough information? (134)

☐ Is this the whole story? (134)

☐ Has the entire process been valid and reliable? (164)

☐ Have I identified all quoted material in my text? (176)

☐ Are all quotes accurate and grammatical? (177)

☐ Are all paraphrases accurate and clear? (178)

☐ Have I documented all sources not considered common knowledge? (189)

☐ Is the documentation complete and correct? (189)

Reasoning

☐ Am I reasoning instead of rationalizing? (183)

☐ Am I confident that my causal reasoning is correct? (183)

☐ Can all the numbers and statistics be trusted? (185)

☐ Have I resolved (or at least acknowledged) any conflicts among my findings? (137)

☐ Do I avoid conclusions that are specious or forced? (187)

☐ Have I decided whether my final answer is definitive, or only probable, or even inconclusive? (182)

☐ Is this the most reasonable conclusion (or merely the most convenient)? (187)

☐ Do I allow for other possible interpretations or conclusions? (137)

☐ Have I accounted for all sources of bias, including my own? (182)

☐ Should I reconsider the evidence? (135)

EXERCISES

1. Assume you are an assistant communications manager for a new organization that prepares research reports for decision makers worldwide. (A sample topic: "What will be the probable impact of the North American Free Trade Agreement on the U.S. computer industry?") These clients expect answers based on the best available evidence and reasoning.

Although your recently hired co-workers are technical specialists, few have experience in the kind of wide-ranging research required by your clients. Training programs in the research process are being developed by your communications division, but will not be ready for several weeks.

Meanwhile, your boss directs you to prepare a one- or two-page memo (pages 498–499) that introduces employees to major procedural and reasoning errors that affect validity and reliability in the research process. Your boss wants this memo to be comprehensive, but not vague.

2. Assume the scenario from Exercise 1. In a memo to colleagues, offer guidelines for avoiding unintentional plagiarism in quoting, paraphrasing, and citing the work of others. Explain what to document and how, using MLA style for a parenthetical reference and a works cited entry. (Illustrate with examples, but not those from the book!)

3. From print or broadcast media or from personal experience, identify an example of each of the following interpretive errors:

- reliance on soft evidence
- overestimating the level of certainty
- biased interpretation
- rationalizing
- faulty causal reasoning
- specious conclusion
- meaningless statistic
- undefined average
- distorted percentage figure
- bogus ranking

Submit to your instructor your examples along with a memo explaining each error, or be prepared to discuss your material in class.

4. Locate the style manual for your discipline. (Ask faculty in your major or a librarian). Redesign Figure 9.4 (page 198) according to the guidelines in this manual. Submit your document along with a memo outlining main differences in the two documentation styles. (If your discipline stipulates no particular style, use the *APA Manual* for this assignment.)

COLLABORATIVE PROJECTS

1. Exercises 1, 2, and 3 from the previous section are well suited for collaborative work.

2. *Evaluating sources.* Figure 9.4 (page 198) lists the final sources for the research project discussed early in Chapter 8. Review pages 180–81 and then return to Figure 9.4, and evaluate the sources on the basis of these criteria: currency, range, balance, relative objectivity, and reputability. Here are more specific questions:

- Should the sources generally have been more current? Why, or why not?
- Which types of sources are represented here: business, science, trade, or general-interest publications; newspapers or news magazines; scholarly journals; government reports; or primary sources? Does the range of sources seem adequate (from general to specialized)? Explain.
- How many different viewpoints are represented here (representatives of the company [County Electric], consumer advocates, independent researchers, print journalists, the media, people in the industry, investors, others)? Do the sources represent a fair balance of views on this controversial issue [electromagnetic radiation hazards]? Explain.
- Which sources seem most likely to be objective or impartial? Explain.
- Which seem most likely to be biased? Explain.
- Which seem most expert or authoritative? Explain.
- Which seem most comprehensive? Explain.
- Overall, do the sources seem adequate for the topic and situation described in Chapter 8? Explain.

Make the explanations brief but informative enough to justify your evaluation.

Appoint a group manager, who will lead the discussion and assign the following tasks: taking notes, reporting the group's evaluation in a memo, editing and revising the memo, orally reporting the evaluation to the class.

Summarizing Information

Purpose of Summaries

Elements of a Summary

Critical Thinking in the Summary Process

A Sample Situation

The Forms Summarized Information Can Take

Placement of Summarized Information in a Document

A summary is a short version of a longer message. An economical way to communicate, a summary saves time, space, and energy.

PURPOSE OF SUMMARIES

Chapter 8 shows how abstracts (a type of summary) aid our research by providing an encapsulated glimpse of an article or other long document. As we record our research findings, in turn, we summarize and paraphrase to capture the main ideas in a compressed form. In addition to this dual role as a research aid, summarized information is vital in the day-to-day transactions among professionals.

On the job, you have to write concisely about your work. You might summarize a news article or report, describe your progress on a project, or introduce the reader to a procedure. A routine assignment for many new employees is to provide superiors (decision makers) with summaries of the latest developments in their field.

Given today's pace and volume of information, summaries are more vital than ever. Some reports and proposals can be hundreds of pages long. Those who must act on this information need to rapidly identify what is most important in a document. From a good summary, busy readers can get enough information to decide whether they should read the entire document, parts of it, or none of it.

Whether you summarize someone else's information or your own, your job is to communicate the *essential message*—to represent the original document accurately and in the fewest words. The principle is simple: include what your readers need and omit what they don't.

The essential message in any well-written document is easy enough to identify, as in the following passage:

> The lack of technical knowledge among owners of television sets leads to their suspicion about the honesty of television repair technicians. Although television owners might be fairly knowledgeable about most repairs made to their automobiles, they rarely understand the nature and extent of specialized electronic repairs. For instance, the function and importance of an automatic transmission in an automobile are generally well known; however, the average television owner knows nothing about the flyback transformer in a television set. The repair charge for a flyback transformer failure is roughly $150—a large amount to a consumer who lacks even a simple understanding of what the repairs accomplished. In contrast, a $450 repair charge for the transmission on the family car, though distressing, is more readily understood and accepted.

The original passage

Three significant ideas comprise the essential message: (1) television owners lack technical knowledge and are suspicious of repair technicians; (2) an owner usually understands even the most expensive automobile repairs; and (3) owners do not understand or accept expenses for television repairs. A possible summary might read like this:

> Because television owners lack technical knowledge about their sets, they often are suspicious of repair technicians. Although consumers may understand expensive automobile repairs, they rarely understand or accept repair and parts expenses for their television sets.

A summarized version

This summary is almost 30 percent of the original length because the original itself is short. With a longer original, a summary might be 5 percent or less. But length is less important than informative value: an effective summary gives readers all they need.

Summaries are vital whenever people have no time to read in detail everything that crosses their desks.[1] For letters, memos, or other short documents that can be read quickly, the only summary needed is usually an opening thesis or topic sentence that previews the contents.

[1] A recent U.S. president reportedly required all significant world news for the last twenty-four hours to be condensed into one typed page and placed on his desk, first thing each morning. Another president had a writer who summarized articles from more than two dozen major magazines.

ELEMENTS OF A SUMMARY

All effective summaries display the elements discussed below.

Essential Message

The essential message is the significant material from the original: controlling ideas (thesis and topic sentences); major findings; important names, dates, statistics, and measurements; and conclusions or recommendations. Significant material does not include background; the author's personal comments or conjectures; introductions; long explanations, examples, or definitions; visuals; or data of questionable accuracy. (These distinctions are illustrated on pages 210–13.)

Nontechnical Style

More people generally read the summary than any other part of a report. Write at the lowest level of technicality. Translate technical data into plain English. "The patient's serum glucose measured 240 mg%" can be translated: "The patient's blood-sugar remained critically high." Of course, if you know all your readers are experts, you won't need to simplify.

Independent Meaning

In meaning as well as style, your summary should stand alone as a self-contained message. Readers should have to read the original only for more detail—not to make sense of your message.

No Added Material

Avoid personal comments ("This interesting report . . . ," or "The author is correct in assuming . . ."). Add nothing to the original.

Conciseness

Conciseness is vital, but never at the expense of clarity and accuracy. Make the summary short enough to be economical, but long enough to be clear and comprehensive.

CRITICAL THINKING IN THE SUMMARY PROCESS

Follow these guidelines for summarizing your own writing or another's.

1. *Read the entire original.* When summarizing another's work, get a complete picture before writing a word.

2. *Reread and underline.* Reread the original, underlining essential material. Focus on the thesis and topic sentences.

3. *Edit the underlined data.* Reread the underlined material and cross out whatever does not advance the meaning.

4. *Rewrite in your own words.* Include all essential material in the first draft, even if it's too long; you can trim later.

5. *Edit your own version.* When you have everything readers need, edit for conciseness.

   a. Cross out all needless words without harming clarity or grammar. Use complete sentences.

The summer internship in journalism gives the ~~journalism~~ student ~~first hand~~ experience ~~at what goes~~ on ~~within~~ a ~~real~~ newspaper.

   b. Cross out needless prefaces such as "The writer argues . . ." or "Also discussed is. . . ."

   c. Use numerals for numbers, except to begin a sentence.

   d. Combine related ideas in order to emphasize relationships (pages 274–277).

6. *Check your version against the original.* Verify that you have preserved the essential message and added no comments.

7. *Rewrite your edited version.* Add transitional expressions to reinforce the connection between related ideas.

8. *Document your source.* If summarizing another's work, cite the source immediately below the summary, and place directly quoted statements within quotation marks. (See Chapter 9 for documentation formats.)

Although the summary is written last by the writer, it is read *first* by the reader. Take the time to do a good job.

A SAMPLE SITUATION

Imagine that you work in the information office of your state's Department of Environmental Management (DEM). In the coming election, citizens will vote on a referendum proposal for constructing municipal trash incinerators. Referendum supporters argue that incinerators would help solve the growing problem of waste disposal in highly populated parts of the state. Opponents argue that incinerators cause air pollution.

To clarify the issues for voters, the DEM is preparing a newsletter to be mailed to each registered voter. You have been assigned the task of researching the recent data, and of summarizing them for newsletter readers. Here is one of the articles you decided to summarize (shown with steps 1, 2, and 3 from The Summary Process completed).

An Article to be Summarized
INCINERATING TRASH: A HOT ISSUE GETTING HOTTER
THE FAILURE OF LANDFILLS; THE HOPE OF RESOURCE RECOVERY

Combine as orienting sentence (controlling idea)

Alarmed by the tendency of landfills to contaminate the environment, both public officials and citizens are vocally seeking alternatives. The most commonly discussed alternative is something called a *resource recovery facility.* Nearly 100 U.S. cities have built such in the last 15 years, and another 150 or so are in various stages of planning.

Include definition

These recovery facilities are a new form of an old technology. Basically, they're incinerators. But, unlike the incinerators of old, they don't just burn waste. They also recover energy. The energy is sold as steam to an industrial customer, or it is converted to electricity and sold to the local utility. (A few facilities, not many, also recover metals or other materials before using the waste as fuel.)

Include major fact

Include major statistic

A ton of trash possesses the energy content of a barrel and a half of oil. This is not a trivial amount. The United States discards 150 million tons of municipal refuse a year. If all of it were converted to energy, we could replace the equivalent of 12 percent of our oil imports.

Include major fact

At the local level, selling energy or materials not only replaces nonrenewable resources; it also provides a source of income that partly offsets the cost of operating the facility.

Include major fact

Delete explanation

The new facilities are, on average, much cleaner than the municipal incinerators of old. Many have two-stage combustion units, in which the second-stage burns exhaust gases at high temperature, converting many potential organic pollutants to less harmful emissions such as carbon dioxide. Some, especially the larger and newer facilities, also come equipped with the latest in pollution control devices.

Delete questionable point

The environmental community is uneasy with this new technology. Environmentalists have argued for many years that the best method of handling municipal trash is to recycle it—i.e., to separate the glass, metal, paper, and other materials and use them again, either without reprocessing or as raw materials in producing new products. The thought of the potential resources in municipal solid waste simply being burned, even with energy recovery, has made many environmentalists opponents of resource recovery.

More recently, opponents have found a stronger reason to oppose burning waste: *dioxin* in the plants' emissions. The amounts present are extremely small, measured in trillionths of a gram per cubic meter of air. But dioxin can be deadly, at least to animals, at very low levels.

Include key finding and explanation

WHAT IS DIOXIN?

Dioxin is a generic term for any of 75 chemical compounds, the technical name for which is poly-chlorinated dibenzo-p-dioxins (PCDDs). A related group of 135 chemicals, the PCDFs or furans, are often found in association with PCDDs.

Include definition

Delete technical details

The most infamous of these substances, 2,3,7,8-TCDD, is often referred to as the "most toxic chemical known." This judgment is based on animal test data. In laboratory tests, 2,3,7,8-TCDD is lethal to guinea pigs at a concentration of *500 parts per trillion*. A part per trillion is roughly equivalent to the thickness of a human hair compared to the distance across the United States.

Include major fact

The effects on humans are less certain, for many reasons: it is difficult to measure the amounts to which humans have been exposed; difficult to isolate the effects of dioxin from the effects of other toxic substances on the same population; and the latency period for many potential effects, such as cancer, may be as long as 20 to 30 years.

Include major point

Delete long explanation

Nevertheless, because of the extreme effects of this substance on animals, known releases of dioxin have generated considerable public alarm. One of the most publicized releases occurred at Seveso, Italy, in July 1976, where a pharmaceutical plant explosion resulted in the contamination of at least 700 acres of fields and affected more than 5,000 people. Dioxin was found in the soil in concentrations of 20 to 55 parts per billion.

Include continuation of major point

Delete long example

The immediate effects on humans were nausea, headaches, dizziness, diarrhea, and an acute skin condition called chloracne, which causes burn-like sores. The effects on animals were more severe: birds, rabbits, mice, chickens, and cats died by the hundreds, within days of the explosion. In response to the explosion, the Italian provincial authorities evacuated 730 people from the zone nearest the plant, and sealed off an area containing another 5,000 people from contact with nonresidents.

In this country, perhaps the best known dioxin contamination incident occurred at Times Beach, Missouri, where used oil, contaminated with dioxin, was

Delete long example

sprayed on roads as a dust suppressant. Soil samples showed dioxin at levels exceeding 100 parts per billion. While no human health effects were documented at Times Beach, a flood, in December, 1982, led to widespread dispersal of the contamination, as a result of which the entire town was condemned, the population evacuated, and over $30 million of Superfund money used to purchase the condemned property.

Dioxin was among the substances of concern at Love Canal. And it was the major contaminant in the chemical defoliant, Agent Orange, the subject of a lawsuit by 15,000 Vietnam veterans and dependents and an out-of-court settlement of those complaints valued at $180 million.

As early as 1978, trace amounts of dioxin were found in the routine emissions of a municipal incinerator. Virtually every incinerator tested since that date has shown traces of dioxin.

THE MEANING OF IT ALL

At the request of Congress, the Agency began in 1984 a major research effort on dioxin, the National Dioxin Study. The study is intended to provide a context in which to place mounting concerns about dioxin. Research for the study was organized into seven "tiers," each tier including a group of sites at which dioxin contamination may be present.

- Tier 1, production sites, includes the 10 sites at which 2,4,5-TCP, a pesticide known to have been contaminated by dioxin, was produced, and additional sites where waste materials from its production were disposed.
- Tier 2, precursor sites, includes 9 sites where 2,4,5-TCP was used as a precursor to make other chemical products, and related waste disposal sites. The chemical products included the herbicides 2,4,5-T and silvex, and hexachlorophene, a disinfectant that was widely used in soaps and deodorants, but was banned from nonprescription uses by the Food and Drug Administration in 1981.
- Tier 3 includes 60 to 70 sites at which 2,4,5-TCP and its derivatives were formulated into herbicide products, and associated waste disposal sites.
- Tier 4 includes a wide range of combustion sources, including internal combustion engines, wood stoves, fireplaces, forest fires, oil burners and other sources burning waste oil, and many others.
- Tier 5 includes 20 to 30 of the thousands of sites at which dioxin-contaminated pesticides have been used, for example, power line rights-of-way, forests, and rice and sugar cane fields.
- Tier 6 includes about 20 of the chemical and pesticide production facilities where improper quality control may have led to the accidental production of dioxin.

Include the most striking and familiar example

Include key findings

Include major fact

Condense list

- And Tier 7 includes samples from sites where the Agency least expected to find dioxin, to determine whether there are background levels of dioxin in the environment. Soil samples have been taken at 500 randomly selected locations across the country—200 in rural areas, 300 in urban areas—and fish have been sampled from over 400 locations.

While the study is not yet complete, data from a variety of sources have already produced disturbing—yet, perhaps, in an odd way, reassuring—results. Dioxin in trace amounts appears to be widely present in the environment, even in remote locations where industrial activity and waste combustion are unlikely to be the source.

Include key finding

Environment Canada, the Canadian EPA, also has an extensive dioxin testing program under way. One of the more startling findings of their research, conducted at a resource recovery facility on picturesque Prince Edward Island, is that the garbage delivered to the plant contained more dioxin than the plant's emissions. The source of the dioxin in this case is not known, though it could include pesticide residues or other products contaminated with dioxin during manufacturing processes.

Delete long example

Delete speculation

In short, dioxin is not just a problem created by burning municipal waste. It is not clear at this time whether municipal waste combustion is even the major source of dioxin in the environment.

Include key conclusion

Ultimately, the dioxin problem is like other toxic substance issues. We know less than we need to know to thoroughly evaluate the risk. The more we find out, the more complex the issues tend to become. There is no risk-free solution, since all the potential disposal methods may result in some release of toxic substances to the environment. Yet those who counsel delay, to allow the collection of more data, are met with the suspicion that their real agenda is to prevent action entirely.

Include conclusion

Delete personal comment

What we do know at present does not seem to suggest that we should stop planning to build resource recovery facilities. What it does suggest is that we proceed cautiously, inform the public of both what is known and what is unknown, install pollution controls if the plants' uncontrolled emissions are significant or if the exposed population wants added protection, and hope that continued examination of all the sources and effects of dioxin will eventually produce a consensus.

Include recommendations

Source: James E. McCarthy. *Congressional Research Service Review* Apr. 1986: 19–21.

Assume that in two early drafts of your summary, you rewrote and edited; for coherence and emphasis, you inserted transitions and combined related ideas. Here is your final draft, with all steps from The Summary Process completed.

A Summary of the Article

INCINERATING TRASH: A HOT ISSUE GETTING HOTTER (A SUMMARY)

Because landfills often contaminate the environment, trash incinerators (resource recovery facilities) are becoming a popular alternative. Nearly 100 are operating in U.S. cities, and 150 more are planned. Besides their relatively clean burning of waste, these incinerators recover energy, which can be sold to offset operating costs. One ton of trash has roughly the energy content of 1.5 barrels of oil. Converting all U.S. refuse to energy could reduce oil imports by 12 percent.

Unfortunately, incinerator emissions contain very small amounts of dioxin (a generic name for any of 75 related chemicals). Even low dioxin levels can be deadly to animals. In fact, animal tests have helped label one dioxin substance "the most toxic chemical known." Although effects on human beings are less certain, news of dioxin in the environment creates public alarm, as evidenced at Love Canal and by the successful Agent Orange lawsuit by 15,000 Vietnam veterans. Almost every municipal incinerator tested since 1978 has shown traces of dioxin.

At Congress's request, the Environmental Protection Agency in 1984 began the National Dioxin Study. Sites included herbicide and pesticide production and waste-disposal facilities, combustion sources such as woodstoves and forest fires, areas of dioxin-contaminated pesticide use such as rice and sugarcane fields, and random soil and fish samples nationwide. Findings indicate that trace amounts of dioxin are widely present in the environment, even in areas remote from industry and waste combustion. Waste incineration is by no means the only source of environmental dioxin—and may not even be the major source.

At this stage, we know too little to evaluate the risk. And no risk-free waste-disposal solution in fact exists. But no evidence so far suggests that we stop planning incinerators. We should, however, move cautiously, fully informing the public, installing pollution controls as needed, and searching for a better solution.

Source: James E. McCarthy. *Congressional Research Service Review* Apr. 1986: 19–21.

The version above is trimmed, tightened, and edited: word count is reduced to roughly 20 percent of the original. A summary this long serves well in many situations, but other audiences might want a briefer and more compressed summary—say, 10 to 20 percent of the original:

A More Compressed Summary

Because landfills often contaminate the environment, trash incinerators (resource recovery facilities) are becoming a popular alternative across the United States. Besides their relatively clean burning of waste, these incinerators recover energy, which can be sold to offset operating costs.

Unfortunately, incinerator emissions contain very small amounts of dioxin, a chemical proven so deadly to animals, even at low levels, that it has been labeled the most toxic chemical known. Although its effects on human beings are less certain, news of dioxin in the environment creates public alarm. Almost every municipal incinerator tested since 1978 has shown traces of dioxin.

Findings of an EPA study begun in 1984 suggest that trace amounts of dioxin are widely present in the environment, even in areas remote from industry and waste combustion. The dioxin source is by no means only waste incineration.

We lack risk-free disposal solutions and know too little to evaluate risks. But no evidence so far suggests that we stop planning incinerators. We should, however, move cautiously, fully informing the public, installing pollution controls as needed, and searching for better solutions.

Notice that the essential message is still intact; related ideas are again combined and fewer supporting details are included. Clearly, length is adjustable according to your audience and purpose.

THE FORMS SUMMARIZED INFORMATION CAN TAKE[2]

In preparing a report, proposal, or other document, we might summarize works of others as part of our presentation. But we often summarize our own presentations as well. For instance, if our document extends to several pages, we usually include, near the end or the beginning, a summary. Depending on its location and its level of detail, this summarized information takes one of these three forms: **closing summary, informative abstract,** or **descriptive abstract.** Figure 10.1 depicts these forms.

[2]Although I take liberties with his classification, David Vaughan's insightful article helped clarify my thinking about the overlapping terminology that perennially seems to confound discussions of these distinctions. The full citation appears in Works Cited, page 675.

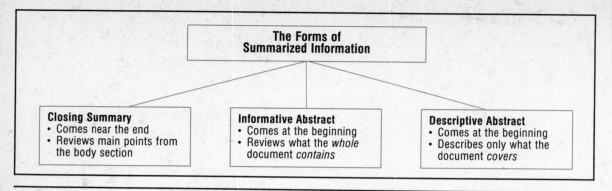

FIGURE 10.1 Summarized Information Assumes Various Forms

The Closing Summary

Summarized information at the end of a document's body section helps readers review and remember the main points or major findings from the presentation. This look back at "the big picture" helps readers appreciate and understand any conclusions and recommendations that follow.

The Informative Abstract[3]

In addition to a closing summary, we might include an opening summary to show readers what the document is all about, to help them decide whether to read it, and to give them an orientation—a framework for understanding what follows. We call this opening summary an **informative abstract,** partly to differentiate it from its closing counterpart and partly because it is different: besides merely reviewing main points or major findings, the informative abstract identifies the issue or need that led to the report and includes condensed conclusions and recommendations.

The Descriptive Abstract

As we have seen above, the informative abstract *explains what the original document contains* (its origins, findings, and conclusions—like the samples on pages 375, 659). But yet another more compressed form of summarized

---

[3]Informative abstracts sometimes are referred to as **executive summaries,** but some writers argue that the executive summary has more of a persuasive emphasis, for readers who are executives rather than technical professionals. Executive summaries are very important to decision-making—in cases when readers *expect* the writer to help guide their thinking. ("Tell me how to think about this" instead of merely "Help me understand this.")

information can precede a document: this version is called the **descriptive abstract,** and it merely *describes what the original is about* (its subject).

A descriptive abstract, then, conveys only the nature and extent of the document. It presents the broadest view, and offers no major facts from the original. Whereas the informative abstract contains the "meat" of the original, the descriptive abstract has only its skeletal structure—a kind of summary of a summary. Compare, for instance, the abstract that follows with the version on page 214:

A Descriptive Abstract

INCINERATING TRASH: A HOT ISSUE GETTING HOTTER

As an alternative to landfill dumps, municipal trash incinerators offer environmental benefits, but they also create the danger of dioxin emissions.

Descriptive abstracts of articles often appear in magazine or journal tables of contents.

On the job, you might write informative abstracts for a boss who needs the information but who has no time to read the original. Or you might write descriptive abstracts to accompany a bibliography of works you are recommending to colleagues or clients (an **annotated bibliography**).

PLACEMENT OF SUMMARIZED INFORMATION IN A DOCUMENT

Placement requirements for various report elements vary from company to company, but these general guidelines serve in many situations:

- Place the closing summary in the concluding section of your report, usually preceding any conclusions and recommendations.
- Place a descriptive abstract single-spaced on the lower-third of your title page.
- Place an informative abstract on its own separate page, immediately following your table of contents.

Reports and proposals in later chapters illustrate the various placement options for summarized information.

REVISION CHECKLIST FOR SUMMARIES

Use this checklist to refine your summaries. (Page numbers in parentheses refer to the first page of discussion.)

Content

☐ Does the summary contain only the essential message? (207)

☐ Does the summary make sense as an independent piece? (208)

☐ Is the summary accurate when checked against the original? (209)

☐ Is the summary free of any additions to the original? (208)

☐ Is the summary free of needless details? (209)

☐ Is the summary economical yet clear and comprehensive? (208)

☐ Is the source documented? (209)

☐ Does the descriptive abstract tell what the original is about? (216)

Organization

☐ Is the summary coherent? (209)

☐ Are there enough transitions to reveal the line of thought? (209)

Style

☐ Is the summary's level of technicality appropriate for its audience? (208)

☐ Is the summary free of needless words? (209)

☐ Are all sentences clear, concise, and fluent? (209)

☐ Is the summary written in correct English? (Appendix A)

EXERCISES

1. Read each of these two paragraphs, and then list the significant ideas comprising each essential message. Write a summary of each paragraph.

> In recent years, ski-binding manufacturers, in line with consumer demand, have redesigned their bindings several times in an effort to achieve a noncompromising synthesis between performance and safety. Such a synthesis depends on what appear to be divergent goals. Performance, in essence, is a function of the binding's ability to hold the boot firmly to the ski, thus enabling the skier to change rapidly the position of his or her skis without being hampered by a loose or wobbling connection. Safety, on the other hand, is a function of the binding's ability both to release the boot when the skier falls, and to retain the boot when subjected to the normal shocks of skiing. If achieved, this synthesis of perfor-

mance and safety will greatly increase skiing pleasure while decreasing accidents.

Contrary to public belief, sewage-treatment plants do not fully purify sewage. The product that leaves the plant to be dumped into the leaching (sievelike drainage) fields is secondary sewage containing toxic contaminants such as phosphates, nitrates, chloride, and heavy metals. As the secondary sewage filters into the ground, this conglomeration is carried along. Under the leaching area develops a contaminated mound through which groundwater flows, spreading the waste products over great distances. If this leachate reaches the outer limits of a well's drawing radius, the water supply becomes polluted. And because all water flows essentially toward the sea, more pollution is added to the coastal regions by this secondary sewage.

2. Attend a campus lecture on a topic of interest and take notes on the significant points. Write a summary of the lecture's essential message.

3. Find an article about your major field or area of interest and write both an informative abstract and a descriptive abstract of the article.

4. Select a long paper you have written for one of your courses; write an informative abstract and a descriptive abstract of the paper.

5. Read the following article and write both a descriptive abstract and an informative abstract, using the steps under The Summary Process as a guide. Define a specific audience and use for your material. (A possible scenario: You are assistant shipping manager for a large wholesaler of roofing and construction supplies to hardware stores and lumber yards. Company executives have asked all managers to provide them with summaries of any news relevant to the building-supply business. After coming across this article, you decide to summarize it for your bosses. As far as you can tell, your intended readers have little or no background on the asbestos issue. Give them the information they need.) Bring your summary to class and exchange it with a classmate's for editing according to the revision checklist. Revise your edited copy before submitting it to your instructor.

MOVING TO RID AMERICA OF ASBESTOS

On January 19, EPA published a proposal in the *Federal Register* to rid the United States of the "miracle" fiber called asbestos.

In the proposed rule, issued under authority of the Toxic Substances Control Act (TSCA), EPA invites public opinion on its intent to immediately ban five major asbestos products and phase out all remaining uses of the substance over the next 10 years. Why is EPA proposing such a measure for a product long considered so commercially important, and still so pervasive throughout American society?

Asbestos is really a common name for a group of natural minerals—silicates—that separate into thin but strong fibers. The fibers are chemically inert and heat-resistant, and they cannot be destroyed or degraded easily.

Since 1900, over 30 million tons of asbestos have been used in hundreds of products. Much of it was sprayed on ceilings and other parts of schools and public and private buildings for fireproofing, sounddeadening, insulation, or decoration.

Unfortunately, some of the characteristics that make this mineral fiber so useful commercially—such as its great stability—also help make it a dangerous killer when it is breathed in. Unless completely sealed into a product, asbestos can easily break into a dust or into tiny fibers. These fibers can then float and be inhaled. Once asbestos gets into the body, it can remain there for many years.

"There can be no debate about the health risks of asbestos," says EPA Administrator Les Thomas. A well-documented cause of lung and other cancers in humans, including mesothelioma (a cancer of the chest and abdominal lining), asbestos is now generating up to 12,000 cancer cases a year in the United States, almost all of which are fatal. Aside from the cancer threat, about 65,000 persons in this country are currently suffering from asbestosis, a chronic scarring of the lungs which makes breathing more and more difficult and eventually causes death. Cigarette smokers exposed to asbestos face extra risk, having a much higher chance of getting lung cancer than exposed nonsmokers.

Asbestos-caused cancers can remain latent and not occur for 15 to 40 years after the first exposure. EPA also believes that even small amounts of asbestos in the air are dangerous.

Asbestos is released into the air throughout its entire life cycle: manufacturing, use, destruction, and disposal. Since substitutes are, or will soon be, available for nearly all uses of the fiber, EPA has no feasible alternative but to phase out asbestos and all its products. This is what the January proposal sets out to do.

Prohibited would be the importing, manufacture, and processing of five products that account for as much as one half of United States asbestos consumption. The bulk of these products are used mainly in the construction and renovation industry. They are:

- *Saturated and unsaturated roofing felt:* This is a product made from paper felt and intended to cover or lie under other roof coverings. Its purpose is to insulate and help prevent corrosion.
- *Flooring felt and asbestos felt-backed sheet flooring:* Used as an underside backing for vinyl sheet flooring, this felt helps maintain original product shape and helps prolong floor life, especially when moisture from below the surface is a problem.

- *Vinyl-asbestos floor-tile:* Especially popular for use in heavy traffic areas such as in stores, kitchens, and entry ways.
- *Asbestos-cement pipe and fittings:* This is used primarily to carry water or sewage, and to a lesser extent, as conduit pipe for the protection of electrical or telephone cable or for air ducts.
- *Asbestos clothing:* Not street clothes, but special occupational garments worn by those needing protection from extreme heat, such as firefighters.

What about the rest of the asbestos products in use? EPA is proposing to get rid of the asbestos in these products indirectly by phasing down all domestic mining and importing of asbestos by a certain percentage each year over the next 10 years. This phasedown would be carried out by allowing a company to mine or import an annually decreasing percentage of the amount of asbestos it mined or imported during the years 1981–1983.

EPA estimates that as a result of what it is proposing, about 1,900 cancer deaths from asbestos will be avoided.

EPA intends that labels be put on all products not immediately banned, warning users that they contain asbestos. EPA hopes that these warnings would encourage users to take steps to reduce their exposure.

For the five products banned under the proposal, EPA is convinced that industry has adequate, readily available substitutes which should minimize the economic impact of this action. The 10-year phaseout should give industry time to develop good alternatives for all remaining asbestos products.

(Public hearings on EPA's proposed rules are tentatively scheduled for mid-May; written public comments must be submitted by April 29.)

Source: Adapted from Dave Ryan. "Moving to Rid America of Asbestos." *EPA Journal* 12.2 (1986): 7–8.

COLLABORATIVE PROJECT

Organize into small groups and choose a topic for discussion: an employment problem, a campus problem, plans for an event, suggestions for energy conservation, or the like. (A possible topic: Should employers have the right to require lie detector tests, drug tests, or AIDS tests for their employees?) Discuss the topic for one class period, taking notes on significant points and conclusions. Afterward, organize and edit your notes in line with the directions for writing summaries. Next, write a summary of the group discussion in no more than 200 words. Finally, as a group, compare your individual summaries for accuracy, emphasis, conciseness, and clarity.

Sequence, Shape, and Style in a Document

Outlining

Types of Outline

Elements of a Formal Outline

An Outline Model

The Importance of Being Messy

The Report Design Worksheet

■   ■   ■

Successful writers spend time thinking and planning so they can produce organized information. A recent survey of workers, full-time students, and students working part-time found that all seventy respondents used outlines to organize their writing, especially in preparing long documents (Roundy and Mair 91).

An outline enables you to move from a random listing of items, as they occurred to you, to a deliberate map that enables readers to locate, understand, and remember your information. The computer is especially useful for rearranging outlines until they reflect the sequence in which you expect readers to approach your message.

TYPES OF OUTLINE

The Informal Outline

An informal outline is a simple list, probably all you need for a short report. For a longer report, an informal outline can serve as a tentative (or working) outline that keeps you on track and allows for alterations as you work.

Following is an informal outline for a report titled "An Analysis of the Advisability of Converting Our Office Building from Oil to Gas Heat."

1. Description of Our Present Heating System
2. Removal of the Oil Burner and Tank
3. Installation of a Gas Pipe from the Street to the Building
4. Installation of a Gas Burner
5. Estimation of Gas Heating Costs

Beginning with a description of the present system, the topics then follow the sequence of the actual conversion process. The writer now has a general plan for gathering data and writing a first draft. She can expand this outline and make it more specific by adding subtopics as shown below. Before the final draft, she will develop a formal outline, to check the organization of her report, and to reveal that organization to readers.

The Formal Topic Outline

The formal topic outline is detailed and systematic, using notation (numbers and letters) to mark divisions and to show how parts relate. Choose between two systems of notation: the roman numeral–letter–arabic numeral system or the decimal system.

Roman Numeral–Letter–Arabic Numeral Notation.   Here are the five topic headings from the earlier informal outline developed into a formal outline, using roman numeral–letter–arabic numeral notation:

II. REPORT BODY (or COLLECTED DATA)
   A. Description of Our Present Heating System
      1. Physical condition
      2. Required yearly maintenance
      3. Fuel supply problems
         a. Overworked distributor
         b. Varying local supply[1]
      4. Cost of operation
   B. Removal of the Oil Burner and Tank
      1. Data from the oil company
      2. Data from the salvage company
         a. Procedure
         b. Cost

The body section
of a formal outline

[1] Any division must yield at least two subparts. You could not logically divide "Fuel supply problems" into "a. Overworked distributor," without other subparts. If you cannot divide your major topic into at least two subtopics, change your original heading.

3. Possibility of private sale
   C. Installation of a Gas Pipe from the Street to the Building
     1. Procedure
     2. Cost of installation
     3. Cost of landscaping
   D. Installation of a Gas Burner
     1. Procedure
     2. Cost of plumber's labor and materials
   E. Estimation of Gas Heating Costs
     1. Rate determination
     2. Required yearly maintenance
     3. Cost data from neighboring facility
     4. Overall cost of operation
       a. Cost of conversion
       b. Cost of maintenance
       c. Cost of gas supply

The writer now adds the introduction and conclusion.

Introductory and concluding sections added

  I. INTRODUCTION
   A. Background
   B. Purpose of the Report
   C. Intended Audience (usually for in-school reports only)
   D. Information Sources
   E. Limitations of the Report
   F. Scope (list of major topics)
 II. BODY (as shown earlier)
III. CONCLUSION
   A. Summary of Findings
   B. Comprehensive Interpretation of Findings
   C. Recommendations

In short reports, the introduction and conclusion still are included, but usually shortened to one or two sentences.

A formal outline easily converts to a table of contents for the finished report, as shown in Chapter 16. (Because they serve mainly to guide the *writer,* minor outline headings may be omitted from the table of contents or the report itself. Needless headings make a document seem fragmented.)

Decimal Notation.   Here is a partial outline in decimal notation:

The style of notation common in technical documents

2.0 Collected Data
   2.1 Description of Our Present Heating System
      2.1.1 Physical condition
      2.1.2. Required yearly maintenance
      2.1.3 Fuel supply problems
         2.1.3.1 Overworked distributor
         2.1.3.2 Varying local supply
      2.1.4 Cost of operation
   2.2 Removal of the Oil Burner and Tank
      2.2.1 Data from the oil company
      2.2.2 (and so on)

The decimal outline makes it easier to refer readers to various sections. But both systems achieve the same organizing objective. Unless readers express a preference, use whichever system you prefer.

The Sentence Outline

The previous outline is a *topic outline* because each division is expressed as a topic phrase. A topic outline may be expanded into a *sentence outline*.

A sentence outline

II. COLLECTED DATA
   A. Our present heating needs are supplied by circulating hot air generated by a Model A-12, electrically fired, Zippo oil burner fed by a 275-gallon fuel tank. Both are twelve years old.
      1. Burner and tank are in good working order and physical condition, for they have been carefully maintained.
      2. The oil burner and associated components require cleaning once yearly at a service charge of $38. The air filter, costing $6.25, is replaced three times yearly at a total cost of $18.75.
      3. (and so on)

Each sentence serves as a topic sentence for a paragraph in the report. Sentence outlines are used mainly in collaborative projects in which various team members prepare different sections of a long document.

ELEMENTS OF A FORMAL OUTLINE

Parallel Construction for Parallel Levels

Make all items at the same level parallel, or grammatically equal, to emphasize connections among related ideas.

Not parallel E. Estimation of Gas Heating Costs
1. Rate determination
2. The system requires yearly maintenance
3. Cost data were obtained from a neighboring facility
4. Overall cost of operation

Parallel E. Estimation of Gas Heating Costs
1. Rate determination
2. Required yearly maintenance
3. Cost data from neighboring facility
4. Overall cost of operation

For a full discussion of parallelism, see Appendix A.

Clear and Informative Headings

Be sure topic headings contain informative words. Under "Description of Our Present Heating System," a heading titled "Fuel" is less informative than "Fuel supply problems."

Avoid repetitive headings that add no information.

Not informative C. Environmental Effects of Strip Mining
1. Effects on land
2. Effects on erosion
3. Effects on water
4. Effects on flooding

Informative C. Environmental Effects of Strip Mining
1. Permanent land scarring
2. Increased erosion
3. Water pollution
4. Increased flood hazards

Each revised heading has a key phrase that summarizes the message.

Parts in a Sequence Logical to Readers

A sequence logical to readers enables them to follow your reasoning, to see relationships. Does your subject suggest a sequence? Specialists in document design suggest listing the questions you could expect from readers, and then arranging these questions in the sequence in which you would expect readers to ask them (Redish et al. 142). Here are some possible sequences.

Chronological Sequence.   Follow the order of events or steps (as in a set of instructions). Also, use this sequence to explain how the parts of a mechanism operate (as in how the heart pumps blood).

Spatial Sequence.   Follow the arrangement of parts (left to right, top to bottom, front to rear), as in describing how an office will be remodeled to accommodate automated equipment.

Reasons For and Against.   Follow the sequence in which both sides of an issue are argued—first one side, then the other—as in weighing the risks and benefits of a flu-vaccination program.

Problem-Causes-Solution.   Follow the sequence of the problem-solving process, from a description of the problem, through diagnosis, to a solution. An analysis of business failures in your area would require this sequence.

Cause and Effect.   Follow actions to their results, as in analyzing the therapeutic benefits of meditation.

Comparison and Contrast.   Evaluate two or more items by considering their similarities and then their differences. An example would be comparison of two proposed sites for dumping low-level nuclear wastes.

Simple to Complex.   Begin with the simple and familiar. An explanation of satellite transmission, for instance, would describe what we see on our television screens *before* explaining how the video signal is transmitted.

Sequence of Priorities.   Arrange items according to their relative importance, as in a proposal for increasing your town's budget.

Documents often require a combination of sequences. Our heating conversion outline fuses the chronological sequence with comparison-contrast and cause-and-effect. A single paragraph generally follows *one* sequence, as shown on pages 244–48.

Research demonstrates that material presented in an organized sequence is easier for readers to understand and remember than unorganized material (Felker et al. 11). Organize in the sequence in which you expect readers to approach the material.

AN OUTLINE MODEL

The following model is adaptable to most reports directed toward reaching a decision. (For information-type reports, the Conclusion becomes simply a Summary.)

GENERAL OUTLINE MODEL
 I. INTRODUCTION
   A. Definition, Description, and History
   B. Statement of Purpose
   C. Target Audience (omitted for workplace audiences)
   D. Information Sources (including research methods and materials)
   E. Working Definitions
   F. Limitations of the Report
   G. Scope of Coverage (sequence of major topics in the body)

 II. BODY
   A. First Major Topic
      1. First subtopic of A
      2. Second subtopic of A
         a. First subtopic of 2
         b. Second subtopic of 2
           (and so on—subdivision carried as far as necessary)
   B. Second Major Topic
     (and so on)

 III. CONCLUSION (where everything is tied together)
   A. Summary of Information in II
   B. Overall Interpretation of Information in II
   C. Recommendations Based on Information in II

Suggestions for developing each section follow.

Introduction

A typical introduction has some combination of these sections:

1. *Definition, description, and history.* Begin by giving background on the origins and significance of your topic.
2. *Statement of purpose.* Tell why you are writing, and what you plan to achieve.
3. *Target audience.* Identify your audience and its intended use of your information. (Omit this section if you submit an audience-and-use-analysis or if you write for a workplace audience.)
4. *Information sources.* Identify any information sources briefly here. (You will document them fully later.)
5. *Working definitions.* If needed, define specialized terms or general terms that have special meanings. If you must define more than five terms, place definitions in a glossary at the end.
6. *Report limitations.* Explain any incomplete information. Perhaps you were unable to locate a key book or interview a key person. Or perhaps your study must be titled "preliminary" instead of "definitive" (the final word). Or perhaps your report treats only one side of a controversial issue, as in studying the *negative* effects of automation on employee morale.
7. *Scope.* List the major topics to be discussed in the body section.

Not all introductions require each of these sections. Keep the introduction as brief as you can—without compromising its informative value. Reports too often waste readers' time with needless background. Know your readers, and give them only what they need.

Body

In the body you present your evidence and explanations. "Show me!" is any reader's implied demand. Trace the steps by which you move from introduction to conclusion.

Give your body section an informative title. For a mechanism description, you might title the body "Description and Function of Parts." For instructions, "Required Steps." For a problem-solving report, "Collected Data." (See section titles in documents in this book.)

Conclusion

The closing section of a report has several purposes: it might evaluate the significance of the report, take a position, predict an outcome, offer a solution, or suggest further study. Whatever the conclusion's specific purpose, readers expect a clear perspective on the whole document.

Conclusions vary with the document. You might conclude a mechanism description by reviewing the mechanism's major parts, and then briefly describing one operating cycle. You might conclude a comparison or feasibility report by offering judgments about the facts you've presented, and then recommending a course of action. Here are the possible elements in your closing section:

1. *Summary of the body.* When your body section is long and involved or your major findings need reemphasis, summarize them briefly.

2. *Overall interpretation.* Tell the readers what your data mean—even when the report does not call for recommendations.

3. *Recommendations.* Base recommendations directly on your findings and interpretations.

A good beginning, middle, and ending are indispensable, but alter your outline as you see fit. No model should be followed slavishly. *The organization of any document ultimately is determined by what your readers need.*

THE IMPORTANCE OF BEING MESSY

The neat and ordered outlines shown earlier represent the *products* of outlining, not the *process*. Beneath any finished outline (or any finished document, for that matter) lie pages of scribbling and things crossed out, lists, arrows, and fragments of ideas. Writing begins in disorder. Messiness is a natural and often essential part of writing in its early stages.

In a recent survey to assess how computers influence workers' writing, traditional formal outlining was found to be giving way to outlining that took the form of "notes on audience, purpose, direction, key content points, tone." These outlines were "flexible, sketchy, punctuated by arrows, numbers or exclamation points; they looked more like lists . . ." (Halpern 179). Outlines needn't be pretty, as long as they help you control your material.

Not until your final draft of a long document do you compose the finished outline, which serves as a model for your table of contents. This final outline serves as a check on your reasoning, and a way of revealing to your readers a logical line of thinking.

REPORT DESIGN WORKSHEET

Preliminary Information

What is to be done? *A report on the feasibility of converting our home office from oil to gas heat*

Whom is it to be presented to, and when? *Charles Jones, company president: April 1*

Audience Analysis	Primary Reader(s)	Secondary Reader(s)
Position and title:	*President, Abco Engineering consultants*	*Company officers engineering staff*
Relationship to author or organization:	*Employer*	*supervisors, colleagues, junior members*
Technical expertise:	*nontechnical (for this subject)*	*nontechnical*
Personal characteristics:	*highly efficient; demands quality and economy*	*all serious-minded professionals*
Attitude toward author or organization:	*is considering me for promotion to assistant V.P.*	*friendly and respectful; officers will vote on my promotion*
Attitude toward subject:	*highly interested because of last winter's inconvenience*	*interested*
Effect of report on readers or organization:	*will be read closely and acted upon*	*will be read and discussed at our next staff meeting*

Reader's Purpose

Why has reader requested it?	*wants to make a practical decision*	_____
What does reader plan to do with it?	*use the data to make the best choice*	*confer with the president about the choice*
What should reader know beforehand to understand it as written?	*nothing special; history of problem is reviewed in report*	*same*
What does reader already know?	*remembers last winter's problems*	*same*
What amount and kinds of detail will reader find significant?	*brief description of conversion procedures and detailed cost analysis*	*same*
What should reader know and/or be able to do after reading it?	*make an educated decision*	*advise the president about his decision*

FIGURE 11.1 Report Design Worksheet

Writer's Purpose

Why am I writing it?   *to communicate my research findings clearly*

What effect(s) do I wish to achieve?   *to have my readers conclude that conversion is not economically feasible; to persuade them to accept my recommendation of an alternative to conversion*

Design Specifications

Sources of data: *gas company, our oil company representative, Tubo Plumbing Corp., Jumbo Salvage Co., Watt Electronics, Inc.*

Tone: *semiformal*

Point of view: *first- and second-person*

Needed visuals and supplements: *title page, letter of transmittal, table of contents, informative abstract, data sheet appendix reviewing the procedure for cost analysis*

Appropriate format (letter, memo, etc.):   *formal report format with full heading system*

Rhetorical mode (description, definition, classification, etc.—or some combination):   *primary mode: analysis; secondary modes: description, process narration*

Basic organization (problem-causes-solution, intro-instructions-summary, etc.):   *questions-answers-conclusions and recommendations*

Main points in introduction: *Background*
Purpose
Intended Audience
Data Sources
Limitations/Scope

Main points in body: *Description of Present System*
Removal of Oil Burner and Tank
Installation of Oil Pipe
Installation of Gas Burner
Estimation of Gas Heating Costs

Main points in conclusion: *Summary of Findings*
Interpretation of Findings
Recommendation

Other Considerations   *no frills: these readers are all engineers interested in hard facts*

FIGURE 11.1 Report Design Worksheet   *Continued*

THE REPORT DESIGN WORKSHEET

As an alternative to the audience-and-use profile sheet (page 60), the worksheet in Figure 11.1 can supplement your outline and help you focus on your audience and purpose. This version is based on a worksheet developed by Professor John S. Harris of Brigham Young University. The figure has been completed for the heating-conversion report outlined earlier.

EXERCISE

For each of the following documents, indicate the most logical sequence. (For example, a description of a proposed computer lab would follow a spatial sequence.)

- a set of instructions for operating a power tool
- a campaign report describing your progress in political fund-raising
- a report analyzing the weakest parts in a piece of industrial machinery
- a report analyzing the desirability of a proposed oil refinery in your area
- a detailed breakdown of your monthly budget to trim excess spending
- a report investigating the reasons for student apathy on your campus
- a report investigating the effects of the ban on DDT in insect control
- a report on any highly technical subject, written for a general reader
- a report investigating the success of a no-grade policy at other colleges
- a proposal for a no-grade policy at your college

COLLABORATIVE PROJECTS

1. Organize into small groups. Choose *one* of these topics, formulate a statement of purpose, and brainstorm to develop a formal outline for the body section of a report. One representative from your group can write the final draft on the board, for class revision.

- job opportunities in your career field
- a physical description of the ideal classroom
- how to organize an effective job search
- how the quality of your higher educational experience can be improved
- arguments for and against a formal grading system
- arguments for and against a college-wide computer literacy requirement

2. Assume your group is preparing a report titled "The Negative Effects of Strip Mining on the Cumberland Plateau Region of Kentucky." After brainstorming and researching your subject, you all settle on these four major topics:

- economic and social effects of strip mining
- description of the strip-mining process

- environmental effects of strip mining
- description of the Cumberland Plateau

Arrange these topics in the most sensible sequence.

When your topics are arranged, assume that subsequent research and further brainstorming produce this list of subtopics:

- method of strip mining used in the Cumberland Plateau region
- location of the region
- permanent land damage
- water pollution
- lack of educational progress
- geological formation of the region
- open-pit mining
- unemployment
- increased erosion
- auger mining
- natural resources of the region
- types of strip mining
- increased flood hazards
- depopulation
- contour mining

Arrange each subtopic (and perhaps some sub-subtopics) under its appropriate topic headings. Use effective notation to create the body section of a formal outline. Appoint one group member to present the outline in class.

Hint: Assume that this is your thesis: "Decades of strip mining (without reclamation) in the Cumberland Plateau have devastated this region's environment, economy, and social structure."

Shaping the Paragraphs

The Standard Paragraph

Paragraph Unity

Paragraph Coherence

Sequence in a Paragraph

Paragraph Length

■   ■   ■

To follow your thinking, readers need a message organized in a way that makes sense to *them*. But thinking rarely occurs in a neat, predictable sequence, and so we don't have the luxury of "reporting" thoughts in the same random order they occur. Instead we have to *shape* this material into an organized unit of meaning. In trying to organize, we face decisions like these:

- *What do I want to emphasize?*
- *What do I say first?*
- *What comes after that?*
- *How do I end?*

Useful writing of any length—a book, chapter, news article, letter, or memo—typically follows a common organizing pattern:

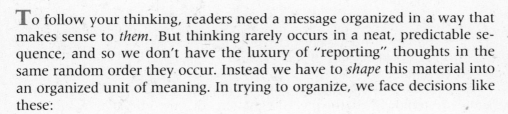

This pattern is best illustrated in the form of a standard paragraph.

Whereas outlines help us shape the whole message into this logical pattern, paragraphs help us shape each part of the whole.

THE STANDARD PARAGRAPH

In a standard paragraph, a group of sentences focuses on one main organizing point. This main point is expressed as a *topic sentence*:

Topic sentences

Computer literacy soon will be a requirement for all "educated" people.

A video display terminal can endanger the operator's health.

Chemical pesticides and herbicides are both ineffective and hazardous.

Each topic sentence introduces an idea or judgment or opinion. But in order to grasp the writer's exact meaning, readers need some explanation. Consider the third statement:

Chemical pesticides and herbicides are both ineffective and hazardous.

Imagine you are a researcher for the Epson Electric Light Company and have been assigned this task: determine whether the company should (1) begin spraying pesticides and herbicides under its power lines, as many other utilities are doing, or (2) continue with its manual (and nonpolluting) ways of minimizing foliage and insect damage to lines and poles. If you simply responded with the preceding statement, your employer would have questions:

- *Why, exactly, are chemical pesticides and herbicides ineffective and hazardous?*
- *What are the problems?*
- *Can you explain?*

By answering these questions in a fully developed paragraph, you provide the supporting details:

Intro. (topic sent.)
Body (2–6)

[1]Chemical pesticides and herbicides are both ineffective and hazardous. [2]Because none of these chemicals has permanent effects, pest populations invariably recover and require respraying. [3]Repeated applications cause pests to develop immunity to the chemicals. [4]Furthermore, most of these products attack species other than the intended pest, killing off its natural predators, and thereby actually increasing the pest population. [5]Above all, chemical residues survive in the environment (and living tissue) for years, often carried hundreds of miles by

wind and water. [6]This toxic legacy includes such biological disasters as birth deformities, reproductive failures, brain damage, and cancer. [7]The ultimate victims of these chemicals would be our customers. [8]I therefore recommend we continue our present control methods.

Conclusion (7–8)

Most paragraphs in technical writing have an introduction-body-conclusion structure. They begin with a clear topic (or orienting) sentence stating a generalization. Details in the body support the generalization.

Writing the Topic Sentence

Readers look to the first sentences in a paragraph to orient themselves, to envision a framework for understanding your message. Without this framework, readers cannot possibly grasp your exact meaning. Consider this paragraph, whose topic sentence has been omitted (read it *once* only):

> Besides containing several toxic metals, it percolates through the soil, leaching out naturally present metals. Pollutants such as mercury invade surface water, accumulating in fish tissues. Any organism eating the fish—or drinking the water—in turn faces the risk of heavy metal poisoning. Moreover, acidified water can release heavy concentrations of lead, copper, and aluminum from metal plumbing, making ordinary tap water hazardous.

A paragraph with its topic sentence omitted

After one reading, can you identify the paragraph's main idea? Probably not, even after a second reading. Without the orientation of a topic sentence, you have no framework for understanding the larger meaning. And you don't know where to place the emphasis: on polluted fish, on metal poisoning, on tap water?

Now, insert this sentence at the beginning, and reread the paragraph:

> Acid rain indirectly threatens human health.

The missing topic sentence

In light of this organizing point, the exact meaning of the message becomes clear. The topic sentence gives us a framework by:

1. Naming the subject of the message (acid rain).
2. Stating the topic—the writer's specific judgment on the subject (that acid rain threatens human health).
3. Forecasting how the message will be developed (by explaining the process) in response to the reader's main question: *How exactly does acid rain threaten human health?*

In most standard paragraphs, readers expect the topic sentence to serve as the key to understanding the whole. The topic sentence therefore should appear *first* in the paragraph, unless you have good reason to place it else-where. Think of your topic sentence as the one sentence you would keep if you could keep only one (U.S. Air Force Academy 11).

Tell readers immediately what to expect. Don't write *Some pesticides are less hazardous and often more effective than others* when you mean *Organic pesticides are less hazardous and often more effective than their chemical counter-parts.* The first topic sentence leads everywhere and nowhere; the second helps us focus, tells us what to expect from the paragraph. Don't write *Acid rain poses a danger,* leaving readers to decipher your meaning of *danger.* If you mean that *Acid rain is killing our lakes and polluting our water supplies,* say so. Uninformative topic sentences keep readers guessing.

Writing the Body

The paragraph body contains the details that explain and expand your main idea. This support material answers the questions you can expect from readers about your topic sentence:

- *Says who?*
- *What proof do you have to support your claim?*
- *Can you give examples?*

Here is how writer Tracy Kidder develops a paragraph to support his topic sentence (in boldface):

> **Obviously, computers have made differences.** They have fostered the development of spaceships—as well as a great increase in junk mail. The computer boom has brought the marvelous but expensive diagnostic device known as the CAT scanner, as well as a host of other medical equipment; it has given rise to machines that play good but rather boring chess, and also, on a larger game board, to a proliferation of remote-controlled weapons in the arsenals of nations. Computers have changed ideas about waging war and about pursuing science, too. It is hard to see how contemporary geophysics or meteorology or plasma physics can advance very far without them now. Computers have changed the nature of research in mathematics, though not every mathematician would say it is for the better. And computers have become a part of the ordinary conduct of businesses of all sorts. (243–44)

With the topic sentence and supporting details on paper, the paragraph is ready for its conclusion.

Writing the Conclusion

Your concluding statement rounds out the paragraph by summarizing, interpreting, or judging the facts. In a longer document, the paragraph conclusion can prepare readers for a subsequent paragraph. Here is Tracy Kidder's conclusion for the preceding paragraph:

| They really help, in some cases.

This conclusion serves as transition to Kidder's next main idea:

> **Not always, though.** One student of the field has estimated that about forty percent of commercial applications of computers have proved uneconomical, in the sense that the job the computer was bought to perform winds up costing more to do after the computer's arrival than it did before. Most computer companies have boasted that they aren't just selling machines, they're selling *productivity*. ("We're not in competition with each other," said a PR man. "We're in competition with labor.") But that clearly isn't always true. Sometimes they're selling paper-producers that require new legions of workers to push that paper around.   (244)[1]

An introduction-body-conclusion structure should serve most of your paragraph needs in technical writing.[2] Begin each support paragraph with a solid topic sentence and you will stay on target.

Structural Variation

Sometimes a main idea calls for a topic statement of two or more sentences (as in the needle description, page 245). Or your main idea might have several distinct parts, which would result in an excessively long paragraph. You might break up the paragraph, making the topic statement a brief introductory paragraph that serves the separate subparts, which are set off as independent paragraphs for the readers' convenience.

> **The most common types of strip-mining procedures are open-pit mining, contour mining, and auger mining. The specific type employed will depend on the type of terrain covering the coal.**

A single topic statement serving multiple paragraphs

---

[1]From *The Soul of a New Machine* by Tracy Kidder. Copyright © 1981 by John Tracy Kidder. By permission of Little, Brown and Co.

[2]Besides support paragraphs, of course, you will write introductory paragraphs, transitional paragraphs, and concluding paragraphs (as illustrated in the sample reports in later chapters).

Open-pit mining is employed in the relatively flat lands in western Kentucky, Oklahoma, and Kansas. Here, draglines and scoops operate directly on the coal seams, producing long, parallel rows of packed spoil banks, ten to thirty feet high, with steep slopes. Between the spoil banks are large pits that fill with water to create pollution and flood hazards.

Contour mining is most widely practiced in the mountainous terrain of the Cumberland Plateau and eastern Kentucky. Here, bulldozers and explosives cut and blast the earth and rock covering a coal seam. Wide bands are removed from the mountain's circumference to reach the embedded coal beneath. The cutting and blasting result in a shelf along with a manmade cliff some sixty feet high, at a right angle to the shelf. The blasted and churned earth is pushed over the shelf to form a massive and unstable spoil bank that creates a danger of mud slides.

Auger mining is employed when the mountain has been cut so thin it can no longer be stripped. It is also used in other difficult-access terrain. Here, large augers bore parallel rows of holes into the hidden coal seams to extract the embedded coal. Among the three strip-mining methods, auger mining causes least damage to the surrounding landscape.

Each paragraph begins with a clear statement about the part of the subtopic to be discussed.

PARAGRAPH UNITY

A paragraph is unified when all parts work toward the same end—when every word, phrase, and sentence explains, illustrates, and clarifies the topic sentence.

A unified paragraph

Solar power offers an efficient, economical, and safe solution to the Northeast's energy problems. To begin with, solar power is highly efficient. Solar collectors installed on fewer than 30 percent of roofs in the Northeast would provide more than 70 percent of the area's heating and air-conditioning needs. Moreover, solar heat collectors are economical, operating for up to twenty years with little or no maintenance. These savings recoup the initial cost of installation within only ten years. Most important, solar power is safe. It can be transformed into electricity through photovoltaic cells (a type of storage battery) in a noiseless process that produces no air pollution—unlike coal, oil, and wood combustion. In sharp contrast to its nuclear counterpart, solar power produces no toxic waste and poses no catastrophic danger of meltdown. Thus, massive conversion to solar power would ensure abundant energy and a safe, clean environment for future generations.

One way to damage unity in the paragraph above would be to discuss the differences between active and passive solar heating, or manufacturers of solar technology, or the advantages of solar power over wind power. Although these matters do *broadly* relate to the general issue of solar energy, none directly advances the meaning of *efficient, economical,* or *safe.*

Every topic sentence has a key word or phrase that carries the meaning. In the pesticide-herbicide paragraph (page 238), the key words are *ineffective* and *hazardous.* Anything that fails to advance their meaning throws the paragraph—and the readers—off track.

PARAGRAPH COHERENCE

A paragraph is coherent when it hangs together and flows smoothly in a clear direction—when all sentences are logically connected like links in a chain, leading toward a definite conclusion.

Three ways to damage paragraph coherence are (1) to use short, choppy sentences, (2) to place sentences in the wrong sequence, and (3) to use insufficient transitions and other connectors (pages 623–626) for linking related ideas. Here is how the paragraph on solar energy might become incoherent:

> Solar power offers an efficient, economical, and safe solution to the Northeast's energy problems. Unlike nuclear power, solar power produces no toxic waste and poses no danger of meltdown. Solar power is efficient. Solar collectors could be installed on fewer than 30 percent of roofs in the Northeast. These collectors would provide more than 70 percent of the area's heating and air-conditioning needs. Solar power is safe. It can be transformed into electricity. This transformation is made possible by photovoltaic cells (a type of storage battery). Solar heat collectors are economical. The photovoltaic process produces no air pollution.

An incoherent paragraph

The paragraph above (only part is shown) lacks coherence in three ways: the sentences are choppy; the sequence of sentences is chaotic; and transitions and connectors are missing.

Here again is the original, coherent paragraph. Notice how this version reveals a clear line of thought:

> [1]Solar power offers an efficient, economical, and safe solution to the Northeast's energy problems. [2]**To begin with,** solar power is highly efficient. [3]Solar

A coherent paragraph

collectors installed on fewer than 30 percent of roofs in the Northeast would provide more than 70 percent of the area's heating and air-conditioning needs. [4]**Moreover,** solar heat collectors are economical, operating for up to twenty years with little or no maintenance. [5]**These savings** recoup the initial cost of installation within only ten years. [6]**Most important,** solar power is safe. [7]**It** can be transformed into electricity through photovoltaic cells (a type of storage battery) in a noiseless process that produces no air pollution—unlike coal, oil, and wood combustion. [8]**In sharp contrast** to its nuclear counterpart, solar power produces no toxic waste and poses no danger of catastrophic meltdown. [9]**Thus,** massive conversion to solar power would ensure abundant energy and a safe, clean environment for future generations.

We can easily trace the sequence of thoughts in this paragraph.

1. The topic sentence establishes a clear direction.
2–3. The first reason is given and then explained.
4–5. The second reason is given and explained
6–8. The third and major reason is given and explained.
9. The conclusion sums up and reemphasizes the main point.

Within this line of thinking, each sentence follows logically from the one before it. Readers know exactly where they are at any point in the paragraph. To reinforce the logical sequence, related ideas are combined in individual sentences, and transitions and connectors (in boldface) signal clear relationships. The whole paragraph sticks together.

SEQUENCE IN A PARAGRAPH

In developing the sequence of a paragraph, you decide on which idea to discuss first, which second, and so on. Some possibilities follow.

Spatial Sequence

A spatial sequence begins at one location and ends at another. It is most useful in describing a physical item or a mechanism. Describe the parts in the sequence in which readers would actually view them or in the order in which each part functions: left or right, inside to outside. This description of a hypodermic needle proceeds from the needle's base (hub) to its point:

A hypodermic needle is a slender, hollow steel instrument used to introduce medication into the body (usually through a vein or muscle). It is a single piece composed of three parts, all considered sterile: the hub, the cannula, and the point. The hub is the lower, larger part of the needle that attaches to the neck-like opening on the syringe barrel. Next is the cannula (stem), the smooth and slender central portion. Last is the point, which consists of a beveled (slanted) opening, ending in a sharp tip. The diameter of a needle's cannula is indicated by a gauge number; commonly, a 24–25 gauge needle is used for subcutaneous injections. Needle lengths are varied to suit individual needs. Common lengths used for subcutaneous injections are $3/8$, $1/2$, $5/8$, and $3/4$ inch. Regardless of length and diameter, all needles have the same functional design.

Product and mechanism descriptions almost always have some type of visual to amplify the verbal description.

Chronological Sequence

A paragraph describing a series of events or giving instructions is most effective when its details are arranged according to a strict time sequence: first step, second step, and so on.

Instead of breaking into a jog too quickly and risking injury, take a relaxed and deliberate approach. Before taking a step, spend at least ten minutes stretching and warming up, using any exercises you find comfortable. (After your first week, consult a jogging book for specialized exercises.) When you've completed your warmup, set a brisk pace walking. Exaggerate the distance between steps, taking long strides and swinging your arms briskly and loosely. After roughly 100 yards at this brisk pace, you should feel ready to jog. Immediately break into a very slow trot: lean your torso forward and let one foot fall in front of the other (one foot barely leaving the ground while the other is on the pavement). Maintain the slowest pace possible, just above a walk. *Do not bolt out like a sprinter!* The biggest mistake is to start fast and injure yourself. While jogging, relax your body. Keep your shoulders straight and your head up, and enjoy the scenery—after all, it is one of the joys of jogging. Keep your arms low and slightly bent at your sides. Move your legs freely from the hips in an action that is easy, not forced. Make your feet perform a heel-to-toe action: land on the heel; rock forward; take off from the toe.

The paragraph explaining how acid rain endangers human health (page 239) is another example of chronological sequence.

Effect-to-Cause Sequence

A paragraph that first identifies a problem and then discusses its causes is typically found in problem-solving reports.

> Modern whaling techniques have brought the whale population to the threshold of extinction. In the nineteenth century, invention of the steamboat increased hunters' speed and mobility. Shortly afterward, the grenade harpoon was invented so that whales could be killed quickly and easily from the ship's deck. In 1904, a whaling station opened on Georgia Island in South America. This station became the gateway to Antarctic whaling for the nations of the world. In 1924, factory ships were designed that enabled round-the-clock whale tracking and processing. These ships could reduce a ninety-foot whale to its by-products in roughly thirty minutes. After World War II, more powerful boats with remote sensing devices gave a final boost to the whaling industry. The number of kills had now increased far beyond the whales' capacity to reproduce.

Cause-to-Effect Sequence

In a cause-to-effect sequence, the topic sentence identifies the cause(s), and the remainder of the paragraph discusses its effects.

> Some of the most serious accidents involving gas water heaters occur when a flammable liquid is used in the vicinity. The heavier-than-air vapors of a flammable liquid such as gasoline can flow along the floor—even the length of a basement—and be explosively ignited by the flame of the water heater's pilot light or burner. Because the victim's clothing frequently ignites, the resulting burn injuries are commonly serious and extremely painful. They may require long hospitalization, and can result in disfigurement or death. *Never, under any circumstances, use a flammable liquid near a gas heater or any other open flame.* (Consumer Product Safety Commission)

Emphatic Sequence

Paragraphs that provide detailed reasons to support a specific viewpoint or recommendation often appear in workplace writing, as in the pesticide-herbicide paragraph on page 238 or the solar energy paragraph on page 242. For emphasis, the reasons or examples usually are arranged in decreasing or increasing order of importance.

> Although strip mining is safer and cheaper than conventional mining, it is highly damaging to the surrounding landscape. Among its effects are scarred mountains, ruined land, and polluted waterways. Strip operations are altering

our country's land at the rate of 5,000 acres per week. An estimated 10,500 miles of streams have been poisoned by silt drainage in Appalachia alone. If strip mining continues at its present rate, 16,000 square miles of U.S. land eventually will be stripped barren.

In this paragraph, the most dramatic example is saved for the end.

Problem-Causes-Solution Sequence

The problem-causes-solution paragraph is commonly used in daily activity or progress reports. After outlining the cause of the problem, the writer of this paragraph explains how the problem has been solved:

On all waterfront buildings, the unpainted wood exteriors had been severely damaged by the high winds and sandstorms of the previous winter. After repairing the damage, we took protective steps against further storms. First, all joints, edges, and sashes were treated with water-repellent preservative to protect against water damage. Next, three coats of nonporous primer were applied to all exterior surfaces to prevent paint from blistering and peeling. Finally, two coats of wood-quality latex paint were applied over the nonporous primer. To keep coats of paint from future separation, the first coat was applied within two weeks of the priming coats, and the second within two weeks of the first. Two weeks after completion, no blistering, peeling, or separation has occurred.

Comparison/Contrast Sequence

A paragraph discussing the similarities or differences (or both) between two or more items often is used in job-related writing.

The ski industry's quest for a binding that ensures good performance as well as safety has led to development of two basic types. Although both bindings improve performance and increase the safety margin, they have different release and retention mechanisms. The first type consists of two units (one at the toe, another at the heel) that are spring-loaded. These units apply their retention forces directly to the boot sole. Thus the friction of boot against ski allows for the kind of ankle movement needed at high speeds over rough terrain, without causing the boot to release. In contrast, the second type has one spring-loaded unit at either the toe or the heel. From this unit extends a boot plate that travels the length of the boot to a fixed receptacle on its opposite end. With this plate binding, the boot has no part in release or retention. Instead, retention force is applied directly to the boot plate, providing more stability for the recreational skier, but allowing for less ankle and boot movement before releasing. Overall, the double-unit binding performs better in racing, but the plate binding is safer.

For comparing and contrasting more specific data on these bindings, two lists would be most effective.

The Salomon 555 offers the following features:

1. upward release at the heel and lateral release at the toe (thus eliminating 80 percent of leg injuries)
2. lateral antishock capacity of 15 millimeters, with the highest available return-to-center force
3. two methods of re-entry to the binding: for hard and deep-powder conditions
4. five adjustments
5. (and so on)

The Americana offers these features:

1. upward release at the toe as well as upward and lateral release at the heel
2. lateral antishock capacity of 30 millimeters, with moderate return-to-center force
3. two methods of re-entry to the binding
4. two adjustments, one for boot length and another for comprehensive adjustment for all angles of release and elasticity
5. (and so on)

Instead of this block structure (in which one binding is discussed and then the other), the writer might have chosen a point-by-point structure (in which points common to all items, such as "Re-entry Methods" are listed together). The point-by-point comparison is favored in feasibility and recommendation reports because it offers readers a meaningful comparison between common points rather than by cataloging items separately.

PARAGRAPH LENGTH

Paragraph length depends on the writer's purpose and the reader's capacity for understanding. A question that guides your decision about length:

- *How much and what kind of support do I need, to convey my exact meaning to these readers?*

When deciding about paragraph length, consider these guidelines:

1. Too many short paragraphs make a message seem choppy and poorly organized. Short paragraphs, however, are effective in step-by-step instructions.

2. Long paragraphs can be hard to follow. Important ideas can get buried in the middle of a long paragraph. Unless your paragraph is in list form (like this one), it should rarely be longer than fifteen lines (or about one-half of a double-spaced page).

3. A short paragraph (even a single-sentence paragraph) can attract attention by setting off an important idea.

> We can prevent further damage from mud slides only by building a retaining wall behind the number 3 construction site immediately.

4. A combination of shorter and longer paragraphs is best for emphasis, if your subject allows this arrangement.

5. Avoid long paragraphs at the beginning or end of a document.

EXERCISE

Locate, copy, and bring to class a paragraph that has the following features:

- an orienting topic sentence
- adequate development
- unity
- coherence
- a recognizable sequence
- appropriate length for its purpose and audience

Be prepared to identify and explain each of these features in a class discussion.

Revising for a Readable Style

Revising for Clarity

Revising for Conciseness

Revising for Fluency

Finding The Right Words

Adjusting Your Tone

Avoiding Reliance on Automated Tools

You might write for a diverse or specific audience, or for experts or non-experts. But no matter how technically appropriate your document, the audience's needs will not be served unless your style is readable.

What is *writing style,* and how does it influence reader response to a document? Your writing style is the product of

- the words you choose
- the way in which you put a sentence together
- the length of your sentences
- the way in which you connect sentences
- the tone you convey.

Efficient writing style is neither fancy nor impressive; it is straightforward, easy to follow and understand—in a word, *readable.*

Efficient style requires much more than correct grammar, punctuation, and spelling. Granted, basic mechanical errors do distract readers; but correctness alone is no guarantee of a readable style. This response to a job applicant is mechanically "correct" but inefficient:

We are in receipt of your recent correspondence indicating your interest in securing the advertised position. Your correspondence has been duly forwarded to the office having employment candidate selection responsibility for consideration. You may expect to hear from the aforementioned office relative to your application as the selection process progresses. Your interest in the position is appreciated.

Notice how hard you have worked to extract information that could be expressed this simply:

Your application for the advertised position has been forwarded to the office that selects candidates. At each stage, we will inform you of your status. Thank you for your interest.

Inefficient style makes readers work harder than they should.

Style can be inefficient for many reasons, but it is especially inept when it

- makes the writing impossible to interpret
- takes too long to make the point
- reads like a Dick-and-Jane story from primary school
- uses imprecise or needlessly big words
- sounds stuffy and impersonal.

Regardless of its cause, inefficient style results in writing that is less informative, less persuasive. Moreover, inefficient style can be unethical—by confusing or misleading readers.

To help your audience spend less time reading, spend more time revising for a style that is *clear, concise, fluent, exact,* and *likable*.

REVISING FOR CLARITY

A clear sentence conveys the writer's exact meaning on the first reading. It signals relationships among its parts, and it emphasizes the main idea. The following guidelines will help you revise for clarity.

Avoid Ambiguous Phrasing

Workplace writing ideally has *one* meaning only, allows for *one* interpretation. Does one's "suspicious attitude," then, mean that one is "suspicious" or "suspect"?

Ambiguous	All managers are not required to submit reports. (Are some or none required?)
Revised	Managers are not all required to submit reports.
	or
	No managers are required to submit reports.
Ambiguous	Our two helicopters crashed, injuring eight people. (Did they crash separately or did they collide?)
Revised	Each of our two helicopters crashed, injuring eight people.
	or
	Our two helicopters collided, injuring eight people.

Make sure your writing conveys the meaning you intend.

Avoid Ambiguous Pronoun References

Whenever you use a pronoun (*he, she, it, their,* etc.), that pronoun must refer to one clearly identified noun. If the pronoun's referent (or antecedent) is vague, readers will be confused.

Ambiguous referent	Our patients enjoy the warm days while they last. (Are the patients or the warm days on the way out?)

Depending on whether the referent for *they* is *patients* or *warm days*, the sentence can be clarified.

Clear referent	While these warm days last, our patients enjoy them.
	or
	Our terminal patients enjoy the warm days.
Ambiguous referent	Jack resents his assistant because he is competitive. (Who's the competitive one—Jack or his assistant?)
Clear referent	Because his assistant is competitive, Jack resents him.
	or
	Because Jack is competitive, he resents his assistant.

Be sure readers can identify the noun your pronoun replaces. (See pages 601–602 for more on pronoun references, and pages 297–98 for advice on avoiding sexist bias in pronoun use.)

Avoid Ambiguous Punctuation

A missing comma, hyphen, or other punctuation mark can obscure your meaning.

Missing hyphen	Replace the trailer's inner wheel bearings. (The inner-wheel bearings or the inner wheel-bearings?)
	Our president is a high fidelity fanatic. (Someone who likes drugs, but refuses to fool around?)
Missing comma	Does your company produce liquid hydrogen? If so, how[,] and where do you store it? (Notice how the meaning changes with a comma after "how.")
	Police surrounded the crowd[,] attacking the strikers. (Without the comma, the crowd appears to be attacking the strikers.)

Although missing hyphens and commas are prime culprits, other omissions can cause ambiguity as well. A missing colon after *kill* yields the headline "Moose Kill 200." A missing apostrophe after *Myers* creates this gem: "Myers Remains Buried in Portland." Punctuation *does* affect meaning.

Avoid Telegraphic Writing

Function words show relationships between the *content words* (nouns, adjectives, verbs, and adverbs) in a sentence. Some examples of function words:

- articles (*a, an, the*)
- prepositions (*in, of, to*)
- linking verbs (*is, has, looks*)
- relative pronouns (*who, which, that*)

Some writers mistakenly try to compress their writing by eliminating these function words.

Ambiguous	Proposal to employ retirees almost dead.
Revised	The proposal to employ retirees **is** almost dead.
Ambiguous	Uninsulated end pipe ruptured. (What ruptured? The pipe or the end of the pipe?)
Revised	**The** uninsulated end **of the** pipe ruptured.

or

The uninsulated pipe **on the** end ruptured.

Ambiguous	The reactor operator told management several times he expected an accident. (Did he tell them once or several times?)
Revised	The reactor operator told management several times **that** he expected an accident.

or

The reactor operator told management **that** several times he expected an accident.

Avoid Ambiguous Modifiers

Modifiers explain, define, or add detail to other words or ideas. If a modifier is too far from the words it modifies, the message can be ambiguous.

Misplaced modifier	**Only** press the red button in an emergency. (Does *only* modify *press* or *emergency*?)
Revised	Press **only** the red button in an emergency.

or

Press the red button in an emergency **only**.

Another problem with ambiguity occurs when the modifier has no word to modify.

Dangling modifier	**Being so well known in the computer industry**, I would appreciate your advice.

The writer intended to say that the *reader* is well known, but with no word to modify, the modifying phrase dangles. We can eliminate the confusion by adding a subject:

Revised	Because **you** are so well known in the computer industry, I would appreciate your advice.

See pages 604–606 for more on modifiers.

Unstack the Modifying Nouns

One noun can modify another noun (as in "software development"). But when two or more nouns modify a noun, the string of densely packed words becomes hard to read and ambiguous.

Stacked	Be sure to leave enough time for a **training session participant** evaluation. (Evaluation of or by participants?)

With no function words (articles, prepositions, verbs, relative pronouns) to break up the string of nouns, readers cannot see the relationships among the nouns. What modifies what?

Stacked nouns also deaden your style. Bring your style *and* your reader to life by using action verbs (*complete, prepare, reduce*) and prepositional phrases.

Revised	Be sure to leave enough time **for** participants **to evaluate** the training session.

or

Be sure to leave enough time **to evaluate** participants **in the** training session.

No such problem with ambiguity occurs when *adjectives* are stacked in front of a noun.

Clear	He was a **nervous**, **angry**, **confused**, but **dedicated** employee.

Readers can readily see that the adjectives modify *employee*.

Arrange Words for Coherence and Emphasis

In coherent writing, everything sticks together; each sentence builds on the preceding sentence, and looks ahead to the following sentence. Sentences generally work best when the beginning looks back at familiar information and the end provides the new information:

Familiar		*Unfamiliar*
My dog	has	fleas.
Our boss	just won	the lottery.
This company	is planning	a merger.

Besides helping a message stick together, the familiar-to-unfamiliar structure emphasizes the new information. Just as every paragraph has a key sentence, every sentence has a key word or phrase that sums up the new information. That key word or phrase usually is best emphasized at the end of the sentence.

Faulty emphasis	We expect a **refund** because of your error in our shipment.
Correct	Because of your error in our shipment, we expect a **refund**.
Faulty emphasis	In a business relationship, **trust** is a vital element.
Correct	In a business relationship, a vital element is **trust**.

One exception to placing key words last occurs with *instructions*. Each step in a list of instructions should contain an action verb (*insert, open, close, turn, remove, press*). To provide readers with a forecast, place the verb in that instruction at the beginning.

Correct	**Insert** the diskette before activating the system.
	Remove the protective seal.

With the key word at the beginning of the instruction, readers know immediately the action they need to take.

Use Active Voice Often, Passive Voice Selectively

The active voice ("I did it") is more direct, concise, and persuasive than the passive voice ("It was done by me"). In the active voice, the agent performing the action serves as subject:

	Agent	*Action*	*Recipient*
Active	Joe	lost	your report.
	Subject	*Verb*	*Object*

The passive voice reverses the pattern, making the recipient of an action serve as subject.

	Recipient	*Action*	*Agent*
Passive	Your report	was lost	by Joe.
	Subject	*Verb*	*Prepositional phrase*

Sometimes the passive eliminates the agent altogether:

Passive	Your report was lost. (Who lost it?)

Some writers mistakenly rely on the passive voice because they think it sounds more objective and important. But the passive voice often makes writing wordy or indecisive or evasive and unethical. Consider the effect when an active statement is recast in the passive voice:

Concise and direct (active)	I underestimated labor costs for this project. (7 words)
Wordy and indirect (passive)	Labor costs for this project were underestimated by me. (9 words)
Evasive (passive)	Labor costs for this project were underestimated.

For economy, directness, and clarity, use the active voice in most of your writing.

Do not evade responsibility by hiding behind the passive voice:

Passive irresponsibles	A **mistake** was made in your shipment.
	It was decided not to hire you.
	A **layoff** is recommended.

Acknowledge responsibility for your actions:

Active	**I** made a mistake in your shipment.
	I decided not to hire you.
	Our committee recommends a layoff.

In reporting errors or bad news, use the active voice. Readers appreciate clarity and sincerity.

But do use the passive voice if the person behind the action has reason for being protected.

| Correct passive | The criminal was identified. |
| | The embezzlement scheme was exposed. |

Here, the passive protects the innocent person who identified the criminal or who exposed the scheme.

Use the passive only when your audience does not need to know the agent.

| Correct passive | Mr. Jones was brought to the emergency room. |
| | The bank failure was publicized statewide. |

Readers do not need to know *who* brought Mr. Jones or *who* publicized the bank failure. Notice again how the passive voice focuses on the *recipient* rather than the *agent*.

Use the passive voice to focus on events or results when the agent is unknown, unapparent, or unimportant.

> **Correct passive**   All memos in the firm are filed in a database.
>
> Fred's article was published last week.

The information that will interest readers here is *how* the memos are filed or *that* Fred's article was published.

Prefer the passive when you deliberately wish to be indirect or inoffensive (as in requesting the customer's payment or the employee's cooperation, or to avoid blaming someone—such as your boss) (Ornatowski 94).

> **Active but**   **You** have not paid your bill.
> **offensive**   **You** need to overhaul our filing system.
>
> **Inoffensive**   **This bill** has not been paid.
> **passive**   **Our filing system** needs overhauling.

By focusing on the recipient rather than the agent, these passive versions help retain the audience's goodwill.

The passive voice is weaker than the active voice, and creates an impersonal tone.

> **Weak and**   An offer will be made by us next week.
> **impersonal**
>
> **Strong and**   We will make an offer next week.
> **personal**

Use the active voice when you want action. Otherwise, your statement will have no power.

> **Weak passive**   If my claim is not settled by May 15, the Better Business
>   Bureau will be contacted, and their advice on legal ac-
>   tion will be taken.

This passive and tentative statement is unlikely to persuade readers to act: *Who* is making the claim? *Who* should settle the claim? *Who* will contact the Bureau? *Who* will take action? Nobody is here! Following is a more direct, concise, and forceful version, in the active voice:

Strong active   If you do not settle my claim by May 15, I will contact the Better Business Bureau for advice on legal action.

Notice how this active version emphasizes the new and significant information by placing it at the end.

Ordinarily, use the active voice for giving instructions.

Faulty passive   The bid should be sealed.

Care should be taken with the dynamite.

Correct active   **Seal** the bid.

Be careful with the dynamite.

Avoid shifts from active to passive voice in the same sentence.

Faulty shift   During the meeting, project members spoke and presentations were given.

Correct   During the meeting, project members spoke and gave presentations.

Unless you have a deliberate reason for choosing the passive voice, prefer the *active* voice for making forceful connections like the one described below:

By using the active voice, you direct the reader's attention to the subject of your sentence. For instance, if you write a job-application letter that is littered with passive verbs, you fail to achieve an important goal of that letter: to show the readers the important things you have done, and how prepared you are to do important things for them. That strategy requires active verbs, with clear emphasis on *you* and what you have done/are doing. (Pugliano 6)

Avoid Overstuffed Sentences

A sentence that crams too many ideas forces readers to struggle over its meaning:

Overstuffed   Publicizing the records of a private meeting that took place three weeks ago to reveal the identity of a manager who criticized our company's promotion policy would be unethical.

Notice how the details are hard to remember, the relationships hard to identify. Clarify by sorting out the relationships.

Revised	In a private meeting three weeks ago, a manager criticized our company's policy on promotion. It would be unethical to reveal the manager's identity by publicizing the records of that meeting. (Other versions are possible here, depending on the writer's intended meaning.)

Give your readers only as much information as they can retain easily in one sentence.

EXERCISES IN REVISING FOR CLARITY

1. Revise each sentence below to eliminate ambiguities in phrasing, pronoun reference, or punctuation.

Ambiguous phrasing	Most city workers strike on Friday.
Revised	Most city workers **are planning to strike** Friday.
	or
	Most city workers **typically strike** on Friday.
Ambiguous pronoun reference	Fred reminded Joe of the contract he had signed.
Revised	Fred reminded Joe of the contract **Joe** had signed.
	or
	Fred reminded Joe of the contract **Fred** had signed.
Ambiguous punctuation	Dial "10" to deactivate the system and sound the alarm.
Revised	Dial "10" to deactivate the system, and sound the alarm.
	or
	Dial "10" to deactivate the system and **to** sound the alarm.

a. Call me any evening except Tuesday after 7 o'clock.
b. The benefits of this plan are hard to imagine.
c. I cannot recommend this candidate too highly.
d. Visiting colleagues can be tiring.
e. Janice dislikes working with Claire because she's impatient.
f. Despite his efforts, Joe misinterpreted Sam's message.

g. Our division needs more effective writers.

h. Tell the reactor operator to evacuate and sound a general alarm.

i. If you don't pass any section of the test, your flying days are over.

2. Revise each sentence below to replace missing function words or to clarify ambiguous modifiers.

Telegraphic writing	Send response to our client advising immediate action.
Revised	Send a response to our client **who** advised immediate action.
	or
	Send a response to our client **to** advise immediate action.
Ambiguous modifier	As a microprocessor expert, I look forward to your reply.
Revised	Because you are a microprocessor expert, I look forward to your reply.

a. The manager claims repeatedly he reported the danger.

b. I want the final Amex report written by your division.

c. Replace main booster rocket seal.

d. The president refused to believe any internal report was inaccurate.

e. Our client failed to understand our recent proposal was preliminary.

f. Only use this phone in a red alert.

g. After offending our best client, I am deeply annoyed with the new manager.

h. Send memo to programmer requesting explanation.

i. He failed completely to explain the malfunction.

j. Do not enter test area while contaminated.

3. Revise each sentence below to unstack modifying nouns or to rearrange the word order for clarity and emphasis.

Stacked modifiers	Sarah's job involves fault analysis systems troubleshooting handbook preparation.
Revised	Sarah prepares handbooks **for** troubleshooting fault analysis systems.
Faulty word order	We have a critical need for technical support.
Revised	Our need for technical support is **critical**.

a. Develop online editing system documentation.

b. We need to develop a unified construction automation design.

c. Install a hazardous materials dispersion monitor system.

d. I recommend these management performance improvement incentives.

e. Our profits have doubled since we automated our assembly line.

f. Education enables us to recognize excellence and to achieve it.

g. In all writing, revision is required.

4. The sentences below are wordy, weak, or evasive because of passive voice. Revise each sentence as a concise, forceful, and direct expression in the active voice, to identify the person or agent performing the action.

Wordy passive	Our test results will be sent to you as soon as verification is completed.
Revised	We will send you our test results as soon as we verify them.
Weak passive	It is believed by us that this contract is faulty.
Revised	We believe that this contract is faulty.
Evasive passive	The decision was made that your request for promotion should be denied.
Revised	I decided to deny your request for promotion.

a. The evaluation was performed by us.

b. The report was written by our group.

c. Unless you pay me within three days, my lawyer will be contacted.

d. Hard hats should be worn at all times.

e. It was decided to reject your offer.

f. Gasoline was spilled on your Ferrari's leather seats.

g. The manager was kissed.

5. The sentences below lack proper emphasis because of active voice. Revise each ineffective active as an appropriate passive, to emphasize the recipient rather than the actor.

Agent is unimportant	The publisher has accepted Mary's story for publication.
Revised	Mary's story has been accepted for publication.
Recipient needs emphasis	A tornado destroyed the barn.

Revised	The barn was destroyed by a tornado.
Active is too blunt	You apparently failed to edit this report.
Revised	This report apparently was not edited.

a. Joe's company fired him.

b. A rockslide buried the mine entrance.

c. Someone on the maintenance crew has just discovered a crack in the nu-clear-core containment unit.

d. A power surge destroyed more than 2,000 lines of our new applications program.

e. Your report confused me.

f. You are paying inadequate attention to worker safety.

g. You are checking temperatures too infrequently.

6. Unscramble this overstuffed sentence by making shorter, clearer sentences:

A smoke-filled room causes not only teary eyes and runny noses but also can alter people's hearing and vision, as well as creating dangerous levels of carbon monoxide, especially for people with heart and lung ailments, whose health is particularly threatened by "second-hand" smoke.

REVISING FOR CONCISENESS

Earlier sections showed that sometimes you have to add words to make meaning clear. But *needless* words can obscure your meaning.

Writing can suffer from two kinds of wordiness: one kind occurs when readers are given information they don't need (think of a typical weather report on local television news). The other kind occurs when too many words are used to convey information readers do need ("a great deal of potential for the future" instead of "great potential").

Every word in the document should advance your meaning. We quote an expert on writing:

Writing improves in direct ratio to the number of things we can keep out of it that shouldn't be there. (Zinsser 14)

A concise message conveys most information in fewest words. It is highly informative, but not cluttered.

| Cluttered | These collar ties weigh 300 pounds apiece, thereby ex-ceeding by a total of 10 percent the load tolerance speci-fied by the building inspector for this project. |

A brief but vague message, on the other hand, is useless.

| Brief but vague | These collar ties are too heavy. |

Here is a concise version:

| Brief but informative | These collar ties weigh 300 pounds apiece, exceeding our specified load tolerance by 10 percent. |

First drafts rarely are concise. Get rid of anything that adds no meaning.

| Cluttered | At this point in time, I would like to say that we are ready to move ahead. |
| Concise | We are ready. |

Use fewer words when fewer will do. But remember the difference between *clear writing* and *compressed writing* that is impossible to decipher.

| Impenetrable | Give new vehicle air conditioner compression cut-off system specifications to engineering manager advising immediate action. |

The following strategies will help you revise for conciseness.

Avoid Needless Phrases

Don't use a whole phrase when one word will do. Instead of *in this day and age,* write *today.* Each needless phrase here can be reduced to one word—without loss in meaning:

at a rapid rate	=	rapidly
due to the fact that	=	because
the majority of	=	most
on a personal basis	=	personally
give instruction to	=	instruct
would be able to	=	obvious

readily apparent	=	obvious
a large number	=	many
prior to	=	before
aware of the fact that	=	know
conduct an inspection of	=	inspect

Eliminate Redundancy

A redundant expression says the same thing twice, in different words, as in *fellow colleagues*. Each boldfaced word below merely adds clutter, because its meaning is included in the other word.

a **dead** corpse	**end** result
completely eliminate	cancel **out**
basic essentials	consensus **of opinion**
enter **into**	**utter** devastation
mental awareness	**the month of** August
mutual cooperation	**utmost** perfection

Avoid Needless Repetition

Unnecessary repetition clutters writing and dilutes meaning.

> **Repetitious** In trauma victims, breathing is restored by **artificial respiration**. Techniques of **artificial respiration** include mouth-to-mouth **respiration** and mouth-to-nose **respiration**.

Repetition in that passage disappears when sentences are combined.

> **Concise** In trauma victims, breathing is restored by artificial respiration, either mouth-to-mouth or mouth-to-nose.

Repetition can, of course, be useful. Don't hesitate to repeat, or at least rephrase, material (even whole paragraphs in a longer document) if you feel that readers need reminders. Effective repetition helps avoid cross-references like this: "See page 23" or "Review page 10."

Avoid *There* Sentence Openers

Save words, add force, and improve emphasis by avoiding *There is* and *There are* to begin sentences—whenever your intended meaning allows.

Weak	**There is** sensitive equipment required for this procedure.
Revised	This procedure requires sensitive equipment.
Weak	**There is** a danger of explosion in Number 2 mineshaft.
Revised	Number 2 mineshaft is in danger of exploding.

Dropping these openers places the key words at sentence end, where they are best emphasized.

Avoid Some *It* Sentence Openers

Don't begin a sentence with *It*—unless the *It* clearly points to a specific referent in the preceding sentence.

Weak	**It** was his bad attitude that got him fired.
Revised	He was fired because of his bad attitude.
Weak	**It** is necessary to complete both sides of the form.
Revised	Please complete both sides of the form.

Delete Needless Prefaces

Don't keep readers waiting for the new information in your sentence. Get right to the point.

Wordy	**I am writing this letter because** I wish to apply for the position of copy editor.
Concise	I wish to apply for the position of copy editor.
Wordy	**As far as artificial intelligence is concerned**, the technology is only in its infancy.
Concise	Artificial-intelligence technology is only in its infancy.

Avoid Weak Verbs

Prefer verbs that express a definite action: *open, close, move, continue, begin.* These strong verbs advance your meaning. Avoid verbs that express no specific action: *is, was, are, has, give, make, come, take*. These weak verbs add words without advancing meaning. All forms of *to be* are weak. This next sentence achieves conciseness because of the strong verb *consider*:

> **Concise** Please **consider** my offer.

Here is what happens with a weak verb in the same sentence:

> **Weak and wordy** Please **take into consideration** my offer.

Don't disappear behind weak verbs, and their baggage of needless nouns and prepositions.

> **Weak** My recommendation **is** for a larger budget.
>
> **Strong** I **recommend** a larger budget.

Strong verbs, or action verbs, suggest a positive, confident, and persuasive writer. Some weak verbs converted to strong:

is in conflict with	=	conflicts
has the ability to	=	can
give a summary of	=	summarize
make an assumption	=	assume
come to the conclusion	=	conclude
take action	=	act
make a decision	=	decide

Delete Needless *To Be* Constructions

The preceding section showed that all forms of *to be* (*is, was, are,* and so on) are weak. Sometimes, the *to be* form itself mistakenly appears behind such verbs as *appears, seems,* and *finds.*

> **Wordy** Your product seems **to be** superior.
>
> I consider this employee **to be** highly competent.

Eliminating *to be* in these examples saves words while preserving meaning.

Avoid Excessive Prepositions

On page 255, we saw how prepositions can increase clarity by breaking up clusters of stacked modifying nouns. But prepositions combined with forms of *to be* can make wordy sentences.

Wordy	The recommendation first appeared **in** the report written **by** the supervisor **in** January **about** that month's productivity.
Concise	The recommendation first appeared in the supervisor's productivity report for January.

Here are some needlessly long prepositional phrases reduced to one or two words—without loss in meaning:

with the exception of	=	except for
in reference to	=	about (or regarding)
in order that	=	so
in the near future	=	soon
in the event that	=	if
at the present time	=	now
in the course of	=	during
in the process of	=	during (or in)

Fight Noun Addiction

Nouns manufactured from verbs (nominalizations) make sentences weak and wordy. Nominalizations often accompany weak verbs and needless prepositions.

Weak and wordy	We ask for the **cooperation** of all employees.
Strong and concise	We ask all employees to **cooperate**.
Weak and wordy	Give **consideration** to the possibility of a career change.
Strong and concise	**Consider** a career change.

Besides being wordy, nominalizations can be vague—by hiding the agent of an action.

Wordy and vague	A **need** for immediate action exists. (Who should take the action? We can't tell.)
Precise	We **must act** immediately.

Here are nominalizations restored to their verb forms:

conduct an investigation of	=	investigate
provide a description of	=	describe
conduct a test of	=	test
make a discovery of	=	discover

Along with weak verbs and needless prepositions, nominalizations drain the life from your style. In cheering for your favorite team, you wouldn't say

Blocking of that kick is a necessity!
>   *instead of*
Block that kick!

Write as you would speak, but avoid colloquialisms and slang.

Make Negatives Positive

A positive expression is more easily understood than a negative one. As one expert on writing points out, "To understand the negative, we have to translate it into an affirmative, because the negative only implies what we should do by telling us what we shouldn't do. The affirmative states it directly" (Williams 54).

Indirect and wordy	Please do not be late in submitting your report.
Direct and concise	Please submit your report on time.

Readers work even harder to translate sentences with two or more negative expressions:

Confusing and wordy	Do **not** distribute this memo to employees who have **not** received a security clearance.

> **Clear and concise** Distribute this memo only to employees who have received a security clearance.

Besides the directly negative words (*no, not, never*), some words are indirectly negative (*except, forget, mistake, lose, uncooperative*). When these indirectly negative words combine with directly negative words, readers are forced to translate.

> **Confusing and wordy** **Do not neglect** to activate the alarm system.
>
> My diagnosis was **not inaccurate**.

The second example above shows how multiple negatives can make the writer seem evasive.

> **Clear and concise** **Be sure** to activate the alarm system.
>
> My diagnosis was **accurate**.

The positive versions are easier to understand *and* more persuasive.

Some negative expressions, of course, are perfectly correct, as in expressing disagreement.

> **Correct negatives** This is **not** the best plan.
>
> Your offer is **unacceptable**.
>
> This project **never** will succeed.

Prefer positives to negatives, however, whenever your meaning allows. Here are negative expressions translated into positive versions:

did not succeed	=	failed
does not have	=	lacks
did not prevent	=	allowed
not unless	=	only if
not until	=	only when
not absent	=	present

Clean Out Clutter Words

Clutter words stretch a message without advancing its meaning. Here are some of the commonest: *very, definitely, quite, extremely, rather, somewhat, really, actually, currently, situation, aspect, factor.*

Cluttered	**Actually**, one **aspect** of a business **situation** that could **definitely** make me **quite** happy would be to have a **somewhat** adventurous partner who **really** shared my **extreme** attraction to risks.
Concise	I seek an adventurous business partner who enjoys risks.

Use such words only when they actually advance your meaning.

Delete Needless Qualifiers

Qualifiers are expressions that soften the tone and impact of a statement. Here are some common qualifiers: *I feel, It seems, I believe, In my opinion*, and *I think*. Use qualifiers to express uncertainty.

Appropriate	Despite Frank's poor grades last year, he will, **I think**, do well in college.
	Your product **seems** to be what we need.

But when you are certain, eliminate the qualifier so as not to seem tentative or evasive.

Needless	**It seems** that I've made an error.
qualifiers	We **appear** to have exceeded our budget.
	In my opinion, this candidate is outstanding.

EXERCISES IN REVISING FOR CONCISENESS

1. Revise each wordy sentence below to eliminate needless phrases, redundancy, and needless repetition.

Needless phrase	I am aware of the fact that Sam is trustworthy.
Revised	I know that Sam is trustworthy.
Redundancy	Through mutual cooperation, we can achieve our goals.
Revised	By cooperation we can achieve our goals.
Needless repetition	This offer is the most attractive offer I've received.
Revised	This is the most attractive offer I've received.

a. I have admiration for Professor Jones.

b. Due to the fact that we made the lowest bid, we won the contract.

c. On previous occasions we have worked together.

d. He is a person who works hard.

e. We have completely eliminated the bugs from this program.

f. This report is the most informative report on the project.

2. Revise each sentence below to eliminate *There* and *It* openers and needless prefaces.

There opener	There is a coaxial cable connecting the antenna to the receiver.
Revised	A coaxial cable connects the antenna to the receiver.
It opener	It is necessary for us to complete this report by Friday.
Revised	We must complete this report by Friday.
Needless preface	I am writing this memo to spell out evacuation procedures.
Revised	This memo spells out evacuation procedures.

a. There was severe fire damage to the reactor.

b. There are several reasons why Jane left the company.

c. It is essential that we act immediately.

d. It has been reported by Bill that several safety violations have occurred.

e. This letter is to inform you that I am pleased to accept your job offer.

f. The purpose of this report is to update our research findings.

3. Revise each wordy and vague sentence below to eliminate weak verbs.

Weak verb	This manual gives instructions to end users.
Revised	This manual instructs end users.
Weak verb	I have a preference for Ferraris.
Revised	I prefer Ferraris.

a. Our disposal procedure is in conformity with federal standards.

b. Please make a decision today.

c. We need to have a discussion about the problem.

d. I have just come to the realization that I was mistaken.

e. We certainly can make use of this information.

f. Your conclusion is in agreement with mine.

4. Revise each sentence below to eliminate needless prepositions and *to be* constructions, and to cure noun addiction.

Needless *to be*	I consider George to be an excellent technician.
Revised	I consider George an excellent technician.
Needless prepositions	Power surges are associated, in a causative way, with malfunctions of computers.
Revised	Power surges cause computer malfunctions.
Noun addiction	The appearance of this problem was just yesterday.
Revised	This problem appeared just yesterday.

a. Igor seems to be ready for a vacation.

b. Our survey found 46 percent of users to be disappointed.

c. In the event of system failure, your sounding of the alarm is essential.

d. These are the recommendations of the chairperson of the committee.

e. Our acceptance of the offer is a necessity.

f. Please perform an analysis and make an evaluation of our new system.

g. A need for your caution exists.

5. Revise each sentence below to eliminate inappropriate negatives, clutter words, and needless qualifiers.

Confusing negative	Do not accept bids that are not signed.
Revised	Accept only signed bids.
Clutter words	Our current situation is that we actually cannot offer you employment.
Revised	We cannot offer you employment.
Needless qualifier	It seems as if I have just wrecked a company car.
Revised	I have just wrecked a company car.

a. Our design must avoid nonconformity with building codes.

b. Never fail to wear protective clothing.

c. Do not accept any bids unless they arrive before May 1.

d. I am not unappreciative of your help.

e. We are currently in the situation of completing our investigation of all aspects of the accident.

f. I appear to have misplaced the contract.

REVISING FOR FLUENCY

Fluent sentences are easy to read because of clear connections, variety, and emphasis. Their varied length and word order eliminate choppiness and monotony. Fluent sentences enhance *clarity*, allowing readers to see what is most important, with no struggle to sort out relationships. Fluent sentences enhance *conciseness*, often replacing several short, repetitious sentences with one longer, economical sentence. The following strategies will help you write fluent sentences.

Combine Related Ideas

A series of short, disconnected sentences is not only choppy and wordy, but unclear as well.

Disconnected	Jogging can be healthful. You need the right equipment. Most necessary are well-fitting shoes. Without this equipment you take the chance of injuring your legs. Your knees are especially prone to injury. (4 sentences)
Clear, concise, and fluent	Jogging can be healthful if you have the right equipment. Shoes that fit well are most necessary because they prevent injury to your legs, especially your knees. (2 sentences)

Never force readers to figure out connections for themselves.

Most sets of information can be combined in different relationships, depending on what you want to emphasize. Imagine that this set of facts describes an applicant for a junior-management position with your company.

- Roy James graduated from an excellent management school.
- He has no experience.
- He is highly recommended.

Assume you are a personnel director, conveying to upper management your impression of this candidate. To convey a negative impression, you might combine the facts in this way:

Strongly negative emphasis	Although Roy James graduated from an excellent management school and is highly recommended, **he has no experience**.

The *independent* idea (in boldface) receives the emphasis. Earlier ideas are made dependent on (or subordinate to) the independent idea by the subordinating word *although*. When a sentence has two or more ideas of unequal importance, the less important idea is signaled by a subordinating word such as *often, as, because, if, unless, until,* or *while*.

To continue with our Roy James example: If you are undecided, but leaning in a negative direction, you might combine the information in this way:

> **Slightly** Roy James graduated from an excellent management
> **negative** school and is highly recommended, **but** he has no expe-
> **emphasis** rience.

In the sentence above, the ideas before and after *but* are both independent. These independent ideas are joined by the coordinating word *but*, which suggests that both sides of the issue are equally important (or "coordinate"). Placing the negative idea last, however, gives it slight emphasis. When a sentence has two or more ideas equal in importance, their equality is signaled by coordinating words such as *and, but, for, nor, or, so,* or *yet*.

Consider once again our Roy James example. To emphasize strong support for the candidate, you could use this combination:

> Although Roy James has no experience, **he graduated from an excellent management school and is highly recommended**.

In the version above, the earlier idea is subordinated by *although*, leaving the two final ideas independent.

Caution: Combine sentences only to advance your meaning, to ease the reader's task. A sentence with too much information and too many connections is impossible to sort out.

> **Overstuffed** Our night supervisor's orders from upper management to
> repair the overheated circuit were misunderstood by
> Harvey Kidd, who gave the wrong instructions to the
> emergency crew, thereby causing the fire.

Readability is affected less by the number of words in a sentence than by the amount of information. Even short sentences can be unreadable if they incorporate too many details:

> **Overstuffed** Send three copies of Form 17-e to all six departments,
> unless Departments A or B or both request Form 16-w.

Effective combining *enhances* and *streamlines* meaning, but overcombining makes meaning impenetrable.

Vary Sentence Construction and Length

Long and short sentences each have their purpose: to express ideas logically or forcefully.[1] We have just seen how related ideas often need to be linked in one sentence, so that readers can grasp the connections:

Disconnected	The nuclear core reached critical temperature. The loss-of-coolant alarm was triggered. The operator shut down the reactor.
Connected	As the nuclear core reached critical temperature, triggering the loss-of-coolant alarm, the operator shut down the reactor.

The ideas above have been combined to show that one action resulted from another. But an idea that should stand alone for emphasis needs a whole sentence of its own:

Correct	Core meltdown seemed inevitable.

Too much of anything loses effect. An unbroken string of long or short sentences can bore and confuse readers; so too can a series with identical openings:

Dreary	There are some drawbacks about diesel engines. **They** are difficult to start in cold weather. **They** cause vibration. **They** also give off an unpleasant odor. **They** cause sulfur dioxide pollution.
Varied	Diesel engines have some drawbacks. Most obvious are their noisiness, cold-weather starting difficulties, vibration, odor, and sulfur dioxide emission.

Opening sentences repeatedly with *The, This, He, She*, or *I* creates monotony. When you write in the first person, overusing *I* makes you appear self-centered.

Do not, however, avoid personal pronouns if they make the writing more readable (say, by eliminating passive constructions). Instead, to avoid repetitious openings, combine ideas and shift word order.

---

[1]My thanks to Professor Edith K. Weinstein, University of Akron, for suggesting this distinction.

Use Short Sentences for Special Emphasis

With all this talk about combining ideas, you might conclude that short sentences have no place in good writing. Wrong.

Whereas long sentences show connections and clarify relationships, short sentences (even one-word sentences) isolate an idea for special emphasis. They stick in a reader's mind.

EXERCISES IN REVISING FOR FLUENCY

1. The sentences sets below lack fluency because they are disconnected, have no variety, or have no emphasis. Combine each set into *one* or *two* fluent sentences.

> **Choppy**　The world's forests are now disappearing. The rate of disappearance is 18 to 20 million hectares a year (an area half the size of California). Most of this loss occurs in humid tropical forests. These forests are in Asia, Africa, and South America.
>
> **Revised**　The world's forests are now disappearing at the rate of 18 to 20 million hectares a year (an area half the size of California). Most of the loss is occurring in the humid tropical forests of Africa, Asia, and South America.[2]

a. The world's population will grow.
It will grow from 4 billion in 1975.
It will reach 6.5 billion in 2000.
This will be an increase of more than 50 percent.

b. In sheer numbers, population will be growing.
It will be growing faster in 2000 than it is today.
It will add 100 million people each year.
This figure compares with 75 million in 1975.

c. Energy prices are expected to increase.
Many less-developed countries will have increasing difficulty.
Their difficulty will be in meeting energy needs.

d. One-quarter of humanity depends primarily on wood.
They depend on wood for fuel.
For them, the outlook is bleak.

[2]Sample sentences are adapted from *Global Year 2000 Report to the President: Entering the 21st Century.* Washington: GPO, 1980.

e. The world has finite fuel sources.
These include coal, oil, gas, oil shale, and uranium.
These resources theoretically are sufficient for centuries.
These resources are not evenly distributed.

f. Already the populations in parts of Africa and Asia have exceeded the carrying capacity of the immediate area.
This overpopulation has triggered erosion.
This erosion has reduced the land's capacity to support life.

2. Combine each set of sentences below into one fluent sentence that provides the requested emphasis.

Sentence set	John is a loyal employee. John is a motivated employee. John is short-tempered with his colleagues.
Combined for positive emphasis	Even though John is short-tempered with his colleagues, he is a loyal and motivated employee.
Sentence set	This word processor has many features. It includes a spelling checker. It includes a thesaurus. It includes a grammar checker.
Combined to emphasize thesaurus	Among its many features, such as spelling and grammar checkers, this word processor includes a thesaurus.

a. The job offers an attractive salary.
It demands long work hours.
Promotions are rapid.
(Combine for negative emphasis.)

b. The job offers an attractive salary.
It demands long work hours.
Promotions are rapid.
(Combine for positive emphasis.)

c. Our office software is integrated.
It has an excellent database management program.
Most impressive is its word processing capability.
It has an excellent spreadsheet program.
(Combine to emphasize the word processor.)

d. Company X gave us the lowest bid.
Company Y has an excellent reputation.
(Combine to emphasize Company Y.)

e. Superinsulated homes are energy efficient.
Superinsulated homes create a danger of indoor air pollution.
The toxic substances include radon gas and urea formaldehyde.
(Combine for a negative emphasis.)

f. Computers cannot *think* for the writer.
Computers eliminate many mechanical writing tasks.
They speed the flow of information.
(Combine to emphasize the first assertion.)

FINDING THE RIGHT WORDS

Too often, language can be a vehicle for *camouflage* rather than communication. People see many reasons to hide behind language, as when they

- speak for their company but not for themselves
- fear the consequences of giving bad news
- are afraid to disagree with company policy
- make a recommendation some readers will resent
- worry about making a bad impression
- worry about being wrong
- pretend to know more than they do
- avoid admitting a mistake or ignorance

Inflated and unfamiliar words, borrowed expressions, and needlessly technical terms are a few of the poor word choices that camouflage meaning. Whether intentional or not, poor word choices have only one result: inefficient and often unethical writing that resists interpretation and frustrates the reader.

Following are strategies for finding words that are *convincing, precise,* and *informative.*

Use Simple and Familiar Words

If technical vocabulary is essential in your work, by all means use it. Don't replace technically precise words with nontechnical words that are vague or imprecise. Don't write *a part that makes the computer run* when you mean *central processing unit.* Use the precise term, and define it for a nontechnical audience:

> **Correct** Central processing unit: the part of the computer that controls information transfer and carries out arithmetic and logical instructions.

While the technical words may be indispensable, the nontechnical words usually can be simplified without any loss in meaning. Instead of *answering in the affirmative, say yes*; or instead of *endeavoring to promulgate* a new policy, try to *announce* it.

Don't use three syllables when one will do. Generally, trade for less:

aggregate	=	total
approximately	=	roughly
demonstrate	=	show
effectuate	=	cause
endeavor	=	effort
eventuate	=	result
frequently	=	often
initiate	=	begin
is contingent upon	=	depends on
multiplicity of	=	many
optimum	=	best
subsequent to	=	after
utilize	=	use

Count the syllables. Trim whenever you can. Most important, choose words you hear and use in everyday speech—words familiar to all of us.

Don't write *I deem* when you mean *I think*, or *keep me apprised* instead of *keep me informed*, or *I concur* instead of *I agree*, or *securing employment* instead of *finding a job*, or *it is cost prohibitive* instead of *we can't afford it*. Experiments have shown that readers have to spend extra time on passages that contain unfamiliar or less familiar words (Bailey 70; Felker et al. 61).

Don't write like the author of a report from the Federal Aviation Administration, who recommended that DC-10 manufacturers re-evaluate *the design of the entire pylon assembly to minimize design factors which are resulting in sensitive and/or critical maintenance and inspection procedures* (25 words, 50 syllables). A plain English translation: *Redesign the pylons so they are easier to maintain and inspect* (11 words, 18 syllables).

Besides the annoyance they cause, needlessly big or unfamiliar words can be *ambiguous*.

Ambiguous   Make an improvement in the clerical situation.

Should we hire more secretaries or better secretaries? Words chosen to impress readers too often confuse them instead. A plain style is more persuasive because "it leaves no one out" (Cross 6).

Of course, now and then the fancier or more impressive word is best—if it expresses your exact meaning. For instance, we would not substitute *end* for *terminate* in referring to something with an established time limit.

> **Correct**   Our trade agreement terminates this month.

If a fancy word can replace a handful of simpler words—and can sharpen your meaning—use the fancy word.

> **Weak**   Six rectangular grooves **around the outside edge** of the steel plate **are needed for** the pressure clamps **to fit into**.
>
> **Informative and precise**   Six rectangular grooves on the steel plate **perimeter accommodate** the pressure clamps.
>
> **Weak**   We need a **one-to-one exchange of ideas and opinions**.
>
> **Informative and precise**   We need a **dialogue**.
>
> **Weak**   Sexist language **contributes to the ongoing prevalence of** gender stereotypes.
>
> **Informative and precise**   Sexist language **perpetuates** gender stereotypes.

Use the fancy word *only* when your meaning and audience demand it.

Avoid Useless Jargon

Every profession has its own "shorthand." Among specialists, technical terms are a precise and economical way to communicate. On page 20, *stat* is medical jargon for *Drop everything and deal with this emergency.* For computer buffs, a *glitch* is a momentary power surge that can erase the contents of internal memory; a *bug* is an error that causes a program to run incorrectly. Such useful jargon conveys clear meaning to a knowledgeable audience.

Technical language, however, can be used appropriately or inappropriately. The latter is useless jargon, meaningless to insiders as well as outsiders.

In the world of useless jargon people don't *cooperate* on a project; instead, they *interface* or *contiguously optimize their efforts*. Rather than *designing a model*, they *formulate a paradigm*. Instead of *observing limits* or *boundaries*, they *function within specific parameters*.

A popular form of useless jargon is adding *-wise* to nouns, as shorthand for *in reference to* or *in terms of*.

> **Useless jargon** **Expensewise** and **schedulewise,** this plan is unacceptable.
>
> **Revised** In terms of expense and scheduling, this plan is unacceptable.

Writers create another form of jargon when they invent verbs from nouns or adjectives by adding an *-ize* ending: Don't invent *prioritize* from *priority*; instead use *to rank priorities*. Don't write *deambiguize* when you mean *clarify*, or *finalize* when you mean to *settle on*.

Jargon's worst fault is that it makes the writer seem stuffy and pretentious:

> **Pretentious** Unless all parties interface synchronously within given parameters, the project will be rendered inoperative.
>
> **Possible translation** Unless we coordinate our efforts, the project will fail.

Beyond reacting with frustration, readers often conclude that useless jargon is intended as camouflage by a writer who has something to hide.

Before using any jargon, think about your specific readers and ask yourself: "Can I find an easier way to say exactly what I mean?" Use jargon only if it *improves* your communication.

Use Acronyms Selectively

Acronyms are another form of specialized shorthand, or jargon. They are formed from the first letters of words in a phrase (as in *LOCA* from *loss of coolant accident*) or from a combination of first letters and parts of words (as in *bit* from *binary digit* or *pixel* from *picture element*).

Computer technology has spawned countless acronyms, including

DOS = disk operating system

RAM = random-access memory

VDT = video display terminal

Acronyms *can* communicate concisely—but only when the audience already knows their meaning, and only when you use the term often in your document. Always define the acronym (in parentheses) on first use. For lay audiences, try to avoid acronyms altogether.

Avoid Triteness

Writers who rely on tired old phrases (clichés) seem either too lazy or careless to find convincing or exact ways to say what they mean. Here are just a few of the countless expressions worn out by overuse:

make the grade	the chips are down
in the final analysis	not by a long shot
close the deal	last but not least
in-depth study	welcome aboard
water under the bridge	over the hill
holding the bag	bite the bullet
up the creek	work like a dog

If it sounds like a "catchy phrase" you've heard often, don't use it.

Avoid Misleading Euphemisms

Euphemisms are expressions aimed at politeness or at making unpleasant subjects seem less offensive. Thus, we *powder our nose* or *use the boys' room* instead of *using the bathroom*; we *pass away* or *meet our Maker* instead of *dying*. Euphemisms make the truth seem less painful.

When euphemisms avoid offending or embarrassing our audience, they are perfectly legitimate. Instead of telling a job applicant he or she is *unqualified*, we might say, *Your background doesn't meet our needs*. In addition, there are times when friendliness and interoffice harmony are more likely to be preserved with writing that is not too abrupt, bold, blunt, or emphatic (Mackenzie 2).

Euphemisms, however, are unethical if they understate the truth when only the truth will serve. In the sugar-coated world of misleading euphemisms, bad news disappears:

- Instead of being *laid off* or *fired,* workers are *surplused* or *deselected,* or the company is *downsized*.

- Instead of *lying* to the public, the government *engages in a policy of disinformation*.

- Instead of *wars* and *civilian casualties*, we have *conflicts* and *collateral damage*.

Language loses all meaning when *criminals* become *offenders*, when *rape* becomes *sexual assault*, and when people who are just plain *lazy* become *underachievers*. Plain talk is always better than deception. If someone offers you a job *with limited opportunity for promotion*, expect a *dead-end job*.

Avoid Overstatement

When they exaggerate to make a point, writers lose credibility. Be cautious when using words such as *best, biggest, brightest, most,* and *worst.*

Overstated	**Most** businesses have **no** loyalty toward their employees.
Revised	**Some** businesses have **little** loyalty toward their employees.
Overstated	You will find our product to be the **best**.
Revised	You will **appreciate the high quality** of our product.

Avoid Unsupported Generalizations

Base your conclusions on sufficient evidence.

Unsupported generalization	In 1983, twenty-one murderers were executed, and the murder rate dropped 8.1 percent for that year. These figures prove that the death penalty should be reinstated.

This isolated piece of evidence does not justify such a sweeping generalization.

Reasonable generalization	In 1983, twenty-one murderers were executed, and the murder rate dropped 8.1 percent for that year. These figures suggest that the death penalty may deter violent crimes.

Unsupported generalizations harm your credibility because they have no way of being proved.

Be aware of the vast differences in meaning among these words:

few	never
some	rarely
many	sometimes

most	often
all	always

Unless you specify *few, some, many,* or *most,* readers can interpret your statement as meaning *all.*

> **Misleading**   Assembly-line employees are doing shabby work.

Unless you mean *all* assembly-line employees, qualify your generalization with *some, most,* or another limiting word—even better, specify *20 percent.*

Avoid Imprecise Words

Poorly chosen words can be offensive, embarrassing, misleading, or ambiguous. Be sure that what you say is what you mean.

Even words listed as synonyms carry different shades of meaning. Do you mean to say *I'm slender, You're thin, She's lean,* or *He's scrawny*? The wrong choice could be disastrous.

Just one wrong word can offend readers, as in this statement by a job applicant:

> **Offensive**   Another attractive feature of your company is its **adequate** training program.

While "adequate" might convey honestly the writer's intended meaning, the word seems inappropriate in this context (i.e., an applicant expressing a judgment about a program). Although the program may not have been highly ranked, our writer could have used any of several alternatives (*solid, respectable, growing*—or no modifier at all) without overstating the point or being offensive.

Poor word choice can be embarrassing, as in this example:

> **Imprecise**   **Chaos** is running this project.

Strictly speaking, *chaos* can't run anything!

Be especially aware of similar words with dissimilar meanings, as in these examples:

affect/effect	farther/further
all ready/already	fewer/less
almost dead/dying	healthy/healthful
among/between	imply/infer

continual/continuous   invariably/inevitably

eager/anxious   uninterested/disinterested

fearful/fearsome   worse/worst

Don't write *Skiing is healthy* when you mean that skiing promotes good health. Healthful things keep us healthy.

Be on the lookout for imprecisely phrased (and therefore illogical) comparisons.

> **Imprecise**   Your bank's interest rate is higher than Citibank. (Can a rate be higher than a bank?)
>
> **Precise**   Your bank's interest rate is higher than Citibank's.

Precision is above all essential to the informative value of your writing. Imprecise language can be misleading. Consider how meanings differ:

> **Differing**   Include **less** technical details in your report.
>
> **meanings**   Include **fewer** technical details in your report.

Imprecise language can result in ambiguous messages as well:

> **Ambiguous**   Loan payments are due **bimonthly**. (Every other month or twice monthly?)
>
> **Clear**   Loan payments are due twice monthly
>
> *or*
>
> Loan payments are due every other month.

Precision ultimately enhances conciseness, when one exact word replaces multiple inexact words.

> **Wordy and less**   I have **gotten together** all the financial information.
>
> **exact**   **Keep doing** this exercise for ten seconds.
>
> **Concise and**   I have **assembled** all the. . . .
>
> **more exact**   **Continue** this exercise. . . .

Be Specific and Concrete

General words name broad classes of things, such as *job, computer,* or *person.* Such words usually need to be clarified by more specific ones.

> job   =   senior accountant for Softbyte Press

```
computer   =   Macintosh SE, with hard disk

person     =   Sarah Jones, production manager
```

The more specific your words, the sharper your meaning.

General	structure
	dwelling
	vacation home
	log cabin
Specific	log cabin in Vermont
	a three-room log cabin on the banks of the Battenkill River in Vermont

Notice how the picture becomes more vivid as we move to lower levels of generality.

Abstract words name qualities, concepts, or feelings (*beauty, luxury, depression*) whose exact meaning has to be nailed down by *concrete* words—words that name things we can know through our five senses.

a **beautiful** view	=	snowcapped mountains, a wilderness lake, pink granite ledge, ninety-foot birch trees
a **luxurious** condominium	=	imported tiles, glass walls, oriental rugs
a **depressed** worker	=	suicidal urge, insomnia, feelings of worthlessness, no hope for improvement

Informative writing *shows* and *tells*.

> **General**   One of our **workers** was **injured** by a **piece of equipment recently**.

The boldface words only *tell* without showing.

> **Specific**   **Alan Hill** suffered a **broken thumb** while working on a **lathe yesterday**.

Choose high-information words that express exactly what you mean. Don't write *thing* when you mean *lever, switch, micrometer,* or *disk*. Instead of evaluating an employee as *good, great, disappointing,* or *terrible*, use informative words such as *reliable, skillful, dishonest,* or *incompetent*—and give examples.

In some instances, of course, you may wish to generalize for the sake of diplomacy. Instead of writing *Bill, Mary, and Sam have been tying up the office phones with personal calls,* you might prefer to generalize: *Some employees have been tying up. . . .* The second version makes the point without accusing anyone in particular.

When you can, provide solid numbers and statistics that get your point across:

General	In 1972, thousands of people were killed or injured on America's highways. Many families had at least one relative who was a casualty. After the speed limit was lowered to 55 miles per hour in late 1972, the death toll began to drop.
Specific	In 1972, 56,000 people died on America's highways; 200,000 were injured; 15,000 children were orphaned. In that year, if you were a member of a family of five, chances are that someone related to you by blood or law was killed or injured in an auto accident. After the speed limit was lowered to 55 miles per hour in late 1972, the death toll dropped steadily to 41,000 in 1975.

Concrete and specific information not only is more informative; it is more persuasive as well.

Use Analogies to Sharpen the Image

Ordinary comparison shows similarities between two things *of the same class* (two computer keyboards, two technicians, two methods of cleaning dioxin-contaminated sites). Analogy, on the other hand, shows some essential similarity between two things of *different classes* (report writing and computer programming, computer memory and post office boxes).

Analogies are good for emphasizing a point *(Some rain is now as acidic as vinegar)*. But they are especially useful in translating something abstract, complex, or unfamiliar, as long as the easier subject is broadly familiar to readers. Analogy therefore calls for particularly careful analyses of audience.

Analogies can save words and convey vivid images. *Collier's Encyclopedia* describes the tail of an eagle in flight as "spread like a fan." The following sentence from a description of a trout feeder mechanism uses this analogy to clarify the positional relationship between two working parts:

Analogy	The metal rod is inserted (and centered, **crosslike**) between the inner and outer sections of the clip. . . .

Without the analogy *crosslike*, we would need something like this to visualize the relationship:

> **Missing analogy**   The metal rod is inserted, **perpendicular to the long plane and parallel to the flat plane**, between the inner and outer sections of the clip. . . .

This second version is doubly inefficient: more words are needed to communicate, and more work is needed to understand the meaning.[3]

Besides naming things vividly, analogies help *explain* things. The following extended analogy from the *Congressional Research Report* helps us understand something unfamiliar (dangerous levels of a toxic chemical) by comparing it to something more familiar (human hair).

> **Analogy**   A dioxin concentration of 500 parts per trillion is lethal to guinea pigs. One part per trillion is roughly equal to the thickness of a human hair compared to the distance across the United States.

EXERCISES IN REVISING FOR EXACTNESS

1. Revise each sentence below for straightforward and familiar language.

> **Big words**   I wish to upgrade my present employment situation.
>
> **Revised**   I want a better job.
>
> **Unfamiliar words**   Acoustically attenuate the food-consumption area.
>
> **Revised**   Soundproof the cafeteria.

a. May you find luck and success in all endeavors.

b. I suggest you reduce the number of cigarettes you consume.

c. Within the copier, a magnetic-reed switch is utilized as a mode of replacement for the conventional microswitches that were in use on previous models.

d. A good writer is cognizant of how to utilize grammar in a correct fashion.

e. I will endeavor to ascertain the best candidate.

f. In view of the fact that the microscope is defective, we expect a refund of our full purchase expenditure.

[3]Analogy is, of course, a form of metaphor. For an inspired discussion of metaphor, see John S. Harris. "Metaphor in Technical Writing." *Technical Writing Teacher* 2 (1975): 9–13.

2. Revise each sentence below to eliminate useless jargon and triteness.

> **Useless jargon**   Intercom utilization will be employed to initiate substitute employee operative involvement.
>
> **Revised**   Employees who are asked to substitute will be notified on the intercom.
>
> **Triteness**   Managers who make the grade are those who can take daily pressures in stride.
>
> **Revised**   Successful managers are those who cope with daily pressures.

    a. For the obtaining of the X-33 word processor, our firm will have to accomplish the disbursement of funds to the amount of $6,000.

    b. To optimize your financial return, prioritize your investment goals.

    c. The use of this product engenders a 50-percent repeat consumer encounter.

    d. We'll have to swallow our pride and admit our mistake.

    e. We wish to welcome all new managers aboard.

    f. Not by a long shot will this plan succeed.

3. Revise each sentence below to eliminate euphemism, overstatement, or unsupported generalizations.

> **Euphemism**   Because of your absence of candor, we can no longer offer you employment.
>
> **Revised**   Because of your dishonesty, you're fired.
>
> **Overstatement**   Igor is the world's best employee.
>
> **Revised**   Igor is the finest employee I've had.
>
> **Unsupported generalization**   Television is making students nothing but illiterates.
>
> **Revised**   Television seems to negatively affect many students' reading ability.

    a. I finally must admit that I am an abuser of intoxicating beverages.

    b. I was less than candid.

    c. This employee is poorly motivated.

    d. Your faulty machinery traumatically amputated my client's arm.

    e. Most entry-level jobs are boring and dehumanizing.

    f. Clerical jobs offer no opportunity for advancement.

4. Revise each sentence below to make it more precise or informative.

> Imprecise   Push the printer connector into the serial socket.
>
> Revised   Insert the printer connector into the serial socket.
>
> Too general   Industrial emissions are causing atmospheric effects on certain bodies of water. Lakes in particular are affected. Many lakes in fact are now dead.
>
> Revised   Sulfur dioxide emissions from coal-burning plants combine with atmospheric water to produce sulfuric acid. The resultant "acid rain" so increases the acidity of lakes that they no longer can support life.

    a. Our outlet does more business than Chicago.

    b. Anaerobic fermentation is used in this report.

    c. Confusion is in control of this office.

    d. Your crew damaged a piece of office equipment.

    e. His performance was admirable.

    f. This thing bothers me.

ADJUSTING YOUR TONE

Your tone is your personal stamp—the personality that takes shape between the lines. The tone you create in any writing depends on (1) the distance you impose between yourself and the reader, and (2) the attitude you show toward the subject.

Assume that a friend is going to take over a job you've held. You've decided to write instructions for your friend. Here is your first sentence:

> Now that you've arrived in the glamorous world of office work, put on your track shoes; this is no ordinary manager-trainee job.   *Informal*

What is the tone in this sentence? First, we notice that the writer imposes little distance between herself and the reader (she uses the direct address, "you," and the humorous suggestion to "put on your track shoes"). And the ironic use of "glamorous" suggests just the opposite—that the job holds little glamour.

For a different reader (say, the recipient of a company training manual), the writer would have chosen some other way to open:

> As a manager trainee at GlobalTech, you will work for many managers. In short, you will spend little of your day seated at your desk.   *Semiformal*

The tone has changed; it is no longer intimate, and the writer expresses no distinct attitude toward the job. For yet another audience (say, those who will read a company pamphlet for clients or investors), the writer might have altered her tone again:

> Manager trainees at GlobalTech are responsible for duties that extend far beyond desk work.

Here the businesslike impersonal tone imposes more distance between writer and audience, especially with the shift from second- to third-person address. The tone is far too impersonal for any document addressed to the trainees themselves.

Your tone changes in response to the situation and audience, even if the subject remains the same. When you are introduced to someone, for example, you respond in a tone that defines your relationship:

- Honored to make your acquaintance. [*formal tone—greatest distance*]

- How do you do? [*formal*]

- Nice to meet you. [*semiformal—medium distance*]

- Hello. [*semiformal*]

- Hi. [*informal—least distance*]

- What's happening? [*informal—slang*]

Your greeting (and tone) will depend on how much distance you decide is appropriate, and, in turn, the tone will determine how you come across. Each of these greetings is appropriate in some situations, inappropriate in others. "Hi" might be okay when you are introduced to a new friend, but not the college or company president.

To decide on an appropriate distance from which to address a particular audience, follow these guidelines:

- Use a formal or semiformal tone in writing for superiors, professionals, or academics (depending on what you think the reader expects).

- Use a semiformal or informal tone in writing for colleagues and subordinates (depending on how close you feel to your readers).

- Use an informal tone when you want your writing to be conversational, or when you want it to sound like a person talking.

Whichever tone you choose, be consistent throughout your document.

Inconsistent tone	My office isn't fit for a pig [*too informal*]; it is ungraciously unattractive [*too formal*].
Revised	The shabbiness of my office makes it an unfit place to work.

In general, lean toward an informal tone without falling into slang.

In addition to setting the distance between writer and reader, your tone implies your *attitude* toward the subject. Consider the different attitudes expressed in these examples:

- We dine at seven.
- Dinner is at seven.
- Let's eat at seven.
- Let's chow down at seven.
- Let's strap on the feedback at seven.
- Let's pig out at seven.

The words we choose tell readers a great deal about where we stand. If readers expect an impartial report, try to keep your own biases out of it. But for situations in which your opinion *is* expected, or in which you perceive some danger or ethics violation, speak up and let readers know where you stand.

Say *I enjoyed the fiber optics seminar* instead of *My attitude toward the fiber optics seminar was one of high approval*. Say *Let's liven up our dull relationship* instead of *We should inject some rejuvenation into our lifeless liaison*. Say *Methane levels in No. 3 mineshaft pose the definite risk of explosion* instead of *Rising methane levels in No. 3 mineshaft should be evaluated*. Make sure your attitude is clear and appropriate for the situation.

Let your attitude reflect your relationship with the reader—and what you think the reader expects. In an upcoming meeting about the reader's job evaluation, does your reader expect to *discuss* the evaluation, *talk it over, have a chat*, or *chew the fat*? If the situation calls for a serious tone, don't use language that suggests a casual attitude—or vice versa. Use the following guidelines for making your tone conversational and appropriate.

Use Occasional Contractions

Unless you have good reason to be formal, use (but do not overuse) contractions to loosen the tone. Balance an *I am* with an *I'm*, a *you are* with a *you're*, an *it is* with an *it's* (as we've done throughout this book).

Generally, use contractions only with pronouns, not with nouns or proper nouns (names). Otherwise, the constructions are awkward or ambiguous.

Awkward contractions	Barbara'll be here soon.
	Health's important.
Ambiguous contractions	The dog's barking.
	Bill's skiing.

These ambiguous contractions can easily be confused with possessive constructions.

Address Readers Directly

Use the personal pronouns *you* and *your* to create contact with readers. Otherwise, your writing sounds impersonal.

Impersonal tone	A writer should use **you** and **your** generously to create contact with his or her readers.

Notice how distance *and* words increase. Direct address creates a more personal tone.

Impersonal tone	Students at our college will find the faculty always willing to help.
Personal tone	As a student at our college, **you** will find the faculty always willing to help.

Research shows that readers relate better to something they consider meaningful to them (Felker 33).

Caution: Use *you* and *your* only to correspond *directly* with the reader, as in a letter, memo, instructions, or some form of advice, encouragement, or persuasion. By using *you* and *your* when your subject and purpose call for first or third person, you might write something wordy and awkward like this:

Wordy and awkward	**When you** are in northern Ontario, **you** can see wilderness lakes everywhere around **you**.
Appropriate	Wilderness lakes are everywhere in northern Ontario.

Use *I* and *We* When Appropriate

Don't disappear behind your writing. Use *I* or *We* when referring to yourself or your organization.

Distant	This writer would like a refund.
Revised	I would like a refund.

A message can become doubly impersonal when both the writer and the reader disappear.

| Impersonal | The requested report will be sent next week. |
| Personal | **We** will send the report **you** requested next week. |

Avoid, of course, opening too many sentences with *I*. Combine ideas and shift word order instead.

Prefer the Active Voice

Because the active voice is more direct and economical than the passive voice, it generally creates a more personal and less formal tone. (Review pages 256–59.)

| Passive and impersonal | Travel expenses cannot be reimbursed unless receipts are submitted. |
| Active and personal | We cannot reimburse your travel expenses unless you submit receipts. |

Emphasize the Positive

Readers respond more favorably to encouragement than to criticism. When you give advice, suggestions, or recommendations, try to emphasize benefits rather than flaws.

| Critical tone | Because of your division's lagging productivity, a management review may be needed. |
| Encouraging tone | A management review might help boost productivity in your division. |

Avoid Bias

Biased people make judgments without examining the facts. Your responsibility as a writer is to report accurately and fairly without distorting the evidence or injecting "loaded" words that reflect personal biases. If you *are* asked to include your interpretations and conclusions, base them on the facts. Even controversial subjects deserve unbiased treatment.

Imagine you have been sent to investigate the causes of an employee-management confrontation at your company's Omaha branch. Your initial report, written for the New York central office, is intended simply to describe what happened. Here is how an unbiased description might read:

> At 9:00 A.M. on Tuesday, January 21, eighty women employees entered the executive offices of our Omaha branch and remained six hours, bringing business

A factual account

to a virtual halt. The group issued a formal statement of protest, claiming their working conditions were repressive, their salary scale was unfair, and their promotional opportunities were limited. The women demanded affirmative action, insisting that the company's hiring and promotional policies and wage scales be revised. The demonstration ended when Garvin Tate, vice president in charge of personnel, promised to appoint a committee to investigate the group's claims and to correct any inequities.

Notice the absence of implied judgments; the facts are presented objectively. A less impartial version of the event, from a protester's point of view, might read:

A biased version

> Last Tuesday, sisters struck another blow against male supremacy when eighty women employees paralyzed the pin-striped world of solidly entrenched sexism for more than six hours. The timely and articulate protest was aimed against degrading working conditions, unfair salary scales, and lack of promotional opportunities. Stunned executives watched helplessly as the group occupied their offices. The women were determined to continue their occupation until their demands for equal rights were met. Embarrassed company officials soon perceived the magnitude of this protest action and agreed to study the group's demands and to revise the company's discriminatory policies. The success of this long overdue confrontation serves as an inspiration to oppressed women employees everywhere.

Judgmental words and qualifiers (*male supremacy, degrading, paralyzed, articulate, stunned, discriminatory*) inject the writer's attitude, even though it isn't called for. In contrast to this bias, the following version patronizingly defends the status quo:

A biased version

> Our Omaha branch was the scene of an amusing battle of the sexes last Tuesday, when a Women's Lib group, eighty strong, staged a six-hour sit-in at the company's executive offices. The protest was lodged against alleged inequities in hiring, wages, working conditions, and promotion for women in our company. The radicals threatened to remain in the building until their demands for "equal rights" were met. Bemused company officials responded to this carnival display with patience and dignity, assuring the militants that their claims and demands—however inaccurate and immoderate—would receive just consideration.

Again, qualifying adjectives and superlatives slant the tone.

Being unbiased, of course, doesn't mean burying your head—and your values—in the sand. Remaining "neutral" about something you know to be wrong or dangerous is unethical (Kremers 59). You have an ethical responsibility to weigh the facts and make your views known. If, for instance, you conclude that the Omaha protest was clearly justified, don't hesitate to say so.

Avoid Sexist Usage

The way we use language as a culture reflects the way we think about ourselves. Usage that gives people in general a male identity allows no room for females. In fact, females become virtually invisible in a world of *policemen, congressmen, firemen, foremen, selectmen,* and *aldermen.*

Sexist usage refers to doctors, lawyers, and other professionals as *he* or *him,* while referring to nurses, secretaries, and homemakers as *she* or *her.* In this traditional stereotype, males do the jobs that really matter and that pay higher wages, whereas females serve only as support and decoration. And when females do invade traditional "male" roles, we might express our surprise at their boldness by calling them *female executives, female sportscasters, female surgeons,* or *female hockey players.* Likewise, to demean males who have settled for "female" roles, we sometimes refer to *male secretaries, male nurses, male flight attendants,* or *male models.*

In the biased reality of sexist usage, the title "Mr." protects the privacy of a male who might be married or unmarried, while "Mrs." and "Miss" announce a female's marital status to the world. Moreover, an unmarried male is fondly referred to as *bachelor,* while his female counterpart is stigmatized as an *old maid* or a *spinster.*

Besides being misleading and demeaning, sexist usage is offensive. Instead of bringing writer and readers close together, sexist usage severs the respect for humans that our writing might otherwise serve to achieve.

To eliminate sexist usage from your writing, follow these guidelines:

- Use neutral expressions:

chair, or chairperson	rather than	chairman
businessperson	rather than	businessman
supervisor	rather than	foreman
police officer	rather than	policeman
letter carrier	rather than	postman
homemaker	rather than	housewife
humanity, or humankind	rather than	mankind
actor	rather than	actor vs. actress

- Rephrase to eliminate the pronoun, if you can do so without changing your original meaning.

Sexist	A writer will succeed if **he** revises.
Revised	A writer who revises succeeds.

- Use plural forms. Instead of *Each doctor . . . he,* use *All doctors . . . they* (but not *Each doctor . . . they*).

Sexist	A writer will succeed if **he** revises.
Revised	Writers will succeed if **they** revise (but *not* A writer will succeed if **they** revise.)

Note: When using a plural form, don't create an error in pronoun-referent agreement by having the *plural* pronoun *they* or *their* refer to a *singular* referent (as in **Each writer** *should do* **their** *best*).

- When possible (as in direct address) use *you:* **You** *will succeed if* **you** *revise.* But use this form *only* when addressing someone directly. (See page 294 for further discussion.)

- Use occasional pairings (*him* or *her, she* or *he, his* or *hers, he/she*): *A writer will succeed if* **she or he** *revises.*
  Note: Overuse of such pairings can be awkward: *A writer should do* **his or her** *best to make sure that* **he or she** *connects with* **his or her** *readers.* Most handbooks now encourage alternating use between the two pronouns, and discourage pairings and *he/she: An effective writer always focuses on* **her** *audience; The writer strives to connect with all* **his** *readers.*

- Drop diminutive endings such as *-ess* and *-ette* used to denote females (*poetess, drum majorette, actress,* etc.). Such endings seem to perpetuate an image of *the little woman.*

- Use *Ms.* instead of *Mrs.* or *Miss,* unless you know that person prefers one of the traditional titles. Or omit titles completely: *Roger Smith* and *Jane Kelly; Smith* and *Kelly.*

Not only do the words you choose reveal your way of seeing, but they also influence your reader's way of thinking. Sexist language carries built-in judgments, and, as renowned linguist S. I. Hayakawa reminds us: "Judgment stops thought." In a world unethically defined by sexist language, everyone remains frozen in her and his place.

EXERCISES IN ADJUSTING TONE

1. The sentences below have inappropriate tone because of pretentious language, missing contractions, or indirect address. Adjust the tone.

Plain English needed	Avoid prolix nebulosity.
Revised	Don't be wordy and vague.
No contractions	Do not be wordy and vague.
Revised	Don't be wordy and vague.
Indirect address	A writer shouldn't be wordy and vague.
Revised	Don't be wordy and vague.

a. Further interviews are a necessity to our ascertaining the most viable candidate.

b. This project is beginning to exhibit the characteristics of a loser.

c. We are pleased to tell you that you are a finalist.

d. Do not submit the proposal if it is not complete.

e. Employees must submit travel vouchers by May 1.

f. Persons taking this test should use the HELP option whenever they need it.

2. These sentences have too few *I* or *We* constructions, too many passive constructions, or express an unclear attitude. Adjust the tone.

No *I*	This writer would like to be considered for your opening.
Revised	Please consider me for your opening.
Passive	Your request will be given our consideration.
Revised	We will consider your request.
Unclear attitude	My disapproval is far more than negligible.
Revised	I strongly disapprove.

a. Payment will be made as soon as an itemized bill is received.

b. You will be notified.

c. Your help is appreciated.

d. Our reply to your bid will be sent next week.

e. I am not unappreciative of your help.

f. My opinion of this proposal is affirmative.

3. The sentences below suffer from negative emphasis, biased expressions, or sexist language. Adjust the tone.

Negative emphasis	Aggressive management of this risky project will help you avoid failure.
Revised	Aggressive management of this risky project will increase the chance for success.
Biased	Compare our product with its lesser competitors.
Revised	Compare our product with its competitors.
Sexist	While the girls played football, the men waved pompoms.
Revised	While the women played football, the men waved pompoms.

a. If you want your workers to like you, show sensitivity to their needs.

b. By not hesitating to act, you prevented my death.

c. The union has won its struggle for a decent wage.

d. The group's spokesman demanded salary increases.

e. Each employee should submit his vacation preferences this week.

f. Each applicant must prove that he or she has received his or her state certification.

AVOIDING RELIANCE ON AUTOMATED TOOLS

Many of the strategies covered in this chapter could be executed rapidly with good word-processing software. By using the *global search-and-replace function* in some programs, you can command the computer to search for ambiguous pronoun references, overuse of passive voice, *to be* verbs, *There* and *It* sentence openers, negative constructions, clutter words, needless prefaces and qualifiers, overly technical language, jargon, sexist language, and so on. With an online dictionary or thesaurus, you can check definitions or see a list of synonyms for a word you have used in your document.

Despite the increasing sophistication of style checkers, diction checkers, and other such editing aids, they can be extremely imprecise. No amount of automation can eliminate the writer's burden of *choice*. None of the "rules" or advice offered in this chapter applies universally. Our language confronts us with almost infinite choices that cannot be programmed into a computer. Ultimately, it is the informed writer's sensitivity to meaning, emphasis, and tone—the human contact—that determines the effectiveness of any document.

Graphic and Design Elements

IV

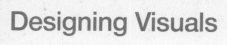

Designing Visuals

Purpose of Visuals

Tables

Graphs

Charts

Illustrations

Computer Graphics

Guidelines for Visuals

■   ■   ■

A visual is any pictorial representation used to simplify or emphasize material.

Besides saving space and words, visuals help audiences process, understand, and remember information. Because they offer powerful new ways of looking at data, visuals reveal trends, problems, and possibilities that might otherwise remain buried in lists of facts and figures. In written reports and oral presentations alike, effective visuals help the writer or presenter appear prepared, credible, and persuasive.

PURPOSE OF VISUALS

We know that readers expect more than just raw information; they want the information "processed" for their understanding. Readers themselves like to feel smart, to understand the message at a glance. More receptive to images than words, today's readers resist uninterrupted pages of printed words. Visuals break down a reader's resistance in several ways:

- *Visuals provide emphasis.* To show that the average personal computer has dropped in price 60 percent in four years, a bar graph would be more vivid than a prose statement.

- *Visuals display abstract concepts in concrete, geometric shapes.* ("How do lasers work?" "How does the AIDS virus work?")

- *Visuals can compare large amounts of data.* ("How do our profits compare with our expenses?" "How do this month's sales figures compare with last month's?")

- *Visuals depict relationships.* ("How much has our productivity increased since employees received a 15 percent pay hike?" "How has our new incentive program affected employees' absences?")

- *Visuals condense information, displaying it in a meaningful way.* A simple table, for instance, can summarize a long and difficult printed passage, as in the example that follows.

- *Visuals can serve as a "universal language."* In the global workplace, carefully designed visuals can transcend cultural and language barriers, and thus facilitate international communication.

Imagine you are researching a link between chemicals and cancer deaths. From various sources, you might collect these data:

1. In 1900, pneumonia and influenza accounted for 11.8 percent of deaths, heart disease for 8.0 percent, cerebrovascular disease for 6.2 percent, all accidents for 4.2 percent, and cancer for 3.7 percent.

2. In 1960, pneumonia and influenza caused 3.5 percent of deaths, heart disease 38.7 percent, cerebrovascular disease 11.3 percent, all accidents 5.5 percent, and cancer 15.6 percent.

3. In 1970. . . .

In the form above, information is repetitious, tedious, and hard to interpret. But arranged in Table 14.1, these statistics are easy to compare.

Besides their value as presentation devices, visuals also help us analyze information. Table 14.1 is just one example of how technical people can use visuals to work with information as part of the critical thinking process.

Translate your prose into visuals whenever they make your point more clearly than the prose. But use visuals only to *clarify* and to enhance your discussion, not merely to *decorate* it.

TABLE 14.1 An Effective Table

Leading Causes of Death in 1900, 1960, and 1970

	Cause of death (as percentage of all deaths)				
Year	Pneumonia and influenza	Heart disease	Cerebrovascular disease	All accidents	Cancer
1900	11.8	8.0	6.2	4.2	3.7
1960	3.5	38.7	11.3	5.5	15.6
1970	3.3	38.3	10.8	6.0	17.2
Average	6.2	28.3	9.4	5.2	12.1

Source: Adapted from *Facts of Life and Death*. Washington: GPO, 1984: 31.

TABLES

Tables are especially useful for displaying exact quantities or for comparing large amounts of information in a small space. Large tables, however, present so much information that readers can find them confusing (Felker et al. 95). An otherwise impressive-looking table (like Table 14.2) can be difficult for some readers to interpret because it presents too much information all at once. We can see how an unethical writer could use a complex table to bury numbers that are questionable or embarrassing (Williams 12).

Use a table only when you are reasonably certain it will enlighten—rather than frustrate—your audience. For nonspecialized readers, use fewer tables, and keep them simple. Readers of any table need to understand how it is organized, where to find what they need, and how to interpret the information they find (Hartley 90).

To construct a table, use tabulating markers and the tab keys on your typewriter or word processor, and follow these suggestions:

- Number the table in its order of appearance (Table 1, Table 2) and give it a title describing exactly what the table compares or measures.
- Compare your data vertically (in columns) instead of horizontally (in rows). Readers find columns easier to scan than rows.
- Begin each column with a heading naming the category for the items listed (*Heart disease, Cancer*), and stipulate the units of measurement (*percentage of deaths, miles per gallon*).
- Convert all fractions into decimals, and align the decimals vertically. For easy comparison, round off insignificant decimals to the nearest whole number.

- Label all parts of the table clearly, so that readers will know what they are looking at.

- Space your listed items so that they are neither cramped nor too far apart for easy comparison.

- Whenever possible, include row or column averages, to give readers reference points for comparing individual figures.

- If the table is too wide for the page, turn it 90 degrees and place its top toward the inside of the binding. Or you might divide the data into two tables.

- Try to hold the table to one page. Otherwise write "continued" at the bottom, and begin the second page with the full title, "continued," and the original column headings.

- Try to avoid footnotes, unless you must explain or define entries (as in Table 14.2). If footnotes are necessary, label them with lowercase letters (a, b, c) in the table.

TABLE 14.2 A Table with Too Much Information

Air Pollutant Emissions, by Pollutant and Source: 1970 and 1983
(In millions of metric tons, except lead in thousands of metric tons. Metric ton = 1.1023 short tons)

Year and pollutant	Total emissions	Controllable emissions						Misc. uncontrollable	Percentage of total		
		Transportation		Total	Electric utilities	Industrial processes	Solid waste disposal		Transportation	Fuel combustion[a]	Industrial
		Total	Road vehicles								
1970: Carbon monoxide	98.3	71.8	62.7	3.9	.2	9.0	6.4	7.2	73.0	4.0	9.2
Sulfur oxides	28.2	.6	.3	21.3	15.8	6.2	(Z)	.1	2.1	75.5	22.0
Volatile organic compounds	27.0	12.3	11.1	.9	(Z)	8.7	1.8	3.3	45.6	3.3	32.2
Particulates[b]	18.0	1.2	.9	4.5	2.3	10.1	1.1	1.1	6.7	25.0	56.1
Nitrogen oxides	18.1	7.6	6.0	9.1	4.5	.7	.4	.3	42.0	50.3	3.9
Lead	203.8	163.6	156.0	9.6	.3	23.9	6.7	(Z)	80.3	4.7	11.7
1983: Carbon monoxide	67.6	47.7	41.2	7.0	.3	4.6	2.0	6.3	70.6	10.4	6.8
Sulfur oxides	20.8	.9	.5	16.8	14.0	3.1	(Z)	(Z)	4.3	80.8	14.9
Volatile organic compounds	19.9	7.2	6.0	2.1	(Z)	7.5	.6	2.5	18.8	29.0	37.9
Particulates[b]	6.9	1.3	1.1	2.0	.5	2.3	.4	.9	45.4	50.0	33.3
Nitrogen oxides	19.4	8.8	7.0	9.7	6.3	.6	.1	.2	45.4	50.0	3.1
Lead	46.9	40.7	38.7	.6	.1	2.5	3.1	(Z)	86.8	1.3	5.3

Z Less than 50,000 metric tons.   [a]Stationary.   [b]See footnote a, table 352.

Source: U.S. Environmental Protection Agency, *National Air Pollutant Emission Estimates, 1940–1983.* Washington, DC: GPO, 1984.

- Cite your sources of data beneath any footnotes, even if you make your own table from borrowed data.
- Include a prose explanation of the comparisons (as in Table 14.3).

Tables work well for displaying exact quantities, but for easier interpretation, readers prefer graphs or charts. Also, shapes (bars, curves, circles) are generally easier to remember than lists of numbers (Cochran et al. 25).

Any visual other than a table usually is categorized as a *figure,* and so titled (*Figure 1   Aerial View of the Panhandle Site*). Figures covered in this chapter include graphs, charts, and illustrations.

Like all other components in the document, visuals are designed with audience and purpose in mind (Journet 3). An accountant doing an audit might need a table listing exact amounts, whereas the average public stockholder reading an annual report would prefer the "big picture" in the easily grasped bar graph or pie chart (Van Pelt 1). The next section shows how the data in Table 14.3 can be presented in a number of displays, depending on the point the writer wants to make.

TABLE 14.3 Data Displayed in a Tabular Version

Death Rates for Heart Disease and Cancer, 1970–1986

	Number of Deaths (per 100,000) population[a]			
	Heart disease		Cancer	
Year	Male	Female	Male	Female
1970	419	309	172	135
1978	371	300	201	160
1986	348	291	219	173
Percentage of change, 1970–1986	−17.6	−5.8	+21.4	+22.0

[a]Figures are approximate.

Source: Adapted from *Statistical Abstract of the United States.* Washington, DC: GPO, 1986: 205.

As Table 14.3 indicates, both male and female death rates from heart disease decreased from 1970 to 1986, but males showed a sizably larger decrease. Cancer deaths during this period increased at roughly equal rates for both males and females.

GRAPHS

Graphs translate numbers into pictures. Plotted as a set of points (a *series*) on a coordinate system, a graph shows the relationship between two variables.[1] Graphs are especially useful for displaying comparisons, changes with passing time, or trends. When you decide to use a graph, choose the best type for your purpose: bar graph or line graph.

Bar Graphs

Easily understood by most readers, bar graphs show discrete comparisons, as on a year-by-year or month-by-month basis. Each bar represents a specific quantity. Use bar graphs to help readers focus on one value or compare values that change over equal time intervals (expenses calculated at the end of each month, sales figures totaled at yearly intervals). Use a bar graph only to compare values that are noticeably different. Otherwise, all the bars will appear almost identical.

Simple Bar Graphs.   The simple bar graph in Figure 14.1 displays one relationship taken from the data in Table 14.3, the rate of male deaths from heart disease. To aid interpretation, you can record exact values above each bar—but only if readers need exact numbers.

Multiple-Bar Graphs.   A bar graph can display as many as two or three relationships simultaneously, with each relationship plotted as a separate series. Figure 14.2 expresses two comparisons from Table 14.3, the rate of male deaths from both heart disease and cancer.

Because Figure 14.2 exhibits more than one relationship (or series), each relationship is represented by a different pattern, and the patterns are identified by a *legend*.

The more relationships your bar graph displays, the harder it will be to interpret. As a rule, plot no more than three bars on one graph.

Horizontal-Bar Graphs.   To make a horizontal-bar graph, turn a vertical-bar graph (and scales) on its side, or 90 degrees to the right. Horizontal-bar graphs are good for displaying a large quantity of bars arranged in order of

---

[1]Graphs have a horizontal and a vertical axis. The horizontal axis carries categories (the independent variables) to be compared, such as years within a period (1970, 1978, 1986). The vertical axis shows the range of values (the dependent variables) for comparing or measuring the categories, such as the number of people who died from heart failure in a specified year. A dependent variable changes according to activity in the independent variable (say, a decrease in quantity over a set time, as in Figure 14.1). In the equation $y = f(x)$, x is the independent variable and y is the dependent variable.

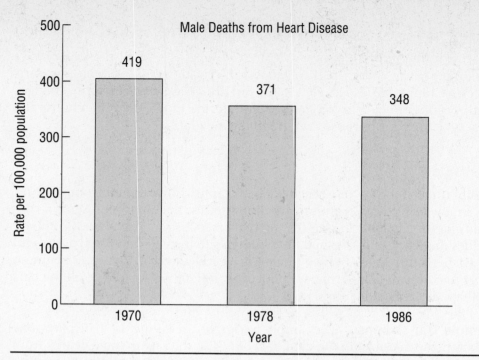

FIGURE 14.1 A Simple Bar Graph

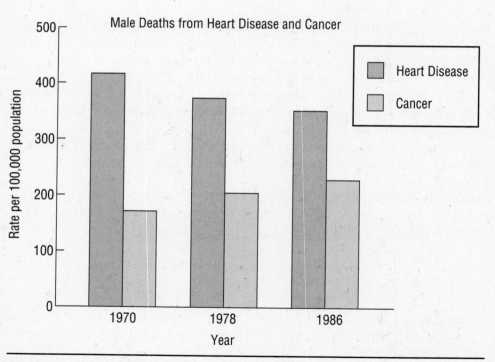

FIGURE 14.2 A Multiple-Bar Graph

increasing or decreasing value, as in Figure 14.3. The horizontal format leaves room for labeling the categories horizontally (*All races, etc.*). A vertical-bar graph would leave no room for horizontal labeling.

As with all visuals, choose the format that presents the clearest and most accurate display.

Stacked-Bar Graphs. Instead of side-by-side clusters of bars, you can display multiple relationships by stacking bars. Stacked-bar graphs are especially useful for showing how much each item contributes to the whole. Figure 14.4 displays other comparisons from Table 14.3.

Percentage of Adults Who Have Completed Four
Years of High School or More: 1950 to 1980

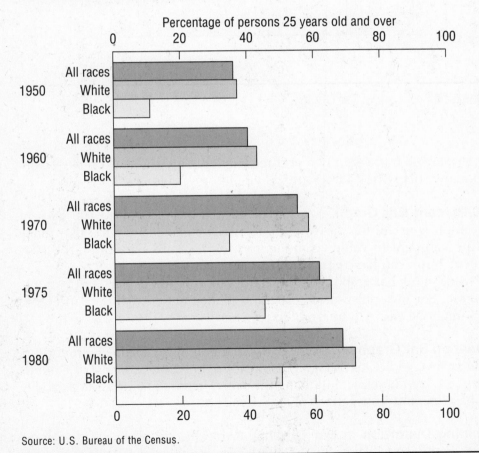

Source: U.S. Bureau of the Census.

FIGURE 14.3 A Horizontal-Bar Graph

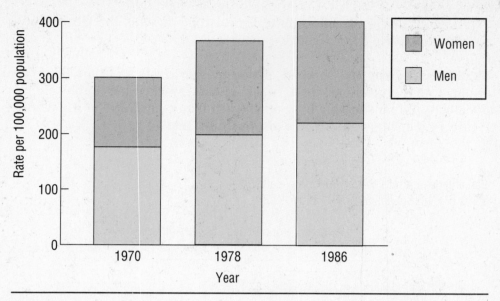

FIGURE 14.4 A Stacked-Bar Graph

Display no more than two or three relationships in a stacked-bar graph. Too many subdivisions and patterns would only confuse your readers.

100-Percent Bar Graph.   A type of stacked bar graph, the 100-percent bar graph is useful for showing the relative values of the parts that make up the 100-percent value, as in Figure 14.5. Like any bar graph, the 100-percent graph can have either horizontal or vertical bars.

Notice how bar graphs become harder to interpret as bars and patterns increase. For unsophisticated readers, the large quantity of data in Figure 14.5 might be easier to interpret in pie charts (pages 319–20).

Deviation Bar Graphs.   The deviation bar graph can display both positive and negative values, as in Figure 14.6. Notice how the vertical axis extends below the zero baseline, following the same incremental division as above the baseline, but in negative values instead.

Avoiding Distortion in Bar Graphs.   Any one set of data can support contradictory conclusions. Even though your numbers may be accurate, the way in which you display them could distort their meaning.

The visual relationships on your graph should always represent the numerical relationships. Never stretch or compress the scales to reinforce your point. Make your vertical scale at least 75 percent as long as your horizontal scale, and begin the vertical scale at zero. Notice how the visual relationships in Figure 14.7 become distorted when the value scale is compressed or when it fails to begin at zero. In version (a) of Figure 14.7, the bars accurately depict the numerical relationships measured from the value scale. In version (b), Z (400) is depicted as three times X (200). In version (c), the scale is overly compressed, causing the shortened bars to understate the differences in quantity. Deliberate distortions are unethical because they imply conclusions contradicted by the actual data.

Graphics software plots the scales of your graphs automatically, eliminating distortion. But if you make graphs by hand, be sure to experiment with scales until you find the most accurate representation.

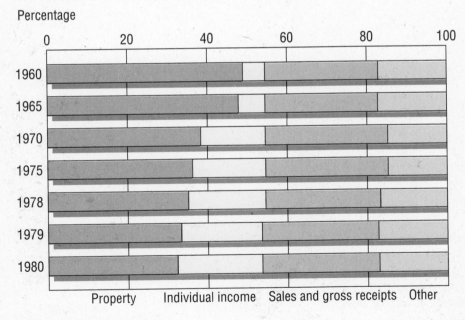

State and Local Government Taxes—
Percentage Distribution by Type, 1960—1980

Source: U.S. Bureau of the Census.

FIGURE 14.5 A 100-percent Bar Graph

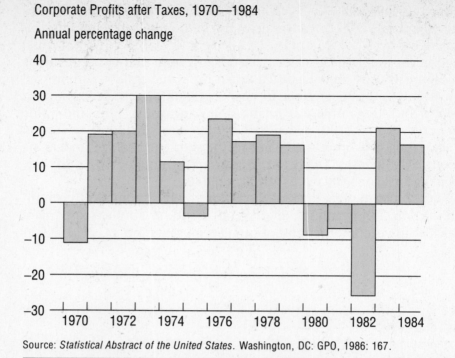

Corporate Profits after Taxes, 1970—1984

Annual percentage change

Source: *Statistical Abstract of the United States*. Washington, DC: GPO, 1986: 167.

FIGURE 14.6   A Deviation Bar Graph

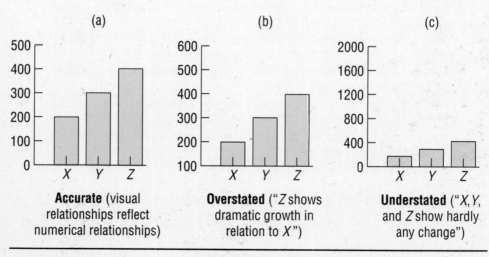

(a)

Accurate (visual relationships reflect numerical relationships)

(b)

Overstated ("*Z* shows dramatic growth in relation to *X*")

(c)

Understated ("*X, Y,* and *Z* show hardly any change")

FIGURE 14.7   An Accurate Bar Graph and Two Distorted Versions

Bar Graph Guidelines. Once you decide on a type of bar graph, follow these suggestions for presenting the graph to your audience:

- Keep the graph simple and easy to read. Avoid plotting more than three types of bars in each cluster.

- Number your scales in units the audience will find familiar and easy to follow. Units of 1 or multiples of 2, 5, or 10 are best (Lambert 45). Space the numbers equally.

- Label both scales to show what is being measured or compared. If space allows, keep all labels horizontal for easier reading.

- Use *tick marks* to show the points of division on your scale. If the graph has many bars, extend the tick marks into *grid lines* to help readers relate bars to values.

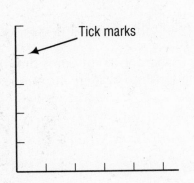

Tick marks

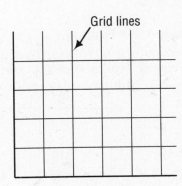

Grid lines

- To avoid confusion, make all bars the same width (unless you are overlapping them, as on page 333). If you must produce your graphs by hand, use graph paper to keep bars and increments evenly spaced.

- In a multiple-bar graph, use a different pattern or color for each bar in a cluster. Be sure to provide a legend identifying each pattern or color.

- If you are trying for emphasis, be aware that darker bars are seen as larger and closer and more important than lighter bars of the same size (Lambert 93).

Many computer graphics programs automatically employ most of these techniques. Anyone producing visuals, however, should know all the conventions.

Line Graphs

A line graph can accommodate many more data points than a bar graph (say, a twelve-month trend, measured monthly). Line graphs help readers synthesize large bodies of information in which exact quantities need not be emphasized. Whereas bar graphs display quantitative differences among items (cities, regions, yearly or monthly intervals), line graphs display data whose value changes repeatedly, as in a trend, forecast, or other change during a specified time (profits, losses, growth). And some line graphs depict cause-and-effect relationships (say, how seasonal patterns affect sales or profits).

Simple Line Graphs.   A simple line graph, as in Figure 14.8, uses one line to plot time intervals on the horizontal scale and values on the vertical scale. The relationship depicted here would be much harder to express in words alone.

Multiple-Line Graphs.   A multiple-line graph displays as many as three relationships simultaneously, as in Figure 14.9.

Because readers usually find line graphs harder to interpret than bar graphs, be sure to explain in prose the relationships readers are supposed to see.

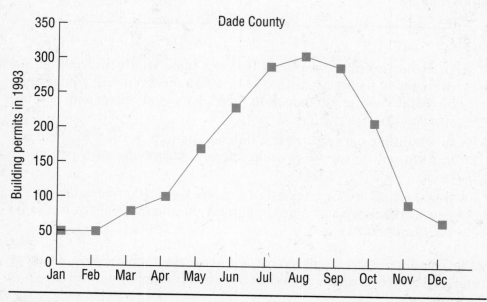

FIGURE 14.8   A Simple Line Graph

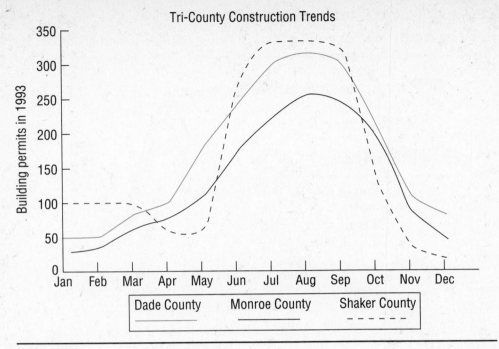

FIGURE 14.9 A Multiple-Line Graph

As displayed in Figure 14.9, building permits in all three counties increased sharply during summer and early fall. Dade and Monroe counties showed a steady increase as the weather grew warmer, but Shaker County's seasonal increase was more erratic. Shaker's permits actually declined in April and May, but then rose to surpass permits in Dade and Monroe counties during June through September.

A prose explanation of visual relationships

Deviation Line Graphs. By extending your vertical scale below the zero baseline, you can display both positive and negative values in one graph, as in Figure 14.10. Be sure to mark the same intervals for values below the baseline.

Band or Area Graphs. A type of line graph, a band or area graph displays changes or trends over a specified period. To attract attention, however, the area beneath each plotted line in a band graph is filled with a pattern. The Figure 14.11 band graph plots the same data as the Figure 14.8 line graph.

A multiple-band graph is especially useful for displaying relationships among sums rather than direct comparisons. Figure 14.12 shows how the criss-crossed line graph in Figure 14.9 would appear as a band graph.

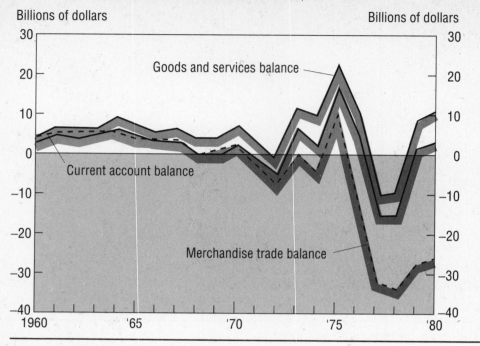

FIGURE 14.10 A Deviation Line Graph

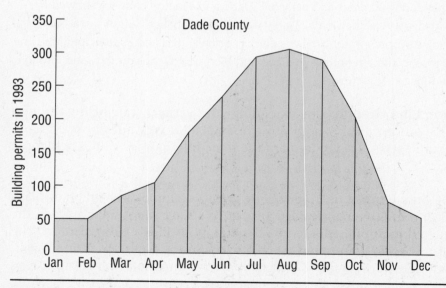

FIGURE 14.11 A Simple Band Graph

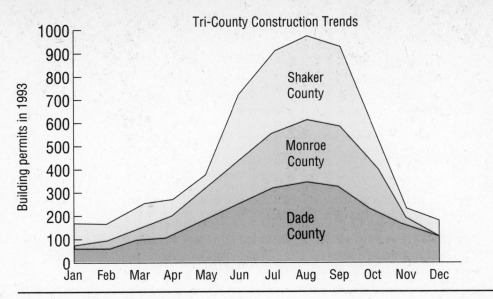

FIGURE 14.12 A Multiple-Band Graph

Although they seem more interesting than line graphs, multiple-band graphs are easy to misinterpret unless they are carefully labeled and explained. We recall that every line in a multiple-line graph represents a distance from the zero baseline. In a multiple-band graph, however, the very top line represents the *total* change or trend; each band below the top line represents a part that contributes to the total (like segments in the stacked-bar graph, Figure 14.4). Always explain these relationships to uninitiated readers.

Avoiding Distortion in Line Graphs.   A poorly constructed line graph can distort visual relationships between numbers that are otherwise accurate. Figure 14.13 shows how one type of distortion can occur, even on a computer, when data that would provide a complete picture are selectively omitted. Version (a) accurately depicts the numerical relationships measured from the value scale. But in version (b), too few points are plotted. If you plot by hand, begin the vertical scale at zero, and avoid stretching or compressing either scale.

Visuals have their own rhetoric and persuasive force, which we can use to advantage—for positive or negative purposes, for the reader's benefit or detriment (Van Pelt 2). Avoiding visual distortion is ultimately a matter of ethics.

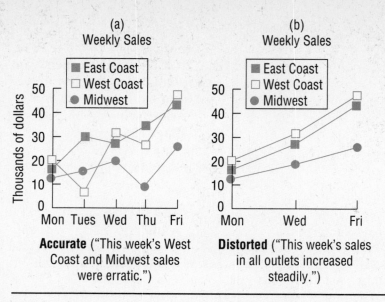

FIGURE 14.13 An Accurate Line Graph and a Distorted Version

Line Graph Guidelines. Line graphs follow the same conventions as bar graphs, with these additions:

- Compare no more than three lines on one graph.
- Mark the individual data points on each line so that readers can see how many points have been used to plot the line.
- In a multiple-line graph, make each line distinct (using colors, dots, dashes, or symbols).
- Label each line clearly so that readers will know what it represents.
- Avoid grid lines; readers might confuse them with the plotted lines.

CHARTS

The terms *chart* and *graph* often are used interchangeably. But a chart is more precisely a figure that displays relationships (quantitative or cause-and-effect) without being plotted on a coordinate system. Commonly used charts include pie charts, organizational charts, flowcharts, tree charts, column charts, and pictorial charts (pictograms).

Pie Charts

Considered easy for readers to understand, a pie chart depicts the relative amounts of the parts that make up a whole. In a pie chart, readers can compare the parts to each other as well as to the whole (to show how much was spent on what, how much income comes from which sources, and so on). Figure 14.14 shows a pie chart. Figure 14.15 shows two other versions of the pie chart in Figure 14.14. Version (a) displays dollar amounts, and version (b) the percentage relationships among these dollar amounts.

For constructing pie charts, follow these suggestions:

- Be sure the parts add up to 100 percent.
- If you must produce your charts by hand, use a compass and protractor for precise segments. Each 3.6-degree segment equals 1 percent. Include any number from two to eight segments. A pie chart containing more than eight segments can be hard to interpret, especially if the segments are small (Hartley 96).
- Combine small segments under the heading "Other."
- Create a sense of spatial logic by locating your first radial line at

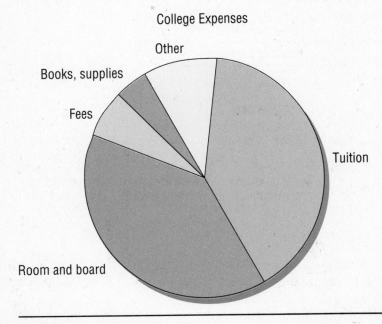

FIGURE 14.14 A Simple Pie Chart

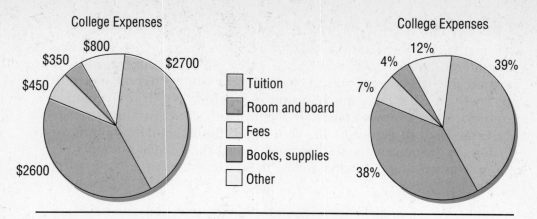

FIGURE 14.15 Two Other Versions of Figure 14.14

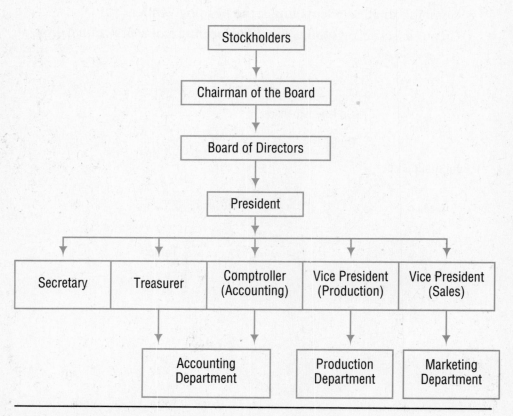

FIGURE 14.16 An Organization Chart for One Corporation

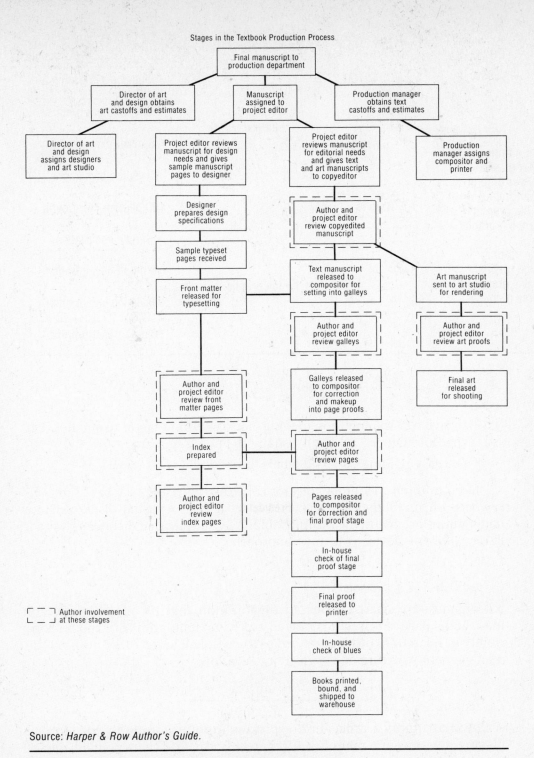

Stages in the Textbook Production Process

Source: *Harper & Row Author's Guide.*

FIGURE 14.17 A Flowchart for Producing a Textbook

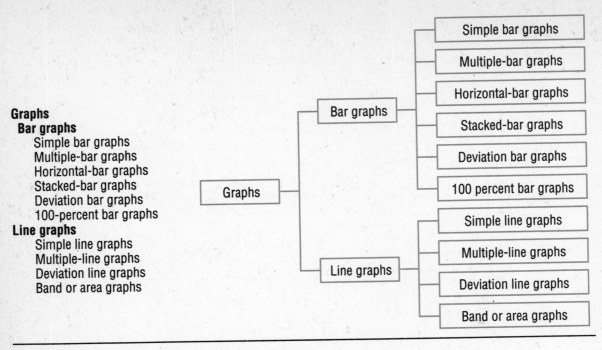

Graphs
  Bar graphs
    Simple bar graphs
    Multiple-bar graphs
    Horizontal-bar graphs
    Stacked-bar graphs
    Deviation bar graphs
    100-percent bar graphs
  Line graphs
    Simple line graphs
    Multiple-line graphs
    Deviation line graphs
    Band or area graphs

FIGURE 14.18 An Outline Converted to a Tree Chart

twelve o'clock and then moving clockwise from larger to smaller (except for "Other," which is usually the final segment).

- For easy reading, keep all labels horizontal.

Keep in mind that guidelines for constructing visuals are based on *conventions*, on methods that have proven effective for visual display. These conventions are by no means inflexible. As your purpose dictates—and accuracy allows—devise your own approach.

Organization Charts

An organization chart divides an organization into its administrative or managerial parts. Each part is ranked according to its authority and responsibility in relation to other parts and to the whole. Figure 14.16 displays the management structure of a typical organization.

Flowcharts

A flowchart traces a procedure or process from beginning to end. In displaying the steps in a manufacturing process, the flowchart would begin at the raw materials and proceed to the finished product. Figure 14.17 traces the procedure for producing a textbook. (Other flowchart examples appear on pages 113, 136, and elsewhere throughout the text.)

Tree Charts

Whereas flowcharts display the steps in a process, tree charts show how the parts of an idea or concept relate to each other. Figure 14.18 displays the parts of an outline for this chapter so that readers can better visualize the relationships. Notice that the tree version seems clearer and more interesting than the prose listing.

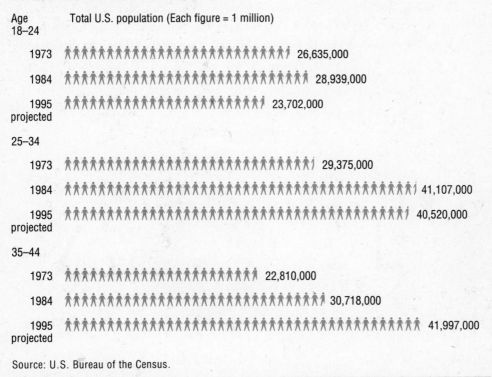

After increasing between 1973 and 1984, the college-age population is projected to decline over the 1984–1995 period.

Source: U.S. Bureau of the Census.

FIGURE 14.19 A Pictogram

Pictorial Charts

Pictorial charts, or pictograms, depict numerical relationships with icons or symbols (cars, houses, smokestacks) of the items being measured, instead of using bars. Each symbol represents a stipulated quantity, as in Figure 14.19. Many graphics programs provide an assortment of predrawn symbols.

Use pictograms when you want to make your information more interesting for nontechnical audiences.

ILLUSTRATIONS

Illustrations consist of diagrams, maps, and photographs depicting relationships that are physical rather than numerical. Good illustrations help readers understand and remember the material (Hartley 82). Consider this information from a government pamphlet, explaining the operating principle of the seat belt:

> The safety-belt apparatus includes a tiny pendulum attached to a lever, or locking mechanism. Upon sudden deceleration, the pendulum swings forward, activating the locking device to keep passengers from pitching into the dashboard.

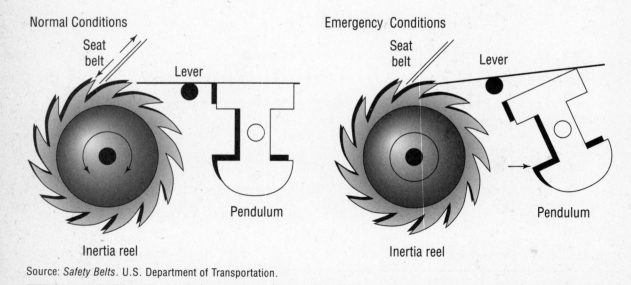

Source: *Safety Belts*. U.S. Department of Transportation.

FIGURE 14.20   A Diagram of a Safety-Belt Locking Mechanism

Without the illustration in Figure 14.20 we have difficulty visualizing the mechanism. Clear and uncluttered, a good diagram eliminates unnecessary details, and is focused only on material useful to the reader. The following pages sample some commonly used diagrams.

Exploded Diagrams

Exploded diagrams, like that of a brace for an adjustable basketball hoop in Figure 14.21, show how the parts of an item are assembled; they often

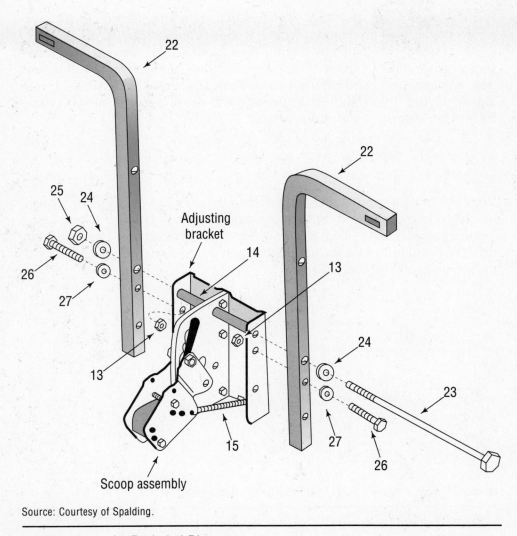

Source: Courtesy of Spalding.

FIGURE 14.21 An Exploded Diagram

appear in repair or maintenance manuals. Notice how all parts are numbered for the reader's easy reference in the written instructions.

Diagrams of Procedures

Diagrams can clarify instructions by illustrating steps or actions, as in Figure 14.22.

Block Diagrams

Block diagrams are simplified sketches that represent the relationship between the parts of an item or process. Because block diagrams are designed to illustrate *concepts* (such as current flow in a circuit), the parts are represented as symbols or shapes. The block diagram in Figure 14.23 illustrates how any process can be controlled automatically through a feedback mechanism. Figure 14.24 shows the feedback concept applied as the cruise-control mechanism on a motor vehicle.

Maps

Besides being interesting and easily remembered, maps are especially useful for showing comparisons and for enabling readers to *visualize* relationships among complex data. Consider how Figure 14.25 conveys important statistical information in a format that is both accessible and understandable. Color enhances the comparisons.

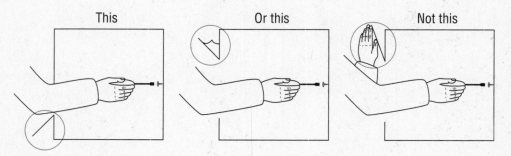

Source: U.S. Dept. of Energy. *Energy and Technology Review* June 1984: 33.

FIGURE 14.22 A Diagram of a Repair Procedure

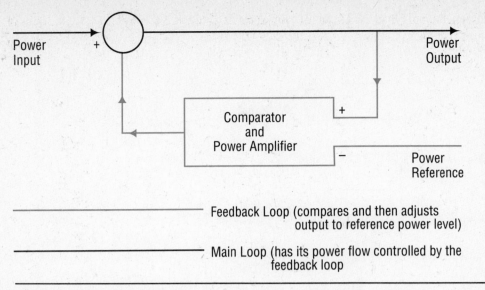

FIGURE 14.23 A Block Diagram Illustrating the Concept of Feedback

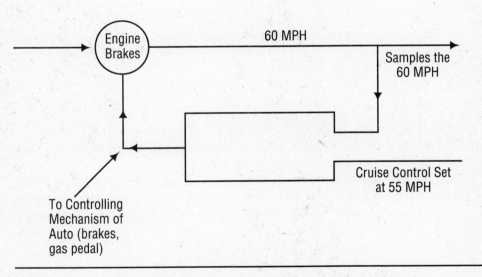

FIGURE 14.24 A Block Diagram Illustrating a Cruise-Control Mechanism

States (shaded) that have lost more than 50 percent of their wetlands between the 1780's and the mid-1980's (in percent loss).

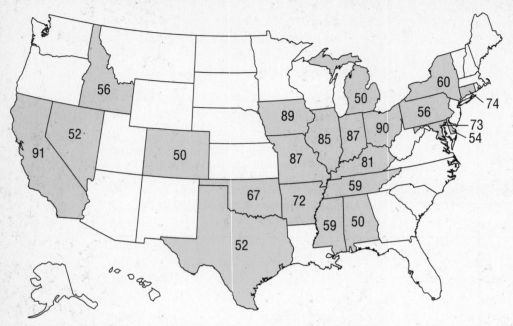

Source: Adapted from Dahl, T.E. and C.E. Johnson, 1991. *Status and Trends of Wetlands in the Coterminous United States*. U.S. Department of the Interior, Fish and Wildlife Service, Washington, D.C.

FIGURE 14.25   A Map Rich in Statistical Significance

Photographs

Photographs provide an accurate overall view, but sometimes they can be too "busy." By showing all details as more or less equal, a photograph sometimes fails to emphasize important areas.

When you use photographs, keep them distinct, well focused, and uncluttered. To emphasize features of a complex mechanism, you probably should rely on diagrams instead of photographs, unless you intend simply to show an overall view. Provide a sense of scale by including a person or a familiar object (such as a hand) in your photo.

An emerging technology known as digital imaging allows photographs to be taken electronically and stored on compact disks instead of film. These stored images can then be retrieved, retouched, and edited. Such capacity for altering photographic content creates unlimited potential for distortion, and raises questions about the ethics of digital manipulation (Callahan 64–65).

COMPUTER GRAPHICS

With one or two commands, today's computer systems can create highly sophisticated, multicolor graphic displays. Simply type your numerical data into the program, and select a type of chart or graph. The system then analyzes, plots, and displays the data in the visual form you selected. The complex visuals possible with computer graphics include these examples:

- With an electronic stylus (a pen with an electronic signal), you can draw pictures on a graphics tablet to be displayed on the monitor, stored, or sent to other computers.

- You can create three-dimensional effects, showing an object from different angles with shading, shadows, or other techniques.

- You can recreate the visual effect of a mathematical model, as in writing equations to explain what happens when high winds strike a tall building. (As the wind deforms the structure, the equations change. Then you can take those new equations and represent them visually.)

- You can create a design, build a model, simulate the physical environment, and let the computer forecast what will happen with different variables.

- You can integrate computer-assisted design (CAD) with computer-assisted manufacturing (CAM), so that the design will direct the machinery that makes the parts themselves (CAD/CAM).

- You can create animations, to see how bodies move (as in a car crash or in athletics).

- You can practice dealing with toxic chemicals, operating sophisticated machines, or making other rapid decisions in medical or technical environments, without the cost or danger in actual situations.

- Through various types of scientific visualization, you can do "what-if" projections and explore countless ways of conceptualizing and understanding your data. Because the computer can generate and evaluate many possibilities rapidly, it enables you to test hypotheses without doing the calculations.

Selecting Design Options

Among all applications for personal computers, graphics production is perhaps the most rapidly expanding (Schmeupe 51). Computer graphics systems allow you to experiment with scales, formats, colors, perspectives, and patterns. Most systems offer WYSIWYG (What You See Is What You Get) packages, in which the on-screen display is almost identical to the hard-copy image that will be printed out. By testing design options on the screen, you can revise and enhance your visual repeatedly until it achieves your exact purpose.

Here is a sampling of design options:

- Update charts and graphs whenever the data change. The software will calculate the new data and plot the relationships.
- Edit your graphics on the screen, adding, deleting, or moving material as needed.
- Create your image at one scale, and later specify a different scale for the same image.
- Annotate and label, and create multiple typefaces within one visual.
- Overlay images in one visual.
- Adjust bar width and line thickness.
- Fill a shape with a color or pattern.

Most of these options call for no more than a single keystroke.

Composing and Enhancing Visuals Electronically

The composing process for visuals is identical to that for written text: you must decide about the purpose, audience, content, arrangement, and style of your visual message, revising until the message conveys your exact meaning. The following short scenario suggests how you might explore options in a typical graphics software package.

Using Graphics Software

Assume that after a few years with Company X (an international producer of communications hardware and software), you have been appointed personnel recruiter. Your job is to visit college campuses across the nation and recruit the finest talent among graduate as well as undergraduate students. To create interest in your company, you decide to prepare a visual presentation that will complement your lectures to student groups during your travels.

Among your most striking data (for this audience) is the dramatic increase in the average salary offered to entry-level people by your company during the past five years; to emphasize the increase, you decide to begin your presentation with a five-year salary comparison. After booting up your graphics software, you get right to work composing your first visual.

To show change at fixed intervals and to focus on specific numbers, you decide to use a bar graph—a visual whose meaning is apparent, even for those who might have little experience interpreting visual messages. You begin by entering average salaries for entry-level personnel with a B.S. degree:

1989	21,500
1990	24,000
1991	26,000
1992	28,500
1993	31,000

After a few commands, your computer processes your data to produce the simple graph in Version A.

Most of your audiences will contain both graduate and undergraduate students, and so you decide to include the salary figures for M.S. and Ph.D. entry-level positions. The computer responds by displaying the multiple-bar graph in Version B.

This second version is neither informative nor engaging enough to present to a varied and unfamiliar audience. As enhancements, you decide to add a legend to identify the various patterns. Then you emphasize the title with bold-face type and a shadowed border. For appeal, you add color. For contrast, you frame the entire graph, as in Version C.

As a further aid to interpretation, you decide to add grid lines, to scale your vertical axis in smaller increments ($5,000 instead of $10,000), and to add major tick marks to the outside of your vertical axis. For easy scanning, you reverse the order of columns in each cluster and overlap them by 50 percent. The finished product in Version D is a "presentation-quality" graph you can

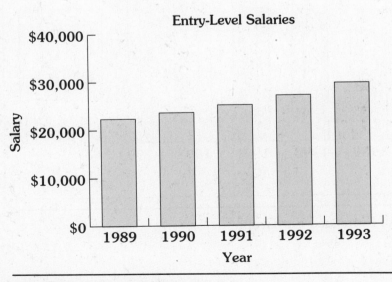

VERSION A   A Simple Bar Graph

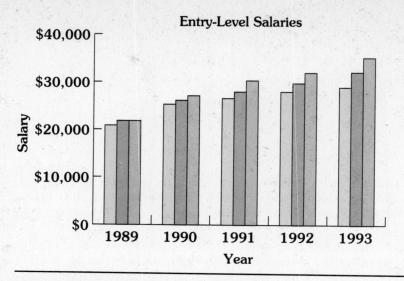

VERSION B A Multiple-Bar Graph

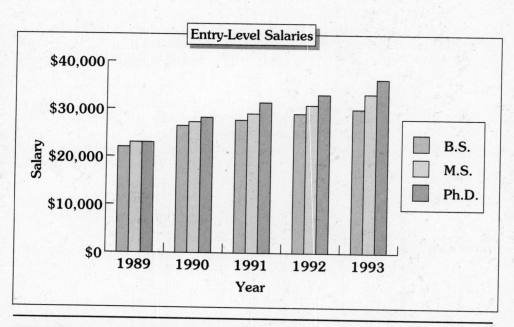

VERSION C An Enhanced Bar Graph

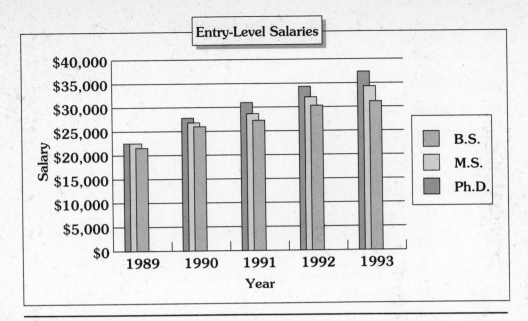

Entry-Level Salaries

VERSION D   A Bar Graph Further Enhanced

project on a large-screen monitor, reproduce as a slide or transparency, or include in a document.

For a different purpose, you might decide to emphasize how salary increases for advanced degrees have outdistanced those for the B.S. One way of achieving this emphasis is to plot the B.S. salary data as a line graph, which would then overlay the original M.S. and Ph.D. bars, as in Version E. This version, of course, is just one among many possibilities for plotting the original data.

We choose visuals according to the kinds of data, the relationships to be illustrated, the needs of our audience, and desired emphasis.

Caution: We have considered only a few among countless options for electronically enhancing visuals. With all the possibilities, we can be tempted to overembellish visuals, as in the unfortunate example in Version F.

Visuals should attract audience attention, but should *not* bombard the senses with a nauseating excess of textures and patterns. Let restraint and good taste guide your selection of enhancement options.

Using Clip Art

Clip art is a generic form for collections of ready-to-use images (of computer equipment, maps, machinery, medical equipment, and so on), all stored on

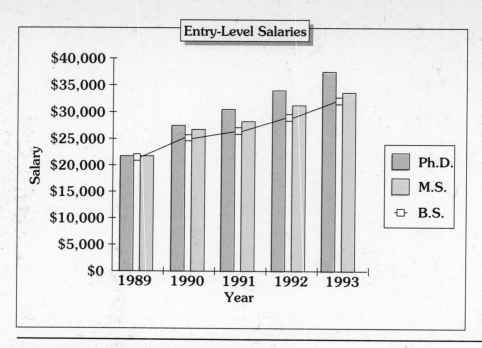

VERSION E An Alternative Display

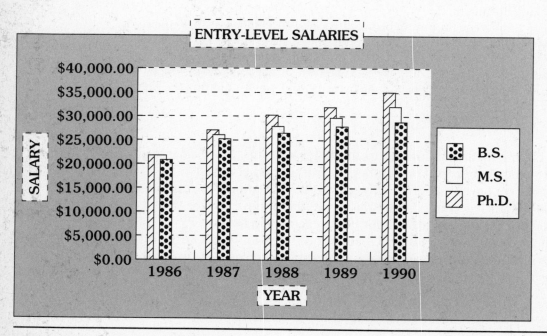

VERSION F An Overdone Display

Source: *Desk Top Art®*; *Business 1*, © Dynamic Graphics, Inc.

FIGURE 14.26 A Clip-art Image

computer disks. Various clip-art packages enable you to "import" into your document countless images like the one in Figure 14.26. By running the image through a drawing program you can enlarge, enhance, or customize it, as in Figure 14.27.

Clip-art software offers images of all kinds from science, architecture, business, and technology (see Figure 14.28).

One form of clip art especially useful in technical writing is the icon (an image with all nonessential background removed). Icons convey a specific idea visually; for instance, icons appear routinely in computer documentation and in other types of instructions because the image provides readers with an immediate signal of the desired action.

Source: Professor R. Armand Dumont.

FIGURE 14.27 A Customized Image

A warning: A direction:

An instruction: A concept:

(safety)

FIGURE 14.28 Icons

Whenever you use an icon, be sure it is "intuitively recognizable" to your readers ("Using Icons" 3). Otherwise, readers are likely to misinterpret its meaning—in some cases with disastrous results.

GUIDELINES FOR VISUALS

An effective visual enables readers to extract the information they need. To simplify your reader's task, observe these general guidelines:

- *Use visuals only when they are needed.* Visual display is warranted in countless situations, but especially these (Dragga and Gong 47–48):
  a. when you want to instruct or persuade
  b. when you want to emphasize something immediately important
  c. when you expect the document to be read in random or selective parts (a manual or other reference work) instead of sequentially (a memorandum or letter)
  d. when you expect readers to be relatively less educated, less motivated, or less familiar with the topic
  e. when you expect readers to be in a distracting environment

  Instead of serving as window dressing, a good visual advances the reader's understanding and and the writer's purpose.

- *Number the visual and give it a clear title and labels.* Your title should tell readers what they are seeing. Label all the important material.

- *Choose the visual most appropriate for your purpose.* To depict the operating parts of a mechanism, for instance, an exploded diagram probably would work better than a photograph.

- *Use prose captions to explain important points made by the visual.* Captions help readers interpret a visual (as in Figure 7.2, 8.11, or 14.7). When possible, make caption typesize smaller than body typesize, so that captions don't "compete" with body type (*Aldus Guide* 35).

- *Match the visual to your audience.* Don't make it too elementary for specialists or too complex for nonspecialists. Be sure your intended audience will be able to interpret the visual correctly.

- *Never include excessive information in one visual.* Any visual that contains too many lines, bars, numbers, colors, or patterns will overwhelm readers, causing them to ignore the visual. In place of one complicated visual, use two or more straightforward ones.

- *Use color with caution.* "Color gains impact when it is used selectively. It loses impact when it is overused" (*Aldus Guide* 39). Ineffective use of color can distort a reader's perception of the relationships (Wickens 117). (Green traditionally signifies safety; red, danger; darker colors seem to make a stronger statement than lighter ones.) Also, readers perceive differently the sizes of variously colored objects. Darker items can seem larger and closer than lighter objects of identical size. The brightest colors will attract the reader's attention (Murch 18–20); subdued colors suggest a dignified tone (*Aldus Guide* 39). If you do use color, be sure to anticipate accurately its effect on the reader's perception.

- *Never mistake distortion for emphasis.* When you want to emphasize a point (a sales increase, a safety record, etc.), be sure your data support the conclusion implied by your visual. For instance, don't use inordinately large visuals to emphasize good news or small ones to downplay bad news (Williams 11). And when using clip art to dramatize a comparison, be sure the relative size of the images or icons reflects the quantities being compared. A visual accurately depicting a 100-percent increase in phone sales at your company might look like this:

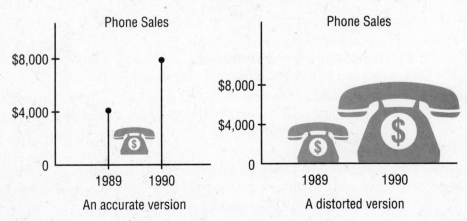

The second version overstates the good news by depicting the larger image four times the size, instead of twice the size, of the smaller. Although the larger image is twice the height, it is also twice the *width,* and so the total area conveys the false impression that sales have *quadrupled.*

Although you are perfectly justified in presenting data in its best light, you are ethically responsible for avoiding misrepresentation.

- *Place the visual where it will best serve your readers.* Avoid clumping visuals in an appendix or at report's end. If the visual clarifies part of your discussion, place it close to that part. But if it will interest only some of your audience, place it in an appendix (discussed on pages 377–79) so that interested readers can refer to it as they wish. Tell readers when to refer to the visual and where to find it.

- *Never refer to a visual that readers cannot easily locate.* In a long document, don't be afraid to repeat a visual if you discuss it again later.

- *Introduce and interpret the visual.* In your introduction, let readers know what to expect.

| Informative | As shown in Table 2, operating costs have increased 7 percent annually since 1980. |
| Uninformative | See Table 2. |

Visuals alone make ambiguous statements (Girill 35); pictures need to be interpreted. Instead of leaving readers to struggle with a page of raw data, explain the relationships displayed.

| Informative | This cost increase means that. . . . |

Always tell readers what to look for and what it means.

- *Be sure the visual's meaning can stand alone.* Even though it repeats or augments information already in the text, the visual should contain everything readers will need to interpret it correctly.

- *Never crowd a visual into a cramped space.* Set your visual off by framing it with plenty of white space (discussed on page 352), and position it on the page for balance.

The Visual Plan Sheet, Figure 14.29, and the Revision Checklist for Visuals on page 340 will help ensure that your visuals observe the above guidelines.

Focusing on Your Purpose

- What is this visual's purpose (to instruct, persuade, create interest)? _____

- What type of information (numbers, shapes, concepts, procedures) will this visual depict? _____

- What kind of relationship(s) will the visual depict (comparison, cause-effect, connected parts, sequence of steps)? _____

- What judgment or conclusion or interpretation is being emphasized (that profits have increased, that toxic levels are rising, that X is better than Y, that time is being wasted)? _____

- Is a visual needed at all? _____

Focusing on Your Audience

- Is this audience accustomed to interpreting visuals? _____

- Is the audience interested in specific numbers or an overall view? _____

- Should the audience focus on one value, compare two or more values, or synthesize a group of values (Wickens 121)? _____

- Which type of visual will be most accurate, representative, accessible, and compatible with the type of judgment the audience is expected to make? _____

- In place of one complicated visual, would two or more straightforward ones be preferable? _____

Focusing on Your Presentation

- What enhancements, if any, will increase audience interest (colors, patterns, legends, labels, varied typefaces, shadowing, enlargement or reduction of some features)? _____

- Which medium—or combination of media—will be most effective for presenting this visual (slides, transparencies, handouts, large-screen monitor, flip chart, report text)? _____

- To achieve the greatest utility and effect, where in the presentation does this visual belong? _____

FIGURE 14.29 A Planning Sheet for Preparing Visuals

REVISION CHECKLIST FOR VISUALS

(Numbers in parentheses refer to the first page of discussion.)

Content

☐ Does the visual serve a legitimate purpose (clarification, not mere ornamentation) in the document? (303)

☐ Is the visual titled and numbered? (336)

☐ Is the level of complexity appropriate for the audience? (337)

☐ Are all patterns in the visual identified by label or legend? (313)

☐ Are all values or units of measurement specified clearly (grams per ounce, millions of dollars)? (304)

☐ Are the numbers accurate and exact? (304)

☐ Do the visual relationships represent the numerical relationships, without distortion? (337)

☐ Are explanatory notes added only as needed? (305)

☐ Are all data sources cited? (306)

☐ Is the visual introduced, discussed, interpreted, and integrated with the text? (338)

☐ Can the visual itself stand alone in meaning? (338)

Arrangement

☐ Is the visual easy to locate? (338)

☐ Are all design elements (title, line thickness, legends, notes, borders, white space) positioned for balance? (338)

☐ Is the visual positioned on the page for balance? (338)

☐ Is the visual set off by adequate white space or borders? (338)

☐ Does the top of a wide visual face the inside binding? (305)

☐ Is the visual in the best report location? (338)

Style

☐ Is this the best type of visual for your purpose and audience? (336)

☐ Are all decimal points in each column vertically aligned? (304)

☐ Is the visual uncrowded and uncluttered? (338)

☐ Is the visual engaging (patterns, colors, shapes), without being too busy? (337)

☐ Is the visual in good taste? (333)

☐ Is the visual ethically acceptable? (317)

EXERCISES

1. The following statistics are based on data from three colleges in a large western city. They give the number of applicants to each college over six years.

- In 1988, X College received 2,341 applications for admission, Y College received 3,116, and Z College 1,807.
- In 1989, X College received 2,410 applications for admission, Y College received 3,224, and Z College 1,784.
- In 1990, X College received 2,689 applications for admission, Y College received 2,976, and Z College 1,929.
- In 1991, X College received 2,714 applications for admission, Y College received 2,840, and Z College 1,992.
- In 1992, X College received 2,872 applications for admission, Y College received 2,615, and Z College 2,112.
- In 1993, X College received 2,868 applications for admission, Y College received 2,421, and Z College 2,267.

Illustrate these data in a line graph, a bar graph, and a table. Which version seems most effective for a reader who (a) wants exact figures, (b) wonders how overall enrollments are changing, or (c) wants to compare enrollments at each college in a certain year? Write a paragraph interpreting one of these versions.

2. Devise a flowchart for a process in your field or area of interest. Include a title and a brief discussion.

3. Devise an organization chart showing the lines of responsibility and authority in an organization where you hold a part-time or summer job.

4. Devise a pie chart to depict your yearly expenses. Title and discuss the chart.

5. Obtain enrollment figures at your college for the past five years by sex, age, race, or any other pertinent category. Construct a stacked-bar graph to illustrate one of these relationships over the five years.

6. Keep track of your pulse and respiration at thirty-minute intervals over a four-hour period of changing activities. Record your findings in a line graph, noting the times and specific activities below your horizontal coordinate. Write a prose interpretation of your graph and give it a title.

7. In textbooks or professional journal articles, locate each of these visuals: a table, a multiple-bar graph, a multiple-line graph, a diagram, and a photograph. Evaluate each according to the revision checklist, and discuss the most effective visual in class.

8. We have discussed the importance of choosing an appropriate scale for your graph, and the most effective form for presenting your data. Study this presentation carefully:

> Strong evidence now indicates that not only nicotine and tar in cigarette smoke can be lethal, but also carbon monoxide. Much of cigarette smoke is carbon monoxide. The bar graph in Figure 14.30 (page 343) lists the ten leading U.S. cigarette brands according to the carbon monoxide given off per pack of inhaled cigarettes.

Is the scale effective? If not, why not? Can these data be presented effectively in a bar graph? Present the same data in some other form that seems most effective.

9. Choose the most appropriate visual for illustrating each of these relationships. Justify each choice in a short paragraph.

    a. A comparison of three top brands of fiberglass skis, according to cost, weight, durability, and edge control.

    b. A breakdown of your monthly budget.

    c. The changing cost of an average cup of coffee, as opposed to that of an average cup of tea, over the past three years.

    d. The percentage of college graduates finding desirable jobs within three months after graduation, over the last ten years.

    e. The percentage of college graduates finding desirable jobs within three months after graduation, over the last ten years—by gender.

    f. An illustration of automobile damage for an insurance claim.

    g. A breakdown of the process of radio wave transmission.

    h. A comparison of five cereals on the basis of cost and nutritive value.

    i. A comparison of the average age of students enrolled at your college in summer, day, and evening programs, over the last five years.

    j. Comparative sales figures for three items made by your company.

10. *Computer graphics:* Using the scenario on pages 330–34 as a model, compose and enhance one or more visuals electronically. You might begin by looking through a recent volume of the *Statistical Abstract of the United States* (in the government documents section of your library). From the *Abstract*, or from a source you

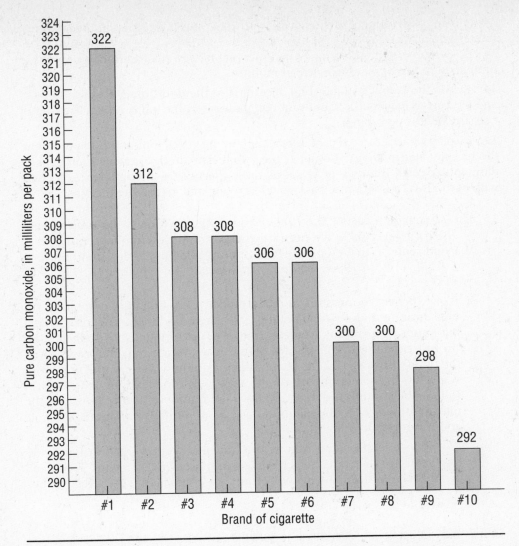

FIGURE 14.30 Ten Leading U.S. Cigarette Brands in Order of CO Content

prefer, select a body of numerical data that will interest your classmates. After completing the Visual Plan Sheet, compose one or more visuals to convey a message about your data, to make a point, as in these examples:

- Consumer buying power has increased or decreased since 1980.
- Defense spending, as a percentage of the federal budget, has increased or decreased since 1980.
- Average yearly temperatures across America are rising or falling.

Experiment with formats and design options, and enhance your visual(s) as appropriate. Add any necessary prose explanations.

Be prepared to present your visual message in class, using either an overhead or opaque projector or a large-screen monitor.

11. Revise the layout of Table 14.4 according to the directions on pages 304–06, and explain to readers the significant comparisons in the table. (*Hint:* The unit of measurement is percentage.)

12. Display each of these sets of information in the visual format most appropriate for the stipulated audience. Complete the Visual Plan Sheet for each visual. Explain why you selected the type of visual as most effective for that audience. Include with each visual a brief prose passage interpreting and explaining the data.

a. (For general readers.) The 1986 *Statistical Abstract of the United States* breaks down energy sources for electrical energy production into these categories, by percentage: In 1970, coal, 46.2; natural gas, 24.3; hydro, 16.2; nuclear, 1.4; oil, 11.9. In 1980, coal, 51.1; natural gas, 15.1; hydro, 12.1; nuclear, 11.0; oil, 10.7.

b. (For experienced investors in rental property.) As an aid in estimating annual heating and air-conditioning costs, here are annual maximum and minimum temperature averages from 1961 to 1990 for five Sunbelt cities (in Fahrenheit degrees): In Jacksonville, the average maximum was 78.7; the minimum was 57.2. In Miami, the maximum was 82.6; the minimum was 67.8. In Atlanta, the maximum was 71.3; the minimum was 51.1. In Dallas, the maximum was 76.9; the minimum was 55. In Houston, the maximum was 77.5; the minimum was 57.4. (From U.S. National Oceanic and Atmospheric Administration.)

c. (For the student senate.) Among the students who entered our university four years ago, here are the percentages of those who graduated, withdrew, or are still enrolled: In Nursing, 71 percent graduated; 27.9 percent with-

TABLE 14.4 Educational Attainment of Persons 25 Years Old and Over

Year	Highest level completed	
	High school (4 years or more)	College (4 years or more)
1970	52.3164	10.7431
1980	66.5432	16.2982
1982	71.0178	17.7341
1984	73.3124	19.1628

Source: Adapted from *Statistical Abstract of the United States*. Washington, DC: GPO, 1986: 164.

drew; 1.1 percent are still enrolled. In Engineering, 62 percent graduated; 29.2 percent withdrew; 8.8 percent are still enrolled. In Business, 53.6 percent graduated; 43 percent withdrew; 3.4 percent are still enrolled. In Arts and Sciences, 27.5 percent graduated; 68 percent withdrew; 4.5 percent are still enrolled.

d. (For the student senate.) Here are the enrollment trends from 1983 to 1992 for two colleges in our university. In Engineering: 1983, 455 students enrolled; 1984, 610; 1985, 654; 1986, 758; 1987, 803; 1988, 827; 1989, 1,046; 1990, 1,200; 1991, 1,115; 1992, 1,075. In Business: 1983, 922; 1984, 1,006; 1985, 1,041; 1986, 1,198; 1987, 1,188; 1988, 1,227; 1989, 1,115; 1990, 1,220; 1991, 1,241; 1992, 1,366.

13. Anywhere on campus or at work, locate at least one visual that needs revision for accuracy, clarity, appearance, or appropriateness. Look in computer manuals, lab manuals, newsletters, financial aid or admissions or placement brochures, student or faculty handbooks, or newspapers—or in your textbooks. Use the Visual Plan Sheet and the Revision Checklist as guides to revise and enhance the visual. Submit to your instructor a copy of the original, along with a memo explaining your improvements. Be prepared to discuss your revision in class.

COLLABORATIVE PROJECT

Assume that your technical writing instructor is planning to purchase five copies of a graphics software package for students to use in designing their documents. The instructor has not yet decided which general-purpose package would be most useful. Your group's task is to test one package and to make a recommendation.

In small groups, visit your school's microcomputer lab and ask for a listing of the graphics packages that are available to students and faculty. Select one package and learn how to use it. Design at least four representative visuals. In a memo or presentation to your instructor and classmates, describe the package briefly and tell what it can do. Would you recommend purchasing five copies of this package for general-purpose use by writing students? Explain. Submit your report, along with the sample graphics you have composed. Appoint one member to present your group's recommendation in class.

Do the same assignment, comparing various clip-art packages. Which package offers the best image selection for writers in your specialty?

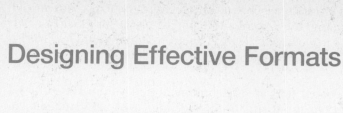

Designing Effective Formats

Formats in Workplace Writing

Automated Formatting

Format Guidelines

Audience Considerations

■   ■   ■

Format is the *look* of a page, the layout of words and graphics. A well-formatted document invites readers in, guides them through the material, and helps them understand it.

Readers' *first* impression of a document tends to be a purely visual, aesthetic judgment. They are attracted by documents that appear inviting and accessible.

FORMATS IN WORKPLACE WRITING

Format decisions become especially significant when we consider these realities about writing and reading in the workplace:

1. *Technical information generally is designed differently from material in novels, news stories, and other forms of writing.* To be accessible, a technical document requires more than just an unbroken sequence of paragraphs. To find their way through complex material, readers may need the help of charts, diagrams, lists, various type sizes and typefaces, different headings, and other page-design elements.

2. *Technical documents rarely get readers' undivided attention.* Readers may be skimming the document while they jot down ideas, talk on the phone, or drink coffee. Or they may refer to sections of the document during a meeting or presentation. Amid frequent dis-

tractions, readers must be able to leave the document and return easily.

3. *People read work-related documents only because they have to.* Novels, newspapers, and magazines are read for relaxation, but work-related documents mean *work*. The more complex the document, the harder readers have to work. If these readers had other ways of getting the information, they probably wouldn't read at all.

4. *As computers generate more and more paper, any document is forced to compete for readers' attention.* Even brilliant writing is useless unless it gets read by its intended audience. Suffering from information

```
                          Sunspaces
Either as an addition to a home or as an integral part of a
new home, sunspaces have gained considerable popularity.
     A sunspace should face within 30 degrees of true south. In
the winter, sunlight passes through the windows and warms the
darkened surface of a concrete floor, brick wall, water-filled
drums, or other storage mass. The concrete, brick, or water
absorbs and stores some of the heat until after sunset, when
the indoor temperature begins to cool. The heat not absorbed
by the storage elements can raise the daytime air temperature
inside the sunspace to as high as 100 degrees Fahrenheit. As
long as the sun shines, this heat can be circulated into the
house by natural air currents or drawn in by a low-horsepower
fan.
     In order to be considered a passive solar heating system,
any sunspace must consist of these parts: a collector, such as
a double layer of glass or plastic; an absorber, usually the
darkened surface of the wall, floor, or water-filled contain-
ers inside the sunspace; a storage mass, normally concrete,
brick, or water, which retains heat after it has been ab-
sorbed; a distribution system, the means of getting the heat
into and around the house by fans or natural air currents; and
a control system, or heat-regulating device, such as movable
insulation, to prevent heat loss from the sunspace at night.
Other controls include roof overhangs that block the summer
sun, and thermostats that activate fans.
```

Source: U.S. Department of Energy. *Sunspaces and Solar Greenhouses*. Washington, DC: GPO, 1984.

FIGURE 15.1 An Ineffective Format

SUNSPACES

Either as an addition to a home or as an integral part of
a new home, sunspaces have gained considerable popularity.

How Sunspaces Work
A sunspace should face within 30 degrees of true south.
In the winter, sunlight passes through the windows and
warms the darkened surface of a concrete floor, brick
wall, water-filled drums, or other storage mass. The
concrete, brick, or water absorbs and stores some of the
heat until after sunset, when the indoor temperature
begins to cool.

The heat not absorbed by the storage elements can raise
the daytime air temperature inside the sunspace to as
high as 100 degrees Fahrenheit. As long as the sun
shines, this heat can be circulated into the house by
natural air currents or drawn in by a low-horsepower fan.

The Parts of a Sunspace
In order to be considered a passive solar heating system,
any sunspace must consist of these parts:

1. A collector, such as a double layer of glass or plastic.
2. an absorber, usually the darkened surface of the wall,
   floor, or water-filled containers inside the sunspace.
3. A storage mass, normally concrete, brick, or water,
   which retains heat after it has been absorbed.
4. A distribution system, the means of getting the heat
   into and around the house (by fans or natural air
   currents).
5. A control system (or heat-regulating device), such as
   movable insulation, to prevent heat loss from the
   sunspace at night. Other controls include roof overhangs that block the summer sun, and thermostats that
   activate fans.

FIGURE 15.2 An Effective Format (Typewritten)

Sunspaces

Either as an addition to a home or as an integral part of a new home, sunspaces have gained considerable popularity.

How Sunspaces Work

A sunspace should face within 30 degrees of true south. In the winter, sunlight passes through the windows and warms the darkened surface of a concrete floor, brick wall, water-filled drums, or other storage mass. The concrete, brick, or water absorbs and stores some of the heat until after sunset, when the indoor temperature begins to cool.

The heat *not* absorbed by the storage elements can raise the daytime air temperature inside the sunspace to as high as 100 degrees Fahrenheit. As long as the sun shines, this heat can be circulated into the house by natural air currents or drawn in by a low-horsepower fan.

The Parts of a Sunspace

In order to be considered a passive solar heating system, any sunspace must consist of these parts:

1. A *collector,* such as a double layer of glass or plastic.
2. An *absorber,* usually the darkened surface of the wall, floor, or water-filled containers inside the sunspace.
3. A *storage mass,* normally concrete, brick, or water, which retains heat after it has been absorbed.
4. A *distribution system,* the means of getting the heat into and around the house (by fans or natural air currents).
5. A *control system* (or heat-regulating device), such as movable insulation, to prevent heat loss from the sunspace at night. Other controls include roof overhangs that block the summer sun, and thermostats that activate fans.

FIGURE 15.3 A Version Prepared on a Computer

overload, today's readers resist any document that appears overwhelming. They want formats that will help them find the information they really need. A "user-friendly" document has an accessible format: at a glance, readers can see how the document is organized, where they are in the document, which items matter most, and how the items relate.

Having decided at a glance whether your document is inviting and accessible, readers will draw conclusions about the value of your information, the quality of your work, and your credibility. For an otherwise useful document, a carefully crafted format becomes your stamp of professionalism.

Even though the information in Figure 15.1 is worthwhile and accurate, its format resists our attention. Without visual cues, we have no way of grouping this information into organized units of meaning. Figure 15.2 shows the same information after a format overhaul. Figure 15.3 shows one version possible on a computer. As readers, we take good format for granted—that is, we hardly notice format *unless* it is offensive.

AUTOMATED FORMATTING

For anyone who writes in the workplace, computer technology has large implications. Planning, drafting, and revising at their workstations, writers themselves are becoming increasingly responsible for all stages in document preparation.

Desktop publishing means less reliance on secretaries, print shops, and graphic artists. Using microcomputers, page-design software, and laser printers, writers at their desks can control the entire production task: designing, illustrating, laying out, and printing the final document.

As automated design continues to improve the look of workplace writing, audiences raise their standards. Now more than ever, readers expect documents to be inviting and accessible. Of course, an electronically produced document—no matter how attractive—is useless if it carries the same old mistakes found in typewriter-produced documents, or if it says nothing worthwhile.

FORMAT GUIDELINES

Whether you write with a typewriter or a computer, approach your formatting decisions from the top down: first, consider the overall look of your pages; next, the shape of each paragraph; and finally, the size and style of

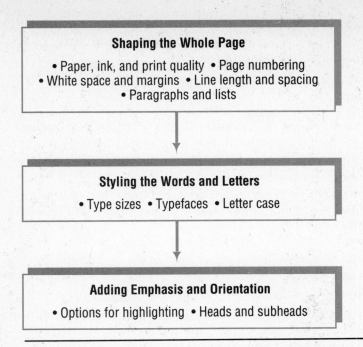

FIGURE 15.4 A Flowchart for Decisions in Format Design

individual letters and words (Kirsh 112). Figure 15.4 shows how format design follows a top-down sequence, moving from large matters to small.[1]

Different audiences may require different formats, but any format must be inviting and accessible. The following guidelines illustrate basic design principles that will enable your formats to satisfy any reader's expectations.

Use the Right Paper and Ink

Type or print in black ink, on 8½-by-11-inch, low-gloss, white paper. Use rag-bond paper (2 pounds or heavier) with a high fiber content (25 percent minimum). Shiny paper produces glare and tires the eyes.

Use High-Quality Type or Print

Keep erasures to a minimum, and retype all smudged pages. Typewriter erasures with opaquing fluid or opaquing film (correction tape) are neatest. Use a fresh ribbon and keep typewriter keys clean. On a computer, print

[1]For advice on formatting specific documents, see the chapters on memos and letters.

your hard-copy on a letter-quality printer, a laser printer, or a dot-matrix printer in the near-letter-quality mode. Many older dot-matrix printers produce copy that is hard to read or looks unprofessional.

Number Pages Consistently

For a long document, count your title page as page i, without numbering it, and number all front-matter pages, including your table of contents and abstract, with lowercase roman numerals (ii, iii, iv). Number the first page of your report and subsequent pages with arabic numerals (1, 2, 3). Place all page numbers in the upper-right corner, or centered in the top or bottom margin. (Some word processors may limit page-number placement; check the manual.)

Use Adequate White Space

White space is all the space not filled by text. White space separates sections in a document, headings and visuals from text, paragraphs on a page, sentences in a paragraph, words in a sentence, letters in a word. White space can be designed to enhance a document's appearance, clarity, and emphasis.

Well-designed white space imparts a shape to the whole document, a shape that orients readers and gives form to the printed matter by

- keeping related elements together
- isolating and emphasizing important elements
- creating a path for the reader's eyes
- providing breathing room between blocks of information

Pages that look uncluttered, inviting, and easy to follow convey an immediate sense of user-friendliness.

Provide Ample and Appropriate Margins

Small margins make a page look crowded and difficult. On your 8½-by-11-inch page, leave margins no smaller than these:

top margin	=	1¼ inches
bottom margin	=	1½ inches
right margin	=	1¼ inches
left margin	=	1¼ inches

For a document that will be bound, widen your left margin to two inches.

If you write using a computer, you can usually choose between *unjustified* text (uneven or "ragged" right margins) and *justified* text (even right margins).

Justified lines

To make the right margin even in justified text, the spaces vary between words and letters on a line, sometimes creating "channels" or rivers of white space. The reader's eyes are then forced to adjust continually to these space variations within a line or paragraph. And because each line ends at an identical vertical space, the eyes must work harder to differentiate one line from another (Felker 85). Moreover, to preserve the even margin, words at line's end are often hyphenated. And frequently hyphenated line endings can be distracting.

Unjustified lines

Unjustified text, on the other hand, uses equal spacing between letters and words on a line, and an uneven right margin (as traditionally produced by a typewriter). For some readers, a ragged right margin makes reading easier. These differing line lengths can prompt the eye to move from one line to another (Pinelli 77). In contrast to justified text, an unjustified page seems to look less formal, less distant, and less "official."

Justified text seems preferable for books, annual reports, brochures, newsletters, and other materials published for a broad readership. Unjustified text seems preferable for more personal forms of communication such as letters, memos, and in-house reports.

Keep Line Length Reasonable

Long lines tire the eyes. The longer the line, the harder it is for readers to return to the left margin and locate the next line (White 25).

Notice how your eye labors to follow the apparently endless message that here seems to stretch in lines that continue long after your eye was prepared to move down to the next line. After reading more than a few of these lines, you begin to feel tired and bored and annoyed, without hope of ever reaching the end.

Short lines force the eyes back and forth (Felker 79). "Too-short lines disrupt the normal horizontal rhythm of reading" (White 25).

Lines that are too
short cause your eye
to stumble from one
fragment to another
at a pace that too
soon becomes
annoying, if not
nauseating.

A reasonable line length is 50–60 characters (or 9 to 12 words) per line for an 8½-by-11-inch single-column page. The number of characters will depend on print size on your typewriter or word processor.

Line length, of course, is affected by the number of columns (vertical blocks of print) on your page. Two-column pages often appear in newsletters and brochures, but research indicates that single-column pages work best for complex, specialized information (Hartley 148).

Keep Line Spacing Consistent

For any document likely to be read completely (letters, memos, instructions), single-space within paragraphs and double-space between. Instead of indenting single-spaced paragraphs, separate them with a line of space.

For longer documents likely to be read selectively (proposals, formal reports), double-space within paragraphs. Indent your double-spaced paragraphs or separate them with a line of space. This open spacing enables readers to scan a long document, and to quickly locate what they need.

Tailor Each Paragraph to Its Purpose

Readers often skim a long document to find what they want. Most paragraphs therefore begin with a topic sentence forecasting the paragraph's content. (See also Chapter 12.)

Use a long paragraph (no more than fifteen lines) for clustering material that is closely related (such as history and background, or any body of information best understood in one block).

Use short paragraphs for making complex material more digestible, for giving step-by-step instructions, or for emphasizing vital information.

Instead of indenting a series of short paragraphs, separate them by inserting an extra line of space (as here).

Avoid "orphan" lines: the paragraph's opening line on the bottom of a page, or the paragraph's closing line on the top line of a page.

Make Lists for Easy Reading

Readers prefer information in list form rather than in continuous prose paragraphs (Hartley 51). Whenever you have a paragraph or long sentence containing a series of related items, consider displaying these items one by one.

Types of items you might list: advice or examples, conclusions and recommendations, criteria for evaluation, errors to avoid, materials and equipment for a procedure, parts of a mechanism, or steps or events in a sequence. Notice how the preceding information becomes easier to grasp and remember when displayed in the list below.

Types of items you might list:

- advice or examples
- conclusions and recommendations
- criteria for evaluation
- errors to avoid
- materials and equipment for a procedure
- parts of a mechanism
- steps or events in a sequence

A list of brief items usually needs no punctuation at the end of each line. But a list of full sentences (page 361) or questions (page 19) requires appropriate punctuation after each item.

Depending on the list's contents, set off each item with some kind of visual or verbal signal. If the items require a strict sequence (as in a series of steps, or parts of a mechanism), use arabic numbers (1,2,3) or the words *First, Second, Third,* and so on. If the items require no strict sequence (as in some examples or advice), use dashes, asterisks, or bullets (heavy dots).

Introduce your list with an explanation. And phrase all listed items in parallel grammatical form (pages 606–607). If the items suggest no strict sequence, try to impose some logical ranking (most important to least important, alphabetical, or some such). Set off the list with extra white space above and below.

Keep in mind that a document with too many lists appears busy, disconnected, and splintered (Felker 55). And long lists could be used by unethical writers to camouflage this or that piece of bad or embarrassing news (Williams 12). As with all format options, use restraint and good judgment.

Use Standard Type Sizes

Word-processing programs offer a variety of type sizes:

9 point
10 point
12 point
14 point
18 point
24 point

The standard type size for most documents is 10 to 12 point. Use larger or smaller sizes only for certain headings, titles, captions (brief explanation of a picture or other visual), or special emphasis. Certain typefaces (see below) are easier to read than others, making smaller type sizes possible.

Select Appropriate Typefaces

Typeface is the style of individual letters and characters. Each typeface can be said to have its own *personality*: "The typefaces you select for . . . [heads], subheads, body copy, and captions affect the way readers experience your ideas" (*Aldus Guide* 24).

Word-processing programs offer a variety of typefaces (or fonts) like the examples below, listed by name. Except for special emphasis, stick to the more conservative typefaces and avoid ornate ones altogether.

All typefaces divide into two broad categories: *serif* or *sans serif*. Serifs are fine lines that extend horizontally from the main strokes of a letter:

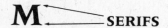

Geneva typeface
Monaco typeface
New York typeface
Chicago typeface
Venice typeface
Athens typeface
London typeface

Serif type makes body copy more readable because the horizontal lines "bind the individual letters" and thereby guide the reader's eyes from letter to letter—as in the type you are now reading (White 14).

In contrast, sans serif type is purely vertical (like this). Clean looking and "businesslike," sans serif is considered ideal for technical material (numbers, equations, etc.), marginal comments, headings, examples, tables, and captions to pictures and visuals, and any other material "set off" from the body copy (White 16).[2]

Avoid Sentences in Full Caps

Sentences or long passages in full capitals (uppercase letters) are hard to read because all uppercase letters and words have the same visual outline (Felker 87).

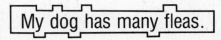

The longer the passage, the harder readers must work to sort things out and grasp your intended emphasis.

Hard   ACCORDING TO THE NATIONAL COUNCIL ON RADIATION PROTECTION, YOUR MAXIMUM ALLOWABLE DOSE OF LOW-LEVEL RADIATION IS 500 MILLIREMS PER YEAR.

Easier   According to the National Council on Radiation Protection, your MAXIMUM allowable dose of low-level radiation is 500 millirems per year.

Lowercase letters take up less space, and the distinctive shapes make each word easier to recognize and remember (Benson 37).

---

[2]European readers generally prefer sans serif type throughout their documents, and other cultures have their own particular preferences as well. Learn all you can about the design conventions of the culture you are addressing.

Use full caps as section headings (INTRODUCTION) or to highlight a word or phrase (WARNING: NEVER TEASE THE ALLIGATOR). As with other highlighting options discussed in the next section, use full caps sparingly in your document.

Highlight for Emphasis and Appeal

Effective highlighting helps readers distinguish the important from the less important elements. Highlighting options include typefaces, type sizes, white space, and other graphic devices that

- emphasize key points
- make headings prominent
- separate sections of a long document
- set off examples, warnings, and notes

On a typewriter, you can highlight with underlining, FULL CAPS, dashes, parentheses, and asterisks.

> You can indent to set off examples, explanations, or any material that should be distinguished from other elements in your document.

Using ruled (or typed) horizontal lines, you can separate sections in a long document:

Using ruled lines, broken lines, or ruled boxes, you can set off crucial information such as a warning or a caution:

— —

> *Caution:* A document with too many highlights can appear confusing, disorienting, and tasteless.

— —

Most software now offers an array of highlighting options that might include **boldface,** *italics,* SMALL CAPS, varying type sizes and typefaces, and color. For specific highlighted items, some options are better than others:

Boldface works well for emphasizing a single sentence or statement, and is perceived by readers as being "authoritative" (*Aldus Guide* 42).

Italics suggest a more subtle or "refined" emphasis than boldface (Aldus Guide 42). Italics can highlight words, phrases, book titles, or anything else you would have underlined on a typewriter. But multiple lines (like these) of italic type are hard to read.

SMALL CAPS WORK FOR HEADINGS AND SHORT PHRASES. BUT ANY LONG STATEMENT ALL IN CAPS IS HARD TO READ.

SMALL TYPE SIZES (USUALLY SANS SERIF) WORK WELL FOR CAPTIONS AND AS LABELS FOR VISUALS OR TO SET OFF OTHER MATERIAL FROM THE BODY COPY.

Large type sizes and dramatic typefaces are both hard to miss and hard to digest. Be conservative—unless you really need to convey a sense of forcefulness.

Color is appropriate only in some documents, and only when used sparingly. (Page 337 discusses how color can influence readers' perception and interpretation of a message.)

Whichever highlights you select for a document, be consistent. Make sure that all headings at one level are highlighted identically, that all warnings and cautions are set off identically, and so on. And *never* combine too many highlights.

Use Headings That Provide Access

Most long documents are read not like novels but like textbooks. Readers look back or jump ahead to sections that interest them most. Without headings, readers don't know how the document is organized, and they can't find what they need. Headings make your writing task easier as well: like guideposts or points on a road map, they keep you on course and provide transitions.

Headings signal readers that something is ending and something else is beginning. They make a document more attractive and less intimidating by dividing it into accessible chunks. An informative heading can help a reader decide whether a section is worth reading (Felker 17). Besides cutting down on reading and retrieval time, headings help readers remember information (Hartley 15).

To be effective, headings should conform to the guidelines that follow.

Make Headings Informative. A heading should be long enough to be informative, without being wordy. Informative headings orient readers, showing them what to expect. Vague or general headings can be more misleading than no headings at all (Redish et al. 144). Whether your heading takes the form of a phrase, a statement, or a question, be sure it advances thought.

Document Formatting

What should we expect here: specific instructions, an illustration, a discussion of formatting policy in general? We can't tell.

Informative
versions

How to Format Your Document

Format Your Document in This Way

How Do I Format My Document?

When you use questions as headings, phrase the questions in the same way as readers might ask them.

Make Headings Specific as Well as Comprehensive.   Focus the heading on a specific topic. Do not preface a discussion of the effects of acid rain on lake trout with a broad heading such as "Acid Rain." Use instead "The Effects of Acid Rain on Lake Trout."

Also, provide enough headings to contain each discussion section. If, say, chemical, bacterial, and nuclear wastes are three *separate* discussion items, provide a heading for each. Do not simply lump them under the sweeping heading "Hazardous Wastes." Adapt major and minor headings from your outline.

Make Headings Grammatically Consistent.   All major topics or minor topics in a document share equal rank; to emphasize this equality, express topics at the same level in identical—or parallel—grammatical form.

Nonparallel
headings

How to Avoid Damaging Your Disks:

1. Clean Disk-Drive Heads
2. Keep Disks Away from Magnets
3. Writing on Disk Labels with a Felt-Tip Pen
4. It is Crucial that Disks Be Kept Away from Heat
5. Disks Should Be Kept Out of Direct Sunlight
6. Keep Disks in Their Protective Jackets

In items 3, 4, and 5, the lack of parallelism (no verbs in the imperative mood) obscures the relationship between individual steps, and confuses readers. This next version emphasizes the equal rank of these items.

3. Write on Disk Labels with a Felt-Tip Pen
4. Keep Disks Away from Heat
5. Keep Disks Out of Direct Sunlight

Parallelism helps make a document readable and accessible.

Make Headings Visually Consistent. To understand how your document is organized, readers consider not only what the headings *say* but also how they *look:* "Wherever heads are of equal importance, they should be given similar visual expression, because that regularity itself becomes an understandable symbol" (White 104). Use identical type size and typeface for all headings at a given rank.

Lay Out Headings by Rank. Like a road map, your headings should identify clearly the large and small segments in your document. A long document ordinarily has section headings to mark its introduction, body, and conclusion. The body section in particular can have several lower ranks of headings: major and minor topic headings, and perhaps subtopic headings as well. (Use the logical divisions from your outline as a model for heading layout.) Think of each heading at a particular rank as an "event in a sequence" (White 95).

Figure 15.5 shows how headings (typewritten) vary in positioning and highlighting, depending on their rank. The layout in the figure embodies these guidelines:

- *Ordinarily, use no more than four levels of heading (section, major topic, minor topic, subtopic).* Excessive heads and subheads can make a document seem cluttered or fragmented.

- *To divide logically, be sure each higher-level heading yields at least two lower-level headings.*

- *Insert one additional line of space above your heading.* For double-spaced text, triple-space before the heading, and double-space after; for single-spaced text, double-space before the heading, and single-space after. The spacing indicates whether the heading belongs with the text preceding it or following it.

- *Never begin the sentence right after the heading with "this," "it," or some other pronoun referring to the heading.* Make the sentence's meaning independent of the heading.

```
                    SECTION HEADING

    Write section headings in full caps, centered horizontally, one
extra line space below any preceding text. Do not underline or
further highlight these heads.

Major Topic Heading
    Begin each word (except articles and prepositions) in major
topic heads with an uppercase letter. Abut the left margin (flush
left), one extra line space below preceding text. Underline.


    Minor Topic Heading
    Begin each important word in minor topic heads with an upper-
case letter. Indent the heads five spaces from the left margin, one
extra line space below preceding text. Underline these heads.

Subtopic Heading. Begin each important word in subtopic heads with
an uppercase letter. Instead of indenting, place the heads flush
left, one extra line space below preceding text, and on the same
line as following text (separated from the text by a period).
```

FIGURE 15.5 Recommended Format for Typewritten Headings

- *Never leave a heading floating on the final line of a page.* Carry the heading over to start the next page.
- *If possible, use different type sizes to reflect the ranks of heads.* Readers tend to equate large type size with importance (White 95). On a typewriter, set major heads in full caps. On a word processor set major heads in a larger type size, and perhaps in boldface (as in Figure 15.6).

When headings show the relationships among all the parts, readers can grasp at a glance how a document is organized.

Section Heading

On a word processor, section headings in boldface and enlarged type are more appealing and readable than heads in full caps. Use a type size roughly 4 points larger than body copy (say, 16-point section heads for 12-point body copy). Avoid *overly large* heads, and use no other highlights. Set these and *all* lower heads one extra line space below any preceding text.

Major Topic Heading

Major topic heads abut the left margin (flush left), and each important word begins with an uppercase letter. Use boldface and a type size roughly 2 points larger than body copy, with no other highlights.

Minor Topic Heading

Minor topic heads also are set flush left. Use boldface, italics (optional), and the same type size as in the body copy, with no other highlights.

Subtopic Heading. Instead of indenting the first line of body copy, place subtopic heads flush left on the same line as following text, and set off by a period. Use boldface and the same type size as in the body copy, with no other highlights.

FIGURE 15.6 Recommended Format for Word-Processed Headings

AUDIENCE CONSIDERATIONS

Like any writing decisions, formatting choices are by no means random. An effective writer designs a document for specific use by a specific audience.

How do you decide on a format? Your best bet is to work from a detailed audience-and-use profile (Wight 11). Know who your readers are and how they will use your information. Design a format to meet their particular needs, as in these examples:

- If readers will use your document for reference only (as in a repair manual), make sure you have plenty of headings.
- If readers will follow a sequence of steps, show that sequence in a numbered list.
- If readers will need to evaluate something, give them a checklist of criteria (as in this book at the end of most chapters).
- If readers need a warning, highlight the warning so that it cannot possibly be overlooked.
- If readers have asked for your one-page résumé, save space by using the 10-point type size.
- If readers will be facing complex information or difficult steps, widen the margins, increase all white space, and shorten the paragraphs.

How effectively you combine your format options will depend mostly on how carefully you have analyzed your audience. But regardless of your audience, never make the document look "too intellectually intimidating" (White 4).

Finally, keep in mind that even the most brilliant formatting cannot redeem a document whose content is worthless, organization chaotic, or style unreadable. The value of a document ultimately depends on elements that are more than skin deep.

REVISION CHECKLIST FOR DOCUMENT FORMATS

Use this checklist to evaluate and revise a document's format. (Numbers in parentheses refer to the first page of discussion.)

- ☐ Is the paper white, low-gloss, rag bond, with black ink? (351)
- ☐ Is all type or print neat and legible? (351)
- ☐ Does the white space guide the reader's eyes? (352)
- ☐ Are the margins ample? (352)

☐ Is line length reasonable? (353)

☐ Is the right margin unjustified? (353)

☐ Is line spacing appropriate and consistent? (354)

☐ Does each paragraph begin with a topic sentence? (354)

☐ Does the length of each paragraph suit its subject and purpose? (354)

☐ Are all paragraphs free of "orphan" lines? (354)

☐ Do parallel items in strict sequence appear in a numbered list? (355)

☐ Do parallel items of any kind appear in a list whenever a list is appropriate? (355)

☐ Are pages numbered consistently? (352)

☐ Is the body type size 10 or 12 point? (356)

☐ Do full caps highlight only words or short phrases? (357)

☐ Is the highlighting consistent and subdued? (358)

☐ Are all format patterns distinct enough so that readers will find what they need? (350)

☐ Are there enough headings so that readers always know where they are in the document? (359)

☐ Are headings informative, comprehensive, specific, parallel, and visually consistent? (359)

☐ Are all headings clearly differentiated according to rank? (361)

☐ Is the overall format inviting without being overwhelming? (364)

EXERCISES

1. Find an example of effective formatting, in a textbook or elsewhere. Photocopy a selection (two to three pages), and attach a memo explaining to your instructor and classmates why this format is effective. Be specific in your evaluation. Now do the same for an example of poor formatting. Bring your examples and explanations to class, and be prepared to discuss why you chose them.

2. These are headings from a set of instructions for listening. Rewrite the headings to make them parallel.

- You Must Focus on the Message
- Paying Attention to Nonverbal Communication

- Your Biases Should Be Suppressed
- Listen Critically
- Listening for Main Ideas
- Distractions Should Be Avoided
- Provide Verbal and Nonverbal Feedback
- Making Use of Silent Periods
- Are You Allowing the Speaker Time to Make His or Her Point?
- Keeping an Open Mind Is Important

3. Using the format checklist on page 364, redesign an earlier assignment or a document you've prepared on the job. Submit to your instructor the revision and the original, along with a memo explaining your improvements. Be prepared to discuss your format design in class.

4. Anywhere on campus or at work, locate a document with a format that needs revision. Candidates include career counseling handbooks, financial aid handbooks, student or faculty handbooks, software or computer manuals, medical information, newsletters, or registration procedures. Redesign the document or a two- to five-page selection from it. Submit to your instructor a copy of the original, along with a memo explaining your improvements. Be prepared to discuss your revision in class.

COLLABORATIVE PROJECT

Working in small groups, redesign the format of a document your instructor provides. Prepare a detailed explanation of your group's revision. Appoint a group member to present your revision to the class, using an opaque or overhead projector, a large-screen computer monitor, or photocopies.

Adding Document Supplements

Purpose of Supplements

Cover

Title Page

Letter of Transmittal

Table of Contents

Table of Tables and Figures

Informative Abstract

Glossary

Appendixes

Documentation

■ ■ ■

Supplements are reference items generally added to a long report or proposal to make the document more accessible. A document's supplements accommodate readers with various interests: The title page, letter of transmittal, table of contents, and abstract give summary information about the content of the document. The glossary, appendixes, and list of works cited can either provide supporting data or help readers follow technical sections. According to their needs, readers can refer to one or more or these supplements or skip them altogether. All supplements, of course, are written only *after* the document itself has been completed.

Some companies and organizations require that a full range of supplements routinely accompany any long document. Others do not. For situations in which your audience has not stipulated its requirements, select only those supplements that *enhance* the informative value of your particular document. Avoid using any supplements merely as decoration.

PURPOSE OF SUPPLEMENTS

Documents have to be *accessible* to varied readers for many purposes. Supplements address these workplace realities:

- *Confronted by a long document, many readers will try to avoid reading the whole thing.* Instead they look for the least information they need to complete the task, make the decision, or take some other action.
- *Different readers often use the same document for different purposes.* Some look for an overview; others want details; others want only conclusions and recommendations, or the "bottom line." Technical personnel might focus on the body of a highly specialized report and on the appendixes for supporting data (maps, formulas, calculations). Executives and managers might read only the transmittal letter and the abstract. If the latter audience reads any of the report proper, they are likely to focus on conclusions and recommendations.

A document with carefully planned supplements accommodates the needs of diverse audiences.

Report supplements can be classified in two groups:

1. *supplements that precede your report* (front matter): cover, title page, letter of transmittal, table of contents (and figures), and abstract
2. *supplements that follow your report* (end matter): glossary, appendix(es), footnotes, endnote pages

COVER

Use a sturdy, plain cover with page fasteners. With the cover on, the open pages should lie flat. Use covers only for long documents.

Center the report title and your name four to five inches from the upper edge:

THE FEASIBILITY OF A TECHNICAL MARKETING CAREER:
AN ANALYSIS
by
Richard B. Larkin, Jr.

4 Spaces

**Feasibility Analysis
of a Career
in Technical Marketing**

10-12 report

Large Font Size
Bold; centered
Contrast Title

4 Spaces

for

Professor J. M. Lannon
Technical Writing Instructor
Southeastern Massachusetts University
North Dartmouth, Massachusetts

Same
(same font in document)

4 Spaces

by

Richard B. Larkin, Jr.
~~English 266 Student~~

All Names
Don't read

May 1, 1993

Bottom Line (centered)

No Page number

FIGURE 16.1 Title Page for a Formal Report

TITLE PAGE

The title page lists the report title, author's name, name of person(s) or organization to whom the report is addressed, and date of submission.

Title.   Your title announces the report's purpose and subject. The previous title (given as an example for the cover) is clear, accurate, comprehensive, and specific. But even slight changes can distort this title's signal.

An unclear   A TECHNICAL MARKETING CAREER
title

The version above is unclear about the report's purpose. Is the report *describing* the career, *proposing* the career, *giving instructions* for career preparation, or *telling one person's career story?* Insert descriptive words ("analysis," "instructions," "proposal," "feasibility," "description," "progress") that accurately state your purpose.

To be sure that your title forecasts what the report delivers, write its final version *after* completing the report.

Placement of Title Page Items.   Do not number your title page, but count it as page i of your prefatory pages. Center the title horizontally, three to four inches below the upper edge, using all capital letters. Place other items in the spacing, order, and typeface shown in Figure 16.1, the title page to a report shown in Chapter 23. Or devise your own system, as long as your page is balanced.

LETTER OF TRANSMITTAL

For college reports, the letter of transmittal usually follows the title page and is bound as part of the report. For workplace reports, the letter usually precedes the title page. Include a letter of transmittal with any formal report or proposal addressed to a specific reader. Your letter adds a note of courtesy and gives you a place for personal remarks. For instance, your letter might

- acknowledge those who helped with the report
- refer to sections of special interest: unexpected findings, key visuals, major conclusions, special recommendations, and the like
- discuss the limitations of your study, or any problems gathering data
- discuss the need and approaches for follow-up investigations
- describe any personal (or "off-the-record") observations

For Thursday

My Address

165 Hammond Way
Hyannis, MA 02457
April 29, 1993

Same space

To: Professor

John Fitton
Placement Director
Southeastern Massachusetts University
North Dartmouth, MA 02747

Ethos oriented

Dear Mr. Fitton:

Intro — where it came from

intro

Here is my analysis to determine the feasibility of a career in technical marketing and sales. In preparing my report, I've learned a great deal about this career, and I believe my information will help other students as well.

Body

Although committed to their specialities, some technical and science graduates seem interested in careers in which they can apply their technical knowledge to customer and business problems. Technical marketing may be an attractive choice of career for those who know their field, who can relate to different personalities, and who are good communicators.

Technical marketing is competitive and demanding, but highly rewarding. In fact, it is an excellent route to upper-management and executive positions. Specifically, marketing work enables one to develop a sound technical knowledge of a company's products, to understand how these products fit into the marketplace, and to perfect sales techniques and interpersonal skills. This is precisely the kind of background that paves the way to top-level jobs.

Concl

I've enjoyed my work on this project, and would be happy to answer any questions.

Sincerely,

4 Spaces

Richard B. Larkin

Richard B. Larkin, Jr.

No Page number

FIGURE 16.2 Letter of Transmittal for a Formal Report

- suggest some special uses for the information
- urge the reader to immediate action

The letter of transmittal can be tailored to a particular reader, as is Richard Larkin's in Figure 16.2. If a report is being sent to a number of people who are variously qualified and bear various relationships to the writer, individual letters of transmittal may vary, within the following basic structure:

Introduction.   Open with reference to the reader's original request. Briefly review the reasons for your report. Maintain a confident and positive tone throughout. Indicate pride and satisfaction in your work. Avoid implied apologies, such as "I hope this report meets your expectations."

Body.   In the letter body, include items from our prior list of possibilities (acknowledgments, special problems). Although your informative abstract will summarize major findings, conclusions, and recommendations, your letter gives a brief and personal overview of the *entire project*.

Conclusion.   State your willingness to answer questions or discuss findings. End positively with something like "I believe that the data in this report are accurate, that they have been analyzed rigorously and impartially, and that the recommendations are sound."

TABLE OF CONTENTS

Your table of contents serves as a road map for readers and a checklist for you. Compose this supplement by assigning page numbers to headings from your outline. Keep in mind, however, that not all levels of outline headings appear in your table of contents or your report. Excessive headings can fragment the discussion.

Follow these guidelines:

- List front matter (transmittal letter, abstract), numbering the pages with small roman numerals. (The title page, though not listed, is counted as page i.) List glossary, appendix, and endnotes; number these pages with arabic numerals, continuing the page sequence of your report proper, in which page 1 is the first page of report text.
- Include no headings in the table of contents not listed as headings or subheadings in the report; the report may, however, contain subheadings not listed in the table of contents.
- Phrase headings in the table of contents exactly as in the report.

Double Space   *1st line*   <u>iii</u>

CONTENTS   *Bold; Caps*
                            Single space

Title Page — not listed   page

FIGURE 16.3   Table of Contents for a Formal Report

FIGURES AND TABLES

FIGURE 16.4 A Table of Tables and Figures for a Formal Report

- To reflect their rank, list headings at various levels in varying typeface and indention.

- Use dots (........) to connect heading to page number. Align rows of dots vertically, each above the other.

Figure 16.3 shows the table of contents for Richard Larkin's feasibility analysis.

TABLE OF TABLES AND FIGURES

Following the table of contents is a table of tables and figures, if needed. When a report has more than four or five visuals, place this table on a separate page. Figure 16.4 shows the table of tables and figures for Larkin's report.

INFORMATIVE ABSTRACT

Some readers have neither time nor willingness to read your entire report. For these readers, the informative abstract is the most important part of the document. The abstract is always written *after* your report proper.

Writing the abstract gives you the chance to measure your own control over the material. Because you are condensing your message, you must

ABSTRACT

[handwritten: — Revise Need]

[handwritten: 1st Paragraph 2-3 sentences]

[handwritten: why you did research {]

The feasibility of technical marketing as a career is based on a college graduate's interests, abilities, and expectations, as well as on possible entry options.

[handwritten: Double Space]

Technical marketing is a feasible career for anyone who is motivated, who can communicate well, and who knows how to get along. Although this career offers job diversity and excellent potential for income, it entails almost constant travel, competition, and stress.

College graduates enter technical marketing through one of four options: entry-level positions *[handwritten: Major Findings]* that offer hands-on experience, formal training programs in large companies, prior experience in one's specialty, or graduate programs. The relative advantages and disadvantages of each option can be measured in resulting immediacy of income, rapidity of advancement, and long-*[handwritten: Recommendations]* term potential.

Anyone considering a technical marketing career should follow these recommendations:

[handwritten: → Indent]

- Speak with people who work in the field.
- Weigh carefully the implications of each entry option.
- Consider combining two or more options.
- Choose options for personal as well as professional benefits.

FIGURE 16.5 Informative Abstract

establish explicit connections. If you cannot effectively summarize your report, it probably needs revision.

Follow these guidelines for your abstract:

- Make your abstract able to stand alone in meaning.
- Write for general readers. Readers of the abstract are likely to vary in expertise, perhaps more than those who read the report itself; translate all technical data.
- Add no new information. Simply summarize the report.
- Present your information in this sequence:

  a. Identify the issue or need that led to the report.

  b. Offer the major findings from the report body.

  c. Include a condensed conclusion and recommendations, if any.[1]

[1]My thanks to Professor Edith K. Weinstein for these suggestions.

The informative abstract in Figure 16.5 accompanies Richard Larkin's report in Chapter 23.

GLOSSARY

A glossary is an alphabetical listing of specialized terms and their definitions, following your report. Specialized reports often contain glossaries, especially when written for both technical and nontechnical readers. A glossary makes key definitions available to nontechnical readers without interrupting technical readers. If fewer than five terms need defining, place them instead in the report introduction as working definitions (as discussed in Chapter 17—or use footnote definitions). If you use a separate glossary, inform readers of its location: "(see the glossary at the end of this report)."

Follow these guidelines for a glossary:

- Define all terms unfamiliar to a general reader (an intelligent layperson).

- Define all terms that have a special meaning in your report (e.g., "In this report, a small business is defined as . . .").

- Define all terms by giving their class and distinguishing features (Chapter 17), unless some terms need expanded definitions.

GLOSSARY

Analgesic: a medication given to relieve pain during the first stage of labor.

Cervix: the neck-shaped anatomical structure that forms the mouth of the uterus.

Dilation: cervical expansion occurring during the first stage of labor.

Episiotomy: an incision of the outer vaginal tissue, made by the obstetrician just before the delivery, to enlarge the vaginal opening.

First stage of labor: the stage in which the cervix dilates and the baby remains in the uterus.

Induction: the stimulating of labor by puncturing the membranes around the baby or by giving an oxytoxic drug (uterine contractant), or both.

FIGURE 16.6 A Glossary (Partial)

- List your glossary and its first page number in your table of contents.
- List all terms in alphabetical order. Underline each term and use a colon to separate it from its single-spaced definition.
- Define only terms that need explanation. In doubtful cases, overdefining is safer than underdefining.
- On first use, place an asterisk in the text by each term defined in the glossary.

Figure 16.6 shows part of a glossary for a comparative analysis of two techniques of natural childbirth, written by a nurse practitioner for expectant mothers and student nurses.

APPENDIXES

An appendix follows the text of your report. The appendix expands items discussed in the report without cluttering the report text. Typical items in an appendix include

- details of an experiment
- statistical or other measurements
- maps
- complex formulas
- long quotations (one or more pages)
- photographs
- texts of laws and regulations
- related correspondence (letters of inquiry, and so on)
- interview questions and responses
- sample questionnaires and tabulated responses
- sample tests and tabulated results
- some visuals occupying more than one full page
- anything more essential to secondary readers than to primary readers

The appendix is a catch-all for items that are important but difficult to integrate within your text. Figure 16.7 shows an appendix to the budget proposal in Chapter 22 (pages 537–546).

Do not stuff appendixes with needless information. Do not use them unethically for burying bad or embarrassing news that belongs in the report proper. Follow the guidelines on page 378.

APPENDIX A

Table 1   Allocations and Performance of Five Massachusetts College Newspapers

	Stonehorse College	Alden College	Simms University	Fallow State	SMU
Enrollment	1,600	1,400	3,000	3,000	5,000
Fee paid (per year)	$65.00	$85.00	$35.00	$50.00	$65.00
Total fee budget	$88,000	$119,000	$105,000	$150,000	$334,429.28
Newspaper budget	$10,000	$6,000	$25,300	$37,000	$21,500
					$25,337.14[a]
Yearly cost per student	$6.25	$4.29	$8.43	$12.33	$4.06
					$5.20[a]
Format of paper	Weekly	Every third week	Weekly	Weekly	Weekly
Average no. of pages	8	12	18	12	20
Average total pages	224	120	504	336	560
					672[a]
Yearly cost per page	$44.50	$50.00	$50.50	$110.11	$38.25
					$38.69[a]

[a]These figures are next year's costs for the SMU *Torch*.

Source: Figures were quoted by newspaper business managers in April 1993.

FIGURE 16.7 An Appendix

- Include only material that is relevant.
- Use a separate appendix for each major item.
- Title each appendix clearly: "Appendix A: Projected Costs."
- Do not use appendixes excessively. Four or five appendixes in a ten-page report would indicate a poorly organized document.

- Limit an appendix to a few pages, unless length is imperative.
- Mention your appendix early in your introduction, and refer readers to it at appropriate points in the report: "(see Appendix A)."

Use an appendix for any material that is essential, but might harm the unity and coherence of your report. Remember that readers should be able to understand your report without having to turn to the appendix. Distill the essential facts from your appendix and place them in your report text.

Improper reference	The whale population declined drastically between 1976 and 1977 (see Appendix B for details).
Proper reference	The whale population declined *by 16 percent* from 1976 to 1977 (see Appendix B for statistical breakdown).

DOCUMENTATION

The endnote or works cited pages list each of your outside references in the same numerical order as they are cited in the report proper. See Chapter 9 for a discussion of documentation.

EXERCISES

1. These titles are intended for investigative, research, or analytical reports. Revise each inadequate title to make it clear and accurate.

a. The Effectiveness of the Prison Furlough Program in Our State

b. Drug Testing on the Job

c. The Effects of Nuclear Power Plants

d. Woodburning Stoves

e. Interviewing

f. An Analysis of Vegetables (for a report assessing the physiological effects of a vegetarian diet)

g. Wood As a Fuel Source

h. Oral Contraceptives

i. Lie Detectors and Employees

2. Prepare a title page, a letter of transmittal (for a definite reader who can *use* your information in a definite way), a table of contents, and an informative abstract for a report you have written earlier.

3. Find a short but effective appendix in one of your textbooks, in a journal article in your field, or in a report from your workplace. In a memo to your instructor and classmates, explain how the appendix is used, how it relates to the main text, and why it is effective. Attach a copy of the appendix to your memo. Be prepared to discuss your evaluation in class.

Specific Documents and Applications

V

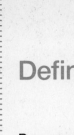

Definitions

Purpose of Definitions

Elements of Definition

Types of Definition

Expansion Methods

Sample Situations

Placement of Definitions

■   ■   ■

To define a term is to explain the precise meaning you intend by using that term. Clear writing depends on definitions that both reader and writer understand. Unless you are sure readers know the exact meaning you intend, always define the term upon first use.

PURPOSE OF DEFINITIONS

Every specialty has its own technical "language." Engineers, architects, or programmers talk about "torque," "tolerances," or "microprocessors"; lawyers, real estate brokers, and investment counselors discuss "easements," "liens," "amortization," or "escrow accounts"—and so on. Whenever such terms are unfamiliar to an audience, they need defining.

For colleagues, you rarely have to define specialized terms (unless the term is new), but reports often are written for the layperson—the client or some other general reader. When you write for nonspecialists, clarify your meaning with definitions.

Most of the specialized terms previously mentioned are concrete and specific. Once "microprocessor" has been defined for the reader, its meaning will not differ appreciably in another context. And when a term is highly technical, a writer can figure out that it should be defined for some readers. Any nonspecialist knows that he or she has no idea what "torque" or "diffraction" means. Readers are less likely to realize, however, that familiar terms like "disability," "guarantee," "tenant," "lease," or "mortgage" acquire very specialized meanings in specialized contexts. Here definition (by all parties) becomes crucial. What "guarantee" means in one situation is not necessarily what it means in another. Contracts are detailed (and legal) definitions of the specific terms of an agreement.

Assume you're shopping for disability insurance to protect your income in case of injury or illness. Besides comparing prices, you want each company to define "physical disability." Although Company A offers the cheapest policy, it might define physical disability as inability to work at *any* job. Therefore, if a neurological disorder prevents you from continuing work as designer of electronic devices, without disabling you for some menial job, you might not qualify as "disabled." In contrast, Company B's policy, although more expensive, might define physical disability as inability to work at your *specific* job. Both companies use the term "physical disability," but each defines it differently. Because you are legally responsible for documents bearing your signature, you need to understand the importance and technique of clear definition.

Definition carries *ethical* requirements, too. Chapter 5, for instance, explains how the shuttle *Challenger* explosion resulted from faulty definition: namely, the definition of "acceptable risk" finally was determined not by the technical facts, but by social pressures to launch on schedule (pages 66–69). And it was this social definition of acceptable risk that was passed forward and influenced the top decision makers to approve the disastrous launch. Agreeing on meaning in such cases may not be a simple affair, but we are ethically bound to communicate definitions based on a legitimate interpretation of the facts as we understand them.

Growth in technology and specialization makes definition increasingly vital. Know for whom you're writing, and why. If unable to pinpoint your audience, assume a general readership and define generously.

ELEMENTS OF DEFINITION

For all definitions, use these guidelines:

Plain English

Your purpose is to clarify meaning, not muddy it. Use language readers understand.

Unclear	A tumor is a neoplasm.
Better	A tumor is a growth of cells that occurs independently of surrounding tissue, and serves no useful function.
Unclear	A solenoid is an inductance coil that serves as a tractive electromagnet. *(A definition appropriate for an engineering manual, but too specialized for general readers.)*
Better	A solenoid is an electrically energized coil that converts electrical energy to magnetic energy capable of performing mechanical functions.

Basic Properties

Definition should convey the properties of an entity that differentiate it from all others. A thermometer has a singular function: it measures temperature; without this essential information, a definition would have no real meaning for uninformed readers. Any other data about thermometers (types, special uses, materials used in construction) are secondary. A book, on the other hand, cannot be defined according to functional properties because books have multiple functions. A book can be used to write in or to display pictures, to record financial transactions, to read, and so on. Also, other items (individual sheets of paper, posters, newspapers, picture frames) serve the same functions. The basic property of a book is physical: it is a bound volume of pages. Readers would have to know this *first*, to understand what a book is.

Objectivity

Unless readers understand that your purpose is to persuade, omit your opinions from a definition. "Bomb" is defined as "an explosive weapon detonated by impact, proximity to an object, a timing mechanism, or other predetermined means." If you define a bomb as "an explosive weapon devised and perfected by hawkish idiots to blow up the world," you are editorializing, *and* ignoring a bomb's basic property.

Likewise, in defining "diesel engine," simply tell what it is and how it works. You might think that diesels are too noisy and sluggish for automobiles, but omit these judgments unless readers ask for them.

TYPES OF DEFINITION

Definitions vary greatly in length and detail: from a few words in parentheses, to one or more complete sentences, to multiple paragraphs or pages.

Your choice of definition type depends on what information readers need, and that, in turn, depends on why they need it. "Carburetor," for instance, could be defined in one sentence, briefly telling readers what it is and how it works. But this definition would be expanded for the student mechanic who needs to know the origin of the term, how the device was developed, what it looks like, how it is used, and how its parts interact. Audience needs should guide your choice.

Parenthetical Definition

A parenthetical definition explains the term in a word or phrase, often as a synonym in parentheses following the term:

> The effervescent (bubbling) mixture is highly toxic.
>
> The leaching field (sievelike drainage area) requires 15 inches of crushed stone.

Parenthetical definitions

Another option is to express your definition as a clarifying phrase:

> The trees on the site are mostly deciduous; that is, they shed their foliage at season's end.

Use parenthetical definitions to convey the general meaning of specialized terms so that readers can follow the discussion where these terms are used. A parenthetical definition of "leaching field" might be adequate in a progress report to a client whose house you are building. But a public-health report titled "Groundwater Contamination from Leaching Fields" would call for expanded definition.

Sentence Definition

A definition may require one or more sentences with this structure: (1) the item or term being defined, (2) the class (specific group) to which the term belongs, and (3) the features that differentiate the term from all others in its class.

Term	Class	Distinguishing features
carburetor	a mixing device	in gasoline engines that blends air and fuel into a vapor for combustion within the cylinders

Elements of sentence definitions

transit	a surveying instrument	that measures horizontal and vertical angles
diabetes	a metabolic disease	caused by a disorder of the pituitary gland or pancreas, and characterized by excessive urination, persistent thirst, and decreased ability to metabolize sugar
liberalism	a political concept	based on belief in progress, the essential goodness of people, and the autonomy of the individual, and standing for the protection of political and civil liberties
brief	a legal document	containing all the facts and points of law pertinent to a specific case, and filed by an attorney before the case is argued in court
stress	an applied force	that strains or deforms a body
laser	an electronic device	that converts electrical energy to light energy, producing a bright, intensely hot, and narrow beam of light
fiber optics	a technology	that uses light energy to transmit voices, video images, and data through hair-thin glass fibers

In this presentation, these elements are combined into one or more complete sentences.

A complete sentence definition

Diabetes is a metabolic disease caused by a disorder of the pituitary gland or pancreas. This disease is characterized by excessive urination, persistent thirst, and decreased ability to metabolize sugar.

Sentence definition is especially useful if you need to stipulate the precise working definition of a term that has several possible meanings. State your working definitions at the beginning of your report:

Classifying the Term.   Be specific and precise in your classification. The narrower your class, the more specific your meaning. "Transit" is correctly classified as a "surveying instrument," not as a "thing" or an "instrument." "Stress" is classified as "an applied force"; to say that stress "takes place when . . ." or "is something that . . ." fails to reflect a specific classification. Be sure to select precise terms of classification: "Diabetes" is precisely classified as "a metabolic disease," not as "a medical term."

Differentiating the Term.   Differentiate the term by separating the item it names from every other item in its class. Make these distinguishing features narrow enough to pinpoint the item's unique identity and meaning, yet broad enough to be inclusive. A definition of "brief" as "a legal document introduced in a courtroom" is too broad because the definition doesn't differentiate "brief" from all other legal documents (wills, written confessions, etc.). Conversely, differentiating "carburetor" as "a mixing device used in automobile engines" is too narrow because it ignores the carburetor's use in all other gasoline engines.

Also, avoid circular definitions (repeating, as part of the distinguishing features, the word you are defining). Thus, "stress" should not be defined as "an applied force that places stress on a body." The class and distinguishing features must express the item's basic property ("an applied force that strains or deforms a body").

Expanded Definition

An expanded definition can include parenthetical and sentence definitions, but it provides greater detail for readers who need it. The sentence definition of "solenoid" on page 385 is good for a general reader who simply needs to know what a solenoid is. But a manual for mechanics or mechanical engineers would define solenoid in detail (as on pages 395–396); these readers need to know also how a solenoid works and how to use it.

The problem with defining an abstract and general word, such as "condominium" or "loan," is different. "Condominium" is a vaguer term than "solenoid" (solenoid A is pretty much like solenoid B) because the former refers to many types of ownership agreements.

Concrete, specific terms such as "diabetes," "transit," and "solenoid" often can be defined in a sentence, and require expanded definition only for certain audiences. But terms such as "disability" and "condominium" require expanded definition for almost any audience. The more general or abstract the term, the more need for expanded definition.

An expanded definition may be a single paragraph (as for a simple tool) or may extend to scores of pages (as for a digital dosimeter—a device for measuring radiation exposure); sometimes the definition itself *is* the whole report.

This excerpt from an automobile insurance policy defines coverage for "bodily injury to others." Its style and detail make this definition clear to general readers. Instead of the "legalese" in some policies, this policy is in plain English.

PART I. BODILY INJURY TO OTHERS

Under this Part, we will pay damages to people injured or killed by your auto in Massachusetts accidents. Damages are the amounts an injured person is legally entitled to collect for bodily injury through a court judgment or settlement. We will pay only if you or someone else using your auto with your consent is legally responsible for the accident. The most we will pay for injuries to any one person as a result of any one accident is $5,000. The most we will pay for injuries to two or more people as a result of any one accident is a total of $10,000. This is the most we will pay as the result of a single accident no matter how many autos or premiums are shown on the Coverage Selections page.

We will *not* pay:

1. For injuries to guest occupants of your auto.
2. For accidents outside of Massachusetts or in places in Massachusetts where the public has no right of access.
3. For injuries to any employees of the legally responsible person if they are entitled to Massachusetts workers' compensation benefits.

EXPANSION METHODS

How you expand a definition depends on the questions you think readers need answered, as shown in Figure 17.1. Begin with a sentence definition, and then use only those expansion strategies which serve your reader's needs.

Etymology

A word's origin (its development and changing meanings) can clarify its definition. *Biological control* of insects is derived from the Greek "bio," meaning *life* or *living organism*, and the Latin "contra," meaning *against* or *opposite*. Biological control, then, is the use of living organisms against in-

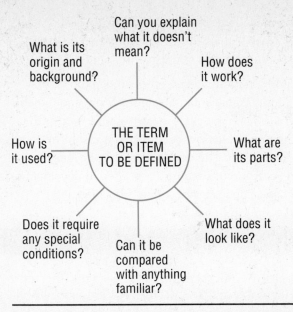

FIGURE 17.1 Directions in Which a Definition Can Be Expanded

sects. College dictionaries contain etymological information, but your best bet is *The Oxford English Dictionary* and encyclopedic dictionaries of science, technology, and business.

Some technical terms are acronyms, derived from the first letters or parts of several words. *Laser* is an acronym for *light amplification by stimulated emission of radiation*.

Sometimes a term's origin can be colorful as well as informative. *Bug* (jargon for *programming error*) is said to derive from an early computer at Harvard that malfunctioned because of a dead bug blocking the contacts of an electrical relay. Because programmers, like many of us, were reluctant to acknowledge error, the term became a euphemism for *error*. Correspondingly, *debugging* is the correcting of errors in a program.

History and Background

The meaning of specialized terms such as "radar," "bacteriophage," "silicon chips," or "X ray" often can be clarified through a background discussion: discovery or history of the concept, development, method of production, applications, and so on. Specialized encyclopedias are a good background source.

The idea of lasers . . . dates back as far as 212 B.C., when Archimedes used a [magnifying] glass to set fire to Roman ships during the siege of Syracuse.   (Gartaganis 22)

The early researchers in fiber optic communications were hampered by two principal difficulties—the lack of a sufficiently intense source of light and the absence of a medium which could transmit this light free from interference and with a minimum signal loss. Lasers emit a narrow beam of intense light, so their invention in 1960 solved the first problem. The development of a means to convey this signal was longer in coming, but scientists succeeded in developing the first communications-grade optical fiber of almost pure silica glass in 1970.   (Stanton 28)

Negation

Readers can grasp some meanings by understanding clearly what the term *does not* mean. The insurance policy on page 388 defines coverage for "bodily injury to others" partly by using negation: "We will *not* pay: 1. For injuries to guest occupants of your auto. . . ." In this next example, negation clarifies the definition of "lasers": "Lasers are not merely weapons from science fiction."

Operating Principle

Most items work according to an operating principle, whose explanation should be part of your definition:

A clinical thermometer works on the principle of heat expansion: as the temperature of the bulb increases, the mercury inside expands, forcing a mercury thread up into the hollow stem.

Air-to-air solar heating involves circulating cool air, from inside the home, across a collector plate (heated by sunlight) on the roof. This warmed air is then circulated back into the home.

Basically, a laser [uses electrical energy to produce] coherent light, light in which all the waves are in phase with each other, making the light hotter and more intense.   (Gartaganis 23)

[A fiber optics] system works as follows: An electrical charge activates the laser . . . , and the resulting light . . . energy passes through the optical fiber. At the other end of the fiber, a . . . receiver . . . converts this light signal back into electrical impulses.   (Stanton 28)

Even abstract concepts or processes can be explained on the basis of their operating principle:

> Economic inflation is governed by the principle of supply and demand: If an item or service is in short supply, its price increases in proportion to its demand.

Analysis of Parts

When your subject can be divided into parts, identify and explain them:

> The standard frame of a pitched-roof wooden dwelling consists of floor joists, wall studs, roof rafters, and collar ties.

"What are its parts?"

> Psychoanalysis is an analytic and therapeutic technique consisting of four parts: (1) free association, (2) dream interpretation, (3) analysis of repression and resistance, and (4) analysis of transference.

In discussing each part, of course, you would further define specialized terms such as "floor joists" and "repression."

Analysis of parts is particularly useful for helping nontechnical readers understand a technical subject. This next analysis helps explain the physics of lasing by dividing the process into three discrete parts:

1. [Lasers require] a source of energy, [such as] electric currents or even other lasers.
2. A resonant circuit . . . contains the lasing medium and has one fully reflecting end and one partially reflecting end. The medium—which can be a solid, liquid, or gas—absorbs the energy and releases it as a stream of photons [electromagnetic particles that emit light]. The photons . . . vibrate between the fully and partially reflecting ends of the resonant circuit, constantly accumulating energy—that is, they are amplified. After attaining a prescribed level of energy, the photons can pass through the partially reflecting surface as a beam of coherent light and encounter the optical elements.
3. Optical elements—lenses, prisms, and mirrors—modify size, shape, and other characteristics of the laser beam and direct it to its target. (Gartaganis 23)

Figure 1 (page 392) shows the three parts of a laser.

Visuals

Well-labeled visuals (such as the laser description) are excellent for clarifying definitions. Always introduce your visual, and explain it. If your vi-

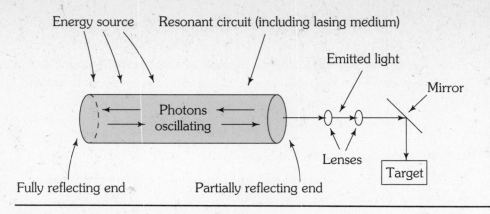

Energy source   Resonant circuit (including lasing medium)

Emitted light

Mirror

Photons oscillating

Lenses

Target

Fully reflecting end   Partially reflecting end

FIGURE 1   **Description of a Simple Laser**

sual is borrowed, credit your source at the bottom left. Unless the visual takes up one whole page or more, do not place it on a separate page. Include the visual near its discussion.

Comparison and Contrast

Comparisons and contrasts help readers understand. Analogies (a type of comparison) to something familiar can help explain the unfamiliar:

"Does it resemble anything familiar?"

> To visualize how a simplified earthquake starts, imagine an enormous block of gelatin with a vertical knife slit through the middle of its lower half. Gigantic hands are slowly pushing the right side forward and pulling the left side back along the slit, creating a strain on the upper half of the block that eventually splits it. When the split reaches the upper surface, the two halves of the block spring apart and jiggle back and forth before settling into a new alignment. Inhabitants on the upper surface would interpret the shaking as an earthquake.   ("Earthquake Hazard Analysis" 8)

> The average diameter of an optical cable is around two-thousandths of an inch, making it about as fine as a hair on a baby's head   (Stanton 29–30).

Here is a contrast between optical fiber and conventional copper cable:

"How does it differ from comparable things?"

> Beams of laser light coursing through optical fibers of the purest glass can transmit many times more information than the present communications systems. . . . A pair of optical fibers has the capacity to carry more than 10,000 times as many signals as conventional copper cable. A ½-inch optical cable can carry as much information as a copper cable as thick as a person's arm. . . .

Not only does fiber optics produce a better signal, [but] the signal travels farther as well. All communications signals experience a loss of power, or attenuation, as they move along a cable. This power loss necessitates placement of repeaters at one- or two-mile intervals of copper cable in order to regenerate the signal. With fiber, repeaters are necessary about every thirty or forty miles, and this distance is increasing with every generation of fiber.   (Stanton 27–28)

Here is a combined comparison and contrast:

Fiber optics technology results from the superior capacity of lightwaves to carry a communications signal. Sound waves, radio waves, and light waves can all carry signals; their capacity increases with their frequency. Voice frequencies carried by telephone operate at 1000 cycles per second, or hertz. Television signals transmit at about 50 million hertz. Light waves, however, operate at frequencies in the hundreds of trillions of hertz.   (Stanton 28)

"How is it both similar *and* different?"

Required Materials or Conditions

Some items or processes need special materials and handling, or they may have other requirements or restrictions. An expanded definition should include this important information.

Besides training in engineering, physics, or chemistry, careers in laser technology require a strong background in optics (study of the generation, transmission, and manipulation of light).

"What is needed to make it work (or occur)?"

Abstract concepts might also be defined in terms of special conditions:

To be held guilty of libel, a person must have defamed someone's character through written or pictorial statements.

Example

Familiar examples showing types or uses of an item can help clarify your definition. This example shows how laser light is used as a heat-generating device:

Lasers are increasingly used to treat health problems. Thousands of eye operations involving cataracts and detached retinas are performed every year by ophthalmologists. . . . Dermatologists treat skin problems. . . . Gynecologists treat problems of the reproductive system, and neurosurgeons even perform brain surgery—all using lasers transmitted through optical fibers.   (Gartaganis 24-25)

"How is it used or applied?"

The next example shows how laser light is used to carry information:

> The use of lasers in the calculating and memory units of computers, for example, permits storage and rapid manipulation of large amounts of data. And audiodisc players use lasers to improve the quality of the sound they reproduce. The use of optical cable to transmit data also relies on lasers.   (Gartaganis 25)

And this final example shows how optical fiber can relay a video signal:

> Acting, in essence, as tiny cameras, optical fibers can be inserted into the body and relay an image to an outside screen.   (Stanton 28)

Examples are a most powerful communication tool—as long as you tailor the examples to the readers' level of understanding.

Whichever expansion strategies you use, be sure to document your information sources. (See Chapter 9 for documentation formats.) Place directly quoted statements within quotation marks.

SAMPLE SITUATIONS

The following definitions employ expansion strategies appropriate to their audiences' needs. Specific strategies are labeled in the margin. Each definition, like a good essay, is unified and coherent: each paragraph is developed around one main idea and logically connected to other paragraphs. Visuals are incorporated. Transitions emphasize the connection between ideas. Each definition is at a level of technicality that connects with the intended audience.

To illustrate the importance of audience analysis in a writer's decision about "How much is enough?" this example, like many throughout the text, is preceded by an audience-and-use profile based on the worksheet on page 32.

An Expanded Definition for Semitechnical Readers

AUDIENCE-AND-USE PROFILE. The intended readers of this material are beginning student mechanics. Before they can repair a solenoid, they will need to know where the term comes from, what a solenoid looks like, how it works, how its parts operate, and how it is used. This definition is designed as merely an *introduction,* and so it offers only a general (but comprehensive) view of the mechanism.

Because the intended readers are not engineering students, they do *not* need details about electromagnetic or mechanical theory (e.g., equations or graphs illustrating voltage magnitudes, joules, lines of force).

EXPANDED DEFINITION: "SOLENOID"

A solenoid is an electrically energized coil that forms an electromagnet capable of performing mechanical functions. The term "solenoid" is derived from the word "sole," which in reference to electrical equipment means "a part of," or "contained inside, or with, other electrical equipment." The Greek word *solenoides* means "channel," or "shaped like a pipe." Formal sentence definition

Etymology

A simple plunger-type solenoid consists of a coil of wire attached to an electrical source, and an iron rod that passes in and out of the coil along the axis of the spiral. A spring holds the rod outside the coil when the current is de-energized, as shown in Figure 1. Description and analysis of parts

When the coil receives electric current, it becomes a magnet and thus draws the iron rod inside, along the length of its cylindrical center. With a lever attached to its end, the rod can transform electrical energy into mechanical force. The amount of mechanical force produced is the product of the number of turns in the coil, the strength of the current, and the magnetic conductivity of the rod. Special conditions and operating principle

The plunger-type solenoid in Figure 1 is commonly used in the starter motor of an automobile engine. This type is 4½ inches long and 2 inches in diameter, with a steel casing attached to the casing of the starter motor. A linkage (pivoting lever) is attached at one end to the iron rod of the solenoid, and at the other end to the drive gear of the starter, as shown in Figure 2. When the ignition key is turned, current from the battery is supplied to the solenoid coil, and the iron rod is drawn inside the coil, thereby shifting the attached linkage. The linkage, in turn, engages the drive gear, activated by the starter motor, with the flywheel (the main rotating gear of the engine). Example and analysis of parts

Explanation of visual

Because of the solenoid's many uses, its size varies according to the work it must do. A small solenoid will have a small wire coil, hence a weak magnetic Comparison of sizes and applications

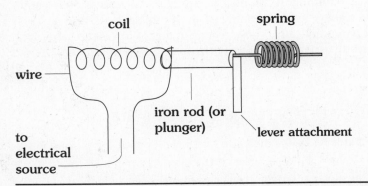

FIGURE 1 Side View of a Plunger-type Solenoid

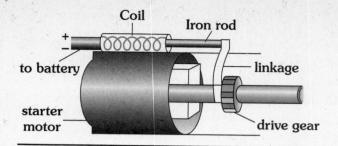

FIGURE 2 Side View of Solenoid and Starter Motor

field. The larger the coil, the stronger the magnetic field; in this case, the rod in the solenoid can do harder work. An electronic lock for a standard door would, for instance, require a much smaller solenoid than one for a bank vault.

The audience for the following definition (an entire community) is too diverse to define precisely, and so the writer wisely addresses the lowest level of technicality—to ensure that all readers will understand.

An Expanded Definition for Nontechnical Readers

AUDIENCE-AND-USE PROFILE. The following definition is written for members of a community whose water supply (all obtained from wells, because the town has no reservoir) is doubly threatened: (1) by chemical seepage from a recently discovered toxic dump site, and (2) by a two-year drought that has severely depleted the water table. This definition forms part of a report analyzing the severity of the problems and exploring possible solutions.

To understand the problems, these readers first need to know what a water table is, how it is formed, what conditions affect its level and quality, and how it figures into town planning decisions. The concepts of *recharge* and *permeability* are vital to readers' understanding of the problem here, and so these terms are defined parenthetically. These readers have no interest in geological or hydrological (study of water resources) theory. They simply need a broad picture.

EXPANDED DEFINITION: WATER TABLE

Formal sentence
definition

Example

The water table is the level below the earth's surface at which the ground is saturated with water. Figure 1 shows a typical water table that might be found in the East. Wells driven into such a formation will have a water level identical to that of the water table.

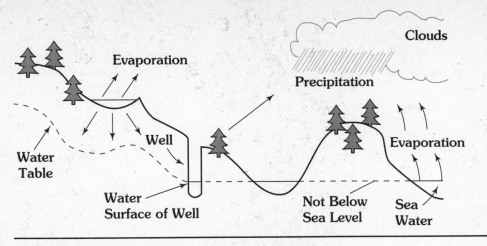

FIGURE 1 A Typical Water Table (Eastern United States)

Operating principle

The world's fresh-water supply comes almost entirely as precipitation that originates with the evaporation of sea and lake water. This precipitation falls to earth, and follows one of three courses: it may fall directly onto bodies of water, such as rivers or lakes, where it is directly used by humans; it may fall onto land, and either evaporate or run over the ground to the rivers or other bodies of water; or it may fall onto land, be contained, and seep into the earth. The latter precipitation makes up the water table.

Comparison

Similar in contour to the earth's surface above, the water table generally has a level that reflects such features as hills and valleys. Where the water table intersects the ground surface, a stream or pond results.

Operating principle

Example

A water table's level, however, will vary, depending on the rate of recharge (replacement of water). The recharge rate is affected by rainfall or soil permeability (the ease with which water flows through the soil). A water table then is never static; rather, it is the surface of a body of water striving to maintain a balance between the forces which deplete it and those which replenish it. In areas of Florida and some western states where the water table is depleted, the earth caves in, leaving sinkholes.

Special conditions and examples

The water table's depth below ground is vital in water resources engineering and planning. It determines an area's suitability for wastewater disposal, or a building lot's ability to handle sewage. A high water table could become contaminated by a septic system. Also, bacteria and chemicals seeping into a water table can pollute an entire town's water supply. Another consideration in water-table depth is the cost of drilling wells. These conditions obviously affect an industry's or homeowner's decision on where to locate.

The rising and falling of the water table give an indication of the pumping rate's effect on a water supply (drawn from wells), and of the sufficiency of the recharge rate in meeting demand. This kind of information helps water resources planners decide when new sources of water must be made available.

PLACEMENT OF DEFINITIONS[1]

Poorly placed definitions interrupt the information flow. If you have only a few parenthetical definitions, place them in parentheses after the terms. Any more than a few definitions per page will be disruptive. Rewrite them as sentence definitions and place them in a "Definitions" section of your introduction, or in a glossary (Chapter 16).

If your sentence definitions are few, place them in a "Definitions" section of your introduction; otherwise, in a glossary. Definitions of terms in the report's title belong in your introduction.

Place expanded definitions in one of three locations:

1. If the definition is essential to the reader's understanding of the *entire* document, place it in the introduction. A report titled "The Effects of Aerosol on the Earth's Ozone Shield" would require expanded definitions of "aerosol" and "ozone shield" early.

2. When the definition clarifies a major part of your discussion, place it in that section of your report. In a report titled "How Advertising Influences Consumer Habits," "operant conditioning" might be defined early in the appropriate section. Too many expanded definitions *within* a report, however, can be disruptive.

3. If the definition serves only as a reference, place it in an appendix (Chapter 16). A report on fire safety in a public building might have an expanded definition of "smoke detectors" in an appendix.

REVISION CHECKLIST FOR DEFINITIONS

Use this list to revise your definition. (Numbers in parentheses refer to the first page of discussion.)

---

[1] Electronic documents pose special problems for placement of definitions. In a hypertext document, for instance, each reader explores the material differently (Chapter 6). One option for making definitions available when they are needed is the "pop-up note": the term to be defined is highlighted in the text, to indicate that its definition can be called up and displayed in a small window on the actual text screen (Horton 25).

Content

☐ Is the type of definition (parenthetical, sentence, expanded) suited to its purpose and readers' needs? (385)

☐ Does the definition convey the basic property of the item? (384)

☐ Is the definition objective? (384)

☐ Is the expanded definition adequately developed? (387)

☐ Are all data sources documented? (394)

☐ Are visuals employed adequately and appropriately? (391)

Arrangement

☐ Does the sentence definition describe features that distinguish the term from other items in the same class? (387)

☐ Is the expanded definition unified and coherent (like an essay)? (394)

☐ Are transitions between ideas adequate? (394)

☐ Does the definition appear in the appropriate location? (398)

Style

☐ Is the definition in plain English? (388)

☐ Will the level of technicality connect with the audience? (394)

☐ Are sentences clear, concise, and fluent? (251)

☐ Is word choice precise? (279)

☐ Is the definition written in grammatical English? (Appendix A)

☐ Is the definition ethically acceptable? (383)

EXERCISES

1. Sentence definitions require precise classification and differentiation. Is each of these definitions adequate for a general reader? Rewrite those which seem inadequate. Consult dictionaries and encyclopedias as needed.

  a. A bicycle is a vehicle with two wheels.

  b. A transistor is a device used in transistorized electronic equipment.

  c. Surfing is when one rides a wave to shore while standing on a board specifically designed for buoyancy and balance.

  d. Bubonic plague is caused by an organism known as *pasteurella pestis.*

  e. Mace is a chemical aerosol spray used by the police.

  f. A Geiger counter measures radioactivity.

  g. A cactus is a succulent.

  h. In law, an indictment is a criminal charge against a defendant.

  i. A prune is a kind of plum.

  j. Friction is a force between two bodies.

  k. Luffing is what happens when one sails into the wind.

  l. A frame is an important part of a bicycle.

  m. Hypoglycemia is a medical term.

  n. An hourglass is a device used for measuring intervals of time.

  o. A computer is a machine that handles information with amazing speed.

  p. A Ferrari is the best car in the world.

  q. To meditate is to exercise mental faculties in thought.

2. Standard dictionaries define for the general reader, whereas specialized reference books define for the specialist. Choose an item in your field and copy the definition (1) from a standard dictionary, and (2) from a technical reference book. For the technical definition, label each expansion strategy. Rewrite the specialized definition for a general reader.

Q. What is cogeneration?

A. It is the simultaneous production of electricity *and* useful thermal energy. This means that you can generate electricity with the same steam you are now using for heating or process. *You can use the same steam twice.* In modern usage cogeneration has also come to mean using waste fuel for in-plant electricity generation.

Q. How does it save money?

A. Cogeneration saves money by allowing you to produce your own electricity for a fraction of the cost of utility power. Cogenerated power is cheaper because cogeneration systems are much more efficient than central utility plants. By using the same steam twice, cogeneration systems can achieve efficiencies of up to 80%, whereas the best utilities can do is about 40%.

Q. Are there other benefits to cogeneration?

A. Yes. For companies using waste fuel, elimination of waste disposal costs can be a very important benefit. Depending upon design, a cogeneration system can provide emergency standby power and can smooth out boiler load swings.

Q. Is cogeneration new?

A. No. It's been done ever since the beginning of the electrification of industrial America. Originally, most electric power was cogenerated by individual manufacturers, not the utilities. In the 1920s and '30s, as cheaper utility electricity became available, cogeneration waned. With cheap oil available, power rates continued to decline through the 1950s and '60s. Then came the 1973–74 Arab Oil Embargo. Everything changed abruptly. Since then electricity prices have risen. Further upward pressure was produced by some utilities' nuclear power plant building programs. Now many companies are getting back to their original source of power: cogeneration.

FIGURE 17.2   Expanded Definition in a Technical Brochure

TECHNICAL CONSIDERATIONS

Turbine generator sets make electricity by converting a steam pressure drop into mechanical power to spin the generator. Conceptually, steam turbines work much the same way as water turbines. Just as water turbines take the energy from water as it flows from a high elevation to a lower elevation, steam turbines take the energy from steam as it flows from high pressure to low pressure. The amount of energy that can be converted to electricity is determined by the difference between the inlet pressure and the exhaust pressure (pressure drop) and the volume of steam flowing through the turbine.

Steam turbines have been used in industry in a variety of applications for decades and are the most common way utilities generate electricity. Exactly how a steam turbine generator can be used in your plant depends upon your circumstances.

IF YOU USE WASTE AS A BOILER FUEL

If you use wood waste or incinerator waste as a boiler fuel you can afford to condense turbine exhaust steam in a condenser. This allows you to convert waste fuel into electricity.

The simplest form of a condensing turbine generator set is the Ewing Power Systems C Series. All surplus steam enters the turbine at high pressure and exhausts to a condenser at a very low pressure, usually a vacuum. Because of the very low exhaust pressure, the pressure drop through the turbine is greater and more energy is extracted from each pound of steam. This is the same basic design as utilities use to produce power. The condenser can be either air or water cooled. In water cooled systems the "cooling" water can be hot enough for use as process hot water or for space heat.

In situations where there is surplus fuel and also a need for low pressure process steam, the Ewing Power Systems CX Series is the system of choice. This arrangement includes a back pressure turbine and a condensing turbine connected to a common generator. Low pressure process or space heating loads are met with the back pressure turbine while surplus steam is directed to the condensing turbine to maximize power production.

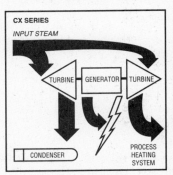

IF YOU PURCHASE BOILER FUEL SUCH AS OIL OR GAS

If oil or gas is used as boiler fuel the best use of a turbine is as a replacement for a steam pressure reducing valve. Many plants produce steam at high pressure and then use some or all of the steam at low pressure after passing it through a pressure reducing valve. Other plants have high pressure boilers but run them at low pressure because they do not need high pressure steam for their process. In either case, a turbine generator can turn the pressure drop energy potential into electricity.

The Ewing Power Systems BP Series turbine generator sets are designed for pressure reducing (back pressure) applications. Very little energy is consumed by the turbine, so most of the inlet steam is available for process. The turbine generator uses about 3631 BTU's

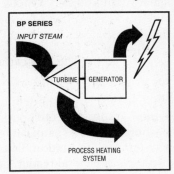

per hour for each kilowatt-hour produced. At 40 cents per gallon for No. 6 fuel oil and 85% boiler efficiency, it will cost about 1.1 cents per kilowatt-hour to generate your own power with a BP Series turbine. Generating costs for gas-fired boilers are similar.

To generate power at this very low cost, all the exhaust steam must be used productively. Generator output is therefore completely governed by process steam demand. For example, if steam is used for space heating you will make more electricity on cold days than on warmer days because more steam will flow through the turbine.

ELECTRICAL CONSIDERATIONS

In most cases the generator will be connected to your plant electrical system and to your utility. This means that you will not give up the security of utility power. It also means that you do not have to generate *all* your power; most cogeneration systems provide only part of the plant load. The more power the generator is making, the less you buy from the utility. If you make more power than you use, you will be able to sell the excess to the utility. If your generator is off-line for any reason, you will be able to buy power from the utility, just as you do now.

There are two primary generator designs: induction and synchronous. Induction generators are similar to induction motors and are much simpler than synchronous generators. Synchronous generators require more elaborate controls and are usually more expensive but offer the advantage of stand-alone capability. Whereas induction generators cannot operate unless they are connected to a utility grid, synchronous sets can be operated in isolation as emergency units or when it is economically advantageous to avoid interconnection with the utility.

All turbine generator sets from Ewing Power Systems include a complete electrical control panel. Our standard panels meet most utility interconnection requirements and we will customize the panel to meet unusual requirements. Synchronous panels can be built for full utility paralleling, stand-alone capability or both.

Courtesy of Ewing Power Systems, So. Deerfield, MA 01373

FIGURE 17.3 Expanded Definition in a Technical Brochure

3. Using reference books as necessary, write sentence definitions for these terms or for terms from your field.

biological insect control	economic inflation	estuary
generator	anorexia nervosa	acid rain
dewpoint	low-impact camping	classical conditioning
microprocessor	hemodialysis	hypothermia
capitalism	gyroscope	thermistor
economic recession	coronary bypass	aquaculture
marsh	oil shale	nuclear fission
artificial intelligence	chemotherapy	modem

4. Select an item from the list in Exercise 3 or from an area of interest. Identify an audience and purpose. Complete an audience-and-use profile sheet (page 32). Begin with a sentence definition of the term. Then write an expanded definition for a first-year student in that field. Next, write the same definition for a layperson (client, patient, or other interested party). Leave a margin at the left side of your page to list expansion strategies (use at least four in each version, and document your sources). Submit, with your two versions, an explanation of your changes from the first version to the second.

5. Figure 17.3 on page 401 shows a page from a brochure titled *Congeneration*. The brochure provides an expanded definition for potential users of fuel conservation systems engineered and packaged by Ewing Power Systems. The intended readers are plant engineers and other technical experts unfamiliar with cogeneration.

Another page of the brochure is designed in a question/answer format. Figure 17.2 shows parts of that page.

Identify each expansion strategy in Figure 17.2 and 17.3. Is the definition appropriate for a technical audience? Why, or why not? Be prepared to discuss your analysis and evaluation in class.

COLLABORATIVE PROJECTS

1. First Class Meeting. Divide into small groups on the basis of academic majors or interests. Appoint one person as group manager. Decide on an item, concept, or process that would require expanded definition for a layperson.

Examples

From computer science: an algorithm, an applications program, artificial intelligence, binary coding, top-down procedural thinking, or systems analysis

From nursing: a pacemaker, coronary bypass surgery, or natural childbirth

Complete an audience-and-use profile (page 32).

2. Beyond informative definition, your group's goal in this next project is to develop a *persuasive* definition.

Assume you work for an organization that has recently formulated a policy to eliminate sexual harassment. Your charge as the communications group is to develop printed material that will publicize this new policy. Your specific task is to develop an expanded definition of *sexual harassment*, to be published in the company newsletter.

At this stage, the company is plagued by confusion, misunderstanding, resentment, and paranoia about the harassment issue. Therefore, beyond compliance with legal requirements, your organization seeks to improve gender relations between co-workers, in the hope of boosting productivity.

Thus, you face an informative and a persuasive and an ethical challenge: to move beyond the usual matters of clarity so that your definition promotes real understanding and reconciliation. In short, you need to amplify the legal definition so that employees are able to recognize harassment, and to understand clearly what does and what does not constitute harassment. But unless you also do something to change their us-against-them *attitude*, you will only create greater tension and overreaction.

In other words, you want your definition not merely to *inhibit* behavior, but to *enlighten* the readers. Insensitive readers, of course, are likely to change their behavior only if they feel coerced (pages 37–39). But appeals to fear almost always have limited success, and people generally tend to be reasonable in the long run. And so you want your definition to have rational appeal, to cause readers to *internalize* the values that underlie the issue. No sermons, please.

Once your group has decided on the appropriate expansion strategies (etymology, negation, etc.), the group manager will assign each member to work on one or two specific strategies as part of the definition.

Second Class Meeting. As a group, edit and incorporate the collected material into an expanded definition, revising as often as needed.

Third Class Meeting. The group manager will assign one member to present the definition in class, using either opaque or overhead projection, a large-screen monitor, or mimeographed copies.

Descriptions and Specifications

Purpose of Description

Specifications

Objectivity in Description

Elements of Description

A General Model for Description

A Sample Situation

■   ■   ■

Description (creating a picture with words) is part of all writing. But technical descriptions convey information about a product or mechanism to someone who will use it, buy it, operate it, assemble it, manufacture it, or to someone who has to know more about it. Any item can be visualized from countless different perspectives. Therefore, *how* you describe— your perspective—depends on your purpose and the needs of your audience.

PURPOSE OF DESCRIPTION

Manufacturers use descriptions to sell products; banks require detailed descriptions of any business or construction venture before approving a loan; medical personnel maintain daily or hourly descriptions of a patient's condition and treatment.

No matter what the subject of description, readers expect answers to as many of these questions as are applicable: *What is it? What does it do? What does it look like? What is it made of? How does it work? How was it put together?* The description in Figure 18.1, part of an installation and operation manual, answers applicable questions for do-it-yourself homeowners.

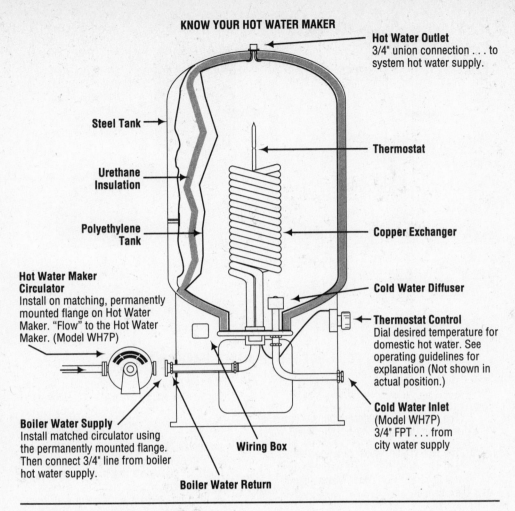

KNOW YOUR HOT WATER MAKER

Hot Water Outlet
3/4" union connection . . . to system hot water supply.

Steel Tank

Thermostat

Urethane Insulation

Polyethylene Tank

Copper Exchanger

Hot Water Maker Circulator
Install on matching, permanently mounted flange on Hot Water Maker. "Flow" to the Hot Water Maker. (Model WH7P)

Cold Water Diffuser

Thermostat Control
Dial desired temperature for domestic hot water. See operating guidelines for explanation (Not shown in actual position.)

Boiler Water Supply
Install matched circulator using the permanently mounted flange. Then connect 3/4" line from boiler hot water supply.

Wiring Box

Boiler Water Return

Cold Water Inlet
(Model WH7P) 3/4" FPT . . . from city water supply

FIGURE 18.1 A Mechanism Description

SPECIFICATIONS

Airplanes, bridges, smoke detectors, and countless other items are produced according to certain *specifications*. A particularly exacting type of description, specifications (or "specs") prescribe standards for performance, safety, and quality. For almost any product, specifications spell out

- the methods for manufacturing, building, or installing the product
- the materials and equipment to be used
- the size, shape, and weight of the product

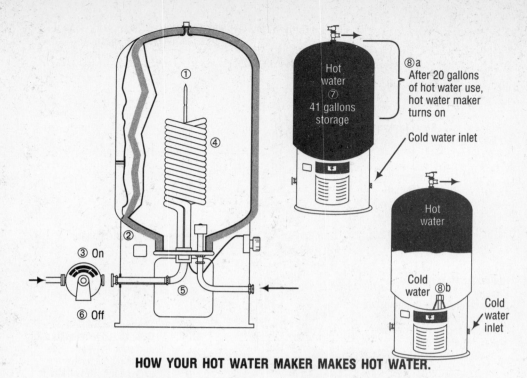

HOW YOUR HOT WATER MAKER MAKES HOT WATER.

1. The thermostat calls for energy to make hot water in your Hot Water Maker.

2. The built-in relay signals your boiler/burner to generate energy by heating boiler water.

3. The Hot Water Maker circulator comes on and circulates hot boiler water through the inside of the Hot Water Maker heat exchanger.

4. Heat energy is transferred, or "exchanged" from the boiler water inside the exchanger to the water surrounding it in the Hot Water Maker.

5. The boiler water after the maximum of heat energy is taken out of it, is returned to the boiler so it can be reheated.

6. When enough heat has been exchanged to raise the temperature in your Hot Water Maker to the desired temperature, the thermostat will de-energize the relay and turn off the Hot Water Maker circulator and your boiler/burner. This will take approximately 23 minutes—when you first start up

the Hot Water Maker. During use, reheating will be approximately 9–12 minutes.

7. You now have 41 gallons of hot water in storage . . . ready for use in washing machines, showers, sinks, etc. This 41 gallons of hot water will stay hot up to 10 hours, if you don't use it, without causing your boiler/burner to come on. (Unless, of course, you need it for heating your home in the winter.)

8. When you do use hot water, you will be able to use approximately 20 gallons, before the Hot Water Maker turns on. Then you will still have 21 gallons of hot water left for use, as your Hot Water Maker "recoups" 20 gallons of cold water. This means, during normal use (3½ GPM Flow), you will never run out of hot water.

9. You can expect substantial energy savings with your Hot Water Maker, as its ability to store hot water and efficiently transfer energy to make more hot water will keep your boiler off for longer periods of time.

Source: Courtesy of AMTROL, Inc.

FIGURE 18.1 A Mechanism Description *Continued*

Because these requirements define an acceptable level of quality, specifications have ethical and legal implications. Any product "below" specifications provides grounds for a lawsuit. And when injury or death results (as in a bridge collapse caused by inferior reinforcement) the contractor, subcontractor, or supplier who "cut corners" is criminally liable.

Federal and state regulatory agencies routinely issue specifications to ensure safety. The Consumer Product Safety Commission specifies that power lawn mowers be equipped with a "kill switch" on the handle, a blade guard to prevent foot injuries, and a grass thrower that aims downward to prevent eye and facial injury. This same agency issues specifications for baby products, as in governing the fire retardancy of pajama material. Passenger airline specifications for aisle width, seat-belt configurations, and emergency equipment are issued by the Federal Aviation Administration. State and local agencies issue specifications in the form of building codes, fire codes, and other standards for safety and reliability.

Government departments (Defense, Interior, etc.) issue specifications for all types of military hardware and other equipment. A set of NASA specifications for spacecraft parts can be hundreds of pages long, prescribing the standards for even the smallest nuts and bolts, down to screw-thread depth and width—in millimeters.

The private sector issues specifications for countless products or projects, to help ensure that customers get *exactly* what they want. Figure 18.2 illustrates partial specifications drawn up by an architect for a building that will house a small medical clinic. This section of the specs covers only the structure's "shell." Other sections detail the requirements for plumbing, wiring, and interior finish work.

Specifications like those in Figure 18.2 (page 408) must be clear enough for *identical* interpretation by the widest possible range of readers (Glidden 258–59):

- *the customer*, who has the "big picture" of what is needed and who wants the best product at the best price
- *the designer* (architect, engineer, computer scientist, etc.), who must translate the customer's wishes into the actual specifications
- *the contractor or manufacturer*, who won the job by making the lowest bid, and so must preserve profit by doing *only* what is prescribed
- *the supplier*, who must provide the exact materials and equipment
- *the workforce*, who will do the actual assembly or construction or installation (managers, supervisors, subcontractors, and workers— some who will work on only one part of the product, such as plumbing or electrical)

Ruger, Filstone, and Grant
Architects

Specifications for the Pownal Clinic Building

Foundation
 footings: 8" x 16" concrete (load-bearing capacity: 3000 lbs. per sq. in.)
 frost walls: 8" x 4' @ 3000 psi
 slab: 4" @ 3000 psi, reinforced with wire mesh over vapor barrier

Exterior walls
 frame: eastern pine #2 timber frame with exterior partitions set inside posts
 exterior partitions: 2" x 4" kiln-dried spruce set at 16" on center
 sheathing: 1/4" exterior-grade plywood
 siding: #1 red cedar with a 1/2" x 6" bevel
 trim: finished-pine boards ranging from 1" x 4" to 1" x 10"
 painting: 2 coats of Clear Wood Finish on siding; trim primed and finished with one
 coat of bone-white, oil base paint

Roof system
 framing: 2" x 12" kiln-dried spruce set at 24" on center
 sheathing: 5/8" exterior-grade plywood
 finish: 240 Celotex 20-year fiberglass shingles over #15 impregnated felt roofing paper
 flashing: copper

Windows
 Andersen casement and fixed-over-awning models, with white exterior cladding,
 insulating glass and screens, and wood interior frames

Landscape
 driveway: gravel base, with 3" traprock surface
 walks: timber defined, with traprock surface
 cleared areas: to be rough graded and covered with wood chips
 plantings: 10 assorted lawn plants along the road side of the building

FIGURE 18.2 Specifications for a Building Project (Partial)

- *the inspectors*, who evaluate how well the product conforms to the specifications (such as building or plumbing or electrical inspectors)

Each of these parties has to understand and agree on exactly *what* is to be done and *how* it is to be done. And in the case of a lawsuit over failure to meet specifications, the readership broadens to include judges, lawyers, and jury. Figure 18.3 shows how a clear set of specifications unifies all readers (their various viewpoints, motives, and levels of expertise) in a shared understanding.

In addition to guiding a product's design and construction, specifications can facilitate the product's use and maintenance. For instance, specifications in a computer manual include the product's performance limits, or *ratings:* its power requirements, work or processing or storage capacity, environment requirements, the makeup of key parts, and so on. Product support literature for appliances, power tools, and other items routinely contains ratings to help readers select a good operating environment or replace worn or defective parts (Riney 186). The ratings in Figure 18.4 are taken from the owner's manual for a dot-matrix printer.

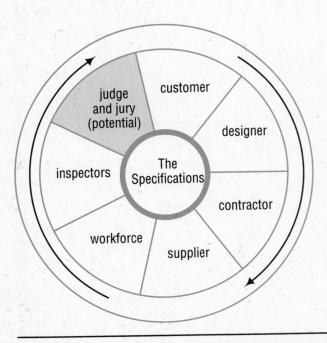

FIGURE 18.3 Readers and Potential Readers of Specifications

TABLE C.1 Continued. Printer Specifications

Paper Types:	Single sheets Pin-feed paper (hole centers 4.0–9.5 inches)
Ribbon:	Cassette containing black inked fabric ribbon 13 mm wide by 13000 mm long, continuous Four color ribbon optional 21 mm wide by 18000 mm long, continuous

Power Options:	American	120 volts AC ± 10%, 60 hertz
	Universal	100 volts AC ± 10%, 50/60 hertz
		120 volts AC ± 10%, 50/60 hertz
		140 volts AC ± 10%, 50/60 hertz
		200 volts AC ± 10%, 50/60 hertz
		220 volts AC ± 10%, 50/60 hertz
		240 volts AC ± 10%, 50/60 hertz

Power Consumption:	Operating: 180 watts maximum Standby: 20 watts maximum
Data Interface:	8-bit serial
Weight:	11.36 kilograms (25 pounds)

Dimensions:	Width	Depth	Height
	431.8	304.8	127.0 millimeters
	17.0	12.0	5.0 inches

Ambient Temperature:
    Operating
    Storage
	
	10 to 40 degrees Celsius (50 to 104 degrees F.)
	−40 to +47 degrees Celcius (−40 to +116 degrees F.)

Humidity:
    Operating
    Storage
	
	20% to 95% relative humidity, noncondensing
	10% to 95% relative humidity, noncondensing

Source: Reprinted by permission of Apple Computer, Inc.

FIGURE 18.4 Specifications for the Imagewriter™ II Printer (Partial)

OBJECTIVITY IN DESCRIPTION

Descriptions are mainly *subjective* or *objective:* based on feelings or fact. Subjective description emphasizes the writer's attitude toward the thing, whereas objective description emphasizes the thing itself. Because they show an *impartial* view, objective descriptions tend to be more precise.

Subjective Description

Essays describing "An Unforgettable Person" or "A Beautiful Moment" express opinions, a personal point of view. Subjective description aims at expressing feelings, attitudes, and moods. You create an *impression* of your subject ("The weather was miserable"), more than communicating factual information about it ("All day, we had freezing rain and gale-force winds").

"Objective" Description

No writing can be strictly objective: the individual writer chooses where to focus and what to ignore. But impartial description filters out—as much as appropriate—personal impressions, and focuses on observable details.

Except for promotional writing, descriptions on the job should be impartial, if they are to be ethical.[1] The six questions on page 404 require objective answers. Notice that "What is your opinion of the item?" is not included.[2] Here are guidelines for remaining impartial.

Record Those Details that Enable Readers to Visualize the Item. Ask these questions: What could any observer recognize? What would a camera record?

> **Subjective**   His office has an *awful* view, *terrible* furniture, and a *depressing* atmosphere.

The italicized words only *tell;* they do not *show*.

> **Objective**   His office has broken windows looking out on a brick wall, a rug with a six-inch hole in the center, chairs with bottoms falling out, missing floorboards, and a ceiling with plaster missing in three or four places.

---

[1]"Pure objectivity" is, of course, humanly impossible. Each writer filters the facts and their meaning through her or his own perspective. Nonetheless, we are ethically bound to communicate the facts as we know them and understand them. One writer offers this useful distinction: "All communication requires us to leave something out, but we must be sure that what is left out is not essential to our [reader's] understanding of what is put in" (Coletta 65).

[2]Of course, "an ethical writer is obligated to express her or his opinions of products, as long as these opinions are based on objective and responsible research and observation" (Mackenzie 3). Being "objective" does not mean forsaking personal evaluation in cases in which a product may be unsafe or unsound.

Use Precise and Informative Language. Use high-information words that enable readers to *visualize*. Name specific parts without calling them "things," "gadgets," or "doohickeys." Avoid judgmental words ("impressive," "poor"), unless your judgment is requested and can be supported by facts. Instead of "large," "long," and "near," give exact measurements, weights, dimensions, and ingredients.

Use words that specify location and spatial relationships: "above," "oblique," "behind," "tangential," "adjacent," "interlocking," "abutting," and "overlapping." Use position words: "horizontal," "vertical," "lateral," "longitudinal," "in cross-section," "parallel."

Indefinite	*Precise*
a late-model car	a 1993 Ford Taurus sedan
an inside view	a cross-sectional, cutaway, or exploded view
next to the foundation	adjacent to the right side
a small red thing	a red activator button with a 1-inch diameter and a concave surface

Do not confuse precise language, however, with overly complicated technical terms or needless jargon. Don't say "phlebotomy specimen" instead of "blood," or "thermal attenuation" instead of "insulation," or "proactive neutralization" instead of "damage control." The clearest writing uses precise but plain language. General readers prefer nontechnical language, as long as the simpler words do the job. Always think about your specific readers' needs.

ELEMENTS OF DESCRIPTION

Clear and Limiting Title

Promise exactly what you will deliver—no more and no less. "A Description of a Velo Ten-Speed Racing Bicycle" promises a complete description, down to the smallest part. If you intend to describe the braking mechanism only, be sure your title so indicates: "A Description of the Velo's Center-Pull Caliper Braking Mechanism."

Overall Appearance and Component Parts

Let readers see the big picture before you describe each part.

The standard stethoscope is roughly 24 inches long and weighs about 5 ounces.

The instrument consists of a sensitive sound-detecting and amplifying device whose flat surface is pressed against a bodily area. This amplifying device is attached to rubber and metal tubing that transmits the body sound to a listening device inserted in the ear.

Seven interlocking pieces contribute to the stethoscope's Y-shaped appearance: (1) diaphragm contact piece, (2) lower tubing, (3) Y-shaped metal piece, (4) upper tubing, (5) U-shaped metal strip, (6) curved metal tubing, and (7) hollow ear plugs. These parts form a continuous unit.

Visuals

Use drawings, diagrams, or photographs generously. Our overall description of the stethoscope is greatly clarified by Figure 1, below.

Function of Each Part

Explain what each part does and how it relates to the whole.

The diaphragm contact piece is caused to vibrate by body sounds. This part is the "heart" of the stethoscope that receives, amplifies, and transmits the sound impulse.

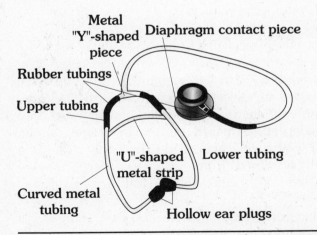

FIGURE 1   Stethoscope with Diaphragm Contact Piece (Front View)

Appropriate Details

Give enough detail for a clear picture, but do not burden readers needlessly. Identify your readers and their reasons for reading your description.

Assume you set out to describe a specific bicycle model. The picture you create will depend on the details you select. How will your reader use this description? What is the reader's level of technical understanding? Is this a customer interested in the bike's appearance—its flashy looks and racy style? Is it a repair technician who needs to know how parts operate? Or is it a helper in your bicycle shop who needs to know how to assemble this bike? If you had designed the bike, you would give the manufacturer detailed specifications.

The description of the hot-water maker on pages 405–06 focuses on what this model looks like, what it's made of, and how it works. Its intended audience of do-it-yourselfers will know already what a hot-water maker is and what it does. That audience needs no background. A description of how it was put together appears with the installation and maintenance instructions later in the manual.

Specifications for readers who will manufacture the hot-water maker would describe each part in exacting detail (say, the steel tank's required thickness and pressure rating as well as required percentages of iron, carbon, and other constituents in the steel alloy).

Clearest Descriptive Sequence

Any item usually has its own logic of organization, based on (1) the way it appears as a static object, (2) the way its parts operate in order, or (3) the way its parts are assembled. We describe these relationships, respectively, in a spatial, functional, or chronological sequence.

Spatial Sequence. Part of all physical descriptions, a spatial sequence answers these questions: *What is it? What does it do? What does it look like? What parts and material is it made of?* Use this sequence when you want readers to visualize the item as a static object or mechanism at rest (a house interior, a document, the Statue of Liberty, a plot of land, a chainsaw, or a computer keyboard). Can readers best visualize this item from front to rear, left to right, top to bottom? (What logical path do the parts create?) A retractable pen would logically be viewed from outside to inside. The specifications in Figure 18.2 proceed from the ground upward.

Functional Sequence. The functional sequence answers: *How does it work?* It is best used in describing a mechanism in action, such as a 35-millimeter camera, a nuclear warhead, a smoke detector, or a car's cruise-control system. The logic of the item is reflected by the order in which its parts function. Like the hot-water maker on page 406, a mechanism usually has only one functional sequence. The stethoscope description on page 413 follows the sequence of parts through which sound travels.

In describing a solar home-heating system, you would begin with the heat collectors on the roof, moving through the pipes, pumping system, and tanks for the heated water, to the heating vents in the floors and walls—from source to outlet. After this functional sequence of operating parts, you could describe each part in a spatial sequence.

Chronological Sequence.   A chronological sequence answers: *How has it been put together?* The chronology follows the sequence in which the parts are assembled.

Use the chronological sequence for an item that is best visualized by its assembly (such as a piece of furniture, an umbrella tent, or a prehung window or door unit). Architects might find a spatial sequence best for describing a proposed beach house to clients; however, they would use a chronological sequence (of blueprints) for specifying to the builder the prescribed dimensions, materials, and construction methods at each stage.

Combined Sequences.   The description of a bumper jack on pages 419–421 alternates among all three sequences; a spatial sequence (bottom-to-top) for describing the overall mechanism at rest, a chronological sequence for explaining the order in which the parts are assembled, and a functional sequence for describing the order in which the parts operate.

A GENERAL MODEL FOR DESCRIPTION

Description of a complex mechanism almost invariably calls for an outline. This model is adaptable to any description.

    I. INTRODUCTION: GENERAL DESCRIPTION[3]
        A. Definition, Function, and Background of the Item
        B. Purpose (and Audience—for classroom only)
        C. Overall Description (with general visuals, if applicable)
        D. Principle of Operation (if applicable)
        E. List of Major Parts

[3]In most descriptions, the subdivisions in the introduction can be combined and need not appear as individual headings in the document.

II. DESCRIPTION AND FUNCTION OF PARTS

    A. Part One in Your Descriptive Sequence

        1. Definition

        2. Shape, dimensions, material (with specific visuals)

        3. Subparts (if applicable)

        4. Function

        5. Relation to adjoining parts

        6. Mode of attachment (if applicable)

    B. Part Two in Your Descriptive Sequence (and so on)

III. SUMMARY AND OPERATING DESCRIPTION

    A. Summary (used only in a long, complex description)

    B. Interrelation of Parts

    C. One Complete Operating Cycle

This outline is tentative, because you might modify, delete, or combine certain parts to suit your subject, purpose, and reader.

Introduction: General Description

Give readers only as much background as they need to get the picture.

A DESCRIPTION OF THE STANDARD STETHOSCOPE

Introduction

Definition and function

The stethoscope is a listening device that amplifies and transmits body sounds to aid in detecting physical abnormalities.

History and background

This instrument has evolved from the original wooden, funnel-shaped instrument invented by a French physician, R. T. Lennaec, in 1819. Because of his female patients' modesty, he found it necessary to develop a device, other than his ear, for auscultation (listening to body sounds).

Purpose and audience

This report explains to the beginning paramedical or nursing student the structure, assembly, and operating principle of the stethoscope. [*Omit this section if you submit to your instructor an audience-and-use profile or if you write for a workplace audience.*]

Finally, give a brief, overall description of the item, discuss its principle of operation, and list its major parts, as in the overall stethoscope description on pages 412–413.

Description and Function of Parts

The body of your text describes each major part. After arranging the parts in sequence, follow the logic of each part. Provide only as much detail as your readers need.

Readers of this description will use a stethoscope daily, and so need to know how it works, how to take it apart for cleaning, and how to replace worn or broken parts.[4]

Diaphragm Contact Piece

The diaphragm contact piece is a shallow metal bowl, about the size of a silver dollar (and twice its thickness), which is caused to vibrate by various body sounds.

Three separate parts make up the piece: hollow steel bowl, plastic diaphragm, and metal frame, as shown in Figure 2.

The stainless steel metal bowl has a concave inner surface, with concentric ridges that funnel sound toward an opening in the tapered base, then out through the hollow appendage. Lateral threads ring the outer circumference of the bowl to accommodate the interlocking metal frame. A fitted diaphragm covers the bowl's upper opening.

The diaphragm is a plastic disk, 2 millimeters thick, 4 inches in circumference, with a molded lip around the edge. It fits flush over the metal bowl, and vibrates sound toward the ridges. A metal frame that screws onto the bowl holds the diaphragm in place.

The stainless steel frame fits over the disk and metal bowl. A ¼-inch ridge between the inner and outer edge accommodates threads for screwing the frame to the bowl. The frame's outside circumference is notched with equally spaced, perpendicular grooves—like those on the edge of a dime—to provide a gripping surface.

The diaphragm contact piece is the "heart" of the stethoscope that receives, amplifies, and transmits sound through the system of attached tubing. The piece attaches to the lower tubing by an appendage on its apex (narrow end), which fits inside the tubing.

Each part of the stethoscope, in turn, is described according to its own logic of organization.

---

[4]Specifications for the manufacturer would require many more technical details (dimensions, alloys, curvatures, tolerances and so on).

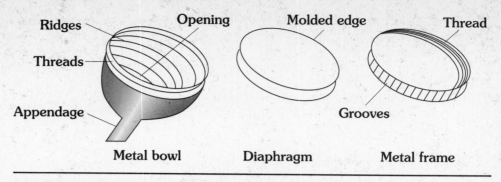

FIGURE 2 Exploded View of a Diaphragm Contact Piece

Summary and Operating Description

Conclude by explaining how the parts work together to make the whole item function.

Conclusion

The seven major parts of the stethoscope provide support for the instrument, flexibility of movement for the operator, and ease in auscultation.

In an operating cycle, the diaphragm contact piece, placed against the skin, picks up sound impulses from the body surface. These impulses cause the plastic diaphragm to vibrate. The amplified vibrations, in turn, are carried through a tube to a dividing point. From here, the amplified sound is carried through two separate but identical series of tubes to hollow ear plugs.

How parts interrelate

One complete operating cycle

A SAMPLE SITUATION

This description of an automobile jack, aimed toward a general audience, follows our outline model.

A Mechanism Description for Nontechnical Readers

AUDIENCE-AND-USE PROFILE. Some readers of this description (written for an owner's manual) will have no mechanical background. Before they can follow instructions for *using* the jack safely, they will have to learn what it is, what it looks like, what its parts are, and how, generally, it works. They will *not* need precise dimensions (e.g., "The rectangular base is 8 inches long and 6½ inches wide, sloping upward 1½ inches from the front outer edge to form a secondary platform 1 inch high and 3 inches square"). The engineer who designed the jack might include such data in specifications for the manufacturer. Laypersons, however, need only the dimensions that will help them recognize specific parts and understand their function, for safe use and assembly.

Also, this audience will need only the broadest explanation of how the leverage mechanism operates. Although the physical principles *(torque, fulcrum)* would interest engineers, they would be of little use to readers who simply need to operate the jack safely.

DESCRIPTION OF A STANDARD BUMPER JACK

Introduction—General Description

The standard bumper jack is a portable mechanism for raising the front or rear of a car through force applied with a lever. This jack enables even a frail person to lift one corner of a 2-ton automobile.

Definition, purpose, and function

The jack consists of a molded steel base supporting a free-standing, perpendicular, notched shaft (Figure 1). Attached to the shaft are a leverage mechanism, a bumper catch, and a cylinder for insertion of the jack handle. Except for the main shaft and leverage mechanism, the jack is made to be dismantled and to fit neatly in the car's trunk.

Overall description (spatial sequence)

The jack operates on a leverage principle, with a human hand traveling 18 inches and the car only 3/8 of an inch during a normal jacking stroke. Such a device requires many strokes to raise the car off the ground, but may prove a lifesaver to a motorist on some deserted road.

Operating principle

Five main parts make up the jack: base, notched shaft, leverage mechanism, bumper catch, and handle.

List of major parts

Description of Parts and Their Function

(chronological sequence)

Base

The rectangular base is a molded steel plate that provides support and a point of insertion for the shaft (Figure 2). The base slopes upward to form a platform containing a 1-inch depression that provides a stabilizing well for the shaft. Stability is increased by a 1-inch cuff around the well. As the base rests on its flat surface, the bottom end of the shaft is inserted into its stabilizing well.

First major part

Definition, shape, and material

Function and mode of attachment

Shaft

The notched shaft is a steel bar (32 inches long) that provides a vertical track for the leverage mechanism. The notches, which hold the mechanism in its position on the shaft, face the operator.

Second major part, etc.

The shaft vertically supports the raised automobile, and attached to it is the leverage mechanism, which rests on individual notches.

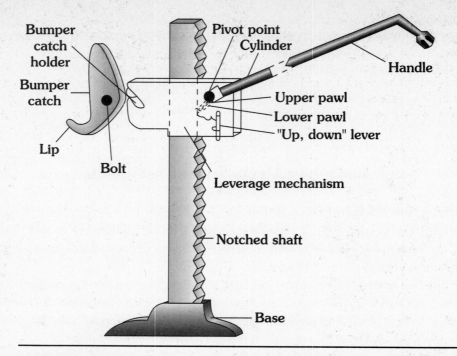

FIGURE 1 A Side View of the Standard Bumper Jack

Leverage Mechanism

The leverage mechanism provides the mechanical advantage needed for its operator to raise the car. It is made to slide up and down the notched shaft. The main body of this pressed-steel mechanism contains two units: one for transferring the leverage and one for holding the bumper catch.

The leverage unit has four major parts: the cylinder, connecting the handle and a pivot point; a lower pawl (a device that fits into the notches to allow forward and prevent backward motion), connected directly to the cylinder; an upper pawl, connected at the pivot point; and an "up-down" lever, which applies or releases pressure on the upper pawl by means of a spring (Figure 1). Moving the cylinder up and down with the handle causes the alternate release of the pawls, and thus movement up or down the shaft—depending on the setting of the "up-down" lever. The movement is transferred by the metal body of the unit to the bumper-catch holder.

The holder consists of a downsloping groove, partially blocked by a wire spring (Figure 1). The spring is mounted in such a way as to keep the bumper catch in place during operation.

Bumper Catch

The bumper catch is a steel device that attaches the leverage mechanism to the bumper. This 9-inch molded plate is bent to fit the shape of the bumper. Its outer ½ inch is bent up to form a lip (Figure 1), which hooks behind the bumper to hold the catch in place. The two sides of the plate are bent back 90 degrees to leave a 2-inch bumper-contact surface, and a bolt is riveted between them. This bolt slips into the groove in the leverage mechanism and provides the attachment between the leverage unit and the car.

Jack Handle

The jack handle is a steel bar that serves both as lever and lug-bolt remover. This round bar is 22 inches long, ⅝ inch in diameter, and is bent 135 degrees roughly 5 inches from its outer end. Its outer end is a wrench made to fit the wheel's lug bolts. Its inner end is beveled to form a blade-like point for prying the wheel covers and for insertion into the cylinder on the leverage mechanism.

Conclusion and Operating Description

One quickly assembles the jack by inserting the bottom of the notched shaft into the stabilizing well in the base, the bumper catch into the groove on the leverage mechanism, and the beveled end of the jack handle into the cylinder. The bumper catch is then attached to the bumper, with the lever set in the "up" position.

Assembly

    As the operator exerts an up-down pumping motion on the jack handle, the leverage mechanism gradually climbs the vertical notched shaft until the car's wheel is raised above the ground. When the lever is in the "down" position, the same pumping motion causes the leverage mechanism to descend the shaft.

One complete operating cycle (functional sequence)

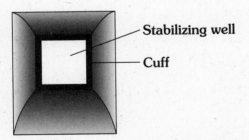

Stabilizing well

Cuff

FIGURE 2   A Top View of the Jack Base

REVISION CHECKLIST FOR DESCRIPTIONS

Use this list to refine the content, arrangement, and style of your description. (Numbers in parentheses refer to first page of discussion.)

Content

☐ Does the title promise exactly what the description delivers? (412)

☐ Have you described the item's overall features, as well as each part? (412)

☐ Is each part defined before it is discussed? (416)

☐ Is the function of each part explained? (413)

☐ Do visuals appear whenever they can provide clarification? (413)

☐ Will readers be able to visualize the item? (414)

☐ Are any details missing, needless, or confusing for this audience? (414)

Arrangement

☐ Does the description follow the clearest possible sequence(s)? (414)

☐ Are relationships among the parts clearly explained? (414)

Style

☐ Is the description sufficiently impartial? (410)

☐ Is the language informative and precise? (412)

☐ Will the level of technicality connect with the audience? (414)

☐ Is the description written in plain English? (412)

☐ Is each sentence clear, concise, and fluent? (451)

☐ Is the description written in grammatical English? (Appendix A)

☐ Is the description ethically acceptable? (411)

EXERCISES

1. Rewrite these subjective statements, making them more impartial:
   a. This classroom is (attractive, unattractive).

b. The weather today is (beautiful, awful, mediocre).

c. (He, she) is the best-dressed person in this class.

d. My textbook is in (good, poor) condition.

e. This room is (too large, too small, just the right size) for our class.

2. Select three possessions (choose simple objects easily described). Write a description of each so that police could identify the items if they were lost or stolen.

3. *a.* Choose an item from this list, or select a device used in your major field. Using the general outline as a model, develop an objective description. Write for a specific use by a specified audience. Attach your written audience profile (based on the worksheet, page 32) to your document.

a soda-acid fire extinguisher	a 100-amp electrical panel
a breathalyzer	a computer printer
a sphygmomanometer	a Skinner box
a ditto machine	a simple radio
a microprocessor	a flat-plate solar collector
a transit	a distilling apparatus
a saber saw	a bodily organ
a drafting table	a Wilson cloud chamber
a hazardous waste site	a programming flowchart
a blowtorch	a brand of woodstove
a photovoltaic panel	a catalytic converter

Remember, you are simply describing the item, its parts, and its function: *do not* provide directions for its assembly or operation.

b. As an optional assignment, describe a place you know well. You are trying to convey a visual image, not a mood; therefore, your description should be impartial, discussing only the observable details.

4. The bumper-jack description in this chapter is aimed toward a general reading audience. Evaluate it by using the revision checklist. In one or two paragraphs, discuss your evaluation, and suggest revisions.

5. Figure 18.5 shows a description from an owner's manual for a telephone answering machine. Study the description and prepare *detailed* answers to these questions:

- Is this description impartial? Explain.
- Compare Figure 18.5 with Figure 18.1. One description focuses on the workings of the *internal* mechanism itself, while the other describes the function of *external* parts. Which is which, and why?
- Are the details in Figures 18.1 and 18.5 adequate and appropriate for the respective audiences? Why?
- What is the descriptive sequence in Figure 18.5? How does it differ from the sequence in 18.1? Why is each sequence appropriate?

Guide To Model 2200 Controls

In the installation and operating instructions that follow, you'll be directed to switches, keys, indicator lights and other controls on your Model 2200. For orientation and easy reference, they're all mapped out here.

1. **Tape Counter**
Three-digit counter keeps track of tape used for incoming messages. Button resets tape counter to zero.

2. **Microphone**
Use it to record your outgoing messages or dictate memos.

3. **POWER key**
Turns the unit on or off. Red light above key indicates "power on."

4. **REW (rewind) key**
Rewinds your message cassette. Green light above key indicates use.

5. **PLAY key**
Plays your incoming or outgoing message cassette. Green light above key indicates use.

6. **FFW (fast forward) key**
Scans the message cassette quickly during playback. Green light above key indicates use.

7. **STOP key**
Stops record, play, rewind or fast forward functions on incoming message cassette.

8. **RECORD key**
Records your announcements on the outgoing cassette. Records memos on the incoming cassette. Red light above key indicates use.

9. **Function Switch**
Set to AUTO to have your calls answered automatically. Set to MESSAGE to play back your incoming messages. Set to ANNOUNCE to record and test your outgoing announcements.

10. **AUTO indicator**
Red indicator glows steadily when your Model 2200 is set to answer calls automatically. It flashes to signal that messages have been received.

11. **VOLUME control**
Controls speaker volume during message playback.

(At the back of unit:)

12. **AC cord**

13. **MESSAGE SELECT switch**
Incoming message length can be set to a 30-second or 270-second limit. The ANS ONLY (answer only) position can be used when you want your callers to hear your outgoing announcement, but you do not want to record any incoming messages.

14. **RING SELECT switch**
Program the unit to answer calls on the second or fourth ring.

15. **Modular telephone line cord**
Connects your Model 2200 to a modular wall jack.

16. **Modular telephone jack**
Accepts modular plug of a telephone or related equipment.

(Not visible in diagrams)

Cassette compartment
Under the unit's cover. The incoming message cassette goes on the left. The outgoing message cassette goes on the right.

Speaker
On the bottom of the unit behind circular grillwork. Plays back messages.

Identification plate
On the bottom of the unit. Includes the Model 2200 FCC registration number, the ringer equivalence number, and the remote control Touch-Tone codes.

Source: Reprinted with Permission of AT&T. Copyright © 1985 AT&T.

FIGURE 18.5 Description of an Answering Machine

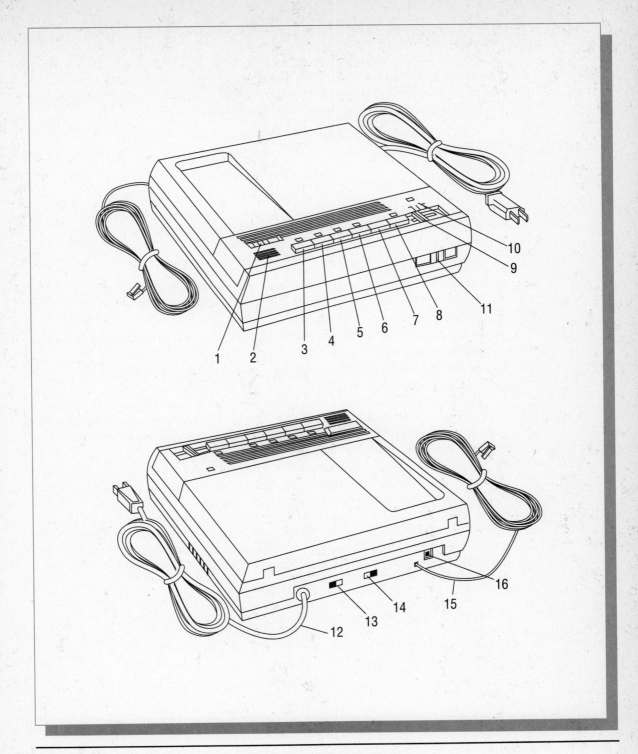

FIGURE 18.5 Description of an Answering Machine *Continued*

COLLABORATIVE PROJECTS

1. Divide into small groups. Assume that your group works in the product-development division of a large and diversified manufacturing company.

Your division has just invented an idea for an inexpensive consumer item with a potentially vast market. (Choose one from the following list.)

- an air pump for bicycle tires
- a flashlight
- nail clippers
- a retractable ball-point pen
- a safety razor
- scissors
- a stapler
- vise grips
- any simple mechanism—as long as it has moving parts

Your group's assignment is to prepare two descriptions of this invention:

a. for company executives who will decide whether to produce and market the item

b. for the engineers, machinists, and so on, who will design and manufacture the item.

Before writing for each audience, be sure to collectively complete audience-and-use profiles (page 418).

Appoint a group manager, who will assign tasks (visuals, typing, etc.) to members. When the descriptions are fully prepared, the group manager will appoint one member to present them in class. The presentation should include explanations of how the two descriptions differ for the different audiences.

2. Assume your group is an architectural firm designing buildings at your college. Develop a set of specifications for duplicating the interior of the classroom in which this course is held. Focus only on materials, dimensions, and equipment (chalkboard, desk, etc.) and use visuals as appropriate. Your audience includes the firm that will construct the classroom, teachers, and school administrators. Use the same format as in Figure 18.2, or design a better one. Appoint one member to present the completed specifications in class. Compare versions from each group for accuracy and clarity.

Procedures and Processes

Purpose of Process-Related Explanation

Elements of Instruction

Measurements of Usability

A General Model for Instructions

A Sample Situation

A General Model for Process Analysis

■   ■   ■

A process is a series of actions or changes leading to a product or result. A procedure is a way of carrying out a process.

PURPOSE OF PROCESS-RELATED EXPLANATION

Writers explain procedures and processes as a way of answering these two questions for readers: (1) *How do I do it?* (2) *How does it happen?* In the workplace, you might instruct a colleague or customer in a specific procedure: how to test a soil sample for chlordane contamination, how to access a database, how to ship radioactive waste. Besides procedural instructions, you might have to explain how something happens: how the budget for various departments is determined, or how the electronic mail system in your company works. This chapter discusses both types of process-related explanation: instructions and process analysis.

ELEMENTS OF INSTRUCTION

Almost anyone with a responsible job writes and reads instructions. The new employee uses instructions for operating office equipment or industrial machinery; the employee going on vacation writes instructions for the person filling in. The person who buys a computer reads the manuals (or *documentation*) for instructions on connecting a printer or running a program.

Instructions carry profound ethical and legal implications. As many as 10 percent of workers are injured yearly on the job (Clement 14). Countless injuries also result from misuse of consumer products such as power tools and car jacks—misuse often caused by defective instructions.

A reader injured because of inaccurate or incomplete or unclear instructions can sue the writer. The court has ruled that a writing defect in product support literature carries liability, as would a design or manufacturing defect in the product itself (Girill 37). Some legal experts argue that defects in the instructions carry even greater liability than product defects because they are more easily demonstrated to a nontechnical jury (Bedford and Stearns 28).

To ensure that your own instructions meet professional and legal requirements for accuracy, completeness, and clarity, observe the following guidelines.

Clear and Limiting Title

Make your title promise exactly what your instructions deliver—no more and no less. The title "Instructions for Cleaning the Carriage and Key Faces of an Electronic Typewriter" tells readers what to expect: instructions for a specific procedure on selected parts. But the title "The Electronic Typewriter" gives no forecast; the document might contain a history of the typewriter, typing instructions, or a description of each part.

Informed Content

Make sure you know *exactly* what you're talking about. Ignorance on your part makes you no less liable for instructions that are faulty or inaccurate:

Never count on ignorance as an excuse

> If the author of [a car repair] manual had no experience with cars, yet provided faulty instructions on the repair of the car's brakes, the home mechanic who was injured when the brakes failed may recover from the author. (Walter and Marsteller 165)

Do not write instructions unless you know the procedure in detail and unless you actually have performed it.

Logically Ordered Steps

Instructions not only divide the procedure into steps; they also guide users through the steps *in chronological order,* to reduce mistakes.

Show how steps are connected

> You can't splice two wires to make an electrical connection until you have removed the insulation. To remove the insulation, you will need. . . .

Try to keep all information for one step close together.

Visuals

Today's readers are image-oriented. Visuals help keep words to a minimum and allow readers to picture what to do, as in Figure 19.1.

Extension Cord Retainer

1. Look into the end of the Switch Handle and you will see 2 slots. The WIDER end of the Retainer goes into the TOP slot (Figure 8).
2. Plug extension cord into Switch Handle and weave cord into Retainer, leaving a little slack (Figure 9).

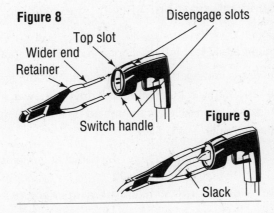

Figure 8
Disengage slots
Top slot
Wider end
Retainer
Switch handle

Figure 9
Slack

Source: Courtesy of Black & Decker® (U.S.), Inc.

FIGURE 19.1 A Visual that Shows "What to Do"

Whenever possible, accompany a specific step with an illustration. Diagrams often are more instructive than photographs because a diagram can selectively omit details unessential to the reader.

Show the same angle of vision the reader will have when doing the activity or using the equipment illustrated—and name the angle (*side view*, *top view*) if you think readers will have trouble figuring it out for themselves.

The less specialized your audience, the more visuals they need. Do not, however, illustrate a step such as "Press the RETURN key," or any action simple enough for readers to visualize on their own.

Appropriate Level of Technicality

Unless you know your readers have the relevant background and skills, write for general readers, and do three things:

1. Give them enough background to understand *why* they need your instructions.
2. Give them enough detail to understand *what* to do.
3. Give them enough examples to visualize the procedure clearly.

Here is how you might adapt instructions titled "How to Copy a Disk Using a Two-Drive Macintosh System" for a general audience.

Background Information.   Begin by explaining the purpose of the procedure.

Tell readers why they are doing this

> You might easily lose information stored on a computer disk if
>
> 1. the disk is damaged by direct sunlight, extreme temperature, or moisture,
> 2. the disk is erased by a faulty disk drive, a power surge, or a user error,
> 3. the stored information is scrambled by a nearby magnet (telephone, computer terminal, or the like).
>
> Always make a backup copy of any disk that contains important material.

Also, state your assumptions about your reader's level of technical understanding.

Tell them what they should know already

> To follow these instructions, you should be able to identify these parts of a Macintosh system: terminal, keyboard, mouse, external drive, and 3.5-inch disk.

Define any specialized terms that appear in your instructions.

WORKING DEFINITION

Initialize: Before you can store or retrieve information on a disk, you must initialize the blank disk. Initializing creates a format the computer can understand—a directory of specific memory spaces (like post office boxes) on the disk, where you can store and retrieve information as needed.

Tell them what each key term means

When your reader understands *what* and *why*, you are ready to explain *how* the reader can carry out the procedure.

Detailed Explanations. Explain the procedure in detail, so your reader knows exactly what to do. Vague instructions result from the writer's failure to consider the readers' needs, as in these unclear instructions for giving first aid to an electrical shock victim:

1. Check vital signs.
2. Establish an airway.
3. Administer external cardiac massage as needed.
4. Ventilate, if cyanosed.
5. Treat for shock.

Give general readers enough background and explanation

These instructions might be clear to medical experts, but not to general readers. Not only are the details inadequate, but terms such as "vital signs" and "cyanosed" are too technical for laypersons. Such instructions posted for workers in a high-voltage area would be useless. The instructions need illustrations and explanations, as in a Red Cross manual.

It's easy to overestimate what people already know, especially when the procedure is almost automatic for you. (Think about when a relative or friend was teaching you to drive a car, or perhaps you tried to teach someone else.) Always assume the reader knows less than you. A colleague will know at least a little less; a layperson will know a good deal less—maybe nothing—about this procedure.

Exactly how much information is enough? These suggestions can help you find an answer:

- Give everything readers need, so the instructions can stand alone.

- Give only what readers need. Don't tell them how to build a computer when they need to know only how to copy a disk.

- Omit steps (*Seat yourself at the computer*) obvious to readers.

- Adjust the information rate ("the amount of information presented in a given page," Meyer 17) to the reader's background and the difficulty of the task. For complex or sensitive steps, slow the information rate. Don't make readers do too much too fast.

- Reinforce the prose with visuals. And don't be afraid to repeat information if it saves readers from flipping pages.
- When writing instructions for consumer products, assume "a barely literate reader" (Clement 151). Simplify.

Examples. Procedures require specific examples (how to load a program, how to order a part), to help readers follow the steps correctly.

Use plenty of examples

To load your program, type this command:

> Load "Style Editor" and press RETURN

Like visuals, examples *show* readers what to do. Examples in fact often appear as visuals.

In the sample procedure that follows, careful use of background, detailed explanation, and visual examples create a "user-friendly" level of technicality for readers just learning to use a computer.

First Step: How to Initialize Your Blank Disk

Before you can copy or store information on a blank disk, you must initialize the disk. Follow this procedure:

1. Switch the computer on.
2. Insert your application disk in the main disk drive.
3. Insert your blank disk in the external disk drive. Unable to recognize this new disk, the computer will respond with a message asking whether you wish to initialize the disk (Figure 1).
4. Using your mouse, place the tip of the on-screen pointer (a small arrow) inside the "Initialize" box.

Begin each instruction with an action verb

Let the visual repeat, restate, or reinforce the prose

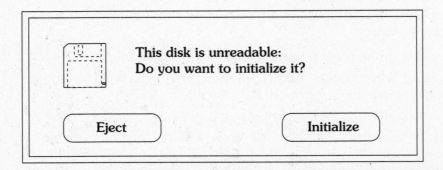

**This disk is unreadable:
Do you want to initialize it?**

Eject   Initialize

FIGURE 1   The "Initialize" Message

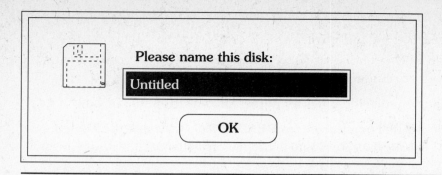

FIGURE 2 The "Disk-naming" Message

5. Press and quickly release the mouse button. Within 15–20 seconds the initializing will be completed, and a message will appear, asking you to name your disk (Figure 2).

Place the visual close to the step

Second Step: How to Name Your Initialized Disk (and so on)

In the sample, notice that instructions and illustrations repeat the same information—they are redundant (Weiss 100). Granted, this prose-visual redundancy makes a longer document, but such repetition helps prevent misinterpretation.[1] When you can't be sure how much is enough, risk overexplaining rather than underexplaining.

Warnings, Cautions, and Notes

Here are the only items that should interrupt the steps in a set of instructions (Van Pelt 3):

- A *note* clarifies a point, emphasizes vital information, or describes options or alternatives.

NOTE: The computer will not initialize a disk that is scratched or imperfect. If your blank disk is rejected, try a new disk.

- A *caution* prevents possible mistakes that could result in injury or damage:

CAUTION: A momentary electrical surge or power failure will erase the contents of internal memory. To avoid losing your work, every few minutes save on disk what you have just typed into the computer.

---

[1]You may recall how Chapter 13 shows how writers avoid *style redundancy* (extra words that give no extra information, page 265). For instructions, however, a writer deliberately seeks *content redundancy*, giving the same information in prose and then in visuals.

- A *warning* alerts users against potential hazards to life or limb:

WARNING: To prevent electrical shock, always disconnect your printer from its power source before cleaning internal parts.

- A *danger* notice identifies an immediate hazard to life or limb:

DANGER: The red canister contains **DEADLY** radioactive material. **Do not break the safety seal** under any circumstances.

In addition to prose warnings, attract readers' attention and help them identify hazards by using symbols, or icons (Bedford and Stearns 128):

Use symbols to alert readers

Do not enter   **Radioactivity**   **Fire danger**

Preview the warnings, cautions, and notes in your introduction, and place them, *clearly highlighted*, immediately before the respective steps.

Use notes, warnings, and cautions only when needed; overuse will dull their effect, and readers may overlook their importance.

Appropriate Words, Sentences, and Paragraphs

Of all communications, instructions have the strictest requirements for clarity, because they lead to *immediate action*. Readers are impatient, and often will not read the entire instructions before plunging into the first step. Poorly phrased and misleading instructions can be disastrous.

Like descriptions, instructions name parts, use location and position words, and state exact measurements, weights, and dimensions. Instructions additionally require your strict attention to phrasing, sentence structure, and paragraph structure.

Active Voice and Imperative Mood.   Use the active voice and imperative mood ("Insert the disk") to address the reader directly. Otherwise, your instructions can lose authority ("You should insert the disk"), or become ambiguous ("The disk is inserted"). In the ambiguous example, we can't tell if the disk is to be inserted or if it already has been inserted.

> Indirect or   The user types in his or her access code.
> confusing   You should type in your access code.
>   It is important to type in the access code.
>   The access code is typed in.

Type in your access code.

The imperative makes instructions more definite and easier to understand because the action verb—the crucial word that tells what the next action will be—comes first. Instead of burying your verb in midsentence, *begin* with an action verb ("raise," "connect," "wash," "insert," "open"), to give readers an immediate signal.

Positive Phrasing.   Research shows that readers respond more quickly and efficiently to instructions phrased positively rather than negatively (Spyridakis and Wenger 205).

| Weaker | Verify that your disk is not contaminated with dust. |
| Stronger | Examine your disk for dust contamination. |

Transitions to Mark Time and Sequence.   Transitional words provide a bridge between related ideas (Appendix A). Some transitions ("in addition," "next," "meanwhile," "finally," "ten minutes later," "the next day," "before") mark time and sequence. They help readers understand the step-by-step process:

> ### Preparing the Ground for a Tent
>
> Begin by clearing and smoothing the area that will be under the tent. This step will prevent damage to the tent floor and eliminate the discomfort of sleeping on uneven ground. *First*, remove all large stones, branches, or other debris within a level 10 × 13-foot area. Use your camping shovel to remove half-buried rocks that cannot easily be moved by hand. *Next*, fill in any large holes with soil or leaves. *Finally*, make several light surface passes with the shovel or a large, leafy branch to smooth the area.

Use transitions to mark time and sequence

Parallel Phrasing.   Like any items in a series, steps should be in identical grammatical form. Parallelism (Appendix A) is important in all writing but in instructions especially, because repeating grammatical forms emphasizes the step-by-step organization of the instructions.

Not parallel To log on to the DEC 20, follow these steps:

1. Switch the terminal to "on."
2. The CONTROL key and C key are pressed simultaneously.

3. Typing LOGON, and pressing the ESCAPE key.

4. Type your user number, and press the ESCAPE key.

Parallel   To log on to the DEC 20, follow these steps:

1. Switch the terminal to "on."
2. Press the CONTROL key and C key simultaneously.
3. Type LOGON, and press the ESCAPE key.
4. Type your user number, and press the ESCAPE key.

Parallelism increases readability and lends continuity to the instructions.

Carefully Shaped Paragraphs and Sentences.   Much of the introductory and explanatory material in instructions takes the form of standard prose paragraphs, with enough sentence variety (page 276) to keep readers interested. But the steps themselves have unique paragraph and sentence requirements. Unless the procedure consists of simple steps (as in "Preparing the Ground for a Tent" on page 435), separate each step by using a numbered list—one step for one activity. If the activity is especially complicated, use a new line (not indented) to begin each sentence in the step.

Instructions ordinarily employ short sentences. But brief is not always best, especially when readers have to fill in the information gaps. Never telegraph your message by omitting articles (*a, an, the*), as shown on page 253.

Unlike many types of documents, instructions call for very little sentence variety. Use sentences with similar structure ("Do this. Then do that.") to avoid confusing readers trying to follow the procedure.

Page 432 shows how the steps usually begin with the action verb. But if a single step covers two related actions, follow the sequence of actions that are required:

Confusing sequence   Insert the disk in the drive before switching on the computer.

The second action required is mistakenly given before the first.

Logical sequence   Before switching on the computer, insert the disk in the drive.

Make your explanations easier to follow by using a familiar-to-unfamiliar sequence (page 255):

Hard   You must initialize a blank disk before you can store information on it.

This sentence is clearer if the familiar material comes first:

> **Easier**   Before you can store information on a blank disk, you must initialize the disk.

Shape every sentence and every paragraph for the reader's access.

Accessible Format

Instructions rarely get undivided attention. The reader, in fact, is doing two things more or less at once: interpreting the instructions and performing the task. *Accessibility,* then, becomes crucial. Readers of instructions have to find immediately what they need. Here are suggestions for designing an accessible instruction format:

- Use informative headings that tell readers what to expect. A heading such as "How to Initialize Your Blank Disk" is more informative than "Disk Initializing." Also, the second version sounds less like a person speaking and more like a robot!
- Arrange all steps in a numbered list.
- Single-space within steps, and double-space between them.
- Double-space to signal a new paragraph, instead of indenting.
- Use white space and highlighting to separate discussion from step.

    Set off your discussion on a separate line (like this), indented or highlighted or both. On a typewriter, you can highlight with underlining, capitals, dashes, parentheses, and asterisks. On most computers, you can use **boldface,** *italics,* varying type sizes, and varying typefaces.

- Set off warnings, cautions, and notes in ruled boxes or use highlighting and plenty of white space.
- Keep the visual and the step close together. If room allows, place the visual right beside the step; if not, right after the step. Set off the visual with plenty of white space.
- Strive for format variety that is appealing but not overwhelming or inconsistent. Readers can be overwhelmed by a page with excessive or inconsistent graphic patterns.

The more accessible the format, the more likely your readers will follow the instructions. Don't be afraid to experiment with formats until you come up with an appealing and accessible design.

MEASUREMENTS OF USABILITY

When we evaluate how well a document achieves its objectives, we measure its **usability.** In measuring the usability of instructions, we ask this question: *How effectively do these instructions enable users to complete the intended task safely, efficiently, and appropriately?*

Companies routinely test the usability of any product—including that product's **documentation** (say, assembly or operating or troubleshooting instructions, warnings, manuals). As a way of avoiding legal action and winning customer goodwill, they go to great lengths to anticipate all the ways a product might be used or misused.

Usability testing is a complex process in which testers have to know (1) how to design tests that are valid and reliable (page 164), and (2) how to interpret their findings accurately (pages 135–36). Primary research methods from Chapter 8 are employed in usability testing (interviews, questionnaires, observation, experiment).

Ideally, usability tests are conducted in a situation that simulates the actual situation, with people who actually will use the document (Redish and Schell 67, Ruhs 8). But even in a classroom setting we can make reasonable assessments of a document's effectiveness on the basis of usability criteria.

Usability criteria for a specific document are defined by its intended use and by characteristics of the user and the environment that affect its use, as depicted in Figure 19.2.

Intended Use

A first step in evaluating usability is to define what we want users to be able to do as a result of reading the document:

- Will users merely carry out a desired action ("What do I do next?"), as in copying a disk correctly? Or will they navigate a complex task that requires decisions ("What option(s) should I choose?") or judgments ("Is this good enough?"), as in diagnosing or treating an ailment? What kinds of knowledge and skills does the task require?

- Besides procedural knowledge (how to do things) what kinds of declarative knowledge (knowing facts) or conceptual knowledge (understanding principles or theories) underly the procedure? How much learning versus performing is required (Mirel 79; Wickens 243)?

- Are we trying to impart perceptual motor skills (swinging a golf club, making a surgical incision), which are less easily forgotten—as opposed to procedural skills (using a computer program, cleaning a fuel-injection system), which are more easily forgotten (Wickens 250)?

INTENDED USE
- Merely follow directions?
- Make judgments?
- Make decisions?

USER CHARACTERISTICS
- Motivation?
- Prior knowledge?
- Skills?
- Ability to process information?
- Experience?
- Limitations?
- Things that can go wrong?

KNOWLEDGE AND SKILLS TO BE IMPARTED
- Declarative knowledge?
- Procedural knowledge?
- Conceptual knowledge?
- Perceptual motor skills?
- Easy material to remember?

WORKPLACE CONSTRAINTS
- Distractions?
- Interruptions?
- Ways the document will be read?
- Things that can go wrong?

DEFINE USABILITY CRITERIA
- Worthwhile content?
- Sensible organization?
- Readable style?
- Accessible design?

FIGURE 19.2 Assessing a Document's Usability

- Will users need to remember this information for a short or long period (Wickens 232)? Will they always have the document in front of them while performing the task?

Human Factors

Those characteristics of the user and the environment that enhance or limit job performance are known as **human factors:** these include the user's attitudes, abilities, and limitations, and the constraints of the work environment (Wickens 3).

- Who are the users? What do they know already? How informed are they in general? How accustomed to such procedures? How educated? How motivated?
- Will readers be scanning the document, studying it or memorizing it, or merely referring to it periodically and randomly?
- Can we anticipate ways in which the document could be misinterpreted or misunderstood (Boiarsky 100)?
- Can we anticipate possible distractions or interruptions from the work environment?

For a closer look at human factors, review the Audience-and-Use Profile Sheet (page 32).

Usability Criteria for the Document

Only after identifying its intended use and the possible human factors can we decide on criteria for evaluating the document's effectiveness.

- Is everything easy to find, appropriate to this reader for this task, easy to understand, and complete and accurate (Haynes 239)?
- Is the document organized in a way the readers are expected to proceed? Is everything easy to read and follow (style, format, visuals)?

The Revision Checklist (pages 455–457) offers one approach for a detailed assessment of usability.

A GENERAL MODEL FOR INSTRUCTIONS

You can adapt the introduction-body-conclusion structure to any instructions. Here are the possible sections to include:

I. INTRODUCTION
    A. Definition, Benefits, and Purpose of the Procedure
    B. Intended Audience (usually omitted for workplace audiences)
    C. Prior Knowledge and Skills Needed by the Audience
    D. Brief Overall Description of the Procedure
    E. Principle of Operation
    F. Materials, Equipment (in order of use), and Special Conditions
    G. Working Definitions (always in the introduction)

    H. Warnings, Cautions, and Notes (mentioned here, and spelled out in the body)

    I. List of Major Steps

II. REQUIRED STEPS

    A. First Major Step

        1. Definition and purpose

        2. Materials, equipment, and special conditions for this step

        3. Substeps (if applicable)

            a.

            b.

    B. Second Major Step (and so on)

III. Conclusion

    A. Review of Major Steps (for a complex procedure only)

    B. Interrelation of Steps

    C. Troubleshooting or Follow-up Advice (as needed)

This outline is only tentative: because you might modify, delete, or combine some elements, depending on your subject, purpose, and audience.

Introduction

Begin with any background readers need to complete the procedure safely and effectively. Discuss the operating principle briefly; most readers are interested primarily in "how to use it or fix it," and need only a general sense of "how it works." You don't want to bury readers in a long introduction, nor to set them loose on the procedure without adequate preparation. Know your audience—what they need and don't need.

Following is an introduction from instructions for users of the college library. Some users will be computer experts; some will be novices—but all will require a detailed introduction to computerized literature searches before they conduct a search on their own.

**HOW TO USE THE OCLC TERMINAL
TO SEARCH FOR A BOOK**

Clear and limiting title

Introduction

Our library's OCLC (On-line Computer Library Center) terminal offers an efficient way to search for books, journals, government publications, and other printed materials. This terminal is connected to the OCLC database in Columbus, Ohio.

Definition of the procedure

The Ohio database contains more than 8 million records of books and other published materials in libraries nationwide. Each record lists the information found on a catalog card, and identifies the libraries holding the work.

These instructions will enable you to determine whether a book you seek can be found in our library or in other libraries throughout the country.

Any library patron can use the OCLC terminal to search for a book. To operate the terminal, you need only be familiar with the keyboard (Figure 1) and to have an operator's manual handy for general reference.

After logging on the system, you can search for a book by TITLE, AUTHOR, or AUTHOR and TITLE. Once you have viewed the catalog entry for the book you seek, you can use the terminal to determine the libraries from which you might borrow the book. These instructions show a search by TITLE only.

The only additional equipment you need is a copy of the *Manual of OCLC Participating Institutions,* to find a listing of library names according to their OCLC symbol displayed on your terminal screen.

Specialized terms that appear in these instructions are here defined:

Cursor: a small, blinking rectangle that indicates the screen position of the next character you will type.
HOME Position: the cursor location at the extreme top left of your screen. HOME position is where you will type most of your messages to the computer, and where the computer will display its messages to you.

Pay close attention to the NOTES that accompany the logging-on step and to the CAUTION preceding the logging-off step.

The major steps in using OCLC to search for a book by its TITLE are (1) logging on the system, (2) initiating the search, (3) viewing the information on your title, (4) locating the book, and (5) logging off.

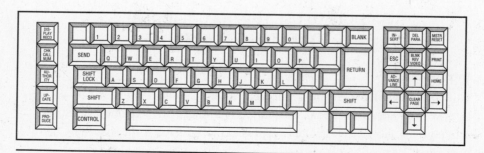

FIGURE 1   The OCLC Terminal Keyboard

Body

In your body section (labeled *Required Steps*), give each step and substep in order. Insert warnings, cautions, and notes as needed. Begin each step with its definition or purpose or both. Readers who understand the reasons for a step will do a better job. A numbered list (like the following) is an excellent way to segment the steps.

Required Steps

Section heading

1. *Logging on the OCLC System*

   Heading previews the step

   To activate the system, you must first log on.

   a. Flick on the red power switch at the keyboard's right edge.

      NOTE: Always press HOME before typing, to make sure your cursor is at HOME position.

      Capitals and spacing set off note

   b. Type the authorization number (07-34-6991).

      NOTE: If you make a typing error, place the cursor over the incorrect character, and type the correct character. To move the cursor one space, press and release one of the four keys marked with an arrow (lower right of keyboard).

   c. Press SEND, and wait for the ⟦ HELLO ⟧ response.

2. *Initiating the Search*

   To initiate a computer search, you must enter your title's exact code name.

   a. Type in the code for your desired title, excluding articles (*a, an, the*) when they appear as the first word of the title. Always type your entry exactly as in the following example:

      Suppose your title is *Presentation Graphics on the Apple Macintosh;* you type

      Explanation begins new line

      pre,gr,on,t

      Example shows what to do

   b. Press DISPLAY/RECD and then SEND.

      The computer will display a numbered list of titles that match the code you have sent (Figure 2).

      Single-spacing within steps: double-spacing between

3. *Viewing the Information on Your Title*

   Now that you have found your title, check to see if our library has the book. To view the full information on your title, follow these steps:

   a. Type in the number of the title you seek.

      All steps have parallel phrasing

   b. Press DISPLAY/RECD and then SEND.

      The computer will display a catalog entry for your title (Figure 3).

      Result of step begins new line

   Because SMU does not have this title, you will have to determine where it can be found.

> 1 Presentation Graphics on the Apple Macintosh: how to
>   use Microsoft Chart to create dazzling graphics for
>   professional and corporate applications / Lambert,
>   Steve. Microsoft Press: Distributed in the U.S. and
>   Canada by Simon & Schuster, c 1984.
>
> 2 Presentation Graphics on the IBM PC and compatibles:
>   how to use Microsoft Chart to create dazzling graph-
>   ics for professional and corporate applications /
>   Lambert, Steve. Microsoft Press: Distributed in the
>   U.S. and Canada by Simon & Schuster, c 1986.

FIGURE 2   The Computer's List of Titles that Match the Code You Typed

4. *Locating the Book*

   To determine which OCLC libraries hold the title you seek, follow these steps:

   a. Type dh and then press DISPLAY/RECD and then SEND. The computer will list symbols of holding libraries by state (Figure 4).

   b. Identify the closest library by looking up symbols in the *Manual of OCLC Participating Institutions*.

      Your title, *Presentation Graphics,* can be found in these Massachusetts libraries: (LDV) Lotus Development Corporation, (LIN) Massachusetts Institute of Technology, and so on.

   c. For reference, obtain a printout of the library listing by pressing PRINT.

> NO HOLDINGS IN SMU—FOR HOLDINGS ENTER dh DEPRESS DIS-
> PLAY RECD. SEND OCLC 10753933
> 3 020 09148511X (pbk. : cover) : $18.95
> 5 050 T385b.L34 1984
> 9 100 Lambert, Steve, 1945—
> 10 245 Presentation graphics on the Apple Macintosh:
>   how to use Microsoft Chart to create dazzling
>   graphics for professional and corporate ap-
>   plications / c Steve Lambert.
> 11 260 Bellevue, Wash.: Microsoft Press; distributed
>   in the U.S. and Canada by Simon & Schuster, New
>   York; c 1984.

FIGURE 3   The Partial Catalog Entry for Your Title

```
REGIONAL LOCATIONS—FOR OTHER HOLDINGS ENTER dhs, dha,
OR dh. DISPLAY RECD. SEND; FOR BIBLIOGRAPHIC RECORD EN-
TER bib. DISPLAY RECD. SEND
STATE        LOCATIONS
  CT           ETN
  MA           LDV LIN MRK
  NY           NYP VYC VYM YQR ZBM ZNC ZQP ZSA
  VT           VTU
```

FIGURE 4 Display of Symbols for Regional Libraries Holding Your Title

5. *Logging off*

   Once you have your printout, you can log off the system.

   CAUTION: Before logging off, press CLEAR PAGE simultaneously with CONTROL. This step will clear any images that might damage the screen's phosphorus coating.

   To log off, type *end* and then press SEND. The computer will respond with GOODBYE.

Conclusion

The conclusion of a set of instructions has several possible functions:

- You might summarize the major steps in a long and complex procedure, to help readers review their performance.
- You might describe the results of the procedure.
- You might offer follow-up advice, about what could be done next or refer the reader to further sources of documentation.
- You might give troubleshooting advice about what to do if anything goes wrong.

You might do all these things—or none of them. If your procedural section has given readers all they need, omit the conclusion altogether.

In the case of our OCLC instructions, these concluding remarks are appropriate:

Conclusion

Once you locate the title you seek, ask a reference librarian for help requesting the book via Interlibrary Loan. (Books usually arrive within two days.)

Follow-up advice

In case the first library is unable to supply the book requested, you initially should choose several libraries from your printout. The Interlibrary Loan system will automatically forward your request to each lender you have specified, until one lender indicates it can supply the book.

A SAMPLE SITUATION

The instructions for *doing something* (felling a tree) in Figure 19.3 are patterned after our general outline, shown earlier. They will appear in a series of brochures for forestry students about to begin summer jobs with the Idaho Forestry Service.

Instructions for Semitechnical Readers

AUDIENCE-AND-USE PROFILE. I'm writing these instructions for partially informed readers who know how to use chainsaws, axes, and wedges but who are approaching this dangerous procedure for the first time. Therefore, I'll include no visuals of cutting equipment (chainsaws and so on) because the audience already knows what these items look like. I can omit basic information (such as what happens when a tree binds a chainsaw) because the audience already has this knowledge also. Likewise, these readers need no definition of general forestry terms such as *culling* and *thinning*, but they *do* need definitions of terms that relate specifically to tree felling (*undercut, holding wood*, and so on).

To ensure clarity, I will illustrate the final three steps with visuals. The conclusion, for these readers, will be short and to the point—a simple summary of major steps with emphasis on safety. I'll experiment with formats until I create a design that is most accessible.

A GENERAL MODEL FOR PROCESS ANALYSIS

An explanation of how things work or happen divides the process into its parts or principles. Colleagues and clients need to know how stock and bond prices are governed, how your bank reviews a mortgage application, how an optical fiber conducts an impulse, and so on. Process analysis emphasizes *the process itself* so that readers can follow it as "observers."

Like instructions, a process analysis must be detailed enough to enable readers to follow the process step by step. Because it emphasizes the process itself, rather than the reader's role, process analysis is written in the third

INSTRUCTIONS FOR FELLING A TREE

INTRODUCTION

Forestry Service personnel fell (cut down) trees to cull or thin a forested plot, to eliminate the hazard of dead trees standing near power lines, to clear an area for construction, and the like.

These instructions explain how to remove sizable trees for personnel who know how to use a chainsaw, axe, and wedge safely.

When you set out to fell a tree, expect to spend most of your time planning the operation and preparing the area around the tree. To fell the tree, you will make two chainsaw cuts, severing the stem from the stump. Depending on the direction of the cuts, the weather, and the terrain, the severed tree will fall into a predetermined clearing. You can then cut the tree into smaller sections for removal.

> **WARNING:** Although these instructions cover the basic procedure, felling is very dangerous. Trees, felling equipment, and terrain vary greatly. Even professionals sometimes are killed or injured because of judgment errors or misuse of tools.
>   Your main concern is safety. Be sure to have an expert demonstrate this procedure before you try it. Also, pay attention to warnings in steps 1 and 4.

To fell sizeable trees, you will need this equipment:
—a 3- to 5-horsepower chainsaw with a 20-inch blade
—a single-blade splitting axe
—two or more 12-inch steel wedges

The major steps in felling a tree are (1) choosing the lay, (2) providing an escape path, (3) making the undercut, and (4) making the backcut.

REQUIRED STEPS

1. *How to Choose the Lay*
   The "lay" is where you want the tree to fall. On level ground in an open field, which way you direct the fall makes little difference. But such ideal conditions are rare.

   Consider ground obstacles and topography, surrounding trees, and the condition of the tree to be felled. Plan your escape path and the location of your cuts depending on surrounding houses, electrical wires, and trees. Then follow these steps:

   a. Make sure the tree still is alive.

FIGURE 19.3   A Set of Instructions

b. Determine the direction and amount of lean.

> **WARNING:** If the tree is dead and leaning substantially, do not try to fell it without professional help. Many dead, leaning trees have a tendency to split along their length, causing a massive slab to fall spontaneously.

c. Find an opening into which the tree can fall in the direction of lean or as close to the direction of lean as possible.

Because most trees lean downhill, try to direct the fall downhill. If the tree leans slightly away from your desired direction, use wedges to direct the fall.

2. *How to Provide an Escape Path*
Falling trees are unpredictable. Avoid injury by planning a definite escape path. Follow these steps:

a. Locate the path in the direction opposite that in which you plan to direct the fall (Figure 1).

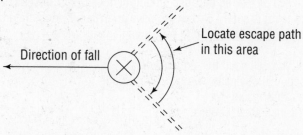

FIGURE 1 Escape-path Location

b. Clear a path 2 feet wide extending beyond where the top of the tree could land.

3. *How to Make the Undercut*
The undercut is a triangular slab of wood cut from the trunk on the side toward which you want the tree to fall. Follow these steps:

a. Start the chainsaw.

b. Holding the saw with the blade parallel to the ground, make the first cut 2 to 3 feet above the ground. Cut into the tree horizontally, to no more than 1/3 its diameter (Figure 2).

FIGURE 19.3 A Set of Instructions *Continued*

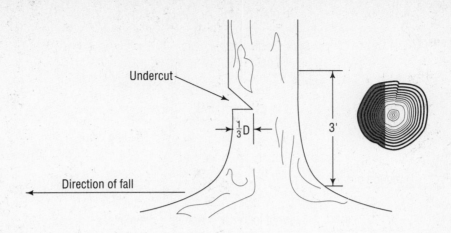

FIGURE 2 Making the Undercut

   c. Make a downward-sloping cut, starting 4 to 6 inches above the first so that the cuts intersect at 1/3 the diameter (Figure 2).

4. *How to Make the Backcut*
After completing the undercut, you make the backcut to sever the stem from the stump. This step requires good reflexes and absolute concentration.

WARNING: Observe tree movement closely during the backcut. If the tree shows any sign of falling in your direction, drop everything and move out of its way.

   Also, do not cut completely through to the undercut; instead, leave a narrow strip of "holding wood" as a hinge, to help prevent the butt end of the falling tree from jumping back at you.

To make your backcut, follow these steps:

   a. Holding the saw with the blade parallel to the ground, start your cut about 3 inches above the undercut, on the opposite side of the trunk. Leave a narrow strip of "holding wood" (Figure 3).

FIGURE 19.3 A Set of Instructions *Continued*

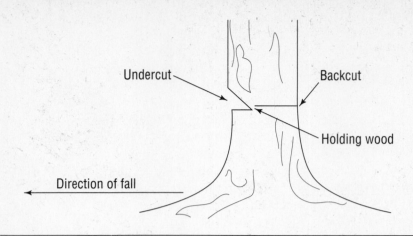

FIGURE 3 Making the Backcut

If the tree begins to bind the chainsaw, hammer a wedge into the backcut with the blunt side of the axe head. Then continue cutting.

b. As soon as the tree begins to fall, turn off the chainsaw, withdraw it, and step back immediately—the butt end of the tree could jump back toward you.

c. Move rapidly down the escape path.

CONCLUSION

Felling is a complex and dangerous procedure. Choosing the lay, providing an escape path, making the undercut, and making the backcut are the basic steps—but trees, terrain, and other circumstances vary greatly.

For the safest operation, seek professional advice and help whenever you foresee *any* complications whatsoever.

FIGURE 19.3 A Set of Instructions *Continued*

person. Instead of leading to immediate action, process analysis helps readers gain understanding. Format and paragraph structure, therefore, resemble those in a traditional essay.

Much of your college writing explains how things happen. Your audience is the professor, who will evaluate what you have learned. Because this informed reader knows *more* than you do about the subject, you often discuss only the main points, omitting details.

But your real challenge comes in analyzing a process for readers who know *less* than you, and who are neither willing nor able to fill in the blank spots; you then become the teacher, and your readers become the students.

For an uninformed audience, introduce your analysis by telling what the process is, and why, when, and where it happens. In the body, tell how it happens, analyzing each stage in sequence. In the conclusion, summarize the stages, and describe one full cycle of the process.

Sections from the following general outline can be adapted to any process analysis:

    I. INTRODUCTION
        A. Definition, Background, and Purpose of the Process
        B. Intended Audience (usually omitted for workplace readers)
        C. Prior Knowledge Needed to Understand the Process
        D. Brief Description of the Process
        E. Principle of Operation
        F. Special Conditions Needed for the Process to Occur
        G. Definitions of Special Terms
        H. Preview of Stages

    II. STAGES IN THE PROCESS
        A. First Major Stage
            1. Definition and purpose
            2. Special conditions needed for the specific stage
            3. Substages (if applicable)
                a.
                b.
        B. Second Stage (and so on)

    III. CONCLUSION
        A. Summary of Major Stages
        B. One Complete Process Cycle

Adapt this outline to the structure of the process you are explaining.

Following is a document patterned after the sample outline. Our writer, Bill Kelly, belongs to an environmental group studying the problem of acid rain in their Massachusetts community. (Massachusetts is among the states most affected by acid rain.) To gain community support, the environmentalists must educate citizens about the problem. Bill's group is publishing and mailing a series of brochures. The first brochure explains how acid rain is formed.

A Process Analysis for Nontechnical Readers

AUDIENCE-AND-USE PROFILE. My audience will consist of general readers. Some already will be interested in the problem; others will have no awareness (or interest). Therefore, I'll keep my explanation at the lowest level of technicality (no chemical formulas, equations). But my explanation needs to be vivid enough to appeal to less aware or less interested readers. I'll use visuals to create interest and to illustrate simply. To give an explanation thorough enough for broad understanding, I'll divide the process into three chronological steps: how acid rain develops, spreads, and destroys.

HOW ACID RAIN DEVELOPS, SPREADS, AND DESTROYS

INTRODUCTION

Definition

Acid rain is environmentally damaging rainfall that occurs after fossil fuels burn, releasing nitrogen and sulfur oxides into the atmosphere. Acid rain, simply stated, increases the acidity level of waterways because these nitrogen and sulfur oxides combine with the air's normal moisture. The resulting rainfall is far more acidic than normal rainfall. Acid rain is a silent threat because its effects, although slow, are cumulative. This analysis explains the cause, the distribution cycle, and the effects of acid rain.

Purpose

Brief description of the process

Most research shows that power plants burning oil or coal are the primary cause of acid rain. The burnt fuel is not completely expended, and some residue enters the atmosphere. Although this residue contains several potentially toxic elements, sulfur oxide and, to a lesser extent, nitrogen oxide are the major problem, because they are transformed when they combine with moisture. This chemical reaction forms sulfur dioxide and nitric acid, which then rain down to earth.

Preview of stages

The major steps explained here are (1) how acid rain develops, (2) how acid rain spreads, and (3) how acid rain destroys.

THE PROCESS

How Acid Rain Develops

Once fossil fuels have been burned, their usefulness is over. Unfortunately, it is here that the acid rain problem begins.

First stage

Fossil fuels contain a number of elements that are released during combustion. Two of these, sulfur oxide and nitrogen oxide, combine with normal moisture to produce sulfuric acid and nitric acid. (Figure 1 illustrates how acid rain develops.) The released gases undergo a chemical change as they combine with atmospheric ozone and water vapor. The resulting rain or snowfall is more acid than normal precipitation.

Acid level is measured by pH readings. The pH scale runs from 0 through 14—a pH of 7 is considered neutral. (Distilled water has a pH of 7.) Numbers above 7 indicate increasing degrees of alkalinity. (Household ammonia has a pH of 11.) Numbers below 7 indicate increasing acidity. Movement in either direction on the pH scale, however, means multiplying by 10. Lemon juice, which has a pH value of 2, is 10 times more acidic than apples, which have a pH of 3, and is 1,000 times more acidic than carrots, which have a pH of 5.

Definition

Because of carbon dioxide (an acid substance) normally present in air, unaffected rainfall has a pH of 5.6. At this time, the pH of precipitation in the northeastern United States and Canada is between 4.5 and 4. In Massachusetts, rain and snowfall have an average pH reading of 4.1. A pH reading below 5 is considered to be abnormally acidic, and therefore a threat to aquatic populations.

How Acid Rain Spreads

Although it might seem that areas containing power plants would be most severely affected, acid rain can in fact travel thousands of miles from its source. Stack gases escape and drift with the wind currents. The sulfur and nitrogen oxides are thus able to travel great distances before they return to earth as acid rain.

Second stage

For an average of two to five days after emission, the gases follow the prevailing winds far from the point of origin. Estimates show that about 50 percent of the acid rain that affects Canada originates in the United States; at the same time, 15 to 25 percent of the U.S. acid rain problem originates in Canada.

The tendency of stack gases to drift makes acid rain a widespread menace. More than 200 lakes in the Adirondacks, hundreds of miles from any industrial center, are unable to support life because their water has become so acidic.

How Acid Rain Destroys

Acid rain causes damage wherever it falls. It erodes various types of building rock such as limestone, marble, and mortar, which are gradually eaten away by the constant bathing in acid. Damage to buildings, houses, monuments, statues, and cars is widespread. Some priceless monuments and carvings already have been destroyed, and even trees of some varieties are dying in large numbers.

More important, however, is acid rain damage to waterways in the affected areas. (Figure 2 illustrates how a typical waterway is infiltrated.) Because of its high acidity, acid rain dramatically lowers the pH in lakes and streams. Although its effect is not immediate, acid rain eventually can make a waterway so acidic it dies. In areas with natural acid-buffering elements such as limestone, the dilute acid has less effect. The northeastern United States and Canada, however, lack this natural protection, and so are continually vulnerable.

The pH level in an affected waterway drops so low that some species cease to reproduce. In fact, a pH level of 5.1 to 5.4 means that fisheries are threatened; once a waterway reaches a pH level of 4.5, no fish reproduction occurs. Because each creature is part of the overall food chain, loss of one element in the chain disrupts the whole cycle.

In the northeastern United States and Canada, the acidity problem is compounded by the runoff from acid snow. During the cold winter months, acid snow sits with little melting, so that by spring thaw, the acid released is greatly

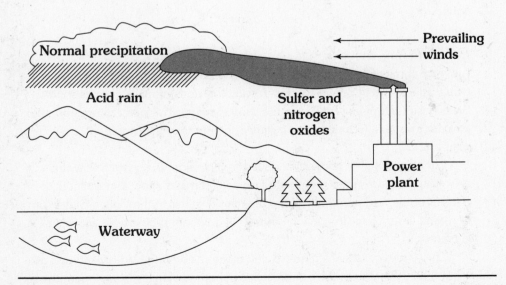

FIGURE 1 How Acid Rain Develops

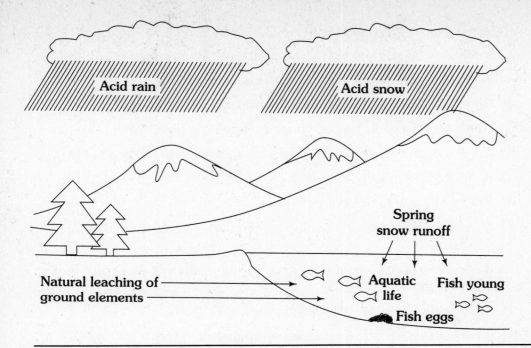

FIGURE 2 How Acid Rain Destroys

concentrated. Aluminum and other heavy metals normally present in soil are also released by acid rain and runoff. These toxic substances leach into waterways in heavy concentrations, affecting fish in all stages of development.

Summary

Acid rain develops from nitrogen and sulfur oxides emitted by industrial and power plants burning fossil fuels. In the atmosphere, these oxides combine with ozone and water to form acid rain: precipitation with a lower-than-average pH. This acid precipitation returns to earth many miles from its source, severely damaging waterways that lack natural buffering agents. The northeastern United States and Canada are the most severely affected areas in North America.

One complete cycle

REVISION CHECKLIST FOR INSTRUCTIONS

Use this checklist to evaluate the usability of instructions. (Numbers in parentheses refer to the first page of discussion.)

Content

☐ Does the title promise exactly what the instructions deliver? (428)

☐ Is the background adequate for this audience? (430)

☐ Do explanations enable readers to understand what to do? (431)

☐ Do examples enable readers to see how to do it correctly? (432)

☐ Are the definition and purpose of each step given as needed? (443)

☐ Is all needless information omitted? (431)

☐ Are all obvious steps omitted? (431)

☐ Do notes, cautions, or warnings appear whenever needed, before the step? (433)

☐ Is the information rate appropriate for the reader's abilities and the difficulty of the task? (431)

☐ Are visuals adequate for clarifying the steps? (429)

☐ Do visuals repeat prose information as needed? (429)

☐ Is everything accurate? (428)

Organization

☐ Is the introduction adequate but not excessive? (441)

☐ Do the instructions follow the exact sequence of steps? (429)

☐ Is all the information for a step close together? (443)

☐ For a complex step, does each sentence begin on a new line? (436)

☐ Is the conclusion necessary and, if necessary, adequate? (445)

Style

☐ Do introductory sentences have enough variety to maintain interest? (436)

☐ Does each step have similar sentence structure? (436)

☐ Do steps generally have short sentences? (436)

☐ Does each step begin with the action verb? (436)

☐ Are all steps in the active voice and imperative mood? (434)

☐ Do all steps have parallel phrasing? (435)

☐ Are transitions adequate for marking time and sequence? (435)

Format

☐ Does each heading clearly tell readers what to expect? (437)

☐ On a typed page, are steps single-spaced within, and double-spaced between? (437)

☐ Do white space and highlights set off discussion from step? (437)

☐ Are notes, cautions, or warnings set off or highlighted? (437)

☐ Are visuals beside or near the step, and set off by white space? (437)

EXERCISES

1. Improve the readability of these instructions by rewriting them in a more appropriate voice and format.

> *What to Do Before Jacking Up Your Car*
> Whenever the misfortune of a flat tire occurs, some basic procedures should be followed before the car is jacked up. If possible, your car should be positioned on as firm and level a surface as is available. The engine has to be turned off; the parking brake should be set; and the automatic transmission shift lever must be placed in "park," or the manual transmission lever in "reverse." The wheel diagonally opposite the one to be removed should have a piece of wood placed beneath it to prevent the wheel from rolling. The spare wheel, jack, and lug wrench should be removed from the luggage compartment.

2. Select part of a technical manual in your field or instructions for a general reader, and make a copy of the material. Using the revision checklist, evaluate the sample's usability. In a memo to your instructor, discuss the strong and weak points of the instructions. Or be prepared to explain to the class why the sample is effective or ineffective.

3. Select a specialized process that you understand well and that has several distinct steps. Using the process analysis on pages 452–455 as a model, explain this process to classmates who are unfamiliar with it. Begin by completing an audience-and-use profile (page 32). Some possible topics: how the body metabolizes alcohol, how economic inflation occurs, how the federal deficit threatens our future, how a lake or pond becomes a swamp, how a volcanic eruption occurs.

Tank must be filled prior to this step (See "Propane Tank" section for details about filling this tank)

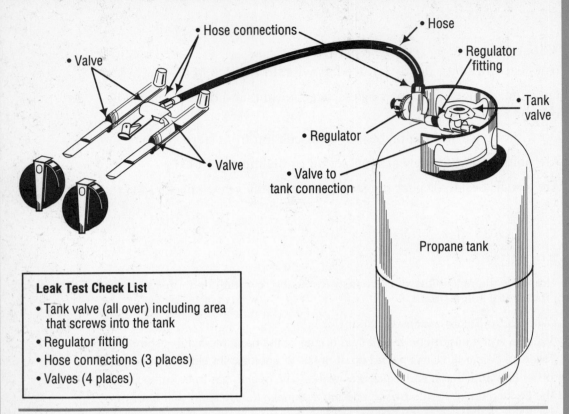

Leak Test Check List
- Tank valve (all over) including area that screws into the tank
- Regulator fitting
- Hose connections (3 places)
- Valves (4 places)

SAFETY! **All models must be Leak Tested! Take the hose, valve, and regulator assembly outdoors in a well ventilated area. Keep away from open flames or sparks. Do Not smoke during this test. Use only a soap and water solution to test for leaks.**

1. Have propane tank filled with propane gas only by a reputable propane gas dealer.
2. Attach regulator fitting to tank valve.
   - Regulator fitting has left-hand threads. Turn counterclockwise to attach.
3. Tighten regulator fitting securely with wrench.
4. Place the two matching control knobs onto the valve stems.
5. Turn the control knobs to the right, clockwise. This is the **"Off"** position.
6. Make a solution of half liquid detergent and half water.

7. Turn gas supply **"ON"** at the tank valve (counterclockwise).
8. Brush soapy mixture on all connections listed in the **Leak Test Check List**.
9. Observe each place for bubbles caused by leaks.
10. Tighten any leaking connections, if possible.
    - If leak cannot be stopped, **Do Not Use Those Parts!** Order new parts.
11. Turn gas supply **"Off"** at the tank valve.
12. Push in and turn control knobs to the left ("HI" position) to release pressure in hose.
13. Disconnect the regulator fitting from the tank valve.
    - Regulator fitting has left-hand threads. Turn fitting clockwise to disconnect.
14. Leave tank outdoors and return to the grill assembly with valve, hose, and regulator.

Source: Instructions reprinted by permission of Thermos® Division.

FIGURE 19.4 Instructions for Leak Testing a Grill

4. Choose a topic from this list, your major, or an area of interest. Using the general outline in this chapter as a model, outline instructions for a procedure that requires at least three major steps. Address a general reader, and begin by completing an audience-and-use profile (page 446). Exchange instructions with a classmate for usability evaluation and revision suggestions based on the checklist.

planting a tree	preparing logs for building a log cabin
hot-waxing skis	removing the rear wheel of a bicycle
hanging wallpaper	adding an electrical outlet
filleting a fish	avoiding hypothermia
warming up before exercise	creating a file on your college's main-
serving a tennis ball	frame computer
hitting a golf ball	

5. Assume you are assistant to the communications manager for a manufacturer of outdoor products. Among the company's best-selling items are its various models of gas grills. Because of fire and explosion hazards, all grills must be accompanied by detailed instructions for safe assembly, use, and maintenance.

One of the first procedures in the manual is the "leak test," to ensure that the gas supply and transport apparatus is leak free. One of the engineers has prepared the instructions in Figure 19.4. Before being published in the manual, they must be approved by communications management. Your boss directs you to evaluate the instructions for accuracy, completeness, clarity, and appropriateness, and to report your findings in a memo. Because of the legal implications here, your evaluation must spell out all positive and negative details of content, organization, style, and format. (Use the Revision Checklist as a guide.) The boss is a busy and impatient reader, and expects your report to be no longer than two pages. Do the evaluation and write the memo.

COLLABORATIVE PROJECTS

1. Select a specialized process you understand well (how gum disease develops, how an earthquake occurs, how steel is made, how a computer compiles and executes a program). Write a brief explanation of the process. Exchange your analysis with a classmate in another major. Study your classmate's explanation for fifteen minutes and then write the same explanation *in your own words,* referring to your classmate's paper as needed. Now, evaluate your classmate's version of your original explanation for accuracy. Does it show that your explanation was understood? If not, why not? Discuss your conclusions in a memo to your instructor, submitted with all samples.

2. Draw a map of the route from your classroom to your dorm, apartment, or home—whichever is closest. Be sure to include identifying landmarks. When your map is completed, write instructions for a classmate who will try to duplicate your map from the information given in your written instructions. Be sure your classmate does not see your map! Exchange your instructions and try to duplicate your classmate's map. Compare your results with the original map. Discuss your conclusions about the usability of these instructions.

3. Divide into small groups and visit your computer center, library, or any place on campus or at work where you can find operating manuals for computers, lab or office equipment, or the like. (Or look through the documentation for your own computer hardware or software.) Locate fairly brief instructions that could use revision for improved content, organization, style, or format. Choose instructions for a procedure you are able to carry out. Make a copy of the instructions, test them for usability, and revise as needed. Submit all materials to your instructor, along with a memo explaining the improvements. Or be prepared to discuss your revision in class.

4. Visit your computer center and ask to borrow a software package that arrived without documentation or with very limited instructions. (Or perhaps you've received such programs as a member of a software club). Run the program, and then prepare instructions for the next users. Test the usability of your instructions by having a classmate use them to run the program. Revise as needed. Remember, you are writing for a user with no experience.

Letters and Employment Correspondence

Elements of Letters

Inquiry Letters

Claim Letters

Résumés and Job Applications

Support for the Application

■   ■   ■

A report may be compiled by a team of writers for multiple readers, but a letter usually is written by one writer for one or more definite readers. Because a letter is more personal than a report and often has a persuasive purpose, proper tone is essential; you want the reader to be on your side. Because your signature certifies the content of your letter (which may serve as a legal document), precision is crucial.

This chapter covers three common letter types: inquiry letters, claim letters, and letters of application, along with résumés. Some other types are discussed in Chapters 16 and 22.

ELEMENTS OF LETTERS

Never send a letter until you feel confident about signing it. Use the guidelines below for judging the adequacy of letters before you sign them.

Introduction-Body-Conclusion Structure

Most letters include (1) a brief *introduction* paragraph (five lines or fewer) that identifies you and your purpose; (2) one or more *body* paragraphs, with the details of your message; (3) a *conclusion* paragraph that sums up and encourages action.

Keep your paragraphs short (usually fewer than eight lines). If your body section is long, divide it into shorter paragraphs, as in Figure 20.1. Notice that this body section is broken down into four questions for easy answering. Figure 20.2 shows the response to these questions.

Standard Parts

Letters typically have six parts, from top to bottom: heading, inside address, salutation, the text (introduction, body, and conclusion), complimentary close, and signature.

Heading. If the stationery has a letterhead, add only the date two lines below the letterhead (Figure 20.2), at the right or left margin. On blank stationery, include your address and the date (but not your name), as in Figure 20.1 and here:

Street Address	154 Sea Lane
City, State Zip Code	Harwich, MA 02163
Month Day, Year	July 15, 1993

Avoid abbreviations (but do use the Postal Service's two-letter state abbreviations on the envelope, and in the heading itself).

Inside Address. Two to six spaces below the heading and abutting the left margin is the inside address.

Reader's Name, Title, and Position	Dr. Ann Mello, Dean
Company Name	Western University
Street Address (if applicable)	Stowe, VT 51350
City, State Zip Code	

Whenever possible, address your letter to a specifically named reader, using that reader's title (Attorney, Major); use Dr. or Ph.D., Dr. or M.D. Abbreviate only titles that are routinely abbreviated (Mr., Ms., Dr.). Titles such as "Captain" are written out in full. (See page 298 for avoiding sexist usage in titles and salutations.)

Salutation. The salutation, two spaces below the inside address, begins with "Dear" and ends with a colon ("Dear Mr. Smith:"). Include the person's full title. If you don't know the person's name or gender, use the position title: "Dear Manager" or, preferably, an attention line (page 466). When appropriate, address your reader by first name—if that is the way you would address this individual in person.

154 Sea Lane
Harwich, MA 02163
July 15, 1993

Mr. A.B. Coolidge
Land Use Manager
Eastern Paper Company
Waldoboro, ME 04572

Dear Mr. Coolidge:

Mr. Melvin Blount, your sales representative, has told me that Eastern Paper Company offers parcels of its lakefront property in northern Maine for public lease. I am interested in a lease, and would appreciate answers to these questions:

1. Does your company have any lakefront parcels in highly remote areas?

2. What is the average size of a leased parcel?

3. How long does a lease remain in effect?

4. What is the yearly leasing fee?

I would welcome any other details you might send along.

Yours truly,

Thomas Stacy

Thomas Stacy

P.S. I am planning a trip to Maine for August 5-11 and would be happy to stop by your office any time you are free.

FIGURE 20.1 Letter of Inquiry

EASTERN PAPER COMPANY
WALDOBORO, MAINE 04572

July 25, 1993

Mr. Thomas Stacy
154 Sea Lane
Harwich, MA 02163

Dear Mr. Stacy:

In answer to your inquiry about leasing lakeside lots in Maine, we have no lands for lease in "highly remote areas." The most remote area is the north shore of Deerfoot Lake, where you can reach the camp lot by boat.

Leases on these 30,000-square-foot parcels are renewable yearly each June. The fee is $250 per year. We have a limited number of leases available, but you would have to visit our office for information about exact location. Come by any weekday morning before 11 o'clock.

The Land Use Regulation Commission in Augusta, Maine regulates campsite leasing, and you need to apply there for a building permit.

Thank you for your inquiry.

Sincerely,

A.B. Coolidge

A.B. Coolidge
Townsite Manager

ABC/de

cc: Melvin Blount

FIGURE 20.2 Response to an Inquiry

Dear. Ms. Jones:

Dear Professor Lexington-Trudeau:

No satisfactory phrase exists for greeting several people at once. Women rightly resent the term "Gentlemen," and "Ladies and Gentlemen" sounds like the beginning of a speech. "Dear Sir or Madam" sounds formal and old-fashioned. And "To Whom It May Concern" is too vague and impersonal. In this case, eliminate the salutation completely by using an attention line (page 466).

Some readers might take offense at the use of their full name. "Dear Mary Smith" or "Dear Joe Brown" can sound too distant and superior and condescending. The implied message in this kind of salutation seems to be, "I don't know you enough to use your first name only, but I don't respect you enough to use your last name only."

Letter Text.   Begin your letter two spaces below the salutation. For letters that fill most of the page, single-space within paragraphs and double-space between. For short letters, double-space within paragraphs and triple-space between, to balance the page.

Complimentary Close.   Place the complimentary close two spaces below the concluding paragraph, aligned with your heading. The closing (polite but not overly intimate) should parallel the level of formality in your salutation, and should reflect your relationship to the reader. These possibilities are ranked in decreasing order of formality:

Respectfully,
Sincerely,
Cordially,
Best wishes,
Warmest regards,
Regards,
Best,

The complimentary close is followed by a comma.

Signature.   Type your full name and title four spaces below the complimentary close. Sign in the space between.

Sincerely yours,

Martha S. Jones
Martha S. Jones
Personnel Manager

If you are writing as a representative of a company or group that bears legal responsibility for the correspondence, type the company's name in full caps two spaces below the complimentary close; place your typed name and title four spaces below the company name, and sign in between.

Yours truly,

HASBROUCK LABORATORIES

Lester Fong

Lester Fong
Research Associate

Specialized Parts

Letters can contain one or more specialized parts.

Attention Line.   Use an attention line when writing to an organization and when you want a specific person (whose name you don't know), title, or department to receive your letter.

ATTENTION: Research and Development Division

ATTENTION: Quality Assurance Supervisor

Drop two spaces below the inside address, and place the attention line either flush with the left margin or centered on the page. The attention line replaces your salutation, as on page 475.

Subject Line.   The subject line forecasts what your letter is about, a good device for getting a busy reader's attention.

SUBJECT: Market Outlook for Absco Portable Computers

Place the subject line two spaces below the inside address (or attention line). Write the subject in caps or underline it.

Typist's Initials.   If someone types your letter, your initials (in caps) and your typist's (in lowercase letters) should appear two spaces below the typed signature, abutting the left margin.

JJ/pl

Because they repeat the signature block, the writer's initials may be eliminated.

Enclosure Notation.   When other documents accompany your letter, add one of these notations one space below the typist's initials, abutting the left margin:

Enclosure

Enclosures 2

Encl. 3

If the enclosures are important documents, name them:

Enclosures: 2 certified checks, 1 set of KBX plans.

Distribution Notation.   If you will distribute copies of your letter to additional readers, so indicate, one space below any enclosure notation.

cc: office file
    Melvin Blount

cc: S. Furlow
    B. Smith

Postscript.   A postscript draws the reader's attention to a point you wish to emphasize. Do not use a postscript for a point you've forgotten in the letter. Rewrite instead. Do use a postscript to add a personal note.

P.S. You will love this model's ease of operation.

Place the postscript two spaces below any other notation, abutting your left margin. Use the postscript sparingly in professional communication. Readers tend to regard postscripts as sales-letter gimmicks.

Appropriate Format

Use uniform margins, spacing, and indention: frame your letter with a 2½-inch top margin and side and bottom margins of 1 to 1¼ inches; single-space within paragraphs and double-space between; avoid hyphens at the ends of lines.

If your letter takes up more than one page, begin each additional page seven spaces from the top, with a line identifying the addressee, date, and page number:

Walter James, June 25, 1987, p. 2

Begin your text two spaces below this line. Place at least two lines of your paragraph at the bottom of page one, and at least two lines of your final text on page two.

On your 9½-by-4-inch envelope, center your reader's address, single-spaced. Use only accepted abbreviations. Place your own single-spaced address in the upper left corner.

Accepted Letter Form

Although several letter forms are acceptable, and your company may have its own, we discuss two common forms: modified block, with paragraphs not indented and the date, complimentary close, and signature aligned slightly right of page center (Figure 20.3); and block, with every line beginning at the left margin (Figure 20.4). The letters of inquiry and response in Figures 20.1 and 20.2 exemplify the modified block and block form, respectively.

Plain English

Workplace correspondence too often suffers from *letterese*, those tired, stuffy, and overblown phrases some writers think they need to sound important. A typically overwritten closing sentence:

> Humbly thanking you in anticipation of your kind cooperation, I remain
>
> Faithfully yours,

Although no one speaks this way, some writers lean on such heavy prose instead of writing in plain English: "We will appreciate your cooperation."

Here are a few of the many old standards that make letters seem unimaginative and boring:

Letterese	*Translation into Plain English*
As per your request	As you requested
Contingent upon receipt of	As soon as we receive
I am desirous of	I want, I would like
Please be advised that my new address is	My new address is
This writer	I
In the immediate future	Soon
In accordance with your request	As you requested

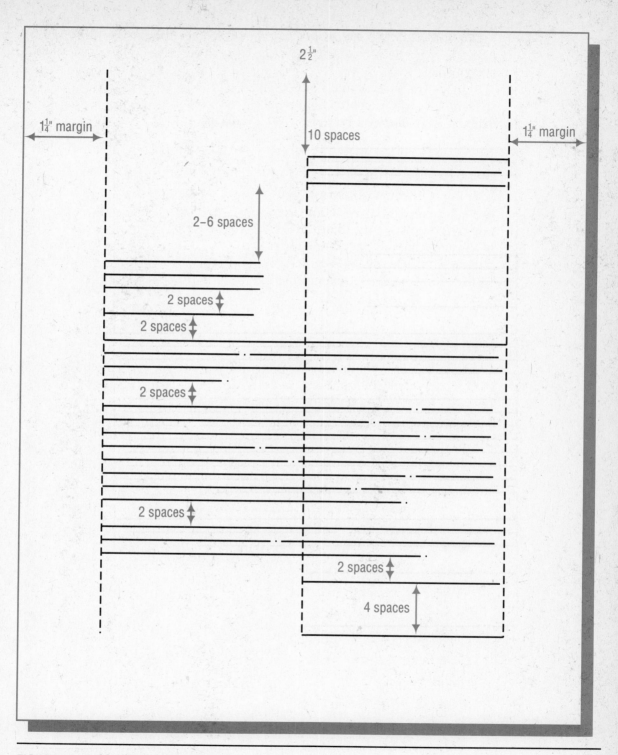

FIGURE 20.3 Modified Block Letter

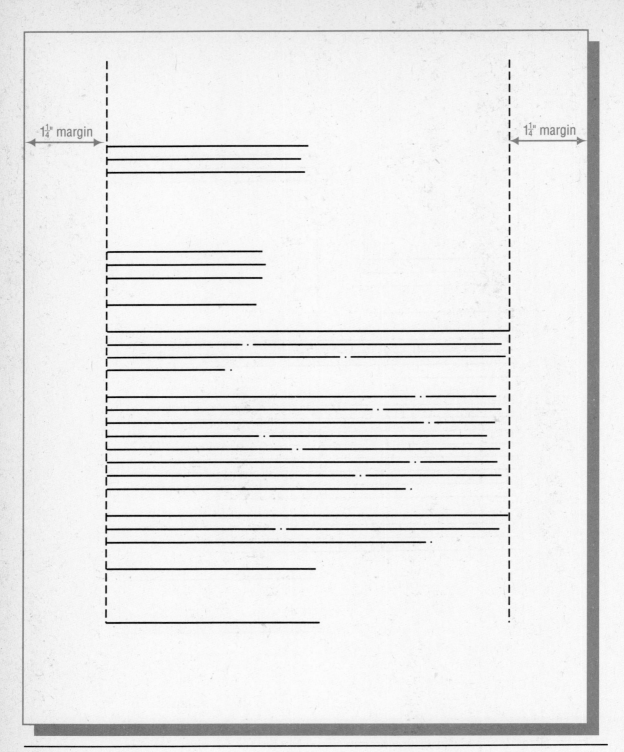

1¼" margin   1¼" margin

FIGURE 20.4 Block Letter

Due to the fact that	Because
I wish to express my gratitude.	Thank you.

Be natural: write as you would speak in a classroom.

"You" Perspective

In speaking face to face, you automatically adjust your delivery as you observe the listener's reactions: a smile, a frown, a raised eyebrow, a nod. Even in a phone conversation a listener can signal approval, anger, and so on. In writing, however, you can easily forget that a flesh-and-blood person will be reacting to what you are saying—or seem to be saying.

The "you" perspective is an element of effective tone; by careful word choice, you show your readers respect. Put yourself in their place; consider how readers will react to what you've written. A letter creates a relationship between writer and reader, and, along with the letter's appearance, the words themselves form the basis for that relationship.

Even a single word, carelessly chosen, can offend. In a letter complaining about the monitor on your new computer, you have the choice of saying, "Although the amber screen causes very little eyestrain, the character resolution is not sharp enough for lengthy word processing use" or "The monitor is crummy for word processing." Clearly, "crummy" is a poor choice because it insults the manufacturer or dealer, and fails to describe the problem (poor resolution).

Clear Purpose

Like any effective writing, good letters do not just "happen." Each is the product of a deliberate process. As you plan, write, and revise, answer these questions:

1. *What do I want the reader to be doing or thinking or feeling after reading this letter?* (offer me a job, give me advice or information, answer my inquiry, follow my instructions, grant me a favor, enjoy good news, accept bad news)

2. *What facts does my reader need?* (measurements, dates, costs, model numbers, enclosures, other details)

3. *To whom am I writing?* (Do you know the reader's name? When possible, write to a person, not a title.)

4. *What is my relationship to my reader?* (Is the reader a potential employer, an employee, a person doing a favor, a person whose products are disappointing, an acquaintance, an associate, a stranger?)

Answer these four questions *before* drafting the letter. After you have a draft, answer the next three questions, which pertain to the *effect* of your letter. Will readers be encouraged to respond favorably?

5. *How will my reader react to what I've written?* (with anger, hostility, pleasure, confusion, fear, guilt, resistance, satisfaction)

6. *What impression of me will my reader get from this letter?* (intelligent, courteous, friendly, articulate, pretentious, illiterate, confident)

7. *Am I ready to sign my letter with confidence?* (Think about it.)

Mail the letter only when you have answered each question to your satisfaction. Revise as often as needed to achieve your purpose.

Direct or Indirect Plan

The reaction you anticipate from your reader (question 5, above) should determine the organizational plan of your letter: either *direct* or *indirect*. The direct plan puts the main point right in the first paragraph, followed by the explanation. Use the direct plan when you expect the reader to react with approval or when you want the reader to know immediately the point of your letter (say, in good-news, inquiry, or application letters—or other routine correspondence).

If you expect the reader to resist or to need persuading, consider an indirect plan: give the explanation *before* the main point (as in refusing a request, admitting a mistake, or requesting a pay raise). An indirect plan might make readers more tolerant of bad news or more receptive to your argument.

Anytime you consider using an indirect plan, think carefully about its *ethical* implications. Never try fooling the reader—or even creating an impression that you have something to hide.

INQUIRY LETTERS

Inquiry letters may be solicited or unsolicited. You often write the first type as a consumer requesting information about an advertised product. Such letters are welcomed because the reader stands to benefit. You can be brief: "Please send me your brochure on. . . ," or some such.

Other inquiries will be unsolicited, that is, not in response to an ad, but requesting information for a report or project. Here, you are asking your reader to spend time reading your letter, considering your request, collecting the information, and writing a response. Always apologize for any imposition, express appreciation, and state a reasonable request clearly and briefly (long, involved inquiries are likely to go unanswered).

In order to ask specific questions, do your homework. Don't expect the respondent to read your mind. A general request ("Please send me all your data on . . .") is likely to be ignored.

A typical inquiry situation: You are preparing an analytical report on the feasibility of harnessing solar energy for home heating in northern climates. You learn that a private, nonprofit research group has been experimenting in solar energy systems. After deciding to write for details, you plan and compose your inquiry.

In your introduction, tell the reader who you are and why you want information. Maintain the "you" perspective with an opening statement that sparks interest and goodwill.

In the body of your letter, write specific and clearly worded questions that are easy to understand and answer. If you have several questions, arrange them in a list. (Lists help readers organize their answers, increasing your chances of getting all the information you want.) Number each question, and separate it from the others, perhaps leaving space for responses right on the page. If you have more than five questions, consider placing them in an attached questionnaire.

Conclude by explaining how you plan to use the information and, if possible, how your reader might benefit. If you have not done so earlier, specify a date by which you need a response. Offer to send a copy of your finished report. Close with a statement of appreciation.

Your letter might resemble the one below (or Figure 20.1).

Inquiry Letter

As a student at Evergreen College, I am preparing a report (April 15 deadline) on the feasibility of solar energy as a viable source of home heating in northern climates.

While gathering data on home solar heating, I encountered references (in *Scientific American* and elsewhere) to your group's pioneering work in solar energy systems. Would you please allow me to benefit from your experience? I specifically would appreciate answers to these four questions:

    States the purpose

    Makes a reasonable request

1. At this stage of development, have you found active or passive solar heating more practical?
2. Do you hope to surpass the 60 percent limit of heating needs supplied by the active system? If so, what level of efficiency do you expect to achieve, and how soon?
3. What is the estimated cost of building materials for your active system, per cubic foot of living space?

    Presents a list of specific questions

4. What metal do you use in collectors to obtain the highest thermal conductivity at the lowest maintenance costs?

Tells how the material will be used, and offers to share findings

Your answers, along with any recent findings you can share, will enrich a learning experience I will put into practice next summer by building my own solar-heated home. I would be glad to send you a copy of my report, along with the house plans I have designed. Thank you.

Sometimes your questions will be too numerous or involved to be answered by letter or questionnaire. You might then request an informative interview (if the respondent is nearby). Karen Granger, our next writer, sought a state representative's "opinions on the EPA's progress" in cleaning up local chemical contamination. Anticipating a complex answer, she requested an interview.

Letter Requesting an Interview

As a technical writing student at the University of Massachusetts, I am preparing a report evaluating the EPA's progress in cleaning up PCB contamination in New Bedford Harbor.

In my research, I have encountered your name repeatedly. Your dedicated work has had a definite influence on this situation, and I am hoping to benefit from your knowledge.

I was surprised to learn that, although this contamination is considered the most extensive anywhere, the EPA still has not moved beyond conducting studies. My own study questions the need for such extensive data gathering. Your opinion, as I can ascertain from *Standard Times* articles, is that the EPA definitely is moving too slowly.

The EPA refutes that argument by asserting they simply do not yet have the information necessary to begin a clean-up operation.

As both a writer and a New Bedford resident, I am very interested in your opinions on the EPA's progress. Could you find time in your busy schedule to grant me an interview? With your permission, I will phone in a few days to arrange an appointment. This interview would be an invaluable aid to me.

Your assistance would be deeply appreciated, and I would gladly send you a copy of my completed report.

Whenever you seek a written response to an unsolicited inquiry, include a stamped, return address envelope.

CLAIM LETTERS

Claim letters request adjustments for defective goods or poor services, or they complain about unfair treatment, or the like. Claims can be routine or arguable. Routine claims follow the direct plan, because the claim is backed by a contract, guarantee, or the company's reputation. Arguable claims, on the other hand, are debatable, and so they call for an indirect plan.

Routine Claims

In a routine claim, make your request or state the problem in your introductory paragraph; then explain in the body section. Close courteously, restating the action you request.

Make the tone courteous and reasonable. Your goal is not to express dissatisfaction, but to achieve results: a refund, replacement, apology. Press your claim objectively yet firmly by explaining it clearly, and by stipulating the *reasonable* action that will satisfy you.

Explain the problem in enough detail for a reader to understand the basis for your claim. Explain that your new alarm clock never rings, instead of merely saying it's defective. Identify the faulty item clearly, giving serial and model numbers, and date and place of purchase. Then propose a fair adjustment. Conclude by expressing goodwill and confidence in the reader's integrity.

The next writer does not ask whether the ski manufacturer will honor his claim; he assumes it will. He asks directly how to return the skis for repair. He uses an attention line to direct his claim to the right department. The subject line, and its re-emphasis in the first sentence, makes clear the nature of the claim.

Routine Claim Letter

Attention: Consumer Affairs Department

Subject: <u>Delaminated Skis</u>

This winter, my Tornado skis began to delaminate. I want to take advantage of your lifetime guarantee to have them relaminated. — *States problem and action desired*

I bought the skis from the Ski House in Erving, Massachusetts, in November 1967. Although I no longer have the sales slip, I did register them with you. The registration number is P9965. — *Provides details*

I'm aware that you no longer make metal skis, but as I recall, your lifetime guarantee on the skis I bought was a major selling point. Only your company and one other were backing their skis so strongly. — *Explains basis for claim*

Would you please let me know how to go about returning my delaminated skis for repair?

Although this is a routine claim, twenty-three years separate the 1967 purchase from the 1990 claim. The claim was honored without question.

Arguable Claims

When your request is in some way unusual, you must *persuade* the firm to grant your claim. Say your parked car is wrecked by a drunk driver, and the insurance company appraises the car at $1,500. But two months earlier, you had the engine rebuilt. By settling for the $1,500, you would lose $1,000. And so you write a claim letter, explaining the circumstances and requesting a fair adjustment.

Use the indirect plan for an arguable claim. Readers are more likely to respond favorably *after* reading your explanation. Begin with a neutral statement both parties can agree to—but which also serves as the basis for your request (e.g., "Customer goodwill is often an intangible quality, but a quality that brings tangible benefits").

Once you've established agreement, explain and support your claim. Include enough information for a fair evaluation: date and place of purchase, order or policy number, dates of previous letters or calls, and background.

Conclude by requesting a *specific action* (a credit to your account, a replacement, a rebate). Ask confidently.

Our next writer employs a tactful, reasonable tone and the indirect plan to achieve her goal.

Arguable Claim Letter

Your company has an established reputation as a reliable wholesaler of office supplies. And for eight years we have counted on that reliability. But a recent episode has left us annoyed and disappointed.

On November 29, 1993, we ordered (#675198) five cartons of Maxell 5¼" diskettes (#A74–866), eight cartons of Scotch 5¼" diskettes (#A74–892), and three cartons of Epson MX 70/80 ribbon cartridges (#A19–556).

On December 5, the order arrived. But instead of double-sided, double-density Maxells ordered, we received single-sided Verbatim diskettes. Instead of Scotch 5¼" diskettes, we received 8" diskettes. And the Epson ribbons were blue, not the black we had ordered. We returned the order the same day.

Also on the 5th, we called John Fitzsimmons at your company to explain our problem. He promised delivery of a corrected order by the 12th. Finally, on the 22nd we did receive an order—the original incorrect one—with a note claiming that the packages had been water damaged while in our possession.

Includes all relevant information

Our warehouse manager insists the packages were in perfect condition when he released them to the shipper. Because we had the packages only five hours, and had no rain on the 5th, we are certain the damage did not occur here.

Sticks to the facts—accuses no one

Responsibility for damages therefore rests with either the shipper or your warehouse staff. What bothers us is our outstanding bill from Hightone ($1,049.50) for the faulty shipment. We insist that the bill be canceled and that we receive a corrected statement. Until this misunderstanding, our transactions with your company were excellent. We hope they can be again.

Requests a specific adjustment

Hightone agreed that the writer had a valid claim, and adjusted the bill.

RÉSUMÉS AND JOB APPLICATIONS

In today's job market, many applicants compete for few openings. Whether you are applying for your first professional job or changing careers, you must market your skills effectively. Your résumé and letter of application must stand out among the competition.

Job Prospecting: The Preliminary Step

Begin your employment search by studying the job market to identify the careers and jobs for which you best qualify.

Being Selective. Focus on openings that suit your skills, aptitudes, and goals:

- What skills have you acquired in schools, in jobs, or from hobbies and interests?
- What can you do well: write, draw, speak other languages, organize, lead, instruct, socialize, sell, solve problems, think creatively?
- Are you good at making decisions?
- Are you better at original thinking or at following directions?
- Are you looking for security, excitement, money, travel, power, prestige, or something else?

Besides helping focus your search, answers to these questions will be handy when you write your résumé.

If your goal is to join some major corporation, consult these library resources for information about prospective employers.

- *Moody's Industrial Index* or Standard and Poor's *Register of Corporations*, for data on plant locations, major subsidiaries, products, executive officers, and corporate assets.
- The *Business Index*, for articles about particular companies in publications such as the *Wall Street Journal, Fortune,* and *Forbes.*
- The yearly issue of *Fortune* Magazine that lists "The Fortune 500" companies.
- Annual reports, for data on a company's assets, innovations, recent performance, and prospects. (Many libraries collect annual reports in a separate file cabinet. Ask your librarian.)

Once you have a general direction, you can begin the actual job search.

Launching Your Search.   Launch your job campaign early. Don't wait for the job to come to you. Take the initiative.

1. Scan the Help Wanted section in major Sunday newspapers for job descriptions, salaries, and qualifications.
2. Ask a reference librarian for occupational handbooks, government publications, newsletters, and job listings in your field.
3. Visit the college placement service; here, openings are posted, interviews are scheduled, and counselors offer job-hunting advice.
4. Ask people in your field for an inside view and practical advice.
5. Sign up at your placement office for interviews with company representatives who visit the campus.
6. Seek the advice of faculty in your major who do outside consulting or who have worked in business, industry, or government.
7. Look for a summer job in your field; this experience may count as much as your education.
8. Establish contacts; don't be afraid to ask for advice. Make a list of names, addresses, and phone numbers of people willing to help.
9. Many professional organizations invite student memberships (at reduced fees). Such affiliations can generate excellent contacts, and they look good on the résumé. If you do join a professional organization, try to attend meetings of the local chapter.

By taking these steps well before your senior year, you may learn that specific courses make you more marketable.

If you're changing jobs or careers, employers will be more interested in what you have accomplished *since* college. Be prepared to show how your experience relates to the new job. Capitalize on the contacts you've made over the years. Collect on favors owed.

Whether you're a beginner or a veteran, you might register with an employment agency. Of course, a fee is payable after you're hired, but many employers pay the fee. Ask about fee arrangements *before* you sign up.

Once you have a picture of where you fit into the job market, you will set out to answer the big question asked by all employers: *What do you have to offer?* Your answer must be a highly polished presentation of yourself, your education, work history, interests, and special skills—your résumé.

The Résumé

The résumé summarizes your experience and qualifications. Written before your application letter, it provides background information to support your letter. In turn, the letter will emphasize specific parts of your résumé, and will discuss how your background is suited to that job.

Employers generally spend less than sixty seconds scanning a résumé. They look for an obvious and persuasive answer to this question: *What can you do for us?* Employers are impressed by a résumé that

1. looks good (conservative, tasteful, uncluttered, on quality paper)
2. reads easily (headings, typeface, spacing, and punctuation that provide clear orientation)
3. provides information the employer needs for making an interviewing decision

Employers generally discard résumés that are mechanically flawed, cluttered, sketchy, or hard to follow. Don't leave readers guessing or annoyed; make your résumé perfect.

Organize your information within these categories:

- name and address
- job and career objectives
- education
- work experience
- personal data
- interests, activities, awards, and skills
- references

Select and organize material to emphasize what you can offer. Don't just list *everything;* be selective. (We're talking about *communicating* instead of merely delivering information.) Don't abbreviate, because some readers may not know the referent. Use punctuation to clarify and emphasize, not to be "artsy." Try to limit your résumé to a single page, as most employers prefer.[1]

Begin your résumé well before your job search. You will need that much time to do a first-class job. Your final version can be duplicated for various similar targets—but each new type of job requires a new résumé that is tailored to fit the advertised demands of that job.

Caution: Never *invent* credentials just to qualify for a job. Your résumé should make you look as good as the facts allow. But distorting the facts is both unethical and counterproductive. Companies routinely investigate claims made in a résumé. And people who have lied are fired.

Name and Address.   Under the first heading, include your full name, mailing address, and phone number (many interview invitations and job offers are made by phone). If your school and summer address differ, include both, indicating dates on which you can be reached at each.

Job and Career Objectives.   From your collected information (steps 1–9, listed earlier), you should have a clear idea of the *specific* jobs for which you *realistically* qualify.[2] Resist the impulse to be all things to all people. The key to a successful résumé is the image of *you* it projects—disciplined and purposeful, yet flexible. State your specific job and career goals:[3]

> Intensive-care nursing in a teaching hospital, with the eventual goal of supervising and instructing.

Do not borrow a trite or long-winded statement from some placement brochure. If applying to a company with various branches, express your willingness to relocate.

Educational Background.   If your education is more impressive than your work experience, place it first. Begin with your most recent school and work backward, listing degrees, diplomas, and schools *beyond* high

---

[1] Of course, if you are changing jobs or careers, or if your résumé looks cramped, you might need a second page.

[2] Be prepared to have different statements of objectives to meet the requirements of different job descriptions.

[3] To save space, you can omit your statement of career objectives from the résumé and include it in your letter instead.

school (unless prestige, program, or your achievement warrants its inclusion). List the courses that have directly prepared you for the job you seek. If your class rank and grade point average are in the upper 30 percent, list them. Include any schools attended or courses completed while you were in military service. If you financed part or all of your education by working, say so, indicating the percentage.

Work Experience.   If you have solid experience, place it before your education. Beginning with your most recent job and working backward, list and clearly identify each job, giving dates and names of employers. Tell whether the job was full-time, part-time (hours weekly), or seasonal. Tell exactly what you did in each job, indicating promotions. If the job was major (and related to this one), describe it in detail; otherwise, describe it briefly. Include military experience. If you have no real experience, show that you have potential by emphasizing your preparation, and by writing an enthusiastic letter.

Do not use complete sentences in your job descriptions; they take up room best left for other items. But do use action verbs throughout (*supervised, developed, built, taught, opened, managed, trained, solved, planned, directed,* and so on). Such verbs emphasize your vitality, and help you stand out.

Personal Data.   An employer cannot legally discriminate on the basis of sex, religion, color, age, national origin, physical features or marital status. Therefore, you aren't required to provide this information or a photograph. But if you believe that any of this information could advance your prospects, by all means include it.

Personal Interests, Activities, Awards, and Skills.   List hobbies, sports, and other pastimes; memberships in teams and organizations; offices held; and any recognition you have received. Include dates and types of volunteer work. Employers know that persons who seek well-rounded lives are likely to take an active interest in their jobs. Be selective in this section. List only items that show the qualities employers seek.

References.   Your list of references names four or five people *who have agreed* to write strong, positive assessments. Often a reference letter is the key to getting an employer to want to meet you; choose your references carefully.

Select references who can speak with authority about your ability and character. Avoid members of your family and close friends not in your field. Choose instead among professors, previous employers, and community figures who know you well enough to write *concretely* on your behalf. In asking for a reference, keep these points in mind:

1. *A mediocre letter of reference is more damaging than no letter at all.* Don't merely ask, "Could you please act as one of my references?" This question leaves the person little chance to refuse. He or she might not know you well or may be unimpressed by your work but, instead of refusing, might write a watery letter that will do more harm than good. Instead, make an explicit request: "Do you feel you know me and my work well enough to write me a strong letter? If so, would you act as one of my references?" This second version gives people the option to decline gracefully or to promise a positive letter.

2. *Letters are time-consuming.* References have no time to write individual letters to every prospective employer. Ask for one letter only, with no salutation. Your reference keeps a copy; you keep the original for your personal dossier (so that you can reproduce it as necessary); and a copy goes to the placement office for your placement dossier. Because the law permits you to read all material in your dossier, this arrangement provides you with your own copy of your credentials.[4] (The dossier is discussed later in this chapter.)

If the people you select as references live elsewhere, you may wish to make your request by letter, like this one:

Letter Requesting a Reference

From September 1988 to August 1990, I worked at Teo's Restaurant as waiter, cashier, and then assistant manager. Because I enjoyed my work, I decided to study for a career in the hospitality field.

In three months I will graduate from San Jose City College with an A.A. degree in Hotel and Restaurant Management. Next month I begin my job search. Do you remember me and my work enough to write me a strong letter of recommendation? If so, would you kindly serve as one of my references?

To save your time, please omit the salutation from your letter. If you could send me the original, I will forward a copy to my college placement office.

To update you on my recent activities, I've enclosed my résumé. Thank you for your help and support.

---

[4]Under some circumstances you may—if you wish to—waive the right to examine your recommendations. Some applicants, especially those applying to professional schools, as in medicine and law, do waive this right. They do so in concession, one supposes, to a general feeling that a letter writer who is assured of confidentiality is more likely to provide a balanced, objective, and reliable assessment of a candidate. In your own case, you might seek the advice of your major adviser or a career counselor.

Opinion is divided about whether names and addresses of references should be in a résumé. If saving space is important, simply state, "References available on request," keeping your résumé only one page long, but if your résumé already takes up more than one page, you probably should include names and addresses of references. (An employer might recognize a name, and thus notice *your* name among the crowd of applicants.) If you are changing careers, a full listing of references is especially important.

General Résumé Models

With data collected and references lined up, you are ready to compose your actual résumé. Imagine you're a twenty-four-year-old student about to graduate from a community college with an A.A. degree in Hotel and Restaurant Management. Before college, you worked at related jobs for more than three years. You now seek a junior management position while you continue your education part time.

You have spent two weeks compiling information for your résumé, and obtaining commitments from four references. Figure 20.5 shows your résumé. Notice that this résumé mentions nothing about salary. Wait until this matter comes up in your interview, or later. The information, specific but concise, describes what you have to offer, and can be read in a moment. (For a résumé composed in search of summer employment, see Figure 20.6.)

When fully satisfied with your résumé, consider having your prototype printed. The prototype should be attractive, neat, and free of mechanical or grammar errors. A perfect copy is your responsibility. *Never* send out carbon, thermofax, or mimeographed copies. If you are applying for various types of jobs, have your résumé prepared on a word processor with a laser printer. This way, you can adapt each résumé to the particular job.

A final suggestion: avoid a résumé-preparation service. Although professionals can use your raw materials to produce an impressive résumé, employers often recognize the source by its style, and might conclude that you are incapable of communicating on your own.

The Job Application Letter

Your Image.   Although it elaborates on your résumé, your application letter must emphasize personal qualities and qualifications convincingly. Your résumé presents raw facts; your application letter relates these facts to the company to which you are applying. The tone and insight you bring to your discussion suggest a good deal about who you are. The letter is your chance to explain how you see yourself fitting into the organization, to interpret

James David Purdy
203 Elmwood Avenue
San Jose, CA 95139
Tel.: 214-316-2419

Career Objective	Customer relations for a hospitality chain, eventually leading to market management.

Education

1990–1993	*San Jose City College, San Jose, CA* Associate of Arts Degree in Hotel/Restaurant Management, June 1993. Grade point average: 3.25 of a possible 4.00. All college expenses financed by scholarship and part-time job (20 hours weekly).

Employment

1990–1993	*Peek-a-Boo Lodge, San Jose, CA* Began as desk clerk and am now desk manager (part-time) of this 200-unit resort. Responsible for scheduling custodial and room service staff, convention planning, and customer relations.
1988–1990	*Teo's Restaurant, Pensacola, FL* Beginning as waiter, advanced to cashier and finally to assistant manager. Responsible for weekly payroll, banquet arrangements, and supervising dining room and lounge staff.
1987–1988	*Encyclopaedia Britannica, Inc., San Jose, CA* Sales representative (part-time). Received top bonus twice.
1986–1987	*White's Family Inn, San Luis Obispo, CA* Worked as bus boy, then waiter (part-time).

Personal	*Awards* Captain of basketball team, 1988; Lion's Club Scholarship, 1990. *Special Skills* Speak French fluently; expert skier. *Activities* High school basketball and track teams (3 years); college student senate (2 years); Innkeepers' Club—prepared and served monthly dinners at the college (2 years). *Interests* Skiing, cooking, sailing, oil painting, and backpacking.
References	Placement Office, San Jose City College, San Jose, CA 90462

[handwritten marginal note: Clubs and Honors / Computer Skills / Interests]

FIGURE 20.5 Résumé for an Entry-level Candidate

Karen P. Granger
82 Mountain Street
New Bedford, MA 02740
Telephone (617) 864-9318

Objective	A summer internship documenting microcomputer software.
Education	Attending Southeastern Massachusetts University, North Dartmouth, MA 02747; will receive B.A. degree in January 1993. Major: English/Writing. Concentration in computer science. Dean's List, five semesters. GPA: 3.54. Class rank 110 of 1792.

Experience

Intern
Technical Writer

Conway Communications, Inc., 39 Wall Street, Marlboro, MA 02864
Learned local area network (LAN) technology and Conway's product line. Wrote and tested five hardware upgrade manuals. Produced a hardware installation/maintenance manual from another writer's work. Specified and approved all illustrations for my projects. Summers 1990, 1991.

Writing
Tutor

Writing/Reading Center, SMU
Tutored writing and word processing for individuals and groups. Edited WRC student newsletter. Trained new tutors. Co-wrote and acted in a video about the WRC. Fall 1989–present.

Reporter
and Editor

Crimson Courier, New Bedford High School
Wrote features, news, and a humorous column titled *The Lone Granger*. As Editor, organized staff meetings, generated story ideas, edited articles, and was responsible for front page layout and pasteup. Fall 1986–Spring 1988.

Computer Skills IBM PC, Apple IIe, and Macintosh microcomputers; DEC 20 mainframe; VAX 11/780 minicomputer; HBJ Writer, Bankstreet Writer, MacWrite, and MASS11 software; and LOGO, Pascal, and BASIC languages.

Achievements Two writing samples published in Dr. John M. Lannon's *Technical Writing,* 6th ed. (HarperCollins, 1994); Massachusetts State Honors Scholarship, 1989–1992; Advanced Placement Literature/Composition (awarded six credits).

Activities Student member, Society for Technical Communication, 1990; student representative, College Curriculum Committee, 1990; SMU Literary Society, 1990–1992.

References Placement Office, Southeastern Massachusetts University, No. Dartmouth, MA 02747

FIGURE 20.6 Résumé for a Summer-internship Candidate

your résumé, and to show how valuable you will be. Your letter's immediate purpose is to secure an interview.

Targets.   Never send a photocopied letter. You can base different letters on one model—with appropriate changes—but prepare each letter fresh.

Sometimes you will apply for jobs advertised in print or by word of mouth (solicited applications). At other times you will write prospecting letters to organizations that have not advertised but might need someone like you (unsolicited applications). Either letter should be tailored to the situation.

The Solicited Letter.   Imagine you are James Purdy. In *Innkeeper's Monthly*, you read this advertisement and decide to apply:

> **RESORT MANAGEMENT OPENINGS**
>
> Liberty International, Inc. is accepting applications for several junior management positions at our new Lake Geneva resort. Applicants must have three years of practical experience, along with formal training in all areas of hotel/restaurant management. Please apply by June 1, 1993, to
>
>   Elmer Borden
>   Personnel Director
>   Liberty International, Inc.
>   Lansdowne, PA 24135

Now plan and compose your letter.

Introduction.   Create a confident tone by directly stating your reason for writing. Name the job, and remember you are talking *to* someone; use the pronoun "you" instead of awkward or impersonal constructions such as "One can see from the enclosed résumé. . . ." If you can, establish a connection by mentioning a mutual acquaintance—but only with that person's permission. Finally, after referring to your enclosed résumé, discuss your qualifications.

Body.   Concentrate on the experience, skills, and aptitudes you can bring to *this* job. Follow these suggestions:

- Don't come across as a jack-of-all-trades. Relate your qualifications specifically to this job.
- Avoid flattery ("I am greatly impressed by your remarkable company").

- Be specific. Replace "much experience," "many courses," or "increased sales" with "three years of experience," "five courses," or "a 35 percent increase in sales between June and October 1991."
- Support your claims with *evidence,* and show how your qualifications will benefit this employer. Instead of saying, "I have leadership skills," say, "I was student senate president during my senior year and captain of the lacrosse team."
- Create a dynamic tone by using *active* voice and action verbs:

Weak	Management responsibilities were steadily given to me.
Strong	I steadily assumed management responsibilities.

- Trim the fat from your sentences:

Flabby	I have always been a person who enjoys a challenge.
Lean	I enjoy a challenge.

- Express self-confidence:

Unsure	It is my opinion that I have the potential to become a successful manager because . . .
Confident	I will be a successful manager because . . .

- Never be vague:

Vague	I am familiar with the 1022 interactive database management system, and RUNOFF, the text processing system.
Definite	As a lab grader for one semester, I kept grading records on the 1022 database management system, and composed lab procedures on the RUNOFF text processing system.

- Avoid letterese. Write in plain English.
- Show enthusiasm. An enthusiastic attitude can be as important as your background, in some instances.

Conclusion. Restate your interest and emphasize your willingness to retrain or relocate (if necessary). If your reader is nearby, request an interview; otherwise, request a phone call, stating times you can be reached. Leave your reader with the impression that you are worth knowing.

Revision.   *Never* settle for a first draft—or a second or third! This letter is your model for letters serving in varied circumstances. Make it perfect.

After several revisions, James Purdy finally signed the letter shown below.

Job Application Letter

203 Elmwood Avenue
San Jose, CA 10462
April 22, 1993

Mr. Elmer Borden
Personnel Director
Liberty International, Inc.
Lansdowne, PA 24135

Dear Mr. Borden:

Writer identifies self and purpose

Establishes a connection

Please consider my application for a junior management position at your Lake Geneva resort. I will graduate from San Jose City College on May 30 with an Associate of Arts degree in Hotel/Restaurant Management. Dr. H. V. Garlid, my nutrition professor, described his experience as a consultant for Liberty International, and encouraged me to apply.

Relates specific qualifications to the job opening

For two years I worked as a part-time desk clerk, and I am now the desk manager at a 200-unit resort. This experience, along with customer relations work described in my résumé, has given me a clear and practical understanding of customers' needs and expectations.

As an amateur chef, I know of the effort, attention, and patience required to prepare fine food. Moreover, my skiing and sailing background might be assets to your resort's recreation program.

Expresses confidence and enthusiasm throughout

I have confidence in my hospitality management skills. My experience and education have prepared me to work well with others and to respond creatively to changes, crises, and added responsibilities.

Makes follow-up easy for the reader

If my background meets your needs, please phone me any weekday after 4 p.m. at 214-555-2419.

Sincerely,

James D. Purdy

James D. Purdy

Enclosure

Purdy wisely emphasizes practical experience because his background is varied and impressive. An applicant with less experience would emphasize education instead, discussing courses and activities.

As an additional example, here is the letter composed by Karen Granger in her quest for a summer internship.

Internship Application Letter

Dear. Mr. White:

I read in Internships 1992 that your company offers a summer documentation internship. Because of my education and previous technical writing employment, I am highly interested in such a position.

Begins by stating purpose

In January 1993, I will graduate from Southeastern Massachusetts University with a B.A. in English/Writing. I have prepared specifically for a computer documentation career by taking computer science, mathematics, and technical writing courses.

Identifies herself and college background

In one writing course, the Computer Documentation Seminar, I wrote three software manuals. One manual uses a tutorial to introduce beginners to the Apple Macintosh and MacWrite. The other manuals describe two IBM PC applications that arrived at SMU with no documentation.

Expands on background

The enclosed résumé describes my work as the intern technical writer with Conway Communications, Inc., for two summers. I learned local area networking (LAN) by documenting Conway's LAN hardware and software. I was responsible for several projects simultaneously and spent much of my time talking with engineers and testing procedures. If you would like samples of my writing, please let me know.

Describes work experience

Although Conway has invited me to return next summer and to work full time after graduation, I would like more varied experience before committing myself to permanent employment. I know I could make a positive contribution to Birchwood Group, Inc. May I telephone you next week to arrange a meeting?

Explains interest in this job

The Unsolicited Letter. Ambitious job seekers do not limit their search to advertised openings. The unsolicited, or "prospecting," letter is a good way to uncover possibilities beyond the Help Wanted section. Such letters have advantages and disadvantages.

Disadvantages.   The unsolicited approach has two drawbacks: (1) you might waste time writing to organizations that have no openings; and (2) because you don't know what the opening is—even if there is one—you cannot tailor your letter to the specific requirements.

Advantages.   For an advertised opening you compete with legions of applicants, but if your timing is lucky your unsolicited letter might arrive just as an opening materializes. Even when no opening exists, most companies welcome unsolicited letters and keep impressive applications on file, or pass them along to a company that has an opening. Unsolicited letters can be a sound investment if your targets are well chosen and your expectations are realistic.

Reader Interest.   Because your unsolicited letter is unexpected, attract attention immediately. Don't begin: "I am writing to inquire about the possibility of obtaining a position with your company." By now, your reader is asleep. If you can't establish a connection through a mutual acquaintance, use a forceful opening:

Opens forcefully

> Does your hotel chain have a place for a junior manager with a degree in hospitality management, proven commitment to quality service, and customer relations experience that extends far beyond mere textbook learning? If so, please consider my application for a position.

Address your letter to the person most likely in charge of hiring. (Consult the business directories listed on page 478 for names of company officers.)

Employers will regard the quality of your application as an indication of the quality of work you will do. Businesses spend much money and time projecting favorable images. The image you project, in turn, must meet their standards.

SUPPORT FOR THE APPLICATION

Your Dossier

Your dossier contains your credentials: college transcript, recommendation letters, and any other items (such as a notice of scholarship award or commendation letter) that document your achievements. An employer impressed by what you say about yourself will want to read what others think, and will request your dossier. By collecting recommendations in one folder, you spare your references from writing the same letter over and over.

Your college placement office will keep your dossier on file and send copies to employers who request them. Always keep your own copy as well. Then, if an employer requests your dossier, you can make a photocopy and mail it, advising your reader that the placement copy is on the way. This is not needless repetition! Most employers establish a specific timetable for (1) advertising an opening, (2) reading letters and résumés, (3) requesting and reviewing dossiers, (4) interviewing, and (5) making an offer. Timing is crucial. Too often, dossier requests from employers sit and gather dust in some "incoming" box in a busy placement office. Weeks can pass before your dossier is mailed. In such a case the only loser is you.

Job Interviews

An employer impressed by your credentials will arrange an interview. The interview's purpose is to confirm the employer's impressions from your application. You are a finalist. But now you must be as impressive in person as you seem on paper—and the best person for the job.

You might meet with one interviewer, a group, or several groups in succession. You might be interviewed alone or with several candidates at once. Interviews can last one hour or less, a full day, or several days. The interview can range from a pleasant chat to grueling quiz sessions. Some interviewers may antagonize you deliberately to observe your reaction.

Take as many interviews as you can—as long as you are genuinely interested in the position. (It would be unethical to waste an employer's time by interviewing just for practice.) Your confidence and competence will increase with each.

Prepare for the interview by learning about the company (its products or services, history, prospects, branch locations) in trade journals and industrial indexes. If time permits, request company literature and annual reports. Prepare specific answers to the obvious questions:

- *Why do you wish to work here?*
- *What do you know about our company?*
- *What do you see as your biggest weakness? Your biggest strength?*
- *What would you like to be doing in ten years?*

Plan informative and direct answers to questions about your background, training, experience, and salary requirements. Prepare your own list of questions about the job and the organization; you will be invited to ask questions, and the questions you ask can be as revealing as the answers you give.

Verify the interview's exact time and location. Come dressed as if you already work for the company. Maintain eye contact most of the time; if you stare at your shoes, the interviewer will not be impressed. Relax but do not slouch. Do not smoke, even if invited. Do not pretend to know more than you do; if you cannot answer a question, say so. Avoid abrupt yes or no answers, as well as life stories. Make your answers concrete but to the point.

Don't be afraid to allow silence. An interviewer simply may *stop* talking, just to observe your reaction to silence. If you really have nothing more to say or ask, don't feel compelled to speak; caught off-guard, many of us are likely to say something stupid. Let the interviewer make the next move.

When your interviewer hints that the meeting is ending (perhaps by checking a wristwatch), don't wear out your welcome. Restate your interest; ask when you can expect further word; thank the interviewer; and leave.

The Follow-Up Letter

A few days after the interview, refresh the employer's memory with a letter restating your interest. James Purdy sent Elmer Borden this follow-up:

> Thank you again for your hospitality during my visit to your Lansdowne offices. The trip and scenery were delightful.
>
> After meeting you and your colleagues and touring the resort, I know I could be a productive staff member.

The Letter of Acceptance

You may receive a job offer by phone or letter. If by phone, request a written offer, and respond with a formal letter of acceptance. This letter may serve as part of your contract; spell out the terms you are accepting. Here is Purdy's letter of acceptance:

> I am happy to accept the position as assistant recreation supervisor at Liberty International's Lake Geneva Resort, with a starting salary of $28,500.
>
> As you requested, I will phone Ms. Druid in your personnel office for instructions on reporting date, physical exam, and employee orientation.
>
> I look forward to a long and satisfying career with Liberty International.

The Letter of Refusal

You may have to refuse offers. Even if you refuse by phone, write a prompt and cordial refusal letter, explaining your reasons, and allowing for future possibilities. Purdy handled one refusal this way:

> Although I was impressed by the challenge and efficiency of your company, I must decline the position as assistant desk manager of your London hotel. I have accepted an offer from Liberty International that will enable me to work full time to complete my degree requirements in hospitality management.

If subsequent openings should materialize, however, I would again appreciate your considering me as a candidate.

Thank you for your interest and courtesy.

REVISION CHECKLIST FOR LETTERS

Use this checklist to refine the content, arrangement, and style of your letters. (Numbers in parentheses refer to the first page of discussion.)

Content

☐ Is the letter addressed to a specifically named person? (462)

☐ Does the letter have all the standard parts? (462)

☐ Does the letter have all needed specialized parts? (466)

☐ Have you given the reader all necessary information? (471)

☐ Have you identified the name and position of your reader? (462)

Arrangement

☐ Does the introduction immediately engage the reader and lead naturally to the body? (461)

☐ Are transitions between letter parts clear and logical? (243)

☐ Does the conclusion encourage the reader to act? (461)

☐ Is the format correct? (467)

☐ Does the letter follow an accepted form? (468)

Style

☐ Is the letter in conversational language (free of letterese)? (468)

☐ Does the letter reflect a "you" perspective? (471)

☐ Does the tone reflect your relationship with your reader? (471)

☐ Is the reader likely to react favorably to this letter? (471)

☐ Are all sentences clear, concise, and fluent? (251)

☐ Is the letter written in grammatical English? (Appendix A)

☐ Does the letter's appearance enhance the writer's image? (472)

EXERCISES

1. Bring to class a copy of a business letter addressed to you or a friend. Compare letters. Choose the most and least effective.

2. Write and mail an unsolicited letter of inquiry about the topic you are investigating for an analytical report or research assignment. In your letter you might request brochures, pamphlets, or other informative literature, or you might ask specific questions. Submit to your instructor a copy of your letter and the response.

3. *a.* As a student in a state college, you learn that your governor and legislature have cut next year's operating budget for all state colleges by 20 percent. This cut will cause the firing of young and popular faculty members, drastically reduce admissions, financial aid, and new programs, and wreck college morale. Write a claim letter to your governor or representative, expressing your strong disapproval and justifying a major adjustment in the proposed budget.

    b. Write a claim letter to a politician about some issue affecting your school or community.

    c. Write a claim letter to an appropriate school official to recommend action on a campus problem.

4. Write a 500- to 700-word essay applying to a college for transfer or for graduate or professional school admission. Cover two areas: (1) what you can bring to this school by way of attitude, background, and talent; and (2) what you expect to gain in personal and professional growth.

5. Write a letter applying for a part-time or summer job, in response to a specific ad. Choose an organization related to your career goal. Identify the exact hours and calendar period during which you are free to work.

6. *a.* Identify a job you hope to have. Using newspaper ads, library (see your reference librarian for assistance), placement office, and personal sources, write your own description of the job: duties, responsibilities, work hours, salary range, requirements for promotion, highest promotion possible, unemployment rate in the field, employment outlook during the next decade, need for further education (advanced degrees, special training), employment rate in terms of geography, optimum age bracket (as in football, is one considered "over the hill" after thirty-five?).

    b. Using these sources, construct a profile of the ideal employee for this job. If you were the personnel director screening applicants, what specific qualifications would you require (education, experience, age, physical ability, appearance, special skills, personality traits, attitude, outside interests, and so on)? Try to locate an actual newspaper ad describing job responsibilities and qualifications. Or assume you're a personnel officer, and compose an ad for the job.

    c. Assess your own credentials against the ideal employee profile. Review the plans you have made to prepare for this job: specific courses, special training, work experience. Assume you have completed your preparation. How do you measure up to the requirements in b? Are your goals realistic? If not, why not? What alternative plan should you create?

    d. Using your list from part c, compose a perfect résumé.

    e. Write a letter applying for the job described in part a.

    f. Write a follow-up letter thanking your fictional employer for your interview and restating your interest.

    g. Write a letter accepting a job offer from this same employer.

    h. Write a letter graciously declining this job offer.

    i. Submit each of these items, in order, to your instructor.

Note: Use the sample letters in this chapter for guidance, but don't borrow specific expressions.

7. Most of these sentences need to be overhauled before being included in a letter. Identify the weakness in each statement, and revise as needed.

    a. Pursuant to your ad, I am writing to apply for the internship.

    b. I need all the information you have about methane-powered engines.

    c. You idiots have sent me a faulty disk drive!

    d. It is imperative that you let me know of your decision by January 15.

    e. You are bound to be impressed by my credentials.

    f. I could do wonders for your company.

    g. I humbly request your kind consideration of my application for the position of junior engineer.

    h. If you are looking for a winner, your search is over!

    i. I have become cognizant of your experiments and wish to ask your advice about the following procedure.

    j. You will find the following instructions easy enough for an ape to follow.

    k. I would love to work for your wonderful company.

    l. As per your request I am sending the county map.

    m. I am in hopes that you will call soon.

    n. We beg to differ with your interpretation of this leasing clause.

    o. I am impressed by the high salaries paid by your company.

8. Write a complaint letter about a problem you've had with goods or service. State your case clearly and objectively, and request a specific adjustment.

COLLABORATIVE PROJECT

Assume this advertisement has appeared in your school newspaper:

STUDENT CONSULTANT WANTED

The Dean of Students invites applications for the position of student consultant to the Dean for the next academic year. Duties: (1) meeting with students as individuals and groups to discuss issues, opinions, questions, complaints, and recommendations about all areas of college policy, (2) presenting oral and written reports to the Dean of Students, and (3) attending college planning sessions as student spokesperson. Time commitment: 15 hours weekly. Salary: $4,000.

Candidates should be full-time students with at least one year of student experience at this college. The ideal applicant will be skilled in report writing and oral communication, will work well in groups, and will demonstrate a firm commitment to our college. Application deadline: May 15.

a. Compose a résumé and a letter of application for this position.

b. Divide your class into screening, interview, and hiring committees.

c. Exchange your group's letters and résumés with those of another group.

d. As an individual committee member, read and evaluate each of the applications you have received. Rank each application privately on paper, according to the criteria in this chapter, before discussing them with your group. *Note:* While screening applicants, you will be competing for selection by another committee, which is reviewing your own application.

e. As a committee, select the three strongest applications, and interview each finalist for ten minutes, after you have prepared a list of standardized questions.

f. On the basis of these interviews, rank your preferences privately, on paper, giving specific reasons for your final choice.

g. Compare your conclusions with those of your colleagues and choose the winning candidate.

h. As a committee, compose a memo to your instructor, justifying your final recommendations. (See pages 501–503 for more on justification reports.)

Memos and Short Reports

Purpose of Memoranda and Short Reports

Recommendation Reports

Justification Reports

Progress Reports

Periodic Activity Reports

Survey Reports

Minutes of Meetings

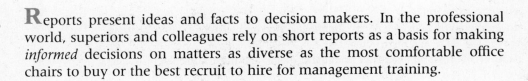

Reports present ideas and facts to decision makers. In the professional world, superiors and colleagues rely on short reports as a basis for making *informed* decisions on matters as diverse as the most comfortable office chairs to buy or the best recruit to hire for management training.

PURPOSE OF MEMORANDA AND SHORT REPORTS

People on the job must communicate rapidly and precisely. Here are some of the kinds of reports you might write on any workday:

- a request for assistance on a project
- a requisition for parts and equipment
- a proposal for a new project
- a report of your progress on a specific assignment
- an hourly or daily account of your work activities
- a report of your inspection of a site, item, or process

NAME OF ORGANIZATION

MEMORANDUM
Date: (also serves as a chronological record for future reference)
To: Name and title (the title also serves as a record for reference)
From: Your name and title (your initials for verification)
Subject: GUIDELINES FOR FORMATTING MEMOS

Subject Line
Announce the memo's purpose and contents, to orient readers to the subject and help them gauge its importance. An explicit title also makes filing by subject easier.

Introductory Paragraph
Unless you have reason for being indirect, state your main point immediately.

Topic Headings
When discussing a number of subtopics, include headings (as we do here). Headings help you organize, and help readers locate information quickly.

Paragraph Spacing
Do not indent the first line of paragraphs. Single space within, and double space between paragraphs.

Second-page Notation
When the memo exceeds one page, begin the second and subsequent pages with recipient's name, date, and page number. For example: Ms. Baxter, June 12, 19xx, page 2. Place this information three lines from the page top and begin your text three lines below.

Memo Verification
Do not sign your memos. Initial the "From" line, after your name.

Copy Notation
When sending copies to people not listed on the "To" line, include a copy notation two spaces below the last line, and list, by rank, the names and titles of those receiving copies. For example,

Copies: J. Spring, V.P., Production
          H. Baxter, General Manager, Production

FIGURE 21.1 Guidelines for Memo Formatting

- a record of a meeting
- a report of your survey to select the best prices or products

Many of these reports take the form of a memorandum.

The major form for internal written communication in organizations, memos leave a "paper trail"[1] of directives, inquiries, instructions, requests, recommendations, and so on. Organizations rely heavily on memos to trace decisions, track progress, recheck data, or identify the person(s) responsible. Therefore, any memo you write can have far-reaching ethical and legal implications. Be sure your memo includes the date, your initials for verification, and that your information is specific, accurate, and unambiguous.

The standard memo has a heading that names the organization and identifies the sender, recipient, subject (often in caps or underlined for emphasis), and date. (Placement of these items may differ among firms.) Other memo elements are shown in Figure 21.1.

Memo reports cover every conceivable topic. Common types include recommendation, justification, progress, periodic, and survey reports.

RECOMMENDATION REPORTS

Recommendation reports interpret data, draw conclusions, and make recommendations, often in response to a specific reader request. The following recommendation report is addressed to the writer's boss. This sample typifies the problem solving carried on daily by professionals and documented in memo form.

A Problem-Solving Recommendation

Assume you are the writer, Bruce Doakes, in the situation that generated the Trans Globe memo below:

You are assistant manager of occupational health and safety for a major airline that employs over two hundred reservation and booking agents. Each agent spends eight hours daily seated at a workstation that has a computer, telephone, and other electronic equipment. Many agents have complained of chronic discomfort from their work: headache, eyestrain and irritation, blurred or double vision, backache, and stiff neck and joints.

Your boss asks you to study the problem and recommend improvements in the work environment. You survey employees and consult ophthalmolo-

---

[1] Electronic-mail memos may leave a "paperless trail" as well. Although E-mail messages tend to be less formal and more hastily written than paper documents, they are saved in both hard copy and online—and can inadvertently be forwarded to someone never intended to receive or read them.

gists, chiropractors, orthopedic physicians, and the latest publications on ergonomics (tailoring work environments for employees' physical and psychological well-being). After completing the study, you compose the report.

Bruce's report needs to be persuasive as well as informative.

Recommendation Memo

TRANS GLOBE AIRLINES

Memorandum

TO: R. Ames, Vice President, Personnel

August 15, 1993

FROM: B. Doakes, Health and Safety *BD*

SUBJECT: **RECOMMENDATIONS FOR REDUCING
AGENTS' DISCOMFORT**

Provides immediate orientation by giving brief background and main point

In our July 20 staff meeting, we discussed physical discomfort among reservation and booking agents, who spend eight hours daily at automated workstations. Our agents complain of headaches, eyestrain and irritation, blurred or double vision, backaches, and stiff joints. This report outlines the apparent causes, and recommends ways of reducing discomfort.

<u>Causes of Agents' Discomfort</u>

Interprets findings and draws conclusions

For the time being, I have ruled out the computer display screens as a cause of headaches and eye problems because of the following reasons:

1. Our new display screens have excellent contrast and no flicker.
2. Research findings about the effects of low-level radiation from computer screens are inconclusive.

The immediate cause of headaches and eye problems seems to be from the excessive glare on display screens from background lighting.

Other discomforts, such as backaches and stiffness, apparently result from the agents' sitting in one position for up to two hours between breaks.

<u>Recommended Changes</u>

Makes general recommendations

We can eliminate much discomfort by improving background lighting, workstation conditions, and work routines and habits.

<u>Background Lighting.</u> To reduce the glare on display screens, these are recommended changes in background lighting:

1. Decrease all overhead lighting by installing lower-wattage bulbs.
2. Keep all curtains and adjustable blinds on the south and west windows at least half-drawn, to block direct sunlight.
3. Install shades to direct the overhead light straight downward, so that it is not reflected by the screens.

Expands on each recommendation

<u>Workstation Conditions.</u> These are recommended changes in the workstations:

1. Reposition all screens so light sources are neither at front nor back.
2. Wash the surface of each screen weekly.
3. Adjust each screen so the top is slightly below the operator's eye level.
4. Adjust all keyboards so they are 27 inches from the floor.
5. Replace all fixed chairs with adjustable, armless, secretarial chairs.

<u>Work Routines and Habits.</u> These are recommended changes in agents' work routines and habits:

1. Allow frequent rest periods (10 minutes each hour instead of 30 minutes twice daily).
2. Provide yearly eye exams for all terminal operators, as part of our routine health-care program.
3. Train employees to adjust screen contrast and brightness whenever the background lighting changes.
4. Offer workshops on improving posture.

These changes will give us time to consider more complex options such as installing hoods and antiglare filters on terminal screens, replacing fluorescent lighting with incandescent, covering surfaces with nonglare paint, or other disruptive procedures.

Discusses benefits of following the recommendations

cc: J. Bush, Medical Director
    M. White, Manager of Physical Plant

For advice on how to think critically as you formulate, evaluate, and refine your recommendations, see pages 555–557.

JUSTIFICATION REPORTS

Justification reports (a type of recommendation report) are in a unique class in that they often are initiated by the writer rather than requested by the readers. As the name implies, such reports *justify* the writer's position on some issue. Justification reports therefore typically begin rather than end with the request or recommendation. Such reports answer this key question for readers: *Why should we?*

Typically, justification reports follow a version of this arrangement:

1. State the problem and your recommendations for solving it.
2. Point out the cost, savings, and benefits of your plan.
3. If needed, explain how your suggestion can be implemented.
4. Conclude by encouraging the reader to act.

Our next writer uses a version of the preceding arrangement: she begins with the problem and recommended solution, spells out costs and benefits, and concludes by re-emphasizing the major benefit. The tone is confident yet diplomatic—appropriate for an unsolicited recommendation to a superior.

Justification Memo

GREENTREE BIONOMICS, INC.
Memorandum

TO: D. Spring, Personnel Director   April 18, 1993
           Greentree Bionomics, Inc. (GBI)
FROM: M. Noll, Biology Division *MN*
SUBJECT: The Need to Hire Additional Personnel

Introduction and Recommendation

Opens with the problem

With twenty-six active employees, GBI has been unable to keep up with scheduled contracts. As a result, we have a contract backlog of roughly $500,000. This backlog is caused by understaffing in the biology and chemistry divisions.

Recommends a solution

To increase production and ease the work load, I recommend that GBI hire three general laboratory assistants.

Expands on the recommendation

The lab assistants would be responsible for cleaning glassware and general equipment; feeding and monitoring fish stocks; preparing yeast, algae, and shrimp cultures; preparing stock solutions; and assisting scientists in various tests and procedures.

Cost and Benefits

Shows how benefits would offset costs

While costing $28,080 yearly (at $4.50/hour), three full-time lab assistants would have a positive effect on overall productivity:

1. Uncleaned glassware no longer would pile up, and the fish-holding tanks could be cleaned daily (as they should be) instead of weekly.

2. Because other employees would no longer need to work more than forty hours weekly, morale would improve.
3. Research scientists would be freed from general maintenance work (cleaning glassware, feeding and monitoring the fish stock, etc.). With more time to perform client tests, the researchers could eliminate our backlog.
4. With our backlog eliminated, clients would no longer have cause for impatience.

Conclusion

Increased production at GBI is essential to maintaining good client relations. These additional personnel would allow us to continue a reputation of prompt and efficient service, thus ensuring our steady growth and development.

Encourages acceptance of recommendation

PROGRESS REPORTS

Large organizations depend on progress (or status) reports to keep track of activities, problems, and progress on various projects. Daily progress reports are vital in a business that assigns crews to many projects. Management uses progress reports to evaluate the project and its supervisor, and to decide how to allocate funds.

Often, a progress report is one of a series. Together, the project proposal (Chapter 22), progress reports (the number varies with the scope and length of the project), and the final report (Chapter 23) provide a record and history of the project.

To give management the answers it needs, progress reports must, at a minimum, answer these questions:

1. *How much has been accomplished since the last report?*
2. *Is the project on schedule?*
3. *If not, what went wrong?*
   a. *How was the problem corrected?*
   b. *How long will it take to get back on schedule?*
4. *What else needs to be done?*
5. *What is the next step?*
6. *Any unexpected developments?*
7. *When do you anticipate completion? Or, on a long project, when do you anticipate completion of the next phase?*

If the report is part of a series, you might also refer to prior problems or developments.

Many organizations have forms for organizing progress reports, and so no one format is best. But each report in a series should follow the same organization. The following memorandum illustrates how one writer organized her report.

Progress Report (On the Job)

SUBJECT: Progress Report: Equipment for New Operations Building

Work Completed

Our training group has met twice since our May 12 report in order to answer the questions you posed in your May 16 memo. In our first meeting, we identified the types of training we anticipate.

Types of Training Anticipated

- Divisional Surveys
- Loan Officer Work Experience
- Divisional Systems Training
- Divisional Clerical Training (Continuing)

- Divisional Clerical Training (New Employees)
- Divisional Management Training (Seminars)
- Special/New Equipment Training

In our second meeting, we considered various areas for the training room.

Training Room

The frequency of training necessitates our having a training room available daily. The large training room in the Corporate Education area (10th floor) would be ideal. Before submitting our next report, we need your confirmation that this room can be assigned to us.

To support the training programs, we purchased this equipment:

- Audioviewer
- 16mm projector
- Videocassette recorder and monitor
- CRT
- Mini/micro computer, for computer-assisted instruction
- Slide projector
- Tape recorder

This equipment will allow us to administer training in a variety of modes, ranging from programmed and learner-controlled instruction to group seminars and workshops.

Work Remaining

To support the training, we need to furnish the room appropriately. Because the types of training will vary, the furniture should provide a flexible environment. Outlined here are our anticipated furnishing needs.

Describes what remains to be done

- Tables and chairs that can be set up in many configurations. These would allow for individual or group training and large seminars.
- Portable room dividers. These would provide study space for training with programmed instruction, as well as allow for simultaneous training.
- Built-in storage space for audiovisual equipment and training supplies. Ideally, this storage space should be multipurpose, providing work or display surfaces.
- A flexible lighting system, important for audiovisual presentations and individualized study.
- Independent temperature control, to ensure that the training room remains comfortable regardless of group size and equipment used.

The project is on schedule. As soon as we receive your approval of these specifications, we will proceed to the next step: sending out bids for room dividers, and having plans drawn for the built-in storage space.

Gives a rough timetable

cc: R.S. Pike, SVP
    G. T. Bailey, SVP

As you work on a longer report or term project, your instructor might require a progress report. In this next memo, Karen Granger documents her progress on her term project: an evaluation of the Environmental Protection Agency's effectiveness in cleaning a heavily contaminated harbor.

Progress Report on Term Project

PROGRESS REPORT

TO: Dr. John Lannon   *Date*
FROM: Karen P. Granger   *To (Professor)*
DATE: April 17, 1993   *From (Me)*
SUBJECT: Evaluation of the EPA's *Remedial Action Master Plan*   *Title of Report*

① Work Completed

February 23: Began general research on the PCB contamination of the New Bedford Harbor.

Summarizes achievements to date

March 8:   Decided to analyze the *Remedial Action Master Plan* (RAMP) in order to determine whether residents are being "studied to death" by the EPA.

March 9–19: Drew a map of the harbor to show areas of contamination. Obtained the RAMP from Pat Shay of the EPA.

Interviewed Representative Grimes briefly by phone; made an appointment to interview Grimes and Sharon Dean on April 13.

Interviewed Patricia Chase, President of the New England Sierra Club, briefly, by phone.

March 24: Obtained *Public Comments on the New Bedford RAMP,* a collection of reactions to the plan.

April 13: Interviewed Grimes and Dean; searched Grimes' files for information. Also searched the files of Raymond Soares, New Bedford Coordinator, EPA.

Work in Progress Don't need

Contacting by telephone the people who commented on the RAMP.

Work to Be Completed ②

April 25: Finish contacting commentators on the RAMP.

April 26: Interview an EPA representative about the complaints that the commentators raised on the RAMP.

Date for Completion: May 3, 1993 Don't need

Complications and comms ③

The issue of PCB contamination is complicated and emotional. The more I uncovered, the more difficult I found it to remain impartial in my research and analysis. As a New Bedford resident, I expected to find that we are indeed being studied to death; because my research seems to support my initial impression, I am not sure I have remained impartial.

Lastly, the people I talk to do not always have the time to find answers for my questions. Everyone, however, has been interested and encouraging, if not always informative.

Page 653 shows another progress report on a term project.

PERIODIC ACTIVITY REPORTS

The periodic activity report is similar to the progress report in that it summarizes activities over a specified period. But unlike progress reports, which summarize specific accomplishments on a given project, periodic reports summarize general activities during a given period. Manufacturers requiring periodic reports often have prepared forms, because most of their tasks are quantifiable (e.g., Units produced). But most white-collar jobs do not lend themselves to prepared-form reports. You will probably have to develop your own format, as our next writer does.

DeWitt's report answers her boss's primary question: *What did you accomplish last month?* Her response has to be detailed and informative.

Periodic Activity Report

DATE: 6/15/93
TO: N. Morgan, Assistant Vice President
FROM: F. C. DeWitt
SUBJECT: Coordinating Meetings for Training

For the past month, I've been working on a cooperative project with the Banking Administration Institute, Computron Corporation, and several banks. My purpose has been to develop training programs, specific to banking, appropriate for computer-assisted instruction (CAI).

Gives overview of recent activities, and their purpose

We have focused on three major areas: Proof/Encoding Training, Productivity Skills for Management, and Banking Principles.

I hosted two meetings for this task force. On June 16, we will discuss Proof/Encoding Training, and on June 17, Productivity Training. The objective for the Proof/Encoding meeting was to compare ideas, information, and current training packages available on this topic. We are now designing a training course.

Gives details

The objective of the meeting on Productivity was to discuss skills that increase productivity in banking (specifically Banking Operations). Discussion included instances in which computer-assisted instruction is appropriate for teaching productivity skills. Computron also discussed computer applications used to teach productivity.

On June 10, I attended a meeting in Washington, D.C., to design a course in basic banking principles for high-level clerical/supervisory-level employees. We also discussed the feasibility of adapting this course to CAI. This type of training, not currently available through Corporate Education, would meet a definite supervisor/management need in the division.

My involvement in these meetings has two benefits. First, structured discussions with trainers in the banking industry provide an exchange of ideas, methods, and experiences. This involvement expedites development of our training programs because it saves me time on research. Second, microcomputers will continue to affect future training. With a working knowledge of these systems and

Explains the benefits of these activities

their applications, I now am able to assist my group in designing programs specific to our needs.

Periodic reports are used by employees to inform management of what they are doing and how well they are doing it. Therefore, accuracy, clarity, and appropriate level of detail are important—as is the persuasive (and ethical) dimension.

SURVEY REPORTS

Brief survey reports often are used to examine conditions that affect an organization (consumer preferences, available markets). The following memo, from the research director for a midwestern grain distributor, gives clear and specific information directly.

Survey Report

SUBJECT: Food-Grain Consumption, U.S., 1989–92

Announces purpose of memo

Here are the data you requested on April 7, as part of your division's annual marketing survey.

U.S. PER CAPITA CONSUMPTION OF FOOD GRAINS IN POUNDS, 1989–92

Presents data in tabular form, for easy comparison

	1989	1990	1991	1992
Corn Products				
Cornmeal and other	16.4	16.4	16.6	16.7
Corn syrup and sugar	22.3	22.7	23.3	23.8
Oat Products	3.8	4.1	4.9	4.8
Barley Products	1.2	1.2	1.2	1.2
Wheat				
Flour	116.2	116.4	117.1	117.8
Breakfast cereals	2.9	2.9	2.9	2.9
Rye, Flour	1.2	1.2	1.2	1.2
Rice, Milled	8.3	6.7	7.7	7.0

Source: U.S. Dept. of Agriculture

Tells how to follow up

If you require additional information, please call Ms. Smith at 316.

cc: C. B. Schultz, Vice President, Marketing

MINUTES OF MEETINGS

There are many team or project meetings on the job that require someone to record the proceedings. Minutes are the records of such meetings. Copies of minutes are distributed to all members and interested parties, to keep track of proceedings. The appointed secretary records the minutes. When you record minutes, answer these questions:

- What group held the meeting? When, where, and why?
- Who chaired the meeting? Who else was present?
- Were the minutes of the last meeting approved (or disapproved)?
- Who said that?
- Was anything resolved?
- Who made which motions, seconded them, and what was the vote?

Minutes are filed as part of an official record, and so must be precise, clear, highly informative, and free of the writer's personal commentary ("As usual, Ms. Jones disagreed with the committee") or judgmental words ("good," "poor," "irrelevant").

Minutes of a Meeting

SUBJECT: Minutes of Managers' Meeting, October 5, 1993

Members Present

Harold Tweeksbury, Jeannine Boisvert, Sheila DaCruz, Ted Washington, Denise Walsh, Cora Parks, Cliff Walsh, Joyce Capizolo.

Tells who came

Agenda

1. The meeting was called to order on Wednesday, October 5, at 10 A.M. by Cora Parks.
2. The minutes of the September meeting were approved unanimously.
3. The first order of new business was to approve the following policies for the Christmas season:
    a. Temporary employees should list their ID numbers in the upper left corner of their receipt envelopes in order to help verification. Discount Clerical assistant managers will be responsible for seeing that this procedure is followed.
    b. When temporary employees turn in their envelopes, personnel from Discount Clerical should spot-check them for completeness and legibility. Incomplete or illegible envelopes should be corrected, completed, or rewritten. Envelopes should not be sealed.

Summarizes discussion of each item

4. Jeannine Boisvert moved that we also hold one-day training workshops for temporary employees in order to teach them our policies and procedures. Denise Walsh seconded the motion. Joyce Capizolo disagreed, saying that on-the-job training (OJT) was enough. The motion for the training session carried 6–3. The first workshop, which Jeannine agreed to arrange, will be held October 25.

5. Joyce Capizolo requested that temporary employees be sent a memo explaining the temporary employee discount procedure. The request was converted to a motion and seconded by Cliff Walsh. The motion passed by a 7–2 vote.

6. Cora Parks adjourned the meeting at 11:55 A.M.

REVISION CHECKLIST FOR SHORT REPORTS

Use this checklist as a guide to refining and revising your short reports. (Numbers in parentheses refer to the first page of discussion.)

☐ Have you chosen the best report form for your purpose? (499)

☐ Does the memo have a complete heading? (498)

☐ Does the subject line forecast the memo's contents? (498)

☐ Are readers given enough information for an *informed* decision? (499)

☐ Are the conclusions and recommendations clear? (501)

☐ Are paragraphs single-spaced within and double-spaced between? (498)

☐ Do headings, charts, or tables appear whenever needed? (498)

☐ If more than one reader is receiving copies, does the memo include a distribution notation (cc:) to identify other readers? (498)

☐ Are all sentences clear, concise, and fluent? (251)

☐ Does the document's appearance create a favorable impression? (499)

☐ Is the document ethically acceptable? (499)

EXERCISES

1. We all would like to see changes in our schools' policies or procedures, whether they are changes in our majors, school regulations, social activities, grading policies, registration procedures, or the like. Find some area of your school that needs obvious changes, and write a justification report to the person who might initiate that change. Explain why the change is necessary and describe the benefits. Follow the format on page 502.

2. Think of an idea you would like to see implemented in your job (e.g., a way to increase productivity, improve service, increase business, or improve working conditions). Write a justification memo, persuading your audience that your idea is worthwhile.

3. Write a memo to your employer, justifying reimbursement for this course. *Note:* You might have written another version of this assignment for Exercise 5 in Chapter 1. If so, compare early and recent versions for content, organization, style, and format.

4. Identify a dangerous or inconvenient area or situation on campus or in your community (endless cafeteria lines, a poorly lit intersection, slippery stairs, a poorly adjusted traffic light). Observe the problem for several hours during a peak-use period. Write a justification report to a *specifically identified* decision maker, describing the problem, listing your observations, making recommendations, and encouraging reader support or action.

5. Assume you have received a $10,000 scholarship, $2,500 yearly. The only stipulation for receiving installments is that you send the scholarship committee a yearly progress report on your education, including courses, grades, school activities, and cumulative average. Write the report.

6. In a memo to your instructor, outline your progress on your term project. Describe your accomplishments, plans for further work, and any problems or setbacks. Conclude your memo with a specific completion date.

7. Keep accurate minutes for one class session (preferably one with debate or discussion). Submit the minutes in memo form to your instructor.

8. Conduct a brief survey (e.g., of comparative interest rates from various banks on a car loan, comparative tax and property evaluation rates in three local towns, or comparative prices among local retailers for an item). Arrange your data and report your findings to your instructor in a memo that closes with specific recommendations for the most economical choice.

9. Although the campus store (or cafeteria) is a convenient place to buy small items (or food), you believe students are overcharged for the convenience. To prove your point, you decide to do a comparative analysis of the cost of five items at the campus store (or cafeteria) and two local stores (or restaurants). Be sure to compare like sizes, weights, brands. Conduct your survey, analyze your findings, and submit the results in memo form to the student senate. Your instructor might want you to include recommendations.

10. Recommendation Report (choose one)

    a. You are legal consultant to the leadership of a large auto-workers' union. Before negotiating its next contract, the union needs to anticipate effects of robotics technology on assembly-line auto workers within ten years. Do the research and write a report recommending a course of action.

b. You are a consulting engineer to an island community of 200 families suffering a severe shortage of fresh water. Some islanders have raised the possibility of producing drinking water from salt water (desalination). Write a report for the town council, summarizing the process and describing instances in which desalination has been used successfully or unsuccessfully. Would desalination be economically feasible for a community this size? Recommend a course of action.

c. You are a health officer in a town less than one mile from a massive radar installation. Citizens are disturbed about the effects of microwave radiation. Do they need to worry? Should any precautions be taken? Find the facts and write your report.

d. You are an investment broker for a major firm. A longtime client calls to ask your opinion. She is thinking of investing in a company that is fast becoming a leader in fiber optics communication links. "Should I invest in this technology?" your client wants to know. Find out, and give her your recommendations in a short report.

e. The buildings in the condominium complex you manage have been invaded by carpenter ants. Can the ants be eliminated by any insecticide *proven* nontoxic to humans or pets? (Many dwellers have small children and pets.) Find out, and write a report making recommendations to the maintenance supervisor.

f. The "coffee generation" wants to know about the properties of caffeine and the chemicals used on coffee beans. What are the effects of these substances on the body? Write your report, making specific recommendations about precautions that coffee drinkers can take.

g. As a consulting dietitian to the school cafeteria in Blandville, you've been asked by the school board to report on the most dangerous chemical additives in foods. Parents want to be sure that foods containing these additives are eliminated from school menus, insofar as possible. Write your report, making general recommendations about modifying school menus.

h. Dream up a scenario of your own in which information and recommendations would make a real difference. (Perhaps the question could be one you've always wanted answered.)

COLLABORATIVE PROJECT

Organize into groups of four or five and choose a topic upon which all group members can take the same position. Here are some possibilities:

- Should your college abolish core requirements?
- Should every student in your school pass a writing proficiency exam before graduating?

- Should courses outside one's major be graded pass/fail at the student's request?
- Should your school drop or institute student evaluation of teachers?
- Should all students be required to be computer literate before graduating?
- Should campus police carry guns?
- Should dorm security be improved?
- Should students with meal tickets be charged according to the type and amount of food they eat, instead of paying a flat fee?

As a group, decide your position on the issue, and brainstorm collectively to justify your recommendation to a stipulated primary audience in addition to your classmates and instructor. Complete an audience-and-use profile (page 446), and compose a justification report. Appoint one member to present the report in class.

Proposals

A proposal is an offer to do something or a recommendation that something be done. A proposal's general purpose is to *persuade* readers to improve conditions, authorize work on a project, accept a service or product (for payment), or otherwise support a plan for solving a problem or doing a job.

Your own proposal may be a letter to your school board to suggest changes in the English curriculum; it may be a memo to your firm's vice president to request funding for a training program for new employees; or it may be a 1,000-page document to the Defense Department to bid for a missile contact (competing with proposals from other firms). You might write the proposal alone or as part of a team. It might take hours or months.

PURPOSE OF PROPOSALS

Whether in science, business, industry, government, or education, proposals are written for decision makers: managers, executives, directors, clients, trustees, board members, community leaders, and the like. Inside or outside your organization, these are the people who decide whether your suggestions are worthwhile, whether your project ever will materialize, whether your service or product is useful. If your job depends on funding from outside sources, proposals might be the most important documents you produce.

THE PROPOSAL PROCESS

The basic proposal process can be summarized simply: someone offers a plan for something that needs to be done. In business and government, this process has three stages:

1. Client *X* needs a service or product.
2. Firms *A, B, C* propose a plan for meeting the need.
3. Client *X* awards the job to the firm offering the best proposal.

The complexity of each phase will of course depend on the situation. Here is a typical situation:

Assume that you manage a mining engineering firm in Tulsa, Oklahoma. On Wednesday, February 19, you spot this announcement in the *Commerce Business Daily:*[1]

> **R—Development of Alternative Solutions to Acid Mine Water Contamination from Abandoned Lead and Zinc Mines** near Tar Creek, Neosho River, Ground Lake, and the Boone and Roubidoux aquifers in northeastern Oklahoma. This will include assessment of environmental impacts of mine drainage followed by development and evaluation of alternate solutions to alleviate acid mine drainage in receiving streams. An optional portion of the contract, to be bid upon as an add-on and awarded at the discretion of the OWRB, will be to prepare an Environmental Impact Assessment for each of three alternative solutions as selected by the OWRB. The project is expected to take 6 months to accomplish, with anticipated completion date of September 30, 19XX. The projected effort for the required task is 30 person-months. Requests for proposals will be issued and copies provided to interested sources upon receipt of a written request. Proposals are due March 1, 19XX. (044)

[1]This daily publication lists the government's latest needs for *services* (salvage, engineering, maintenance) and for *supplies, equipment, and materials* (guided missiles, engine parts, building materials, and so on). The *Commerce Business Daily* is an essential reference tool for anyone whose firm seeks government contracts.

The *Grantsmanship Center News*, published six times yearly, contains a wealth of information about federal grants for nonprofit organizations in areas such as rural development, media, science, energy and environment, and education. This publication also lists notices of training programs in proposal writing, and offers books on proposal writing and evaluation, along with bibliographies of publications about the proposal process.

Oklahoma Water Resources Board
P.O. Box 53585
1000 Northeast 10th Street
Oklahoma City, OK 73152
(405) 555-2541

Your firm has the personnel, experience, and time to do the job, and so you decide to compete for the contract. Because the March 1 deadline is fast approaching, you write immediately for a copy of the request for proposal (RFP). The RFP will give you the guidelines for developing the proposal—guidelines for spelling out your plan to solve the problem (methods, timetables, costs).

You receive a copy of the RFP on February 21—only one week before deadline. You get right to work, with the two staff engineers you have appointed to your proposal team. Because the credentials of your staff could affect the client's acceptance of the proposal, you ask team members to update their résumés (for inclusion in an appendix to the proposal).

Several other firms will be competing for this project. The client will award it to the firm submitting the best proposal, based on the following criteria (and perhaps others):

- understanding of the client's needs, as described in the RFP
- soundness of the firm's technical approach
- quality of the project's organization and management
- ability to complete the job by the deadline
- ability to control costs
- specialized experience of the firm in this type of work
- qualifications of staff to be assigned to the project
- the firm's record for similar projects

A client's specific evaluation criteria often are listed (in order of importance or on a point scale) in the RFP. Although these criteria may vary, every client expects a proposal that is *clear, informative,* and *realistic.*

Some clients hold a preproposal conference for the competing firms. During this briefing, the firms are informed of the client's needs, expectations, specific start-up and completion dates, criteria for evaluation, and any other details that might help guide proposal development.

In most cases, the merits of your proposed plan are judged *solely* by what you have *on paper*. This reality is summed up in the advice below from a government publication that solicits proposals.

[Proposal] reviewers will base their conclusions only on information contained in the proposal. It cannot be assumed that readers are acquainted with the firm or key individuals or any referred-to experiments.[2]

Make any proposal able to stand alone in meaning and persuasiveness.

PROPOSAL TYPES

Despite their variety, proposals are classified in three ways: according to *origin, audience,* or *intention.* Based on its origin, a proposal is either *solicited* or *unsolicited*—that is, requested by an employer or potential client or initiated on your own because you have recognized a need. Business and government proposals usually are solicited, and they originate from a customer's request (as shown in the sample situation on pages 517–519).

Based on its audience, a proposal may be *internal* or *external*—written for members of your organization or for clients and funding agencies. (The situation on pages 522–523 calls for an external proposal.)

Based on its intention, a proposal may be a *planning, research,* or *sales* proposal. These last categories by no means account for all variations among proposals. Some proposals may in fact fall within all three categories, but these are the types you most likely will have to write. A discussion of each type follows.

The Planning Proposal

A planning proposal suggests ways of solving a problem or bringing about improvement. It might be a request for funding to expand the campus newspaper, an architectural plan for new facilities at a ski area, or a plan to develop energy alternatives to fossil fuels. The successful planning proposal always answers this main question for readers:

- *What are the benefits of following your suggestions?*

The planning proposal that follows is external and solicited. The XYZ Corporation has contracted a team of communication consultants to design in-house writing workshops. The consultants need to persuade the reader that their methods are likely to succeed. In their proposal, addressed to the company's education officer, the consultants offer concrete and specific solutions to clearly identified problems.

[2]*Small Business Innovation Research Program* (Washington: U.S. Department of Defense, 1983): 9.

After a brief introduction summarizing the problem, our writers develop their proposal under two headings, "Assessment of Needs" and "Proposed Plan."

Under "Proposed Plan," subheadings offer an even more specific forecast. The "Limitations" section shows that our writers are careful to promise no more than they can deliver.

Because this proposal is external, it is cast in letter form. Notice, however, that the word choice ("thanks," "what we're doing," "Jack and Terry") creates an informal, familiar tone. Such a tone is appropriate in this external document because the writers and reader have spent many hours in conferences, luncheons, and phone conversations.

Planning Proposal

Dear Mary:

States purpose

Thanks for sending the writing samples from your technical support staff. Here is what we're doing to design a realistic approach.

Assessment of Needs

Identifies problem

After conferring with technicians in both Jack's and Terry's groups, and analyzing their writing samples, we identified this hierarchy of needs:

- improving readability
- achieving precise diction
- summarizing information
- organizing a set of procedures
- formulating various memo reports
- analyzing audiences for upward communication
- writing persuasive bids for transfer or promotion
- writing persuasive suggestions

Proposed Plan

Proposes solution

Based on the needs listed above, we have limited our instruction package to eight carefully selected and readily achievable goals.

Course Outline. Our eight 2-hour sessions are structured as follows:

Details what will be done

1. achieving sentence clarity
2. achieving sentence conciseness
3. achieving fluency and precise diction
4. writing summaries and abstracts
5. outlining manuals and procedures
6. editing manuals and procedures

7. designing various reports for various purposes
8. analyzing the audience and writing persuasively

<u>Classroom Format</u>. The first three meetings will be lecture-intensive with weekly exercises to be done at home and edited collectively in class. The remaining five weeks will combine lecture and exercises with group editing of work-related documents. We plan to remain flexible so we can respond to needs that arise.

Details how it will be done

<u>Limitations</u>

Given our limited contact time, we cannot realistically expect to turn out a batch of polished communicators. By the end of the course, however, our students will have begun to appreciate writing as a deliberate process.

Sets realistic expectations

If you have any suggestions for refining this plan, please let us know.

Encourages reader response

The Research Proposal

Research (or grant) proposals request approval (and often funding) for a research project. A chemistry professor might address a research proposal to the Environmental Protection Agency for funds to identify toxic contaminants in local groundwater. Research proposals are solicited by many government and private agencies: National Science Foundation, National Institutes of Health, and others. Each granting agency has its own requirements for proposal format and content, but any successful research proposal answers these questions:

- *Why is this project worthwhile?*
- *What qualifies you to undertake the project?*
- *What are its chances of succeeding?*

In college, you might submit proposals for independent study, field study, or a thesis project. Here is the title of a research proposal submitted to the thesis committee in a geology department:

PROPOSAL FOR A MASTER'S THESIS PROJECT
TO INVESTIGATE THE TERTIARY GEOLOGY
OF THE ST. MARIES RIVER DRAINAGE
FROM ST. MARIES TO CLARKIA, IDAHO

A technical writing student might submit an informal proposal requesting the instructor's approval for a term project (which, in turn, may be a formal proposal). The introduction of the next proposal describes the problem and justifies the need for the study. The body outlines the scope, method, and sources for the proposed investigation. The conclusion describes the goal of the investigation and encourages reader support. This proposal is convincing because it answers questions about *what, why, how, when,* and *where.* Because this proposal is internal, it is cast informally as a memo.

Research Proposal *← center*

March 16, 1993

TO: Dr. John Lannon
FROM: T. Sorrells Dewoody
SUBJECT: Proposal for Determining the Feasibility of Marketing Dead Western
          Precise White Pine

No Greeting   Logos

Introduction

Over the past four decades, huge losses of western white pine have occurred in the northern Rockies, primarily attributable to white pine blister rust and the attack of the mountain pine beetle. Estimated annual mortality is 318 million board feet. Because of the low natural resistance of white pine to blister rust, this high mortality rate is expected to continue indefinitely.

If white pine is not harvested while the tree is dying or soon after death, the wood begins to dry and check (warp and crack). The sapwood is discolored by blue stain, a fungus carried by the mountain pine beetle. If the white pine continues to stand after death, heart cracks develop. These factors work together to cause degradation of the lumber and consequent loss in value.

Statement of Problem

White pine mortality reduces the value of white pine stumpage, because the commercial lumber market will not accept it. The major implications of this problem are two: first, in the face of rising demand for wood, vast amounts of timber lie unused; second, dead trees are left to accumulate in the woods, where they are rapidly becoming a major fire hazard here in northern Idaho and elsewhere.

Proposed Solution   *Simple precise*

One possible solution to the problem of white pine mortality and waste is to search for markets other than the conventional lumber market. The last few years have seen a burst of popularity and growing demand for weathered barn boards and wormy pine for interior paneling. Some firms around the country are marketing defective wood as specialty products. (These firms call the wood from which their products come "distressed," a term I will use hereafter to refer to dead and defective white pine.) Distressed white pine quite possibly will find a place in such a market.

Opens with background and causes of the problem

Describes problem

Describes one possible solution

Scope

To assess the feasibility of developing a market for distressed white pine, I plan to pursue six areas of inquiry:

1. What products are presently being produced from dead wood, and what are the approximate costs of production?
2. How large is the demand for distressed-wood products?
3. Can distressed white pine meet this demand as well as other species meet it?
4. Does the market contain room for distressed white pine?
5. What are the costs of retrieving and milling distressed white pine?
6. What prices for the products can the market bear?

Defines scope of the proposed study

Methods

My primary data sources will include consultations with Dr. James Hill, Professor of Wood Utilization, and Dr. Sven Bergman, Forest Economist—both members of the College of Forestry, Wildlife, and Range. I will also inspect decks of dead white pine at several locations, and visit a processing mill to evaluate it as a possible base of operations. I will round out my primary research with a letter and telephone survey of processors and wholesalers of distressed material.

Secondary sources will include publications on the uses of dead timber, and a review of a study by Dr. Hill on the uses of dead white pine.

Describes how study will be done

My Qualifications

I have been following Dr. Hill's study on dead white pine for two years. In June of this year I will receive my B.S. in forest management. I am familiar with wood milling processes and have firsthand experience at logging. My association with Drs. Hill and Bergman gives me the opportunity for an in-depth feasibility study.

Gives the writer's qualifications for this project

Conclusion   *Purpose ; Persuasion*

Clearly, action is needed to reduce the vast accumulations of dead white pine in our forests. The land on which they stand is among the most productive forests in northern Idaho. By addressing the six areas of inquiry mentioned earlier, I can determine the feasibility of directing capital and labor to the production of distressed white pine products. With your approval I will begin research at once.

Encourages reader acceptance

The Sales Proposal

A sales proposal offers a service or product. Sales proposals may be solicited or unsolicited. If they are solicited, several firms may compete with proposals of their own. Because sales proposals are addressed to readers outside your organization, they are cast as letters (if they are brief). But long sales proposals, like long reports, are formal documents with supplements (cover letter, title page, table of contents).

The sales proposal, a major marketing tool in business and industry, will be successful if it answers this question:

- *How will you serve our needs better than your competitors:*
  or
- *Why should we hire you instead of someone else?*

The following solicited proposal offers a service. Because the writer is competing with other firms, he explains specifically *why* his machinery is best for the job, *how* the job can best be completed, *what* his qualifications are for getting the job done, and *how much* the job will cost. He will be legally bound by his estimate. To protect himself, he points out possible causes of increased costs. *Never underestimate costs by failing to account for all variables—* a sure way to lose money or clients.

The introduction describes the subject and purpose of the proposal. The conclusion reinforces the confident tone throughout and encourages readers' acceptance by ending with—and thus emphasizing—two vital words: "economically" and "efficiently."

Sales Proposal

SUBJECT: <u>Proposal to Dig a Trench and Move Boulders at Site Ten Miles West of Bliss</u>

Dear Mr. Haver:

Offers to do the job

I've inspected your property and would be happy to undertake the landscaping project necessary for the development of your farm.

Gives the writer's qualifications

The backhoe I use cuts a span 3 feet wide and can dig as deep as 18 feet— more than an adequate depth for the mainline pipe you wish to lay. Because this backhoe is on tracks rather than tires, and is hydraulically operated, it is particularly efficient in moving rocks. I have more than twelve years of experience with backhoe work and have completed many jobs similar to this one.

Explains how job will be done

After examining the huge boulders that block access to your property, I am convinced they can be moved only if I dig out underneath and exert upward pressure with the hydraulic ram while you push forward on the boulders with your D-9 Caterpillar. With this method, we can move enough rock to enable you to farm that now inaccessible tract. Because of its power, my larger backhoe will save you both time and money in the long run.

This job should take 12 to 15 hours, unless we encounter subsurface ledge formations. My fee is $100 an hour. The fact that I provide my own dynamiting crew at no extra charge should be an advantage to you because you have so much rock to be moved.

Gives a *qualified* cost estimate

Please phone me anytime for more information. I'm sure we can do the job economically and efficiently.

Encourages reader acceptance

The proposal categories (planning, research, and sales) discussed in this section are neither exhaustive nor mutually exclusive. A research proposal, for example, may request funds for a study that will lead to a planning proposal. The Vista proposal partially shown on pages 524–528 is a combined planning and sales proposal: if clients accept the writer's preliminary plan, they will hire the firm to install the automated system.

PROPOSAL GUIDELINES

Readers will evaluate your proposal according to how clearly, informatively, and realistically you answer these questions:

- *What are you proposing?*
- *What problem will you solve?*
- *Why is your plan worthwhile?*
- *What is unique about your plan?*
- *What are your (or your firm's) credentials?*
- *How will the plan be implemented?*
- *How long will the project take to complete?*
- *How much will it cost?*
- *How will we benefit if we accept your plan?*

In addition to answering the questions above, successful proposal writers adhere to the following guidelines.

Design an Accessible and Appealing Format

Format is the *look* of a document, including such features as

- the layout of words and graphics
- typeface, type size, and white space
- highlights and lists
- headings

See Chapter 15, "Designing Effective Formats," for a discussion of these features.

A poorly designed proposal suggests to readers the writer's careless attitude toward the project.

Signal Your Intent with a Clear Title

Decision makers are busy people who have no time for guessing-games. The title should clearly signal the proposal's purpose and content.

> **Unclear**   PROPOSED OFFICE PROCEDURES FOR VISTA
>   FREIGHT, INC.

What kinds of office procedures are being proposed? This title is too broad.

> **Revised**   A PROPOSAL FOR AUTOMATING VISTA'S FREIGHT
>   BILLING SYSTEM

Don't write "Recommended Improvements" when you mean "Recommended Wastewater Treatment." A specific and comprehensive title signals the proposal's intent.

Include Supporting Material and Appropriate Supplements

Both short and long proposals may include supporting materials (maps, blueprints, specifications, calculations, and so forth). Place supporting material in an appendix (pages 377–379), to avoid interrupting the discussion.

Depending on your readers, appropriate supplements (pages 367–379) for a long proposal might include a title page, cover letter, table of contents, summary, abstract, and appendixes. Readers with various responsibilities will be interested in different parts of your proposal: Some know about the problem and will read only your plan; some look only at the summary; others will study recommendations or costs; and still others need all the details. If you're unsure as to which supplements to include in an internal proposal, ask the intended reader(s) or study other proposals. For a solicited proposal (one written for an outside agency) follow the agency's instructions *exactly.*

Focus on the Problem and the Objective

Readers want specific suggestions for filling specific needs. Their biggest question is "What's in this for me?" Show them you understand their problem and offer a plan for improving their products, sales, or services.

Notice how proposal writer Gerald Beaulieu focuses on Vista's inefficient office procedures, and then outlines specific solutions.

Statement of the Problem

Vista provides two services. (1) It locates freight carriers for its clients. The carriers, in turn, pay Vista a 6 percent commission for each referral. (2) Vista handles all shipping paperwork for its clients. For this auditing service, clients pay Vista a monthly retainer.

Gives background

Although Vista's business has increased steadily for the past three years, record-keeping, accounting, and other paperwork still are done *manually*. These inefficient procedures have caused a number of problems, including late billings, lost commissions, and poor account maintenance. Unless its office procedures are updated, Vista stands to lose clients.

Describes problem and its effects

Objective

This proposal offers a realistic and efficient plan for Vista to streamline office procedures. We first identify the burden imposed on your staff by the current system, and then we show how to reduce inefficiency, eliminate client complaints, and improve your cash flow by automating most office procedures.

Enables readers to visualize results

Treat Contingencies and Limitations Realistically

Do not underestimate the project's complexity. Identify contingencies (occurrences subject to chance) readers might not anticipate, and propose realistic methods for dealing with the unexpected. Here is how the Vista proposal qualifies its promises:

As outlined below, Vista can realize tangible benefits by automating office procedures. But, as countless firms have learned, *imposing* automated procedures on a staff can create severe morale problems—particularly among senior staff who feel coerced. To diminish employee resistance, invite your staff to comment on this proposal. To help avoid hardware and software problems once the system is operational, we have included recommendations and a budget for staff training. (Firms have learned that inadequate training is counterproductive to the automation process.)

Assesses contingencies realistically

If the best available solutions have limitations, let readers know. Otherwise, you and your firm could be liable in the case of project failure. Avoid overstatement. Notice how the above solutions are qualified ("diminish" and "help avoid" instead of "eliminate") so as not to promise more than the writer can deliver. Ethical communication is essential.

Explain the Benefits of Implementing Your Plan

A persuasive proposal shows readers how they (or their organization) will benefit by adopting your plan.

The Vista proposal specifies the following benefits. (Each benefit will be described at length in the body section.)

> Once your automated system is operational, you will be able to
>
> - identify cost-effective carriers
> - coordinate shipments (which will ensure substantial discounts for clients)
> - print commission bills
> - track shipments by weight, miles, fuel costs, and destination
> - send clients weekly audit reports on their shipments
> - bill clients on a 25-day cycle
> - produce weekly or monthly reports
>
> Additional benefits include eliminating repetitive tasks, improving cash flow, and increasing productivity.

Provide Concrete and Specific Information

Vagueness is a fatal flaw in a proposal. Before you can persuade readers, you must inform them; therefore, you need to *show* as well as *tell*. Instead of writing, "We will install state-of-the-art equipment," write,

> To meet your automation requirements, we will install 12 IBM PS/2 computers with 20 megabyte hard drives. The IBM's will be connected (networked) for rapid file transfer between offices. The plan also includes interconnection with 4 Epson high-speed dot matrix printers, and 1 Hewlett-Packard laser printer.

To avoid any misunderstanding and to reflect your ethical commitment, a proposal must elicit *one* interpretation only.

Include Effective Visuals

If they enhance your proposal, use visuals, properly introduced and discussed. Following is one of the visuals from the Vista proposal.

> As the flowchart illustrates (Figure 22.1), your routing and billing system creates a good deal of redundant work for your staff. The routing sheet alone is handled at least six times. Such extensive handling leads to errors, misplaced sheets, and late billing.

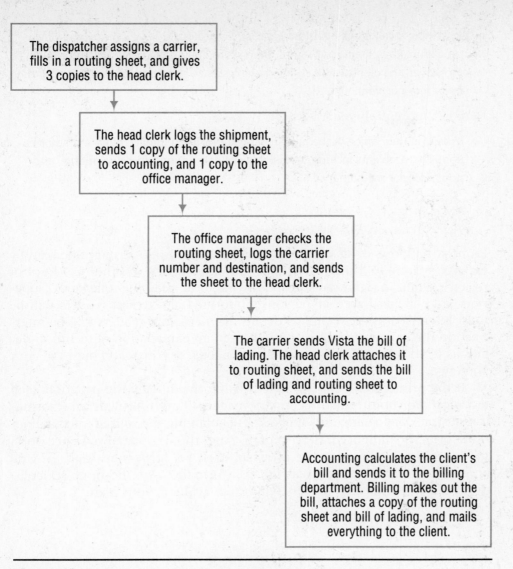

FIGURE 22.1 Flowchart of Vista's Manual Routing and Billing System

Use a Tone That Connects with Readers

Your tone is the voice and personality that appear between the lines. Make the tone confident, encouraging, and diplomatic. Show readers you believe in your plan; urge them to act, and anticipate how they will react to your suggestions. But do not come across with a bossy or insulting tone, as in this example:

> *Had Vista's managers been better trained*, they could have streamlined office procedures by eliminating some of the more obvious paper-shuffling problems. Moreover, the clerical staff should have lessened the number of errors *had they been more vigilant*.

Here is a more diplomatic version:

Avoids blaming anyone

> Vista's manual office procedures have resulted in an inefficient flow of information. The problems include repetitive tasks, late billings, lost documents, and poor account maintenance.

Analyze Audience Needs

Proposals address diverse audiences. A research proposal might be read by experts, who would then advise the granting agency whether to accept or reject it. Planning and sales proposals might be read by colleagues, superiors, and clients (often laypersons). Informed and expert readers will be most interested in the technical details. Nontechnical readers will be interested in the expected results, but will need an explanation of technical details as well. Learn all you can about the needs, interests, and biases of your audience.

If the primary audience is expert or informed, keep the proposal itself technical. For uninformed secondary readers (if any), provide an informative abstract, a glossary, and appendixes explaining specialized information. If the primary audience has no expertise and the secondary audience does, follow this pattern: write the proposal itself for laypersons, and provide appendixes with the technical details (formulas, specifications, calculations) that the informed readers will use to evaluate your plan.

A GENERAL MODEL FOR PROPOSALS

Like all informative writing, proposals are organized into these main sections: introduction, body, and conclusion. Depending on project complexity, each section has some or all of the subsections listed in the following general outline:

I. INTRODUCTION
   A. Statement of Problem and Objective
   B. Background
   C. Need
   D. Benefits

    E. Qualifications of Personnel

    F. Data Sources

    G. Limitations and Contingencies

    H. Scope

II. BODY

    A. Methods

    B. Timetable

    C. Materials and Equipment

    D. Personnel

    E. Available Facilities

    F. Needed Facilities

    G. Cost

    H. Expected Results

    I. Feasibility

III. CONCLUSION

    A. Summary of Key Points

    B. Request for Action

Headings can be rearranged, combined, divided, or deleted as needed. Not every proposal has all subsections, but each major section must answer specific readers' questions, as illustrated next.

Introduction

The introduction answers all these questions—or all those that apply to the situation:

- *What problem do you propose to solve?*
- *In general, what solution are you proposing?*
- *Why are you proposing it?*
- *What are the benefits?*
- *What are your qualifications for this project?*

From the beginning, your goal is to sell your idea, to persuade readers the job needs doing and you are the one to do it. If your introduction is long-winded, evasive, or vague, readers might stop reading. Make it concise, specific, and clear.

Spell out the problem to make it clear to the audience—and to show you understand it fully. Explain the benefits of solving the problem or undertaking the project. Identify any sources of data. In a research or sales proposal, state your qualifications for doing the job. If your plan has limitations or contingencies, explain them. Finally, give the scope of your plan by listing the subsections to be discussed in the body section.

Following is the introduction for a planning proposal titled "A Proposal for Solving the Noise Problem in the University Library." Jill Sanders, a library work-study student, addresses her proposal to the chief librarian and the administrative staff. Because this proposal is unsolicited, it must first make the problem vivid through details that arouse concern and interest. This introduction is longer than it would be in a solicited proposal, whose readers would already agree on the severity of the problem.

INTRODUCTION

Statement of Problem

Concise descriptions of problem and objective immediately alert readers

During the October 19XX Convocation at Margate University, students and faculty members complained about noise in the library. Soon afterward, areas were designated for "quiet study," but complaints about noise continue. To create a scholarly atmosphere, the library should take immediate action to decrease noise.

Objective

This proposal examines the noise problem from the viewpoint of students, faculty, and library staff. It then offers a plan to make areas of the library quiet enough for serious study and research.

Sources

This section comes early because it is referred to in next section

My data come from a university-wide questionnaire, interviews with students, faculty, and library staff, inquiry letters to other college libraries, and my own observations for three years on the library staff.

Details of the Problem

Details enable readers to understand problem

This subsection examines the severity and causes of the noise.

Severity. Since the 19XX Convocation, the library's fourth and fifth floors have been reserved for quiet study, but students hold group-study sessions at the large tables and disturb others working alone. The constant use of computer terminals on both floors adds to the noise, especially when students converse. Moreover, people often chat as they enter or leave study areas.

On the second and third floors, designed for reference, staff help patrons locate materials, causing constant shuffling of people and books, as well as loud conversation. At the computer service desk on the third floor, conferences between students and instructors create more noise.

The most frequently voiced complaint from the faculty members interviewed was about the second floor, where people using the Reference and Government Documents services converse loudly. Students complain about the lack of a quiet spot to study, especially in the evening, when even the "quiet" floors are as noisy as the dorms.

More than 80 percent of respondents (530 undergraduates, 30 faculty, 22 graduate students) to a university-wide questionnaire (Appendix A) insisted that excessive noise discourages them from using the library as often as they would prefer. Of the student respondents, 430 cited quiet study as their primary reason for wishing to use the library.

The library staff recognizes the problem but has insufficient personnel. Because all staff members have assigned tasks, they have no time to monitor noise in their sections.

Causes. Respondents complained specifically about these causes of noise (in descending order of frequency):

1. Loud study groups that often lapse into social discussions.
2. General disrespect for the library, with some students' attitudes characterized as "rude," "inconsiderate," or "immature."
3. The constant clicking of computer terminals on all five floors, and of typewriters on the first three.
4. Vacuuming by the evening custodians.

All complaints converged on lack of enforcement by library staff.

Because the day staff works on the first three floors, quiet-study rules are not enforced on the fourth and fifth floors. Work-study students on these floors have no authority to enforce rules not enforced by the regular staff. Small, black-and-white "Quiet Please" signs posted on all floors go unnoticed, and the evening security guard provides no deterrent.

Needs

Excessive noise in the library is keeping patrons away. By addressing this problem immediately, we can help restore the library's credibility and utility as a campus resource. We must reduce noise on the lower floors and eliminate it from the quiet-study floors.

Scope

The proposed plan includes a detailed assessment of methods, costs and materials, personnel requirements, feasibility, and expected results.

Body

The body (or plan) section of your proposal will receive most attention from readers. It answers all these questions that are applicable:

- *How will it be done?*
- *When will it be done?*
- *What materials, methods, and personnel will it take?*
- *What facilities are available?*
- *How long will it take?*
- *How much will it cost, and why?*
- *What results can we expect?*
- *How do we know it will work?*
- *Who will do it?*

Here you spell out your plan in enough detail for readers to evaluate its soundness. If this section is vague, your proposal stands no chance of being accepted. Be sure your plan is realistic and promises no more than you can deliver. The main goal of this section is to prove that your plan will work.

PROPOSED PLAN

This plan takes into account the needs and wishes of our campus community, as well as the available facilities in our library.

Methods

Tells how plan will be implemented

Noise in the library can be reduced in three complementary phases: (1) improving publicity, (2) shutting down and modifying our facilities, and (3) enforcing the quiet rules.

Describes first phase

Improving Publicity. First, the library must publicize the noise problem. This assertive move will demonstrate the staff's interest. Publicity could include articles by staff members in the campus newspaper, leaflets distributed on campus, and a freshman library orientation acknowledging the noise problem and asking cooperation from new students. All forms of publicity should detail the steps being taken by the library to solve the problem.

Describes second phase

Shutting Down and Modifying Facilities. After notifying campus and local newspapers, you should close the library for one week. To minimize disruption, the shutdown should occur between the end of summer school and the beginning of the fall term.

During this period, you can convert the fixed tables on the fourth and fifth floors to cubicles with temporary partitions (six cubicles per table). You could later convert the cubicles to shelves as the need increases.

Then you can take all unfixed tables from the upper floors to the first floor, and set up a space for group study. Plans already are under way for removing the computer terminals from the fourth and fifth floors.

Enforcing the Quiet Rules. Enforcement is the essential, long-term element in this plan. No one of any age is likely to follow all the rules all the time—unless the rules are enforced.

Describes third phase

First, you can make new "Quiet" posters to replace the present, innocuous notices. A visual-design student can be hired to draw up large, colorful posters that attract attention. Either the design student or the university print shop can take charge of poster production.

Next, through publicity, library patrons can be encouraged to demand quiet from noisy people. To support such patron demands, the library staff can begin monitoring the fourth and fifth floors, asking study groups to move to the first floor, and revoking library privileges of those who refuse. Patrons on the second and third floors can be asked to speak in whispers. Staff members should set an example by regulating their own voices.

Costs and Materials

- The major cost would be for salaries of new staff members who would help monitor. Next year's library budget, however, will include an allocation for four new staff members.
- A design student has offered to make up four different posters for $200. The university printing office can reproduce as many posters as needed at no additional cost.
- Prefabricated cubicles for 26 tables sell for $150 apiece, for a total cost of $3,900.
- Rearrangement on various floors can be handled by the library's custodians.

Estimates costs and materials needed

The Student Fee Allocations Committee and the Student Senate routinely reserve funds for improving student facilities. A request to these organizations presumably would yield at least partial funding for the plan.

Personnel

The success of this plan ultimately depends on the willingness of the library administration to implement it. You can run the program itself by committees

Describes personnel needed

made up of students, staff, and faculty. This is yet another area where publicity is essential to persuade people that the problem is severe and that you need their help. To recruit committee members from among students, you can offer Contract Learning credits.

The proposed committees include an Antinoise Committee overseeing the program, a Public Relations Committee, a Poster Committee, and an Enforcement Committee.

Feasibility

On March 15, 19XX, I mailed survey letters to 25 New England colleges, inquiring about their methods for coping with noise in the library. Among the respondents, 16 stated that publicity and the administration's attitude toward enforcement were main elements in their success.

Improved publicity and enforcement could work for us as well. And slight modifications in our facilities, to concentrate group study on the busiest floors, would automatically lighten the burden of enforcement.

Benefits

Publicity will improve communication between the library and the campus. An assertive approach will show that the library is aware of its patrons' needs and willing to meet those needs. Offering the program for public inspection will draw the entire community into improvement efforts. Publicity, begun now, will pave the way for the formation of committees.

The library shutdown will have a dual effect: it will dramatize the problem to the community, and also provide time for the physical changes. (An antinoise program begun with carpentry noise in the quiet areas would hardly be effective.) The shutdown will be both a symbolic and a concrete measure, leading to reopening of the library with a new philosophy and a new image.

Continued strict enforcement will be the backbone of the program. It will prove that staff members care enough about the atmosphere to jeopardize their friendly image in the eyes of some users, and that the library is not afraid to enforce its rules.

Conclusion

The conclusion reaffirms the need for the project and persuades readers to act. It answers the questions readers will ask:

- *How badly do we need this change?*

Assesses probability of success

Offers a realistic and persuasive forecast of benefits

- *Why should we accept your proposal?*
- *How do we know this is the best plan?*

End on a strong note, with a conclusion that is assertive, confident, and encouraging—and keep it short.

CONCLUSION AND RECOMMENDATION

The noise in Margate University library has become embarrassing and annoying to the whole campus. Forceful steps are needed to restore the academic atmosphere.

Aside from the intangible question of image, close inspection of the proposed plan will show that it will work if you take the recommended steps and—most important—if daily enforcement of quiet rules becomes a part of the library's services.

Re-emphasizes need and feasibility and encourages action

In long proposals, especially those beginning with a comprehensive abstract, the conclusion can be omitted.

A SAMPLE SITUATION

The formal planning proposal in Figure 22.2 typifies the kind of specialized proposal that justifies a funding request.

The situation: A university newspaper is struggling to meet rising costs. The paper's yearly budget is funded by a college allocations committee that disburses money to all student organizations. Because of revenue cuts throughout the university, the newspaper has received no funding increase in three years.

Bill Trippe is the newspaper's business manager. His task is to justify a requested increase of seventeen percent for the coming year's budget. Before drafting his proposal, Bill constructs a detailed profile of his audience (based on the worksheet, page 60).

Audience-and-Use Profile for a Formal Proposal

AUDIENCE IDENTITY AND NEEDS

My primary audience includes all members of the allocations committee. My secondary audience is the newspaper staff, who will implement the proposed plan—if it is approved by the allocations committee.

The primary audience will use my document as perhaps the sole basis for deciding whether to grant the additional funds. Most of these readers have overseen the newspaper budget for years, and so they already know quite a bit about our overall operation. But they still need an item-by-item explanation of the conditions created by our problems with funding and ever-increasing costs. Probable questions I can anticipate:

- Why should the paper receive priority over other campus organizations?
- Just how crucial is the problem?
- Are present funds being used efficiently?
- Can any expenses be reduced?
- How would additional funds be spent?
- How much will this increase cost?
- Will the benefits justify the cost?

ATTITUDE AND PERSONALITY

My primary audience often has expressed interest in this topic. But they are likely to object to any request for more money by arguing that everyone has to economize in these difficult times. I guess I could characterize their attitude as both receptive and hesitant. (Almost every campus organization is trying to make a case for additional funds.)

I do know most committee members pretty well, and they seem to respect my management skills. But I still need to spell out the problem and propose a realistic plan, showing that the newspaper staff is sincere in its intention to eliminate nonessential operating costs. At a time when everyone is expected to make do with less, I need to make an especially strong case for salary increases (to attract talented personnel).

EXPECTATIONS ABOUT THE DOCUMENT

My audience has requested (solicited) this proposal, and so I know it will be carefully read—but also scrutinized and evaluated for its soundness! Especially in a budget request, my audience expects no shortcuts; I'll have to itemize every expense. The Costs sections then could be the longest part of the proposal.

And to further justify the requested budget, I can demonstrate just how well the newspaper manages its present funds. Under Feasibility, I'll give a detailed comparison of funding, expenditures, and size of newspaper in relation to the four other local colleges. This section should be the "clincher" because these facts are most likely to persuade the committee that my plan is cost-effective. To avoid clutter, I'll add an appendix with a table of figures for the comparison above (page 378).

To organize my document, I will (1) identify the problem, (2) establish need, (3) propose a solution, (4) show that the plan is cost-effective, and (5) conclude with a request for action. My audience here expects a confident and businesslike—but not stuffy—tone. I want to be sure that everything in this proposal encourages readers to support our budget request.

A Budget Proposal
for
the SMU *Torch*
(1993–94)

Prepared for
The Student Fee Allocation Committee
Southeastern Massachusetts University
North Dartmouth, Massachusetts

by
William Trippe
Torch Business Manager

May 1, 1993

FIGURE 22.2 A Budget Proposal

The SMU *Torch*

Old Westport Road
North Dartmouth, Massachusetts 02747

May 1, 1993

Charles Marcus, Chair
Student Fee Allocation Committee
Southeastern Massachusetts University
North Dartmouth, MA 02747

Dear Dean Marcus:

No one needs to be reminded about the effects of increased costs on our campus community. We are all faced with having to make do with less.

Accordingly, we at the *Torch* have spent long hours devising a plan to cope with increased production costs—without compromising the newspaper's tradition of quality service. I think you and your colleagues will agree that our plan is realistic and feasible. Even the "bare-bones" operation that will result from our proposed spending cuts, however, will call for a $3837.14 increase in our 1993–94 budget.

We have received no funding increase in three years. Our present need is absolute. Without additional funds, the *Torch* simply cannot continue to function as a professional newspaper. I therefore submit the following budget proposal for your consideration.

Respectfully,

William Trippe

William Trippe
Business Manager, SMU *Torch*

FIGURE 22.2 A Budget Proposal *Continued*

TABLE OF CONTENTS

FIGURE 22.2 A Budget Proposal *Continued*

INFORMATIVE ABSTRACT

The student newspaper at Southeastern Massachusetts University is crippled by inadequate funding, having received no budget increase in three years. Increased costs and inadequate funding are the major problems facing the *Torch*. Increases in costs of printing, layout, and photographic supplies have called for a decrease in production. Moreover, our low salaries are inadequate to attract and retain qualified personnel. A nominal increase would make salaries more competitive.

Our staff plans to cut costs by reducing page count, saving on photographic paper, reducing circulation, and hiring a new printer. The only proposed cost increase (for staff salaries) is essential.

A detailed breakdown of projected costs establishes the need for a $3837.14 budget increase to keep the paper a weekly with adequate page count and distribution to serve our campus.

Compared with other college newspapers, the *Torch* makes much better use of its money. A detailed comparison illustrates the cost effectiveness of our proposal.

FIGURE 22.2 A Budget Proposal *Continued*

INTRODUCTION

Our campus newspaper faces the contradictory challenge of surviving ever-increasing production costs while maintaining its quality. The following proposal offers a realistic plan for meeting the crisis. The plan's ultimate success, however, depends on the Fee Allocation Committee's willingness to approve a long-overdue increase in our 1993–94 budget.

In its ten years, the *Torch* has grown in size, scope, and quality. Roughly 6000 copies (24 pages/issue) are printed weekly each 14-week semester.

With much of its staff about to graduate, the *Torch* faces next year with rising costs in every phase of production, and the need to replace outdated and worn equipment.

Our newspaper also suffers from a lack of student involvement. Few students can be expected to work without some kind of salary. Most staff members do receive minimal weekly salaries: from $10 for the distributor to $45 for the Editor-in-Chief, but salaries averaging less than $2 per hour cannot possibly compete with the minimum wage. Since most SMU students must work part-time, the *Torch* will have to make salaries more competitive.

The newspaper's operating expenses can be divided into four categories: composing costs, salaries, printing costs, and miscellaneous (office supplies, mail, etc.). The first three categories account for nearly 90 percent of the budget. In the past year, costs in all categories have increased from as little as 3 percent for darkroom chemicals to as much as 60 percent for photographic paper and film. Printing costs (roughly one-third of our total budget) rose by 9 percent in the past year, and another price hike of 10 percent has just taken effect.

Despite the steady increase in production costs, the *Torch* has received no increase in its yearly allocation ($21,500) in 3 years.

The plan following includes:
1. methods for reducing production costs while maintaining the quality of our staff
2. projected costs for equipment, material, salaries, and services in 1993–94
3. a demonstration of feasibility, showing our cost effectiveness
4. a summary of the attitudes shared by our staff

FIGURE 22.2 A Budget Proposal *Continued*

PROPOSED PLAN

The following plan is designed to trim operating costs without compromising quality.

Methods

We can overcome our budget and staffing crisis by taking these steps:

Reducing Page Count.   By condensing free notices for campus organizations, abolishing "personals," and limiting press releases to one page, we can reduce page count per issue from 24 to 20, saving 17 percent in production costs.

Saving on Photographic Paper.   A newspaper with fewer pages will call for fewer photographs. Also, we can save money by purchasing a year's supply of photographic paper within the next month, before the upcoming 15 percent increase.

Reducing Circulation.   Reducing circulation from 6000 to 5000 copies will barely cover the number of full-time students, but will save 17 percent in printing costs.

Hiring a New Press.   We can save money by hiring Sadus Press for printing. Other presses (including our present printer) bid at least 25 percent higher than Sadus's price. Also, no other company offers the delivery service we will get from Sadus.

Increasing Staff Salaries.   Although we seek talented students who expect little money and much experience, salaries for all positions *must* increase by an average of $5 weekly. Otherwise, any of our staff could make as much money elsewhere by working only one-third the time. In fact, many students could make more than the minimum wage by working for local newspapers. To illustrate: *The Standard Times* pays $20 to $30 for a news article and $10 for a photo, while the *Torch* pays nothing for articles and $2 for a photo.

A striking example of low salaries is the $3.35 we pay typesetters. Outside, trained typesetters like ours make from $8 to $12 an hour. Our present typesetting cost of $3038 easily could be as much as $7000 or even higher, if we had typesetting done by an outside firm, as many colleges do.

Without this nominal salary increase, we cannot possibly attract qualified personnel.

Budget Request for 1993–94

Our proposed budget is itemized below, but the main point is clear: if the *Torch* is to remain viable, increased funding is essential for meeting projected costs.

FIGURE 22.2   A Budget Proposal   *Continued*

Projected Costs   **Yearly Total**

Equipment Leasing

(From Chase Manhattan Leasing Corporation, for Compugraphic equipment.)

Execuwriter	$106.25/month		
7200 L and hardware	127.38/month		
Execuwriter II	138.63/month		
	372.26/month	=	$3,350.34

Composing Chemicals

Activator	$105.00		
Stabilizer	180.00		
Processor Cleaner	55.00		
	$340.00	=	$340.00

Photo Paper for Execuwriters

3" x 150'	$711.00		
7" x 150'	556.37		
Headline paper	732.63		
8" x 150'	227.00		
	$2,227.00	=	$2,227.00

Layout Supplies

Font strips	$160.00
Layout boards	160.00
Exacto knives and blades	80.00
Non-repro pens	30.00
Rulers	12.00
Scotch tape	50.00
Folders	30.00
Reduction wheels	20.00
Wax	45.00

FIGURE 22.2 A Budget Proposal *Continued*

Projected Costs			Yearly Total
Construction paper	10.00		
Border tape	50.00		
	$647.00	=	$647.00
Photography Supplies			
Paper	$730.00		
Film	250.00		
Darkroom supplies	100.00		
	$1,080.00	=	$1,080.00
Typing Staff			
35 hrs. at $3.35/hr.	$117.25/wk	=	$3,283.00
Salaries			
Editor-in-Chief	$1,400.00/year		
News Editor	840.00		
Asst. News Editor	420.00		
Features Editor	840.00		
Head Writer (Features)	420.00		
Sports Editor	840.00		
Head Writer (Sports)	420.00		
Advertising Manager	1,050.00		
Advertising Designer	700.00		
Free Ad Designer	280.00		
Layout Editor	840.00		
Art Director	560.00		
Photo Editor	840.00		
Business Manager	840.00		
	$10,290.00	=	$10,290.00
Fixed Printing Costs			
5000 copies/week × 28	$12,399.80	=	$12,399.80

FIGURE 22.2 A Budget Proposal *Continued*

Projected Costs		Yearly Total
Miscellaneous Costs		
Graphics by SMU art students:		
3/wk. at $5 each	$420.00	
Mail costs	550.00	
Telephone costs	500.00	
Print shop costs	200.00	
Copier fees	50.00	
	$1,720.00 =	$1,720.00
TOTAL YEARLY COSTS		$35,337.14
Expected Ad Revenue		
(1000/mo. x 10)		$10,000.00
Total Costs minus Ad Revenue		$25,337.14
TOTAL BUDGET REQUEST		$25,337.14

Feasibility

Beyond exhibiting our need, we feel that the feasibility of this proposal can be measured through an objective assessment of our cost effectiveness: Compared with other school newspapers, how well does the *Torch* use its funds?

In a survey of the four area colleges, we found that the *Torch*—by a sometimes huge margin—makes the best use of its money. Table 1 in Appendix A [page 378] shows that, of the five newspapers, the *Torch* costs the students least, runs the most pages weekly, and spends the least money per page, *despite a circulation two to three times the size of the other papers.*

The most striking comparison is between the *Torch* and the newspaper at Fallow State College (Appendix A). Each student at FSC pays $12.33 yearly for a paper averaging 12 pages per issue. Here at SMU, each student pays $4.06 yearly for a paper averaging 20 pages an issue. Thus, for 33 percent of FSC's cost, SMU students are receiving 66 percent more newspaper.

The *Torch* has the lowest yearly costs of all five papers, despite the largest circulation. With the budget increase requested above, the cost would rise only by $0.44, for a yearly cost of $4.50 to each student. Although Alden College's paper costs each student $4.29, it is

---

FIGURE 22.2 A Budget Proposal *Continued*

published only every third week, averages 12 pages per issue, and costs nearly $50.00 yearly per page to print—as opposed to our yearly printing cost of $38.25 per page.

As the figures in Appendix A demonstrate, our cost management is responsible and effective.

Personnel

Students on the *Torch* staff are unanimous in their determination to maintain the highest professionalism. Many are planning careers in journalism, writing, editing, advertising, photography, or public relations. In any *Torch* issue, the balanced, enlightened coverage is evidence of our judicious selection and treatment of articles and our shared concern for quality.

CONCLUSION

As a forum for ideas and opinions, the *Torch* continues to reflect a seriousness of purpose and a commitment to free expression. Its place in the campus community is more vital than ever in these troubled times.

There are increases and decreases in student allocations every year. Last year, for example, eight allocations were increased by an average of $2,166. The *Torch* has received no increase since 1989–90. Presumably, various increases materialize as priorities change and as special circumstances arise. The *Torch* staff urges the Allocation Committee to respond to the paper's legitimate and proven needs by increasing our 1993–94 allocation to $25,337.14.

FIGURE 22.2 A Budget Proposal *Continued*

REVISION CHECKLIST FOR PROPOSALS

Use this list to revise your proposal. (Numbers in parentheses refer to the first page of discussion.)

Format

- ☐ Is the short internal proposal in memo form and the short external proposal in letter form? (517)

- ☐ Does the long proposal have adequate supplements to serve the different needs of different readers? (524)

- ☐ Is the format professional in appearance? (523)

- ☐ Are headings logical and adequate? (523)

- ☐ Does the title forecast the proposal's subject and purpose? (524)

Content

- ☐ Is the problem clearly identified? (524)

- ☐ Is the objective clearly identified? (524)

- ☐ Does everything in the proposal support its objective? (524)

- ☐ Does the proposal *show* as well as *tell?* (526)

- ☐ Is the proposed plan, service, or product beneficial? (526)

- ☐ Are the proposed methods practical and realistic? (525)

- ☐ Are all limitations and contingencies identified? (525)

- ☐ Is the proposal free of overstatement? (525)

- ☐ Are visuals used effectively and whenever appropriate? (526)

- ☐ Is the proposal's length appropriate to the subject? (514)

Arrangement

- ☐ Is there an introduction, body, and conclusion? (528)

- ☐ Are all *relevant* headings from the general outline included? (528)

- ☐ Does the introduction explain *what* and *why?* (528)

- ☐ Does the body explain *how, where,* and *how much?* (532)

□ Does the conclusion encourage acceptance of the proposal? (534)

□ Are there clear transitions between related ideas? (Appendix A)

Style

□ Is the level of technicality appropriate for primary readers? (528)

□ Do supplements follow the appropriate style guidelines? (524)

□ Will the informative abstract be understood by all readers? (528)

□ Does the tone connect with readers? (527)

□ Is each sentence clear, concise, and fluent? (251)

□ Is the language convincing and precise? (279)

□ Is the proposal in grammatical English? (Appendix A)

□ Is the proposal ethically acceptable? (525)

EXERCISES

1. Assume the head of your high school English Department has asked you, as a recent graduate, for suggestions about revising the English curriculum to prepare students for writing. Write a proposal, based on your experience since high school. (Primary audience: the English Department head and faculty; secondary audience: school committee.) In your external, solicited proposal, identify problems, needs, and benefits and spell out a realistic plan. Review the outline on page 528 before selecting specific headings.

2. After identifying your primary and secondary audience, compose a short planning proposal for improving an unsatisfactory situation in the classroom, on the job, or in your dorm or apartment (e.g., poor lighting, drab atmosphere, health hazards, poor seating arrangements). Choose a problem or situation whose resolution is more a matter of common sense and lucid observation than of intensive research. Be sure to (a) identify the problem clearly, give brief background, and stimulate the readers' interest; (b) state clearly the methods proposed to solve the problem; and (c) conclude with a statement designed to gain readers' support for your proposal.

3. Write a research proposal to your instructor (or an interested third party) requesting approval for the final term project (an analytical report or formal proposal). Identify the subject, background, purpose, and benefits of your planned inquiry, as well as the intended audience, scope of inquiry, data sources, methods of inquiry, and a task timetable. Be certain that adequate primary and secondary sources are available. Convince your reader of the soundness and usefulness of the project.

4. As an alternate term project to the formal analytical report (see Chapter 23), develop a long proposal for solving a problem, improving a situation, or satisfying a need in your school, community, or job. Choose a subject sufficiently complex to justify a formal proposal, a topic requiring research (mostly primary). Identify an audience (other than your instructor) who will use your proposal for a specific purpose. Compose an audience-and-use profile, using the sample on pages 535–536 as a model. Here are possible subjects for your proposal:

- improving living conditions in your dorm or fraternity/sorority
- creating a student-oriented advertising agency on campus
- creating a day-care center on campus
- creating a new business or expanding a business
- saving labor, materials, or money on the job
- improving working conditions
- supplying a product or service to clients or customers
- improving campus facilities for the handicapped
- increasing tourism in your town
- eliminating traffic hazards in your neighborhood
- reducing energy expenditures on the job
- improving security in dorms or in the college library
- improving in-house training or job orientation programs
- creating a one-credit course in job hunting or stress management for students
- improving tutoring in the learning center
- making the course content in your major more relevant to student needs
- creating a new student government organization (say, to handle ethnic affairs)
- finding ways for an organization to raise money
- improving faculty advising for students
- purchasing new equipment
- improving food service on campus
- easing freshman students through the transition to college
- making word processing available to all advanced writing students
- establishing a college-wide computer fluency program
- changing the grading system at your school
- establishing more equitable computer terminal use

Analytical Reports

Purpose of Analysis

Typical Analytical Problems

Elements of Analysis

A General Model for Analytical Reports

A Sample Situation

■   ■   ■

The formal analytical report, like the short recommendation reports in Chapter 21, leads to recommendations. The formal report replaces the memo when the topic requires lengthy discussion.

A form of critical thinking in which we separate a whole into its parts, analysis is essential to workplace problem solving.

Say, for example, you receive this assignment from your supervisor:

> Recommend the best method for removing the heavy-metal contamination from our company dump site.

To make the best recommendation, you will have to analyze the situation to learn everything you can about the nature and extent of the problem and the feasibility of various solutions. This is where you apply critical thinking skills from Chapter 2 along with the research process discussed in Chapters 8 and 9. As part of your research, you might do a literature search to discover whether anyone has been able to solve a problem like yours, or to learn about the newest technologies for toxic cleanup. You then have to decide how much, if any, of what others have done can apply to your situation.

PURPOSE OF ANALYSIS

As an employee you may be asked to evaluate a new assembly technique on the production line, or to locate and purchase the best equipment at the best price. You might have to identify the cause behind a monthly drop in sales, the reasons for low employee morale, the causes of an accident, or the reasons for equipment failure. You might need to assess the feasibility of a proposal for a company's expansion or investment. The list is endless but the procedure remains the same: (1) asking the right questions, (2) searching for information, (3) evaluating and interpreting your findings, and (4) drawing conclusions and making recommendations.

TYPICAL ANALYTICAL PROBLEMS

Far more than an encyclopedia presentation of information, the analytical report shows how you arrived at your conclusions and recommendations. Here are some typical analytical problems.

"Will X Work for a Specific Purpose?"

Analysis can answer practical questions. Say your employer is concerned about the effects of stress on employees. She asks you to investigate the claim that low-impact aerobics has therapeutic benefits—with an eye toward such a program for employees. You design your analysis to answer this question: *Do low-impact aerobics programs significantly reduce stress?* The analysis follows a *questions-answers-conclusions* sequence. Because the report could lead to action, you include recommendations based on your conclusions.

"Is X or Y Better for a Specific Purpose?"

Analysis is essential in comparing machines, processes, business locations, computer systems, or the like. Assume that you manage a ski lodge and need to answer this question: *Which of the two most popular ski bindings is best for our rental skis?* In a comparative analysis of the Salomon 555 and the Americana bindings, you would assess the strengths and weaknesses of each binding on the basis of specific criteria (safety, cost, ease of repair, ease of adjustment, dependability, and so on) which you would rank in order of importance.

The comparative analysis follows a *questions-answers-conclusions* sequence and is designed to help the reader make a choice. Examples appear in magazines such as *Consumer Reports* and *Consumer's Digest*.

"Why Does X Happen?"

The causal analysis is designed to answer questions like this: *Why do small businesses have a high failure rate?* This kind of analysis follows a variation of the questions-answers-conclusions structure: namely, *problem-causes-solution*. Such an analysis follows this sequence:

1. identifying the problem
2. examining possible and probable causes, and isolating definite ones
3. recommending solutions

An analysis of low employee morale would investigate causal relationships.

"How Can X Be Improved or Avoided?"

Another form of problem solving focuses on desired results, and recommends methods of achieving these results. This type of analysis answers questions like these: *How can we improve safety procedures in our plant? How can we operate our division more efficiently? How can we improve campus security?* The sample report in Appendix C tackles this question: How can we avoid errors in producing our computer magazine?

"What Are the Effects of X?"

An analysis of the consequences of an event or action would answer questions like these: *How has air quality been affected by the local power plant's change from burning oil to coal? Does electromagnetic radiation pose a significant health risk?*

Another kind of problem-solving analysis is done to predict an effect: "What are the consequences of my changing majors?" Here, the sequence is *proposed action–probable effects–conclusions and recommendations*.

"Is X Practical in This Situation?"

The feasibility analysis assesses the practicality of an idea or plan: *Will the consumer interests of Hicksville support a microcomputer store?* In a variation of the questions-answers-conclusions structure, a feasibility analysis presents *reasons for–reasons against*, with both sides supported by evidence. Business owners often use this type of analysis.

Combining Types of Analyses

Types of analytical problems overlap considerably. Any one study may in fact require answers to two or more of the previous questions. The sample report on pages 570–578 is both a feasibility analysis and a comparative analysis. It is designed to answer these questions: Is technical marketing the right career for me? If so, how do I enter the field?

ELEMENTS OF ANALYSIS

The analytical report incorporates many writing strategies from earlier assignments, along with the guidelines that follow.

Clearly Identified Problem or Question

Know what you're looking for. If your car's engine fails to turn over when you switch on the ignition, you would wisely check battery and electrical connections before dismantling parts of the engine. Apply a similar focus to your report.

On page 551, a hypothetical employer posed this question: *Will a low-impact aerobics program significantly reduce stress among my employees?* The aerobics question obviously requires answers to three other questions: *What are the therapeutic claims for aerobics? Are they valid? Will aerobics work in this situation?* How aerobic exercise got established, how widespread it is, who practices, and other such questions are not relevant to this problem (although some questions about background might be useful in the report's introduction). Always begin by defining the main questions and thinking through any subordinate questions they may imply. Only then can you determine the data or evidence you need.

With the main questions identified, the writer of the aerobics report can formulate her statement of purpose:

> The purpose of this report is to examine and evaluate claims about the thera-   **Definite**
> peutic benefits of low-impact aerobic exercise.[1]

The writer might have mistakenly begun instead with this statement:

> This report examines low-impact aerobic exercise.   **Vague**

[1]Words such as *examine* and *evaluate* (or *compare, identify, determine, measure, describe,* and so on) enable readers to understand the specific analytical activity that is the subject of the report.

Notice how the first version sharpens the focus by expressing the precise subject of the analysis: not aerobics (a huge topic), but the alleged *therapeutic benefits* of aerobics.

Define your purpose by condensing your approach to a basic question: Does low-impact aerobic exercise have therapeutic benefits? or Why have our sales dropped steadily for three months? Then restate the question as a declarative sentence in your statement of purpose.

Subordination of Personal Bias

Interpret evidence impartially. Throughout your analysis, stick to your evidence. Do not force viewpoints on your material that are not substantiated by dependable evidence. (For more on personal bias, see pages 182–183.)

Accurate and Adequate Data

Do not distort the original data by excluding vital points. Say you are asked to recommend the best chainsaw for a logging company. Reviewing test reports, you come across this information.

Give the whole story

> Of all six brands tested, the Bomarc chainsaw proved easiest to operate. It also had the fewest safety features, however.

If you cite these data, present *both* findings, not simply the first—even though you may prefer the Bomarc brand. *Then* argue for the feature you think should receive priority.

As space permits, include the full text of interviews or questionnaires in appendixes.

Fully Interpreted Data

Explain the significance of your data. Interpretation is the heart of the analytical report. You might interpret the chainsaw data in this way:

Explain what the data mean

> Our cutting crews often work suspended by harness, high above the ground. And much work is in remote areas. Safety features therefore should be our first requirement in a chainsaw. Despite its ease of operation, the Bomarc saw does not meet our safety needs.

By saying "therefore . . ." you engage in analysis—not mere information-sharing. *Merely listing your findings is not enough.* Tell readers what your findings mean.

Clear and Careful Reasoning

Each stage of your analysis requires decisions about what to record, what to exclude, and where to go next. As you evaluate your data *(Is this reliable and important?)*, interpret your evidence *(What does it mean?)*, and make recommendations based on your conclusions *(What action is needed?)*, you might have to alter your original plan. Remain flexible enough to revise your thinking if contradictory new evidence appears.

Appropriate Visuals

Use visuals generously (Chapter 14). Graphs are especially useful in an analysis of trends (rising or falling sales, radiation levels). Tables, charts, photographs, and diagrams work well in comparative analyses.

Valid Conclusions and Recommendations

Along with the informative abstract, conclusions and recommendations are the sections of a long report that receive most attention from readers. The goal of analysis is to reach a valid *conclusion*—an overall judgment about what all the material means (that *X* is better than *Y*, that *B* failed because of *C*, that *A* is a good plan of action). Here is the conclusion of a report on the feasibility of installing an active solar heating system in a large building:

1. Active solar space heating for our new research building is technically feasible because the site orientation will allow for a sloping roof facing due south, with plenty of unshaded space.

   Offer a final judgment

2. It is legally feasible because we are able to obtain an access easement on the adjoining property, to ensure that no buildings or trees will be permitted to shade the solar collectors once they are installed.

3. It is economically feasible because our sunny, cold climate means high fuel savings and a faster payback (fifteen years maximum) with solar heating. The long-term fuel savings justify our short-term installation costs (already minimal because the solar system can be incorporated during the building's construction—without renovations).

Conclusions are valid when they are logically derived from accurate interpretation (as discussed on pages 135–137).

Having explained *what it all means*, you then recommend *what should be done*. Taking into account all possible alternatives, your recommendations urge specific action (to invest in *A* instead of *B*, to replace *C* immediately, to follow plan *A*, or the like). Here are the recommendations based on the previous interpretations:

Consider All the Details

- What exactly should be done?
- How exactly should it be done?
- When should it begin and be completed?
- Who will do it, and how willing are they?
- Any equipment, material, or resources needed?
- Any special conditions required?
- What will this cost, and where will money come from?
- What consequences are possible?
- Whom do I have to persuade?
- How should I order my list (priority, urgency, etc.)?

Locate the Weak Spots

- Is anything unclear or hard to follow?
- Is it unrealistic?
- Is it risky or dangerous?
- Is it too complicated or confusing?
- Is anything about it illegal or unethical?
- Will it cost too much?
- Will it take too long?
- Could anything go wrong?
- Who might object or be offended?
- What objections might be raised?

Make Improvements

- Can I rephrase anything?
- Can I change anything?
- Should I consider alternatives?
- Should I reorder my list?
- Can I overcome objections?
- Should I get advice or feedback before I submit this?

Source: Questions adapted from Ruggiero, Vincent R. *The Art of Thinking*, 3rd ed. New York: Harper, 1991: 162–65.

FIGURE 23.1 How to Think Critically About Your Recommendations

1. I recommend we install an active solar heating system in our new research building. Tell what should be done

2. We should arrange an immediate meeting with our architect, building contractor, and solar-heating contractor. In this way, we can make all necessary design changes before construction begins in two weeks.

3. We should instruct our legal department to obtain the appropriate permits and easements immediately.

Recommendations are valid when they propose an appropriate response to the problem or question.

Because they culminate your research and analysis, recommendations challenge your imagination, your creativity, and—above all—your critical thinking skills. Having reached a valid conclusion about *what is*, you now must decide *what ought to be done*. But what strikes one person as a brilliant idea might be seen by others as idiotic or offensive. Depending on whether recommendations are carefully thought out or "off the wall," writers earn an audience's respect or its scorn. (Figure 23.1 depicts the kinds of decisions writers encounter in formulating, evaluating, and refining their recommendations.)

When you do achieve definite conclusions and recommendations, express them with assurance and authority. Unless you have reason to be unsure, avoid noncommittal statements ("It would seem that . . ." or "It looks as if . . ."). Be direct and assertive ("The earthquake danger at the reactor site is acute," or "I recommend an immediate investment"). Let readers know where you stand.

If, however, your analysis yields nothing definite, do not force a simplistic conclusion on your material. Instead, explain your position ("The contradictory responses to our consumer survey prevent me from reaching a definite conclusion. Before we make any decision about this product, I recommend a full-scale market analysis"). The wrong recommendation is far worse than no recommendation at all.

A GENERAL MODEL FOR ANALYTICAL REPORTS

Whether you outline earlier or later, the finished report depends on a good outline. This model outline can be adapted to most analytical reports:

   I. INTRODUCTION
      A. Definition, Description, and Background
      B. Purpose of the Report, and Intended Audience

    C. Sources of Information

    D. Working Definitions (here or in a glossary)

    E. Limitations of the Study

    F. Scope of the Inquiry (topics listed in order of importance)

II. COLLECTED DATA

    A. First Topic for Investigation

        1. Definition

        2. Findings

        3. Interpretation of findings

    B. Second Topic for Investigation

        1. First subtopic

            a. Definition

            b. Findings

            c. Interpretation of findings

        2. Second subtopic (and so on)

III. CONCLUSION

    A. Summary of Findings

    B. Overall Interpretation of Findings (as needed)

    C. Recommendations (as needed and appropriate)

This outline is only tentative. Modify if necessary.

Three sample reports in this text follow our model outline. The text of the first report, "Children Exposed to Electromagnetic Radiation: A Risk Assessment," follows (minus the document supplements that ordinarily accompany a long report). The second report, "The Feasibility of a Technical Marketing Career," appears in Figure 23.2. The third, "Quality-Control Recommendations for *Megacrunch* Magazine," appears in Appendix C. (Shown with the two latter reports are their supplements.)

Each report responds to a slightly different question. The first tackles this question: *What are the effects of X?* The second tackles two questions: *Is X feasible, and which version of X is better for our purposes?* The third: *Why does X happen, and what can we do about it?* At least one of these reports should serve as a model for your own analysis.

Introduction

The introduction engages and orients readers, and provides background—as briefly as possible. Identify the origin and significance of the topic, define or describe the problem or issue, and explain the report's purpose. Identify your audience only in a report for your instructor, and only if you don't attach an audience-and-use profile. Identify briefly your data sources (which you will later document fully), and explain any omissions of data (say, person unavailable for an interview or research still in progress). List working definitions, unless you have so many that a glossary is needed. If you do include a glossary and appendixes, notify readers early. Finally, preview the scope of your analysis by listing the major topics you will discuss in the body section.

Not all reports require every element. Give readers only what they need and expect.

As you read the following introduction, think about the elements designed to engage and orient readers (i.e., local citizens), and evaluate their effectiveness. (Review pages 132–137 for the situation that gave rise to this report.)

Children Exposed to Electromagnetic Radiation: A Risk Assessment

—Laurie A. Simoneau

INTRODUCTION

Wherever electricity flows—through the largest transmission line or the smallest appliance—it emits varying intensities of charged waves known as an electromagnetic field (EMF). Medical research has linked human exposure to EMFs with definite physiological changes and with possible illness, including cancer, miscarriage, and depression.

Definition and background

Some experts disagree over the extent of health risk, if any, from EMFs. Others question whether EMF risk is greater from high-voltage transmission lines, the smaller distribution lines strung on utility poles, or household appliances. Conclusive research may take years; meanwhile, concerned citizens worry about avoiding potential risks.

In Bocaville, four sets of transmission lines (two at 115,000 volts and two at 345,000) cross residential neighborhoods and public property. The Adams elementary school is less than 100 feet from this power-line corridor. EMF risks—whatever they may be—increase with proximity of exposure.

Description of problem

On the basis of examination of recent research and interviews with local authorities, this report assesses whether potential health risks from EMF exposure seem significant enough for Bocaville to (a) increase public awareness, (b) divert the transmission lines that run adjacent to the elementary school, and (c) implement widespread precautions in the transmission and distribution of electrical power throughout Bocaville.

Purpose and sources

Five major topics are covered in this report: what we know about the various sources of EMFs, what research indicates about the physiological and health effects from EMF exposure, how experts differ in their interpretations of the research, what the local power company has to say, and what is being done locally and nationwide to avoid risk.

Body

The body section describes and explains your findings. Present a clear and detailed picture of the evidence, interpretations, and reasoning upon which you will base your conclusion. Divide topics into subtopics, and use clear, informative headings that show readers exactly where they are at any point in the discussion. Remember your ethical responsibility for presenting a *fair* and *balanced* treatment of the material, instead of "loading" the report with only those findings that support your viewpoint.

As you read the following body section, evaluate how effectively it informs readers, keeps them on track, reveals a clear line of reasoning, and presents an impartial analysis.

DATA SECTION

Sources of EMF Exposure

First topic

Although the 2 million miles of power lines crisscrossing the United States have been the focus of the EMF controversy, potentially harmful waves also are emitted by household wiring, appliances, electric blankets, and computer terminals.

Definition

Electromagnetic intensity is measured in *milligauss* (a unit of electrical measurement). The higher the milligauss reading, the stronger the electromagnetic field. Consistent exposure above 2.5 milligauss is thought to increase cancer risk significantly (Abelson 24), but no scientific evidence concludes that exposure even below 2.5 milligauss is safe.

Findings
Interpretation

Table 1 compares approximate intensities of various sources.

EMF intensity from certain appliances tends to be higher than from transmission lines because of the amount of current involved.

Definitions

Voltage measures the speed and pressure of electricity in wires, but *current* measures the volume of electricity passing through wires. Current is what produces electromagnetic fields, and current is measured in amperage. The current flowing through a transmission line typically ranges from 200 to 400 amps. Most homes have a 200-amp service. This means that if every electrical item in the house were turned on at the same time, the house could run about 200 amps—almost the same as the transmission line. Consumers then have the ability to put 200 amps of current-flow into their homes, while transmission lines carrying 200 to 400 amps are at least 50 feet away (Miltane).

Finding

TABLE 1 Approximate Emissions from Selected EMF Sources (in milligauss)

Source	Range[a,b]
Earth's magnetic field	0.1–2.5
Beneath distribution line	10–50
Beneath high-voltage transmission line	15–200
Edge of transmission line right-of-way	3–40
Blow-dryer	60–1400
Four in. from TV screen	40–100
Four ft. from TV screen	0.7–9
Flourescent lights	10–12
Electric razor	1200–1600
Electric blanket	2–25
Computer terminal (12 inches away)	3–15
Toaster	10–60

[a]Data from Brodeur, Paul. "ANNALS OF RADIATION: The Cancer at Slater School." *The New Yorker* 7 Dec. 1992: 118; Castleman, Michael. "Electromagnetic Fields." *Sierra* Jan./Feb. 1992: 21–22; Miltane, John, Interview. 5 Apr. 1993; "What Utilities Can Do about the EMF Threat." *Electrical World* Nov. 1988: 37; Yakutchik M. "A New Jolt of Concern." *USA WEEKEND* 1 Jan. 1993: 5.

[b]Readings are done with a gaussmeter, and tend to vary greatly, depending on technique, the proximity of the gaussmeter to the source, its direction of aim, and other random factors.

Proximity and duration of exposure, however, are other factors in determining risk. People are exposed to EMFs from home appliances at close proximity, but appliances run only periodically: exposure is therefore sporadic, and intensity diminishes sharply within a few feet. Power line exposure, on the other hand, is at a greater distance (usually 50 feet or more), but is constant. Moreover, intensity can remain strong for up to 100 feet (Castleman 21; Miltane). In fact, buildings as far as 150 feet from high-current lines routinely show constant readings of 2 to 3 milligauss (Brodeur, "ANNALS" 87).

Research has not yet determined which type of exposure might be more harmful: to higher intensities for short intervals, or to lower intensities constantly. But in either case, proximity seems most significant because EMF intensity diminishes rapidly with distance.

Finding

Interpretation

Physiological Effects and Health Risks from EMF Exposure

Research on EMF exposure falls into two categories: epidemiologic studies and laboratory studies. The findings are sometimes controversial and inconclusive, but nonetheless disturbing.

Second Topic

First subtopic

Definition

General findings

Epidemiologic Studies   Epidemiologic studies look for statistical correlations between EMF exposure and certain diseases and physical disorders. Several such studies suggest that EMF exposure increases the incidence of the following conditions (Black 44–45; Lee et al. 12–28):

- cancer, especially leukemia and brain tumors
- miscarriage
- stress and depression
- learning disabilities

In one study, for example, pregnant women working at computer terminals at least 20 hours weekly had a risk of miscarriage twice that of other pregnant clerical workers (Toufexis 71).

Following are summaries of four noted epidemiologic studies implicating EMFs in childhood cancer:

Detailed findings

A Landmark Study of the EMF/Cancer Connection.   A Denver, Colorado study by Wertheimer and Leeper in 1979 was the first to implicate EMFs as a cause of cancer. These researchers compared hundreds of homes in which children had developed cancer with similar homes in which children were cancer free. Children with cancer were found to be two to three times as likely to live in "high-current homes" (within 130 feet of a transmission line or within 50 feet of a distribution line).

This study, however, was criticized because (1) it was not "blind" (researchers knew which homes cancer victims were living in), and (2) they never took gaussmeter readings to verify their designation of "high-current" homes (Kirkpatrick 82; Noland 64).

Detailed findings

Follow-up Studies.   At least three major studies of the EMF/cancer connection have confirmed Wertheimer's earlier findings:

- In 1988, Savitz studied hundreds of Denver houses and found that children with cancer were 1.7 times as likely to live in high-current homes. Unlike his predecessors, Savitz did not know whether a cancer victim lived in the home he was measuring, and he took gaussmeter readings to verify that houses could be designated "high-current" (Noland 66).
- In 1990, London and Peters found that Los Angeles children had 2.5 times more risk of developing leukemia if they lived near power lines (Brodeur, "ANNALS" 115).
- In 1992, a massive Swedish study found that children in houses with average intensities of greater than one milligauss had twice the normal risk of developing leukemia; at greater than two milligauss, the risk nearly tripled; at greater than three milligauss, it nearly quadrupled (Brodeur 115).

Recent Workplace Studies. Studies in 1988, 1989, 1990, and—most notably—a 1992 Swedish study found that electricians, electrical engineers, and power-line workers constantly exposed to an average of 1.5 to 4.0 milligauss had 7 to 13 times the normal cancer risk (Brodeur, "ANNALS" 115; Kirkpatrick 82–83).

Detailed findings

None of the above studies can be said to "prove" a direct cause-effect relationship, but their strikingly similar results suggest a probable link between long-term EMF exposure and cancer.

Interpretation

Laboratory Studies Laboratory studies assess the physiological effects of EMFs on humans and animals at cellular, metabolic, and behavioral levels. Such studies show that EMFs directly cause the following physiological changes (Black 46–48; Brodeur, "ANNALS" 88; Toufexis 71):

Second subtopic

- reduced heart rate
- altered brain waves
- impaired immune system
- interference with the synthesis of genetic material
- disrupted regulation of cell growth
- interaction with the biochemistry of cancer cells
- altered hormonal activity

General findings

These changes are documented in the following summaries of several significant laboratory studies.

EMF Effects on Cell Growth and Division. A 1986 study by Aday, Byus, and Pieper found that cancer cells from humans, rats, and mice grew and divided more rapidly when exposed to electromagnetic radiation. The researchers point out, however, that growth of normal cells also increases with EMF exposure (Hecht 28).

Detailed findings

A 1988 study by Phillips showed that the genetic offspring of human cancer cells exposed to electromagnetic radiation grew more rapidly and were highly resistant to attack from white blood cells that normally attack cancer cells. This researcher notes that "cellular effects don't necessarily show up in animals or humans, because a healthy immune system might simply be able to gear itself up and kill off the additional cancer cells" (Noland 65–66).

EMF Effects on Hormones. A 1988 study by Wilson found that EMF exposure from electric blankets inhibits the production of melatonin, a hormone that fights cancer, stimulates the immune system, and regulates natural bodily rhythms

Detailed findings

(Noland 66–67). (*Note:* Only self-regulating blankets, those that switch on and off automatically and therefore require more current, affected melatonin levels.)

Detailed findings

EMF Effects on Behavior. A 1988 study by Liboff and Thomas found that EMF exposure decreased the amount of lithium in the brain cells of rats and produced erratic behavior. Insufficient levels of lithium are implicated in human depression and other emotional disorders (Noland 67–68).

Interpretation

Although laboratory studies seem more conclusive than the epidemiologic studies, what these findings *mean* is debatable.

Third Topic

Conflicting Interpretations of EMF Research

Experts differ over the meaning of EMF research findings largely because of the following limitations attributed to various studies:

Findings

- The epidemiologic studies are criticized for ignoring those physiological effects of EMFs that are beneficial (McDonald 5, 8). These studies are also accused of mistaking *coincidence* for *correlation,* without exploring other "confounding factors" (toxic substances or adverse conditions to which subjects might have been exposed)—including the earth's natural electromagnetic field. But the sheer volume of evidence seems overwhelming (Kirkpatrick 81, 83). Moreover, the 1992 Swedish studies cited earlier seem to invalidate such criticisms (Brodeur, "ANNALS" 115).
- Laboratory studies are criticized, even by the scientists who conduct them, because effects on an isolated culture of cells are not always equal to effects on the total human or animal organism (Jauchem 190–94).

Interpretation

While most scientists do agree that EMFs have effects on the human body, they disagree about whether a real hazard exists. Their only point of agreement seems to be on the need for more and better research.

Fourth topic

Perspective from Local Power Company Representatives

What does our local power company know and think about the alleged EMF risk? Marianne Halloran-Barney, Energy Service Advisor for County Electric, expressed this view:

Findings

> There are definitely some links, but we don't know, really, what the effects are or what to do about them. . . . There are so many variables in EMF research that it's a question of whether the studies were even done correctly. . . . Maybe in a few years there will be really definite answers.

Echoing Halloran-Barney's views, John Miltane, Chief Engineer for County Electric, added this political perspective:

> The public needs and demands electricity, but in regard to the negative effects of generation and transmission, the pervasive attitude seems to be "not

in my back yard!" Utilities in general are scared to death of the EMF issue, but at County Electric we're trying to do the best job we can while providing reliable electricity to 24,000 customers.

These local views seem to parallel the national perspective: informed people are genuinely concerned, but remain unsure about what level of anxiety is warranted or what exactly should be done. Interpretation

Steps Being Taken Toward Risk-Avoidance
Fifth topic

Although conclusive answers about EMFs may require another decade of research, concerned citizens already are taking action against potential hazards.

Risk-Avoidance Nationwide   Communities around the country have taken steps, such as the following, to protect schoolchildren from EMF exposure: First subtopic

- Parents in Boca Raton, Florida are suing to close an elementary school that is very near transmission lines (Kirkpatrick 85). Findings
- Houston schools "forced a utility company to remove a transmission line that ran within 300 feet of three schools. Cost: $8 million" (Kirkpatrick 85).
- California parents and teachers are pressuring reluctant school and public-health officials to investigate cancer rates in the roughly 1000 schools located within 300 feet of transmission lines, and to close at least one school (within 100 feet) in which cancer rates far exceed normal (Brodeur, "ANNALS" 118).

These are just a few examples of the widespread expressions of concern about EMF exposure.

Risk Avoidance Locally   Local awareness of the EMF issue seems very low. The major public concern seems to be with property values of homes near power lines. According to Energy Service Advisor Halloran-Barney, County Electric receives one or two calls monthly from concerned customers, including people buying homes near power lines. The lack of public awareness adds another dimension to the EMF problem: people can't avoid a health threat they don't know exists. Second subtopic

Findings

According to Chief engineer John Miltane, County Electric takes the EMF issue very seriously: Whenever possible, new distribution lines are run underground and configured to diminish EMF intensity:

Although EMFs are impossible to eliminate altogether, we design anything we build to emit minimal intensities. . . . Also, we are considering underground cable (at $1200 per ft.) to replace 8000 feet of transmission lines, but the bill would ring up to nearly $10 million, for the cable alone, without labor. You don't get environmental stuff for free, which is one of the problems.

Before risk avoidance can be implemented on a broader community level, the public must first be informed about EMFs and the associated risks of exposure.

Conclusion

The conclusion is likely to be of most interest to readers because it answers the questions that sparked the analysis in the first place. (Many workplace reports, therefore, are submitted with the conclusion *preceding* the introduction and body sections.)

Here you summarize, interpret, and recommend. Although you have interpreted evidence at each step in your analysis, your conclusion pulls the strands together in a broader interpretation and suggests a course of action, where appropriate. This final section must be consistent in three ways:

1. Your summary must reflect accurately the body of your report.
2. Your overall interpretation must be consistent with the findings in your summary.
3. Your recommendations must be consistent with the purpose of the report, the evidence presented, and the interpretations given.

The summary and interpretations should lead logically to your recommendations.

As you read the following conclusion, evaluate how effectively it provides a clear and consistent perspective on the whole document.

CONCLUSION

Summary and Overall Interpretation of Findings

Electromagnetic fields exist wherever electricity flows; the stronger the current the higher the EMF intensity. While no "safe" EMF level has been identified, long-term exposure to intensities greater than 2.5 milligauss is considered dangerous. Although home appliances can generate high EMFs during use, power lines can generate constant EMFs, typically causing readings of 2 to 3 milligauss in buildings within 150 feet. Our elementary school is less than 100 feet from a high-voltage line corridor.

Notable epidemiologic studies implicate EMFs with increased rates of medical disorders such as cancer, miscarriage, stress, depression, and learning disabilities—all directly related to intensity and duration of exposure. And laboratory studies show that EMFs cause the kinds of cellular, metabolic, and behavioral changes that could produce the above medical disorders.

Though still controversial and inconclusive, the various findings are strikingly similar and they underscore the need for more research and for risk-avoidance, especially as far as children are concerned.

Concerned citizens nationwide are beginning to prevail over resistant school and health officials and utility companies in reducing EMF risk to schoolchildren. And even though our local power company is taking reasonable risk-avoidance steps, our community can do more to learn about the issues and diminish potential risk.

Recommendations

Given the limitations of our present knowledge, drastic and enormously expensive actions (such as burying all the town's power lines) seem inadvisable because these might turn out to be the wrong actions. However, the following inexpensive steps can immediately address EMF risk:

- A version of this report should be distributed to all Bocaville residents.
- The school board should hire a qualified contractor to take milligauss readings throughout the elementary school, to determine the extent of the problem.
- The Town Council should meet with County Electric Company representatives, to explore options and costs for rerouting or burying that segment of the power lines near the school.
- A town meeting should then be held, to answer citizen's questions and to solicit opinions.
- A committee (consisting of at least one physician, one engineer, and other experts) should be appointed to review emerging research as it relates specifically to our school and town.

A reasonable and realistic course of action

As we await conclusive answers, we need to learn all we can about the EMF issue, and to do all we can to diminish this potentially significant health risk.

A closing call to action

The "Works Cited" section for the preceding report appears on page 198. Our author employs MLA documentation style (discussed on pages 189–196).

Supplements

Submit your completed report with these supporting documents, in order:

- title page
- letter of transmittal
- table of contents
- table of figures
- abstract
- **report text (introduction-body-conclusion)**
- glossary (as needed)

- appendixes (as needed)
- works cited page (or alphabetical or numbered list of references)

Each of these supplements is discussed in Chapter 16. Forms of documentation are discussed in Chapter 9.

A SAMPLE SITUATION

The report in Figure 23.2, patterned after our model outline, combines a feasibility analysis with a comparative analysis.

The situation: Richard Larkin, author of the following report, has a work-study job fifteen hours weekly in his school's placement office. Larkin knows that his boss, John Fitton (placement director), likes to keep abreast of trends in various fields. Larkin himself, an engineering major, has become interested in technical marketing and sales, a recent and rapidly growing choice of career for graduates in various specialized fields.

In need of a report topic for his writing course, Larkin offers to analyze the feasibility of a technical marketing and sales career, both for himself and for technical and science graduates in general. Fitton accepts Larkin's offer, looking forward to having the final report in his reference file for use by students choosing careers. Larkin wants his report to be useful in three ways: to satisfy a course requirement, to help him in choosing his own career, and to help other students with their career choices.

With his topic approved, Larkin begins gathering his primary data, using interviews, letters of inquiry, telephone inquiries, and lecture notes. He supplements these primary sources with articles in recent publications. He will document his findings in APA (author-date) style (shown and discussed on pages 196–201).

As a guide for designing his final report, Larkin completes the following audience-and-use profile (based on the worksheet, page 60).

Audience-and-Use Profile for a Formal Report

AUDIENCE IDENTITY AND NEEDS

My primary audience consists of John Fitton, Placement Director, and the students who will be referring to my report as they choose careers. The secondary audience is my writing instructor. And, of course, the data I've uncovered will help me make my own career choice.

Fitton is highly interested in this project, and he has promised to study my document carefully and to make copies available to interested students. Because he already knows something about the technical marketing field, Fitton will need very little background to understand my report. Many student readers, however, may know little or nothing about technical marketing, and so will need background, definitions, and detailed explanations. Here are the questions I can anticipate from my collective audience:

- What, exactly, is technical marketing and sales?
- What are the requirements for this career?
- What are pros and cons of this career?
- Could this be the right career for me?
- If so, how do I enter the field?
- Is there more than one option for entering the field? If so, which option would be best for me?

ATTITUDE AND PERSONALITY

Readers likely to be most affected by my document are students who will be making career choices. And I would expect my readers' attitudes to vary widely:

1. Some readers will approach my document with great interest, especially those seeking a career that is more people-oriented than technology-oriented.
2. Some readers will be only casually interested in my document, as they investigate a range of possible careers.
3. Some readers are likely to feel overwhelmed by the variety and importance of the career choices they face. Members of this group might be looking for easy answers to the problem of choosing a career.
4. Other readers may approach my document skeptically, perhaps unwilling to re-examine their earlier decisions about a more traditional career.

To connect with this array of readers, I will need to persuade them that my conclusions are based on dependable data and careful reasoning.

EXPECTATIONS ABOUT THE DOCUMENT

Although I initiated this document, Fitton has been greatly supportive, and eager to read the final version. All my readers will expect me to spell things out. But I know that my readers are busy and impatient, and so I'll want to make this report concise enough to be read in no more than fifteen or twenty minutes.

Essential information will include an expanded definition of technical marketing and sales, the skills and attitudes needed for success, the career's advantages and drawbacks, and a description of various paths for entering the career. And throughout, I'll relate my material to many technical and science majors, not just engineers.

The body of this report combines a feasibility analysis with a comparative analysis. Therefore, I'll use a reasons-for and reasons-against structure in the feasibility section. In the comparison section, I'll use a block structure (page 248), followed by a table that presents a point-by-point comparison of the four entry paths. Because I want this report to lead to informed decisions, I will include concrete recommendations that are based solidly on my conclusions.

[*This profile continues on page 579.*]

Feasibility Analysis of a Career in Technical Marketing

INTRODUCTION

Technical marketing and sales is a most challenging and lucrative career for science and technology graduates. Specifically, this career involves identifying, reaching, and selling to customers a technical product or service.

Customer orientation is an ever-growing part of today's business and manufacturing climate. According to Richard Marks, sales manager for Global Industries, Inc., much of U.S. industry no longer is "manufacturing-driven" (where customers will buy any products that are available). Instead, companies are "market driven" (customers demand that products be designed to exact specifications). These customer demands are the reason nearly 80 percent of top managers at Global Industries have sales and marketing experience beyond their technical background
(Basta, 1988, p. 84).

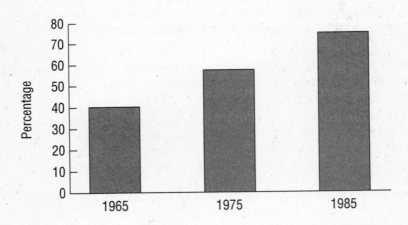

Source: Data from Campbell (1988).

FIGURE 1 Top Managers with Sales and Marketing Backgrounds

FIGURE 23.2 An Analytical Report

The emphasis on sales and maketing at Global Industries is part of a larger trend. Nationwide, roughly 75 percent of today's top managers have worked their way up through the sales and marketing ranks (Campbell, 1988, p. 28). Figure 1 shows how marketing background has become part of the career track to top-management. In the product-oriented industries of 1965, technical marketing accounted for only 39 percent of top-management background. But the number had nearly doubled by 1985 because of growth in customer orientation. And in 1992, listings of government and corporate openings for recent graduates showed nearly one hundred major companies offering positions in technical marketing and sales (*CPC annual*, 1992, p. 389).

Demand for technical salespeople is stimulating interest among technical graduates. Of the more than 40,000 engineering graduates in 1983, nearly 4 percent sought sales and marketing positions (Campbell, p. 20). Today, a growing number of high school seniors identify technical marketing as their career goal (Basta, 1988, p. 85).

But the many undergraduates interested in this career need answers to these basic questions:
—Is this the right career for me?
—If so, how do I enter the field?

To help answer these questions, this report analyzes information gathered from professionals and from the literature.

The following analysis includes these areas: definition of technical marketing, career requirements, advantages and disadvantages, and comparison of four career options.

COLLECTED DATA

Key Factors In Technical Marketing As a Career
Anyone considering technical marketing needs to assess whether this career is practical, considering the individual's interests, abilities, and expectations.

The Technical Marketing Process
Although the terms *marketing* and *sales* usually occur interchangeably, technical marketing involves more than sales work. The process (identifying, reaching, and selling to customers) entails six major activities (Cornelius & Lewis, 1983, p. 43).

1. *Market research:* gathering information about the size and character of the market for a product or service.
2. *Product development and management:* producing the goods to fill a specific market need.

FIGURE 23.2 An Analytical Report *Continued*

3. *Cost determination and pricing:* measuring every expense in the production, distribution, advertising, and sale of the product, to determine its price.
4. *Advertising and promotion:* developing and implementing all strategies for reaching customers.
5. *Product distribution:* coordinating all elements of a technical product or service, from its conception through its final delivery to the customer.
6. *Sales and technical support:* creating and maintaining customer accounts, and servicing and upgrading products.

Fully engaged in all these activities, the marketing professional gains good understanding of the industry, the product, and the customer's needs.

Career Requirements

Besides a strong technical background, technical marketing requires a blend of the traits summarized in Figure 2.

Motivation is essential for marketing work. Professionals must be energetic and able to function with minimal supervision. College recruiter Norman Lance claims that ideal candidates are creative people who can plan and program their own tasks, who can manage their time, and who are not afraid of hard work (personal interview, February 22, 1993). Leadership potential, as demonstrated by extracurricular activities, is an asset.

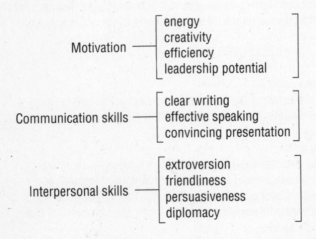

FIGURE 2 Requirements for a Technical Marketing Career

FIGURE 23.2 An Analytical Report *Continued*

But motivation alone is no guarantee of success. Marketing professionals are paid to communicate the virtues of their products or services to customers. This career, therefore, calls for skill in communication, both written and oral. Writing for readers outside the organization includes advertising copy, product descriptions, sales proposals, sales letters, and instructions for customers. In-house writing includes recommendation reports, feasibility studies, progress reports, and memos (*The job outlook*, 1992, p. 89).

Skilled presentation is vital to any sales effort. According to Abco Electronics president Elaine Carto, technical marketing professionals need to speak persuasively—to represent their products and services in the best light (personal interview, February 8, 1993). And sales presentations often require speaking before groups (at conventions, trade shows, and so on).

Besides motivation and communication skills, interpersonal skills are the ultimate requirement for success in marketing (Splaver, 1987, p. 114). Consumers are more likely to buy a product or service if they *like* the person selling it. Marketing professionals are extroverted and friendly; they enjoy meeting people. Because they understand diplomacy, they can motivate customers without alienating them.

Advantages of the Career

A technical marketing career offers diverse experience in every phase of a company's operation, from the design to the sale of a product. And many companies encourage employees to rotate between marketing and their technical specialties (Basta, 1988, p. 84). Because of the broad exposure, a marketing position can provide a direct path to upper-management jobs. In fact, sales engineers with solid experience often open their own firms as "manufacturers' agents" representing a variety of companies ("*Engineering careers*," 1990, p. 232).

Another benefit is the possibility of a high salary. Most marketing professionals receive base pay plus commission. According to John Turnbow, manager of National Electric's Technical Marketing Program, some of NE's new marketing engineers earn more than $60,000 their first year. Many salaries reach six figures, sometimes higher than a company's executive salaries (Basta, p. 87). Figure 3 compares average salaries for engineers with five years of experience to those of marketing engineers. Recent salaries of marketing engineers have accelerated more than engineering salaries.

This career is particularly attractive for its geographic mobility; companies nation-wide seek recent graduates, but especially in the Southeast and on east and west coasts ("*Electronics sales positions*," 1993, pp. 1126–29).

FIGURE 23.2 An Analytical Report *Continued*

Drawbacks of the Career

Technical marketing is not a career for everyone. Personnel might spend 50 to 75 percent of work time traveling. And success requires long hours and occasional weekends (Campbell, 1988, p. 24). Above all, the job is stressful because of pressure to sell and to meet quotas. Anyone considering this career should enjoy competition.

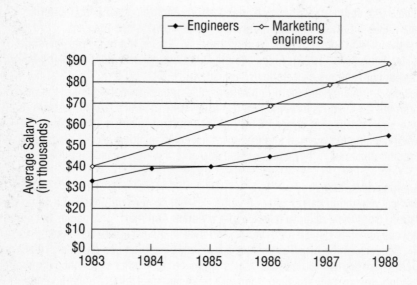

Source: Based on data from Basta (1988).

FIGURE 3 A Comparison of Average Salaries for Engineers and Marketing Engineers with Five Years' Experience

A Comparison of Entry Options

Engineers and other technical graduates enter technical marketing through one of four options. Some graduates join small companies and begin marketing work immediately Others join companies that offer formal training programs. Some begin by acquiring experience in their specialty. And others earn a graduate degree beforehand. Each option is described below, and then evaluated on the basis of specific criteria.

FIGURE 23.2 An Analytical Report *Continued*

Option 1: Entry-Level Marketing Without Training
Direct positions in marketing are available with smaller companies and with firms that represent client manufacturers. Because this option offers no formal training, candidates must be motivated and enterprising. Elaine Carto, president of Abco Electronics, believes small companies offer a unique opportunity; entry-level sales-persons learn about all facets of an organization, with the possibility for rapid advancement (personal interview, February 8, 1993). Career counselor Phil Hawkins says, "It depends on whether you want to be a big fish in a small pond or vice versa" (personal interview, February 10, 1993).

Manufacturers' representatives constitute another entry-level position. These professionals represent products for manufacturers who have no marketing staff of their own. Manufacturers' representatives are, in effect, their own bosses, choosing, from among many offers, the products they wish to represent (Tolland, 1993).

Entry-level marketing offers immediate income and a chance for early promotion. A disadvantage, however, might be the gradual loss of any technical edge that one might have acquired in college.

Option 2: A Marketing and Sales Training Program
Formal training programs are the most popular route into marketing. Large to mid-size companies offer two formats: (a) a product-specific program, in which trainees learn about a product or line of products, or (b) a rotational program, in which trainees learn about various products and work in various positions. Programs last from several weeks to several months, covering products, markets, competition, and customer relations.

Former trainees speak highly of such programs. Ralph Lang, of Allied Products, enjoyed the constant interaction with company personnel (phone interview, April 10, 1993). Bill Collins, sales engineer with Intrex Computers, values his broad knowledge of Intrex's product line, instead of being narrowly focused on one technical area. Sarah Watts, also with Intrex, enjoys applying her training to a variety of sales challenges (Campbell, 1988, p. 29).

Like direct entry, this option offers immediate income. A possible disadvantage, however, is that trainees may find their technical expertise compromised because they have had no chance to practice in their specialty.

Option 3: Practical Experience in One's Specialty
Instead of directly entering marketing, some candidates first gain practical experience in their specialty.

FIGURE 23.2 An Analytical Report *Continued*

This option provides direct exposure to the workplace, and a chance to sharpen technical ability in practical applications. Moreover, some companies offer marketing and sales positions to their outstanding engineers as a first step toward upper-management ("*Engineering careers*," 1990, p. 171).

Many industry experts prefer the practical experience option. Jane Doser, recruitment manager for Trans Electric, has this view of on-the-job experience: "Until you've done some work in your specialty, you really don't know yourself" (Schranke, 1985, p. 15).

This option ensures technical expertise, but not without a delay in entering technical marketing.

Option 4: Graduate Program
Instead of direct entry, some people choose to pursue either an M.S. in their specialty, an M.B.A., or both. According to engineering professor Mary McClane, master's degrees usually are unnecessary for technical marketing unless a sale is highly complex (lecture, April 8, 1993). But the M.B.A. can be a real advantage in terms of general business and marketing skills. According to Kline Jorgenson, manager of college recruiting at Hayes-Proctor, a product-marketing specialist at HP would use much of the knowledge acquired in an M.B.A. program (Schranke, p. 15).

A motivated student might combine degrees. Dora Anson, president of Susimo Cosmic Systems, sees the M.S. and M.B.A. as the ideal combination for technical marketing (lecture, January 15, 1993). To employers, the advanced degree means greater technical competence, stronger motivation, and better performance (13).

One disadvantage of a full-time graduate program is lost salary, compounded by school expenses. But these costs must be balanced against the prospect of promotion and monetary rewards later in one's career.

An Overall Comparison of Advantages
Table 1 compares the four entry options on the basis of three criteria: immediate income, advancement through marketing ranks, and long-term potential. Any entry can have major implications for one's career.

FIGURE 23.2 An Analytical Report *Continued*

Table 1 Relative Advantages among Four Technical Marketing Entry Options

| | Relative Advantage | | |
Option	Immediate income	Early advancement in marketing	Greatest long-term potential
Entry level, no training	yes	yes	no
Training program	yes	yes	no
Practical experience	yes	no	yes
Graduate program	no	no	yes

CONCLUSION

Summary of Findings

Technical marketing and sales involves identifying, reaching, and selling the customer a technical product or service. Besides a solid technical background, anyone entering the field will need motivation and communication and interpersonal skills. This career offers job diversity and excellent potential income, balanced against hard work and the pressure to perform.

College graduates who seek a career in technical marketing and sales have four entry options:

1. Direct entry with no formal training;
2. A formal training program;
3. Practical experience in a technical specialty prior to entry; or
4. Graduate programs.

Each option has advantages and disadvantages in immediacy of income, rate of advancement, and long-term potential.

Interpretation of Findings

For graduates with a strong technical background and the right skills and motivation, technical marketing and sales offers great prospects for income and advancement. Anyone contemplating this field, however, needs to enjoy customer-contact work and to be able to thrive in a highly competitive environment.

Those who decide that technical marketing is for them can choose from among four entry options. Although any of the options might be adequate, one option or another might best suit a person's needs and inclinations:

FIGURE 23.2 An Analytical Report *Continued*

—If immediate income is unimportant, graduate school is an attractive option.
—For hands-on experience, an entry-level job is the logical option.
—For sharpening technical skills, prior work in one's specialty is invaluable.
—For sophisticated sales training, a formal program with a large company is best.

Recommendations

If your interests and abilities match the requirements outlined earlier, follow these suggestions in planning for a technical marketing career.

1. To get a firsthand view, seek the opinions and advice of people in the field.
2. Before settling on an entry option, consider all its advantages and disadvantages, and decide whether this option best coincides with your goals.
3. Remember that you are never committed to one entry option only. For a fuller understanding of technical marketing, you might pursue any combination of options during your professional life.
4. When making any vital career decision, consider career counselor Warren Polgar's advice: "listen to your brain and your heart" (lecture, November 19, 1992). Choose an option or options that offer not only professional advancement, but comfort and happiness as well.

REFERENCES

[The complete list of references is shown in Chapter 9, page 200.]

FIGURE 23.2 An Analytical Report *Continued*

To address various readers who may not wish to read the entire report, I will include an informative abstract.

My tone throughout should be conversational. Because I am writing for a mixed audience (placement director, students, and writing instructor), I will use a third-person point of view.

———————

This report's front matter (title page and so on) and end matter are shown and discussed in Chapter 16, pages 369–376. Its list of references appears on page 200.

REVISION CHECKLIST FOR ANALYTICAL REPORTS

Use this list to refine the content, arrangement, and style of your report. (Numbers in parentheses refer to the first page of discussion.) For evaluating your research methods and reasoning, refer also to the checklist on page 203.

Content

- ☐ Does the report grow from a clear statement of purpose? (553)
- ☐ Is the report's length adequate and appropriate? (550)
- ☐ Are all limitations of the analysis spelled out? (558)
- ☐ Is each topic defined before it is discussed? (558)
- ☐ Are visuals used whenever possible to aid communication? (555)
- ☐ Are all data accurate? (554)
- ☐ Are all data unbiased? (554)
- ☐ Are all data complete? (554)
- ☐ Are all data fully interpreted? (554)
- ☐ Is the documentation adequate and correct? (188)
- ☐ Do the recommendations constitute an appropriate response to the question or problem? (555)

Arrangement

- ☐ Is there an introduction, body, and conclusion? (557)
- ☐ Are headings appropriate and adequate? (560)

□ Are there enough transitions between related ideas? (Appendix A)

□ Does the report have all needed front matter? (567)

□ Does the report have all needed end matter? (568)

Style

□ Is the level of technicality appropriate for the stated audience? (569)

□ Are all sentences clear, concise, and fluent? (251)

□ Is the language convincing and precise? (279)

□ Is the report written in grammatical English? (Appendix A)

EXERCISES

1. These statements are followed by false or improbable conclusions. In order to prevent specious generalizations, what specific supporting data or evidence would be needed to justify each conclusion?

    a. Eighty percent of minority voters in our state voted for John Jones as governor. Therefore, he is not a racist.

    b. Fifty percent of last year's college graduates did not find desirable jobs. Therefore, college is a waste of time and money.

    c. Only 60 percent of incoming students eventually graduate from this college. Therefore, the college is not doing its job.

    d. He never sees a doctor. Therefore, he is healthy.

    e. This house is expensive. Therefore, it must be well built.

2. Prepare an analytical report, using some sequence of these guidelines:

    a. Choose a subject for analysis from the list at the end of this exercise, from your major, or from a subject of interest.

    b. Identify the problem or question so that you will know exactly what you are looking for.

    c. Restate the main question as a declarative sentence in your statement of purpose.

    d. Identify an audience—other than your instructor—who will use your information for a specific purpose.

    e. Hold a private brainstorming session to generate major topics and subtopics.

    f. Use the topics to make an outline based on the model outline in this chapter. Divide as far as necessary to identify all points of discussion.

    g. Make a tentative list of all sources (primary and secondary) that you will investigate. Verify that adequate sources are available.

h. Write your instructor a proposal memo (pages 519–521), describing the problem or question and your plan for analysis. Attach a tentative bibliography.

i. Use your working outline as a guide to research and observation. Evaluate sources and evidence, and interpret all evidence fully. Modify your outline as needed.

j. Submit a progress report to your instructor (pages 503–506), describing work completed, problems encountered, and work remaining.

k. Compose an audience-and-use profile. (Use the samples on pages 535 and 568 as models, along with the profile worksheet on page 60.)

l. Write the report for your stated audience. Work from a clear statement of purpose, and be sure that your reasoning is shown clearly. Verify that your evidence, conclusions, and recommendations are consistent. Be especially careful that your recommendations observe the critical-thinking guidelines in Figure 23.1 (p. 556).

m. After writing your first draft, make any needed changes in the outline and revise your report according to the revision checklist. Include all necessary supplements.

n. Exchange reports with a classmate for further suggestions for revision.

o. Prepare an oral report of your findings for the class as a whole.

Here are some possible subjects for analysis:

- the long-term effects of a vegetarian diet
- the causes of student disinterest in campus activities
- the student transportation problem to and from your college
- comparison of two or more brands of equipment
- noise pollution from nearby airport traffic
- the effects of toxic chemical dumping in your community
- the adequacy of veterans' benefits
- the (causes, effects) of acid rain in your area
- the influence of a civic center or stadium on your community
- the best microcomputer to buy for a specific need
- the qualities employers seek in a job candidate
- the feasibility of opening a specific business
- the best location for a new business
- causes of the high dropout rate in your college
- the pros and cons of condominium ownership for you
- the feasibility of moving to a certain area of the country
- job opportunities in your field
- the effects of budget cuts on public higher education in your state
- effect of population increase on your local water supply
- the feasibility of biological pest control as an alternative to pesticides
- the feasibility of large-scale desalination of sea water as a source of fresh water

- effective water conservation measures that can be used in your area
- the effects of legalizing gambling in your state
- the causes of low morale in the company where you work part-time
- effects of thermal pollution from a local power plant on marine life
- the feasibility of converting your home to solar heating
- adequacy of police protection in your town
- the best energy-efficient, low-cost housing design for your area
- measures for improving productivity in your place of employment
- reasons for the success of a specific business in your area
- the feasibility of operating a campus food co-op
- the advisability of home birth (as opposed to hospital delivery)
- the adequacy of the evacuation plan for your area in a nuclear emergency, hurricane, or other disaster
- the feasibility and cost of improving security in the campus dorms
- the advisability of pursuing a graduate degree in your field, instead of entering the work force with a bachelor's degree.
- major causes of the wage gap between women and men in a particular field

COLLABORATIVE PROJECTS

1. Divide into small groups. Choose a subject for group analysis—preferably, a campus issue—and partition the topic by group brainstorming. Next, select major topics from your list and classify as many items as possible under each major topic. Finally, draw up a working outline that could be used for an analytical report on this subject.

2. Prepare a questionnaire based on your work above, and administer it to members of your campus community. List the findings of your questionnaire and your conclusions in clear and logical form. (Review pages 162–169, on questionnaires and surveys.)

Oral Reports

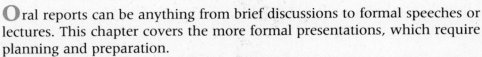

Choosing the Best Delivery

Preparing Extemporaneous Reports

Presenting the Report

■   ■   ■

Oral reports can be anything from brief discussions to formal speeches or lectures. This chapter covers the more formal presentations, which require planning and preparation.

Your own oral presentations may include convention speeches, reports at national meetings, reports given through a teleconferencing network, technical briefings for colleagues, and speeches to community groups. The higher your status, on the job or in the community, the more you will give oral presentations.

In the words of two experts, "most persons in most presentational settings do not perform well" (Goodall and Waagen, 14–15). Despite their vital importance in the workplace, oral presentations often are boring, confusing, unconvincing, or too long. And many are poorly delivered, with the presenter losing her or his place, fumbling through notes, apologizing for forgetting something, or generally seeming unprofessional. How then does a presenter demonstrate *professionalism*? By coming across as purposeful, knowledgeable, prepared, organized, and engaging.

CHOOSING THE BEST DELIVERY

An oral report's effectiveness depends greatly on *how* it is delivered. Not all deliveries are equally effective.

The Impromptu Delivery

The impromptu ("off-the-cuff") delivery is ineffective for formal presentations. Being effective without preparation is nearly impossible. Don't fool yourself into thinking "It's all in my head." Get your plan down on paper.

The Memorized Delivery

In the memorized delivery, you write your report and memorize it. The danger here is that you might sound like a parrot, and forgetting a line can be disastrous.

The Reading Delivery

Certain kinds of complex technical information may need to be read aloud to an audience, but listeners are bored by presentations read verbatim. If you *do* plan to read the report, study it, and practice aloud to decrease reliance on the text.

The Extemporaneous Delivery

An extemporaneous delivery is carefully planned, practiced, and based on notes that keep you on track. In this natural form of presentation, you glance at notes in sentence-outline form, and you speak in a conversational style. Extemporaneous delivery is based on key ideas (represented by topic sentences to refresh your memory) rather than fully developed paragraphs to be read or memorized. One danger in extemporaneous delivery is that you may get off track, unless you are thoroughly prepared.

PREPARING EXTEMPORANEOUS REPORTS

Plan your report step by step, to stay in control and build confidence. Some suggestions follow.

Know Your Subject

Do your homework, *exhaustively*. Be prepared to explain and defend each assertion and opinion with fact. Your audience expects to hear a knowledgeable speaker. Don't disappoint them.

Identify Your Audience

Adjust the amount of detail and the level of technicality to your audience. Audiences often include people with varied technical backgrounds. Unless you know the background of each person, speak to a general audience (Chapter 3), as in a classroom of mixed majors.

As you plan the report, answer these questions:

1. What points do I want to make?
2. How can I make each point interesting and understandable?

If your subject is controversial, consider the average age, political views, education, and socioeconomic status of most audience members. Then decide how to speak candidly and persuasively without offending anyone.

Plan Your Report

Statement of Purpose. Formulate, on paper, a statement of purpose, no more than two or three sentences long. Why are you speaking on this subject? Who is your audience? What effect do you wish to have?

Sentence Outline. If the oral presentation is a version of your written report, you need substantially less preparation. Simply expand your outline for the written report into a sentence outline. For our limited purposes, we will assume that your oral report is based on a written report. Here is a typical sentence outline for a fifteen-minute oral report derived from a written report twenty pages long.

Oral Report Outline

Arnold Borthwick

Purpose: By informing Cape Cod residents about the dangers to the Cape's freshwater supply posed by rapid population growth, this report is intended to increase local interest in the problem.

I. INTRODUCTION
   A. Do you know what you are drinking when you turn on the tap and fill a glass?
   B. The quality of our water is good, but not guaranteed to last forever.
   C. The somewhat unique natural storage facility for our water supply creates a dangerous situation.
   D. Cape Cod's rapidly increasing population could pollute our water.
   E. In fact, pollution in some towns already has begun.

II. BODY

A. The groundwater is collected and held in an aquifer.

    1. This water-bearing rock formation forms a broad, continuous arch under the entire Cape (Visual #1).

    2. The lighter fresh water flows on top of the heavier salt water.

B. With increasing population, sewage and solid waste from landfill dumps invade the aquifer.

    1. As wastes flow naturally toward the sea, they can invade the drawing radii of town wells (Visual #2).

    2. The Cape's sandy soil causes rapid seepage of wastes into the groundwater in the aquifer.

C. Increased population also causes overdraw on some town wells, resulting in saltwater intrusion.

D. Salt and calcium used in snow removal add to the problem by entering the aquifer from surface runoff.

E. The effects of continuing pollution of the Cape's water table will be far-reaching.

    1. Drinking water will have to be piped in more than 100 miles from Quabbin Reservoir.

    2. The Cape's beautiful freshwater ponds will be unfit for swimming.

    3. Aquatic and aviary marsh life will be threatened.

    4. The sensitive ecology of Cape Cod will be destroyed.

F. Such damage would, in turn, lead to economic disaster for Cape Cod's major industry—tourism.

III. CONCLUSION

A. This problem is becoming more real than theoretical.

B. The conclusion is obvious: if the Cape is to survive ecologically and financially, immediate steps must be taken to preserve our *only* water supply.

C. These recommendations offer a starting point for effective action:

    1. Restrict population density in all Cape towns by creating larger building-lot requirements.

    2. Keep strict watch on proposed high-density apartment and condominium projects.

    3. Create a committee in each town to educate residents about conserving water in order to reduce the draw on town wells.

    4. Prohibit salt, calcium, and other additives in sand spread on snow-covered roads.

    5. Identify alternatives to landfill dumps for solid-waste disposal.

D. This crucial issue deserves the attention of every Cape resident.

E. Question-and-answer session.

Each sentence in this outline is a topic sentence for a paragraph the speaker will develop in detail. (Review outlining in Chapter 11.)

Identify areas in your outline where visuals are needed (Chapter 14). Flip charts and transparencies are especially effective for small (classroom-size) audiences.

Transfer your outline to notecards, which you can hold in one hand and shuffle as needed. Or insert the outline pages in a looseleaf binding for easy flipping. For either method, type or print clearly, leaving enough white space so that you can locate material at a glance.

Visuals. Well-chosen visuals increase audience interest. But visuals are no substitute for your report. Select visuals that will clarify and enhance your talk—without making you fade into the background.

Use visuals whenever possible to emphasize a point, when *showing* would be more effective than merely *telling*. As you plan visuals for your presentation, ask these questions:

1. What points do I want to highlight? What visual will provide the best emphasis in each case?

2. How large should the visual(s) be, to be seen by everyone?

3. What·hardware is available (slide projector, opaque projector, overhead projector, film projector, videotape player, terminal with large-screen monitor)? What graphics programs are available? How far in advance do I have to request this equipment?

4. Can I make (or have made) drawings, charts, graphs, or maps as needed? Can transparencies (for overhead projection) be made or slides collected? Do handouts have to be typed and reproduced?

Follow these suggestions for using visuals in your oral report:

- Keep visuals simple. The audience has no time to study each one. Adjust amount and type of detail to the audience's level of technicality.

- Limit their number (no more than four or five in a twenty-minute talk). Otherwise, your presentation may seem like a media event. Be selective about what you choose to highlight with visuals.

- Try not to display your visuals until you refer to them during your presentation; otherwise, the audience may be distracted.

- Interpret each visual as you present it.

- Stand aside when discussing a visual, so that everyone can see it. Don't turn your back on the audience.

- After discussing the visual, remove it, to refocus attention on you.

- If you plan to do drawings on a chalkboard, do them beforehand (in multicolored chalk). Otherwise, your audience will be twiddling their thumbs while you draw away.

- If you want the audience to remember material, prepare handouts for distribution *after* your talk.

- Check the room beforehand to make sure you have adequate space, electrical outlets, and furniture for your equipment. If you will be addressing a large audience by microphone, and plan to point to features on your visuals, be sure the microphone is movable. Don't forget a pointer if you need one.

Length of Delivery.   Unless you have reason to do otherwise, aim for a maximum length of twenty minutes. Longer talks may cause your audience to lose attention. Time yourself in practice sessions, and trim as needed.

Practice Your Delivery

Hold several practice sessions to learn the geography of your report. Then you won't fumble your actual delivery.

Feedback.   Try to practice at least once before friends. Or use a full-length mirror and a tape recorder. Assess your delivery from listeners' comments or from your taped voice (which will sound high to you). Revise as needed.

Organization.   Is your delivery unified, coherent, and logical? Does it hang together? Will the audience be able to follow your reasoning? Do you use enough transitional statements to reinforce the connections between ideas?

Tone.   Maintain a conversational tone. Speaking as you do in the classroom should be appropriate.

Anticipate Audience Questions

Consider those parts of your presentation that listeners might question or challenge. You might need to clarify or justify information that is new, controversial, disappointing, or surprising. Be prepared to support all conclusions, recommendations, assertions, and opinions. Repeat every question, to ensure that all audience members have heard it. If someone tries to engage you in lengthy debate, offer to continue the discussion *after* the presentation.

PRESENTING THE REPORT

Present your report according to the following guidelines:

Use Natural Movements and Postures

Move and gesture as you normally would in conversation, and maintain reasonable postures.

Speak with Conviction and Authority

Show your audience you believe what you are saying. Avoid needless qualifiers ("I suppose," "I'm not sure," "maybe"). Clean up verbal tics ("er," "ah," "uuh,"). Speaking with authority, however, is not the same as speaking like an authoritarian; no sermons, please.

Adjust Volume, Pronunciation, and Rate

When using a microphone, try not to speak too loudly. Without a microphone, try not to speak too softly. Ask your audience about the sound and speed of your delivery after a few sentences.

Because nervousness can cause rapid speech and unclear or slurred pronunciation, pay close attention to your pace and pronunciation. Usually, the rate you feel is a bit slow will be just about right for your audience.

Maintain Eye Contact

Eye contact is vital in relating to your audience. With a small audience, eye contact is one of your best connectors. As you speak, establish eye contact with as many members of your audience as possible. With a large group, maintain eye contact with those in the first rows.

Read Audience Feedback

Assess your audience's feedback continually and make adjustments as needed. If you are laboring through a long list of facts or figures and people begin to doze or fidget, you might summarize. Likewise, if frowns, raised

Oral Report Peer Evaluation for (name/topic) _____

Comments

Content

☐ Began with a clear purpose.   _____

☐ Showed command of the material.   _____

☐ Supported assertions with evidence. _____

☐ Used adequate and appropriate visuals

☐ Used material suited to this audience's
   needs, knowledge, concerns, and
   interests.   _____

☐ Acknowledged opposing views.   _____

☐ Gave the right amount of information. _____

Organization

☐ Presented a clear line of reasoning. _____

☐ Used transitions effectively.   _____

☐ Avoided needless digressions.   _____

☐ Summarized before concluding.   _____

☐ Was clear about what the audience
   should think or do.   _____

Style

☐ Seemed confident, relaxed, and
   likable.   _____

☐ Seemed in control of the speaking
   situation.   _____

☐ Showed appropriate enthusiasm.   _____

☐ Pronounced, enunciated, and spoke
   well.   _____

☐ Used appropriate gestures, tone,
   volume, and delivery rate.   _____

☐ Had good posture and eye contact.   _____

☐ Answered questions concisely and
   convincingly.   _____

Overall professionalism: Superior _____ **Acceptable** _____ **Needs work** _____

Evaluator's signature: _____

FIGURE 24.1 Peer Evaluation Sheet for Oral Reports

eyebrows, or questioning looks indicate confusion, skepticism, or indignation, you can backtrack with a specific example or explanation. By tuning in to your audience's reactions, you can avoid leaving them confused, hostile, or simply bored.

Be Concise

Say what you came to say, then summarize and close—politely and on time. Don't punctuate your speech with "clever" digressions that pop into your head. Unless a specific anecdote was part of your original plan to clarify a point or increase interest, avoid excursions. We often tend to be more interested than our listeners in what we have to say!

Conclude Emphatically

Before ending, summarize the major points and re-emphasize anything of special importance. Tell the audience clearly what you want them to be doing or thinking or feeling at the end. Are they supposed to remember something, have a different attitude, take a specific action? Let them know! As you conclude, thank your listeners.

Allow Time for Questions and Answers

As you begin, inform your audience that a question-and-answer period will follow. Announce a specific time limit (such as ten minutes), to avoid public debate. Then you can end the session gracefully, without making anyone feel cut off or excluded. Don't be afraid to admit ignorance. If you can't answer a question, say so, and move to the next question. End the session by saying, "We have time for one more question," or the like.

PEER EVALUATION SHEET FOR ORAL REPORTS

Use the oral report checklist in Figure 24.1 to evaluate your own practice sessions and presentations by your classmates.

EXERCISES

1. In a memo to your instructor, identify and discuss the kinds of oral reporting duties you expect to encounter in your career.

2. Design an oral report for your writing class. (Base it on a written report.) Make a sentence outline, and include at least two visuals. Practice with a tape recorder or a friend. Use the Figure 24.1 checklist to evaluate your delivery.

3. Observe a lecture or speech, and evaluate it according to the checklist (Figure 24.1). Write a memo to your instructor (without naming the speaker), identifying strong and weak areas and suggesting improvements.

4. In an oral report to the class, present your findings, conclusions, and recommendations from the analytical report assignment in Chapter 23, #2.

Appendix A
Review of Grammar, Usage, and Mechanics

■   ■   ■

No matter how vital and informative a message may be, its credibility is damaged by basic errors. Any of these errors—an illogical, fragmented, or run-on sentence; faulty punctuation; or a poorly chosen word—stands out and mars otherwise good writing. Not only do such errors annoy the reader, but they speak badly for the writer's attention to detail. Your career will make the same demands for good writing that your English classes do. The difference is that evaluation (grades) in professional situations usually shows in promotions, reputation, and salary.

COMMON SENTENCE ERRORS

Any piece of writing is only as good as each of its sentences. Here are common sentence errors, with suggestions for easy repairs.

Sentence Fragment

A sentence expresses a logically complete idea. Any complete idea must have a subject and a verb and must not depend on another complete idea to make sense. Your sentence *might* contain several complete ideas, but it *must* contain at least one!

Although Mary was nervous, she grabbed the line, and she saved the sailboat.
(incomplete idea)   *(complete idea)*   *(complete idea)*

Symbol	Meaning	Page	Symbol	Meaning	Page
ab	abbreviation	626	– – /	dashes	621
agr p	pronoun/referent agreement	601	. . . /	ellipses	619
agr sv	subject/verb agreement	600	! /	exclamation point	609
			– /	hyphen	621
amb	ambiguity	251	*ital*	italics	619
appr	inappropriate diction	292	( ) /	parentheses	620
bias	biased tone	295	. /	period	609
ca	pronoun case	602	? /	question mark	609
cap	capitalization	629	" / "	quotation	618
chop	choppy sentences	274	; /	semicolon	609
cl	clutter word	270	*qual*	needless qualifier	271
coh	paragraph coherence	243	*red*	redundancy	265
cont	contraction	617	*rep*	needless repetition	265
coord	coordination	597	*ref*	faulty reference	602
cs	comma splice	596	*ro*	run-on sentence	597
dgl	dangling modifier	604	*seq*	sequence of development in a paragraph	244
euph	euphemism	283			
exact	inexact word	285	*sexist*	sexist usage	297
frag	sentence fragment	592	*shift*	sentence shift	607
gen	generalization	284	*st mod*	stacked modifiers	254
jarg	needless jargon	281	*str*	paragraph structure	238
len	paragraph length	248	*sub*	subordination	598
lev	level of technicality	20	*th op*	"th" sentence openers	266
mng	meaning unclear	251	*trans*	transition	623
mod	misplaced modifier	604	*trite*	triteness	283
noun ad	noun addiction	268	*un*	paragraph unity	242
om	omitted word	253	*v*	voice	256
over	overstatement	284	*var*	sentence variety	276
par	parallelism	606	*w*	wordiness	263
pct	punctuation	608	*wo*	word order	604
ap/	apostrophe	616	*ww*	wrong word	285
[ ]/	brackets	620	#	numbers	630
: /	colon	610	¶	begin new paragraph	238
, /	comma	611	*ts*	topic sentence	239

TABLE A-1 Correction Symbols

If the idea is not complete—if your reader is left wondering what you mean—you probably have omitted some essential element (the subject, the verb, or another complete idea). Such a piece of a sentence is a *fragment*.

> Grabbed the line. *(a fragment because it lacks a subject)*
>
> Although Mary was nervous. *(a fragment because—although it has a subject and a verb—it needs to be joined with a complete idea)*

The only exception to the sentence rule applies when we give a command (Run!), in which the subject (you) is understood. Logically complete, this statement is properly called a sentence. So is this one:

> Sam is an electronics technician.

Readers cannot miss your meaning: somewhere is a person; the person's name is Sam; the person is an electronics technician. Suppose instead we write:

> Sam an electronics technician.

This statement is not logically complete, therefore not a sentence. The reader is left asking, "What about Sam the electronics technician?" The verb—the word that makes things happen—is missing. By adding a verb we can easily change this fragment to a complete sentence.

Simple verb	Sam **is** an electronics technician.
Adverb	Sam, an electronics technician, **works hard**.
Dependent clause, verb, and subjective complement	**Although he is well paid**, Sam, an electronics technician, **is not happy**.

Do not, however, mistake the following statement—which seems to have a verb—for a complete sentence:

> Sam being an electronics technician.

Such "ing" forms do not function as verbs unless accompanied by such other verbs as **is, was,** and **will be**. Again, readers are confused unless you complete your idea with an independent clause.

> **Sam**, being an electronics technician, checked all **circuitry**.

Likewise, remember that the "to + verb" form does not function as a verb.

> To become an electronics technician.

The meaning is unclear unless you complete the thought.

> To become an electronics technician, **Sam had to pass an exam**.

Sometimes we inadvertently create fragments by adding certain words (**because, since, if, although, while, unless, until, when, where,** and others) to an already complete sentence, transforming our independent clause (complete sentence) to a dependent clause.

> **Although** Sam is an electronics technician.

Such words subordinate the words that follow them so that an additional idea is needed to make the first statement complete. That is, they make the statement dependent on an additional idea, which must itself have a subject and a verb and be a complete sentence. (See "Faulty Subordination.") Now we have to round off the statement with a complete idea (independent clause).

> Although Sam is an electronics technician, **he hopes to be an engineer**.

Note: Be careful not to use a semicolon or a period, instead of a comma, to separate elements in the preceding sentence. Because the dependent clause depends on the independent clause for its meaning, you need only a *pause* (symbolized by a comma), not a *break* (symbolized by a semicolon), between these ideas. In fact, many fragments are created when too strong a mark of punctuation (period or semicolon) severs the connection between a dependent and an independent clause. (See the later discussion of punctuation.)

Here are some fragments from technical documents. Each is repaired in more than one way. Can you think of other ways of making these statements complete?

Fragment	She spent her first week on the job as a researcher. **Compiling information from digests and journals**.
Correct	She spent her first week on the job as a researcher, compiling information from digests and journals.
	She spent her first week on the job as a researcher. She compiled information from digests and journals.

Fragment	**Because the operator was careless**. The new computer was damaged.
Correct	Because the operator was careless, the new computer was damaged.
	The operator's carelessness resulted in damage to the new computer.

Comma Splice

In a comma splice, two complete ideas (independent clauses), which should be *separated* by a period or a semicolon, are incorrectly *joined* by a comma:

Comma splice	Sarah did a great job, she was promoted.

You can choose among several possibilities for correcting this error:

1. Substitute a period followed by a capital letter:
   Sarah did a great job. She was promoted.

2. Substitute a semicolon to signal the relationship:
   Sarah did a great job; she was promoted.

3. Use a semicolon with a connecting adverb (a transitional word):
   Sarah did a great job; **consequently**, she was promoted.

4. Use a subordinating word to make the less important sentence incomplete, thereby dependent on the other:
   Because Sarah did a great job, she was promoted.

5. Add a connecting word after the comma:
   Sarah did a great job, **and** she was promoted.

Your choice of construction will depend, of course, on the exact meaning or tone you wish to convey. The following comma splice can be repaired in the ways described above.

Comma splice	This is a fairly new technique, therefore, some people don't trust it.
Correct	This is a fairly new technique. Some people don't trust it.
	This is a fairly new technique; therefore, some people don't trust it.

> **Because** this is a fairly new technique, some people don't trust it.

> This is a fairly new technique, and **so** some people don't trust it.

Run-on Sentence

The run-on sentence, a cousin to the comma splice, crams too many ideas without needed breaks or pauses.

> **Run-on** The hourglass is more accurate than the water clock because water in a water clock must always be at the same temperature to flow at the same speed since water evaporates and must be replenished at regular intervals, thus being less effective than the hourglass for measuring time.

Like a runaway train, this statement is out of control. Here is a corrected version:

> **Revised** The hourglass is more accurate than the water clock because water in a water clock must always be at the same temperature to flow at the same speed. Also, water evaporates and must be replenished at regular intervals. These temperature and volume problems make the water clock less effective than the hourglass for measuring time.

Faulty Coordination

Give equal emphasis to ideas of equal importance by joining them with coordinating conjunctions: **and**, **but**, **or**, **nor**, **for**, **so** and **yet**.

> This course is difficult **but** worthwhile.

> My horse is old **and** gray.

> We must decide to support **or** reject the dean's proposal.

But do not confound your meaning by coordinating excessively.

> **Excessive coordination** The climax in jogging comes after a few miles **and** I can no longer feel stride after stride **and** it seems as if I am

floating **and** jogging becomes almost a reflex **and** my arms **and** legs continue to move **and** my mind no longer has to control their actions.

Revised   The climax in jogging comes after a few miles when I can no longer feel stride after stride. By then I am jogging almost by reflex, nearly floating, my arms and legs still moving, my mind no longer having to control their actions.

Avoid coordinating ideas that cannot be sensibly connected:

Faulty   John had a drinking problem and he dropped out of school.

Revised   John's drinking problem depressed him so much that he couldn't study and so he quit school.

Faulty   I was late for work and wrecked my car.

Revised   Late for work, I backed out of the driveway too quickly, hit a truck, and wrecked my car.

Instead of *try and,* use *try to.*

Faulty   I will try and help you.

Revised   I will try to help you.

Faulty   Bill promised to try and be on time.

Revised   Bill promised to try to be on time.

Faulty Subordination

Proper subordination shows that a less important idea is dependent on a more important idea. By using subordination you can combine simple sentences into complex sentences, and emphasize the most important idea. Consider these two ideas:

Joe studies hard. He has a learning disability.

Because these ideas are expressed as simple sentences, they appear to be coordinate (equal in importance). But if you wanted to indicate your opinion of Joe's chances of succeeding, you would need a third sentence: "His disability probably will prevent him from succeeding"; or "His willpower will help him succeed." An easier and more concise way to communicate the intended meaning is by combining ideas and subordinating the one that deserves less emphasis:

> Despite his learning disability *(subordinate idea)*, Joe studies hard *(independent idea)*.

This first version implies that Joe will succeed. Below, subordination is used to imply the opposite meaning:

> Despite his diligent studying *(subordinate idea)*, Joe has a learning disability *(independent idea)*.

A dependent (or subordinate) clause in a sentence is signalled by a subordinating conjunction: **because**, **so**, **if**, **unless**, **after**, **until**, **since**, **while**, **as**, and **although**, among others. Be sure to place the idea you want emphasized in the independent clause; don't write

> Although Mary is receiving excellent medical treatment, she is seriously ill.

if you mean to suggest that Mary has a good chance of recovering.

Do not coordinate when you should subordinate:

> **Weak**   Television viewers can relate to an athlete they idolize and they feel obliged to buy the product endorsed by their hero.

Of the two ideas in the sentence above, one is the cause, the other the effect. Emphasize this relationship through subordination:

> **Revised**   Because television viewers can relate to an athlete they idolize, they feel obliged to buy the product endorsed by their hero.

When combining several ideas within a sentence, decide which is most important, and subordinate the other ideas to it—don't simply coordinate:

> **Faulty**   This employee is often late for work, and he writes illogical reports, and he is a poor manager, and he should be fired.

> **Revised**   Because this employee is often late for work, writes illogical reports, and has poor management skills, **he should be fired**. *(The last clause is independent.)*

Do not overstuff sentences by subordinating excessively:

Overstuffed This job, which I took when I graduated from college, while I waited for a better one to come along, which is boring, where I've gained no useful experience, makes me anxious to quit.

Revised Upon college graduation, I took this job while waiting for a better one to come along. Because I find it boring, and have gained no useful experience, I am eager to quit.

Faulty Agreement—Subject and Verb

The subject should agree in number with the verb. We are not likely to use faulty agreement in short sentences, where subject and verb are not far apart. Thus, we are not likely to say "Jack eat too much" instead of "Jack eats too much," but in more complicated sentences—those in which the subject is separated from its verb by other words—we sometimes lose track of the subject-verb relationship.

Faulty The lion's **share** of diesels **are** sold in Europe.

Although **diesels** is closest to the verb, the subject is **share**, a singular subject that must agree with a singular verb.

Correct The lion's **share** of diesels **is** sold in Europe.

Agreement errors are easy to correct when subject and verb are identified.

Faulty A **system** of lines **extend** horizontally to form a grid.

Correct A **system** of lines **extends** horizontally to form a grid.

A second problem with subject-verb agreement occurs when we use indefinite pronouns such as **each**, **everyone**, **anybody**, and **somebody**. They function as subjects and usually take a singular verb.

Faulty **Each** of the crew members **were** injured.

Correct **Each** of the crew members **was** injured.

Faulty **Everyone** in the group **have** practiced long hours.

Correct **Everyone** in the group **has** practiced long hours.

Agreement problems can be caused by collective nouns such as **herd, family, union, group, army, team, committee**, and **board**. They can call for a singular or plural verb—depending on your intended meaning. When denoting the group as a whole, use a singular verb.

Correct   The **committee meets** weekly to discuss new business.

The editorial **board** of this magazine **has** high standards.

To denote individual members of the group, however, use a plural verb.

Correct   Not all members of the editorial **board are** published authors.

Yet another problem occurs when two subjects are joined by **either . . . or** or **neither . . . nor**. Here, the verb is singular if both subjects are singular, and plural if both subjects are plural. If one subject is plural and one is singular, the verb agrees with the subject closer to the verb.

Correct   Neither **John** nor **Bill works** regularly.

Either apples or **oranges are** good vitamin sources.

Either Felix or his **friends are** crazy.

Neither the boys nor their **father likes** the home team.

If, on the other hand, two subjects (singular, plural, or mixed) are joined by **both . . . and,** the verb will be plural. Whereas **or** suggests "one or the other," **and** announces a combination of the two subjects, thereby requiring a plural verb.

Correct   **Both** Joe and Bill **are** resigning.

The **book and** the **briefcase appear** expensive.

A single *and* between singular subjects makes for a plural subject.

Faulty Agreement—Pronoun and Referent

A pronoun can make sense only if it refers to a specific noun (its referent or antecedent), with which it must agree in gender and number.

Correct   **Joe** lost **his** blueprints.

The **workers** complained that **they** were treated unfairly.

Some instances, however, are not so obvious. When an indefinite pronoun such as **each**, **everyone**, **anybody**, **someone**, and **none** serves as the pronoun referent, the pronoun itself is singular.

Correct	**Anyone** can get **his** degree from that college.
	Anyone can get **his or her** degree from that college.
	Each candidate described **her** plans in detail.

Faulty or Vague Pronoun Reference

Whenever a pronoun is used, it must refer to one clearly identified referent; otherwise, your message will be confusing.

Ambiguous	**Sally** told **Sarah** that **she** was obsessed with her job.
Correct	Sally told Sarah, "I'm obsessed with my job."
	Sally told Sarah, "I'm obsessed with your job."
	Sally told Sarah, "You're obsessed with [your, my] job."
	Sally told Sarah, "She's obsessed with [her, my, your] job."

Avoid using **this**, **that**, or **it**—especially to begin a sentence—unless the pronoun refers to a specific antecedent (referent).

Vague	He drove away from his menial **job**, boring **lifestyle**, and damp **apartment**, happy to be leaving **it** behind.
Correct	He drove away, happy to be leaving behind his menial job, boring lifestyle, and damp apartment.
Vague	The problem with our **defective machinery** is compounded by the **operator's incompetence**. **That** annoys me!
Correct	I am annoyed by the problem with our defective machinery as well as by the new operator's incompetence.

Faulty Pronoun Case

A pronoun's case (nominative, objective, or possessive) is determined by its role in a sentence: as subject, object, or indicator of possession.

If the pronoun serves as the subject of a sentence (**I**, **we**, **you**, **she**, **he**, **it**, **they**, **who**), its case is *nominative*.

She completed her graduate program in record time.

Who broke the chair?

When a pronoun follows a version of **to be** (a linking verb), it explains (complements) the subject, and so its case is nominative.

It was **she**.

The chemist who perfected our new distillation process is **he**.

If the pronoun serves as the object of a verb or a preposition (**me, us, you, her, him, it, them, whom**), its case is *objective*.

Object of the verb	The employees gave **her** a parting gift.
Object of the preposition	Several colleagues left with **him**.
	To **whom** do you wish to complain?

If a pronoun indicates possession (**my, mine, ours, your, yours, his, her, hers, its, their, theirs, whose**), its case is *possessive*.

The brown briefcase is **mine**.

Her offer was accepted.

Whose opinion do you value most?

Here are some frequent errors in pronoun case:

Faulty	**Whom** is responsible to **who**? *(The subject should be nominative and the object should be objective.)*
Correct	**Who** is responsible to **whom**?
Faulty	The debate was between Marsha and **I**. *(As object of the preposition, the pronoun should be objective.)*
Correct	The debate was between Marsha and **me**.
Faulty	**Us** board members are accountable for our decisions. *(The pronoun accompanies the subject, "board members," and thus should be nominative.)*
Correct	**We** board members are accountable for our decisions.
Faulty	A group of **we** managers will fly to the convention. *(The pronoun accompanies the object of the preposition, "managers," and thus should be objective.)*
Correct	A group of **us** managers will fly to the convention.

Hint: By deleting the accompanying noun from the two latter examples, we can easily identify the correct pronoun case ("We . . . are accountable . . . "; "A group of us . . . will fly . . . ").

Faulty Modification

A sentence's word order (syntax) helps determine its effectiveness and meaning. Words or groups of words are modified (i.e., explained or defined) by adjectives, adverbs, phrases, or clauses. Modifiers explain, define, or add detail to other words or ideas. Prepositional phrases, for example, usually define or limit adjacent words:

> the foundation **with the cracked wall**
>
> the repair job **on the old Ford**
>
> the journey **to the moon**

As do phrases with "-ing" verb forms:

> the student **painting the portrait**
>
> **Opening the door**, we entered quietly.

Phrases with "to + verb" form limit:

> **To succeed**, one must work hard.

Some clauses limit:

> the person **who came to dinner**

Problems with word order occur when a modifying phrase begins a sentence, and has no word to modify.

> **Dangling modifier**   **Dialing the phone**, the cat ran out the open door.

The cat obviously did not dial the phone. But because the modifier **Dialing the phone** has no word to modify, the word order suggests that the noun beginning the main clause (*cat*) names the one who dialed the phone. Without any word to join itself to, the modifier *dangles*. By inserting a subject, we can repair this absurd message.

> **Correct**   As Joe dialed the phone, the cat ran out the open door.

A dangling modifier also can obscure your meaning.

| Dangling modifier | **After completing the student financial aid application form**, the Financial Aid Office will forward it to the appropriate state agency. |

Who completes the form—the student or the financial aid office?

Here are some other dangling modifiers that make the message confusing, inaccurate, or downright absurd:

Dangling modifier	**While walking**, a cold chill ran through my body.
Correct	While **I** walked, a cold chill ran through my body.
Dangling modifier	Impurities have entered our bodies **by eating chemically processed foods**.
Correct	Impurities have entered our bodies by **our** eating chemically processed foods.
Dangling modifier	**By planting different varieties of crops**, the pests were unable to adapt.
Correct	By planting different varieties of crops, **farmers** prevented the pests from adapting.

The order of adjectives and adverbs in a sentence is as important as the order of modifying phrases and clauses. Notice how changing word order affects the meaning of these sentences:

I **often** remind myself of the need to balance my checkbook.
I remind myself of the need to balance my checkbook **often**.

Be sure that modifiers and the words they modify follow an order that reflects your meaning.

Misplaced modifier	Joe typed another memo on our computer **that was useless**. *(Was the typewriter or the memo useless?)*
Correct	Joe typed another useless memo on our computer.
	or
	Joe typed another memo on our useless computer.
Misplaced modifier	He read a report on the use of nonchemical pesticides **in our conference room**. *(Are the pesticides to be used in the conference room?)*
Correct	In our conference room, he read a report on the use of nonchemical pesticides.

Misplaced modifier	She volunteered **immediately** to deliver the radioactive shipment. *(Volunteering immediately, or delivering immediately?)*
Correct	She immediately volunteered to deliver . . . *or* She volunteered to deliver immediately . . .

Faulty Parallelism

To reflect relationships among items of equal importance, express them in identical grammatical form:

Correct	We here highly resolve . . . that government **of the people, by the people, for the people** shall not perish from the earth.

The statement above describes the government with three modifiers of equal importance. Because the first modifier is a prepositional phrase, the others must be also. Otherwise, the message would be garbled, like this:

Faulty	We here highly resolve . . . that government **of the people, which the people created and maintain, serving the people** shall not perish from the earth.

If you begin the series with a noun, use nouns throughout the series; likewise for adjectives, adverbs, and specific types of clauses and phrases.

Faulty	The new apprentice is **enthusiastic, skilled**, and **you can depend on her**.
Correct	The new apprentice is **enthusiastic, skilled**, and **dependable**. *(all subjective complements)*
Faulty	In his new job, he felt **lonely** and **without a friend**.
Correct	In his new job, he felt **lonely** and **friendless**. *(both adjectives)*
Faulty	She plans **to study** all this month and **on scoring well** in her licensing examination.
Correct	She plans **to study** all this month and **to score well** in her licensing examination. *(both infinitive phrases)*
Faulty	She **sleeps** well and **jobs** daily, **as well as eating** high-protein foods.

Correct	She **sleeps** well, **jogs** daily, and **eats** high-protein foods. *(all verbs)*

To improve coherence in long sentences, repeat words that introduce parallel expressions:

Faulty	Before buying this property, you should decide whether you will settle down and raise a family, travel for a few years, or pursue a graduate degree.
Correct	Before buying this property, you should decide whether **to settle** down and raise a family, **to travel** for a few years, or **to pursue** a graduate degree.

Be sure that all headings at a given level are parallel in your table of contents, your formal outline, and your report.

Faulty

A. Picking the Fruit
 1. When to pick
 2. Packing
 3. Suitable temperature
 4. Transport with care

The logical connection between steps in that sequence is obscured because each heading is phrased in a different grammatical form.

Correct

A. Picking the Fruit
 1. Choose the best time
 2. Pack the fruit loosely
 3. Store at a suitable temperature
 4. Transport with care

Other forms of phrasing also would be correct here, as long as each item is expressed in parallel form.

Sentence Shifts

Shifts in point of view damage coherence. If you begin a sentence or paragraph with one subject or person, do not shift to another.

Shift in person	When **you** finish the job, **one** will have a sense of pride.
Correct	When **you** finish the job, **you** will have a sense of pride.

Shift in number	**One** should sift the flour before **they** make the pie.
Correct	**One** should sift the flour before **one** makes the pie. (Or better: Sift the flour before making the pie.)

Do not begin a sentence in the active voice and then shift to passive.

Shift in voice	**He delivered** the plans for the apartment complex, and the building site **was also inspected by him**.
Correct	**He delivered** the plans for the apartment complex and also **inspected** the building site.

Do not shift tenses without good reason.

Shift in tense	She **delivered** the blueprints, **inspected** the foundation, **wrote** her report, and **takes** the afternoon off.
Correct	She **delivered** the blueprints, **inspected** the foundation, **wrote** her report, and **took** the afternoon off.

Do not shift from one mood to another (as from imperative to indicative mood in a set of instructions).

Shift in mood	**Unscrew** the valve and then steel wool **should be used** to clean the fitting.
Correct	**Unscrew** the valve and then **use** steel wool to clean the fitting.

Do not shift from indirect to direct discourse within a sentence.

Shift in discourse	Jim wonders **if he will get the job** and **will he like it**?
Correct	Jim wonders **if he will get the job** and **if he will like it**.
	Will Jim get the job, and will he like it?

EFFECTIVE PUNCTUATION

Punctuation marks are like road signs and traffic signals. They govern reading speed and provide clues for navigation through a network of ideas; they mark intersections, detours, and road repairs; they draw attention to points of interest along the route; and they mark geographic boundaries. In short, punctuation marks give us a simple way of making ourselves understood.

End Punctuation

The three marks of end punctuation—period, question mark, and exclamation point—work like a red traffic light by signaling a complete stop.

Period. A period ends a sentence. Periods end some abbreviations.

Ms.	Assn.	Dr.
M.D.	Inc.	B.A.

Periods serve as decimal points for figures.

$15.95
2.14%

Question Mark. A question mark follows a direct question.

> Where is the balance sheet?

Do not use a question mark to end an indirect question.

> **Faulty** He asked if all students had failed the test?
> **Correct** He asked if all students had failed the test.
> *or*
> He asked, "Did all students fail the test?"

Exclamation Point. Because exclamation points symbolize strong feeling, don't overuse them. Otherwise you might seem hysterical or insincere.

> **Correct** Oh, no!
> Pay up!

Use an exclamation point only when expression of strong feeling is appropriate.

Semicolon

A semicolon usually works like a blinking red traffic light at an intersection by signaling a brief but definite stop.

Semicolon Separating Independent Clauses.   Semicolons separate independent clauses (logically complete ideas), whose contents are closely related and are not connected by a coordinating conjunction.

> The project was finally completed; we had done a good week's work.

The semicolon can replace the conjunction-comma combination that joins two independent ideas.

> The project was finally completed, and we were elated.
>
> The project was finally completed; we were elated.

The second version emphasizes the sense of elation.

Semicolons Used with Adverbs as Conjunctions and Other Transitional Expressions.   Semicolons accompany conjunctive adverbs and other expressions that connect related independent ideas (**besides**, **otherwise**, **still**, **however**, **furthermore**, **consequently**, **therefore**, **in contrast**, **in fact**, or the like).

> The job is filled; however, we will keep your résumé on file.
>
> Your background is impressive; in fact, it is the best among our applicants.

Semicolons Separating Items in a Series.   When items in a series contain internal commas, semicolons provide clear separation between items.

> We are opening branch offices in the following cities: Santa Fe, New Mexico; Albany, New York; Montgomery, Alabama; and Moscow, Idaho.
>
> Members of the survey crew were John Jones, a geologist; Hector Lightweight, a draftsman; and Mary Shelley, a graduate student.

Colon

Like a flare in the road, a colon signals you to stop and then proceed, paying attention to the situation ahead, the details of which will be revealed as you move along. Usually, a colon follows an introductory statement that requires a follow-up explanation.

> We need the following equipment immediately: a voltmeter, a portable generator, and three pairs of insulated gloves.
>
> She is an ideal colleague: honest, reliable, and competent.
>
> Two candidates clearly are superior: John and Marsha.

With the exception of **Dear Sir:** and other salutations in formal correspondence, colons follow independent (logically and grammatically complete) statements. Because colons, like end punctuation and semicolons, signal a full stop, they never are used to fragment a complete statement.

> **Faulty**   My plans include: finishing college, traveling for two years, and settling down in Boston.

No punctuation should follow "include."

Colons can introduce quotations.

> The supervisor's message was clear enough: "You're fired."

A colon normally replaces a semicolon in separating two related, complete statements when the second statement explains or amplifies the first.

> His reason for accepting the lowest-paying job offer was simple: he had always wanted to live in the Northwest.

The statement following the colon explains the "reason" mentioned in the statement preceding the colon.

Comma

The comma is the most frequently used—and abused—punctuation mark. Unlike the period, semicolon, and colon, which signal a full stop, the comma signals a *brief pause*. The comma works like a blinking yellow traffic light, for which you slow down without stopping. Never use a comma to signal a *break* between independent ideas; it is not strong enough.

Comma as a Pause between Complete Ideas.   In a compound sentence where a coordinating conjunction (**and**, **or**, **nor**, **for**, **but**) connects equal (independent) statements, a comma usually precedes the conjunction.

> This is a high-paying job, but the stress is high.
> This vacant shop is just large enough for our boutique, and the location is excellent for walk-in customer traffic.

Without the conjunction, these statements would suffer from a comma splice, unless the comma were replaced by a semicolon or period.

Comma as a Pause Between an Incomplete and a Complete Idea.   A comma usually appears between a complete and an incomplete statement in a complex sentence to show that the incomplete statement depends for its meaning on the complete statement. (The incomplete statement cannot stand alone, separated by a break such as a semicolon, colon, or period.)

Because he is a fat cat, Jack diets often.

When he eats too much, Jack gains weight.

Above, the first idea is made incomplete by a subordinating conjunction (**since**, **when**, **because**, **although**, **where**, **while**, **if**, **until**), which here connects a dependent with an independent statement. The first (incomplete) idea depends on the second (complete) for wholeness. When the order is reversed (complete idea followed by incomplete), the comma usually is omitted.

Jack diets often **because he is a fat cat**.

Jack gains weight **when he eats too much**.

Because commas take the place of speech signals, reading a sentence aloud should tell you whether or not to pause (and use a comma).

Commas Separating Items (Words, Phrases, or Clauses) in a Series.
Use a comma to separate items in a series.

Jane, Joe, Marsha, and **John** are joining us on the hydroelectric project.

The office was **yellow, orange,** and **red.**

She works hard **at home, on the job,** and even **during her vacation.**

The employee claimed **that the hours were long, that the pay was low, that the work was boring,** and **that the foreman was paranoid**.

Use no commas when *or* or *and* appears between all items in a series.

She is willing to work in San Francisco or Seattle or even in Anchorage.

Add a comma when *or* or *and* is used only before the final item in the series.

Our luncheon special for Thursday will be rolls, steak, beans, and ice cream.

Without the comma, the sentence might cause readers to conclude that beans and ice cream is an exotic new dessert.

Comma Setting off Introductory Phrases.   Infinitive, prepositional, or verbal phrases introducing a sentence usually are set off by commas.

Infinitive phrase	**To be or not to be,** that is the question.
Prepositional phrase	**In Rome,** do as the Romans do.
Participial phrase	**Moving quickly,** the army surrounded the enemy.

When an interjection introduces a sentence, it is set off by a comma.

Oh, is that the final verdict?

When a noun in direct address introduces a sentence, it is set off by a comma.

Mary, you've done a great job.

Commas Setting off Nonrestrictive Elements. A restrictive phrase or clause limits or defines the subject in such a way that deleting the modifier would change the meaning of the sentence.

All candidates **who have work experience** will receive preference.

The clause, **who have work experience**, defines **candidates** and is essential to the meaning of the sentence. Without this clause, the meaning would be entirely different.

All candidates will receive preference.

This next sentence also contains a restriction:

All candidates **with work experience** will receive preference.

The phrase, **with work experience**, defines **candidates** and thus specifies the meaning of the sentence. Because this phrase *restricts* the subject by limiting the category, **candidates**, it is essential to the sentence's meaning and so is not separated from the sentence by commas.

A nonrestrictive phrase or clause does not limit or define the subject; a nonrestrictive element could be deleted without changing the basic meaning of the sentence.

Our draftsperson, **who has little experience,** is highly competent.

This house, **riddled with carpenter ants,** is falling apart.

In each of those sentences, the modifying phrase or clause does not restrict the subject; each could be deleted:

> Our new draftsperson is highly competent.

> This house is falling apart.

A nonrestrictive clause or phrase is set off from the sentence by commas.

Commas Setting off Parenthetical Elements.   Items that interrupt sentence flow are called parenthetical and are enclosed by commas. Expressions such as **of course**, **as a result**, **as I recall**, and **however** are parenthetical and may denote emphasis, afterthought, clarification, or transition.

Emphasis	This deluxe model, **of course,** is more expensive.
Afterthought	Your report format, **by the way,** was impeccable.
Clarification	The loss of my job was, **in a way,** a blessing.
Transition	Our warranty, **however,** does not cover tire damage.

Direct address is parenthetical.

> Listen, **my children,** and you shall hear . . .

A parenthetical expression at the beginning or the end of a sentence is set off by a comma.

> **Naturally,** we will expect a full guarantee.

> **My friends,** I think we have a problem.

> You've done a good job, **Jim**.

> **Yes,** you may use my name in your advertisement.

Commas Setting off Quoted Material.   Quoted items included within a sentence are often set off by commas.

> The customer said, **"I'll take it,"** as soon as he laid eyes on our new model.

Commas Setting off Appositives.   An appositive, a word or words explaining a noun and placed immediately after it, is set off by commas.

> Martha Jones, **our new president,** is overhauling all personnel policies.

Alpha waves, **the most prominent of the brain waves,** are typically recorded in a waking subject whose eyes are closed.

Please make all checks payable to Sam Sawbuck, **company treasurer.**

Commas Used in Common Practice.   Commas set off the day of the month from the year, in a date.

> May 10, 1984

They set off numbers in three-digit intervals.

> 11,215
> 6,463,657

They set off street, city, and state in an address.

> The bill was sent to John Smith, 184 Sea Street, Albany, NY 01642.

When the address is written vertically, however, the omitted commas are those which would otherwise occur at the end of each address line.

> John Smith
> 184 Sea Street
> Albany, NY 01642

Commas set off an address or date in a sentence.

> Room 3C, Margate Complex, is the site of our newest office.
> December 15, 1997, is my retirement date.

They set off degrees and titles from proper names.

> Roger P. Cayer, M.D.
> Gordon Browne, Jr.
> Sandra Mello, Ph.D.

Commas Used Erroneously.   Avoid needless or inappropriate commas. You are probably safer using too few commas than too many. Reading sentences aloud is one way to identify inappropriate pauses.

> Faulty   As I opened the door, he told me, that I was late. *(separates the indirect from the direct object)*

The universal symptom of the suicide impulse, is depression. *(separates the subject from its verb)*

This has been a long, difficult, project. *(separates the final adjective from its noun)*

John, Bill, and Sally, are joining us on the design phase of this project. *(separates the final subject from its verb)*

An employee, who expects rapid promotion, must quickly prove her or his worth. *(separates a modifier that should be restrictive)*

I spoke in a conference call with John, and Marsha. *(separates two words linked by a coordinating conjunction)*

The room was, eighteen feet long *(separates the linking verb from the subjective complement)*

We painted the room, red. *(separates the object from its complement)*

Apostrophe

Apostrophes serve three purposes: to indicate the possessive, a contraction, and the plural of numbers, letters, and figures.

Apostrophe Indicating the Possessive. At the end of a singular word, or of a plural word that does not end in *s,* add an apostrophe plus *s* to indicate the possessive. Single-syllable nouns that end in *s* take the apostrophe before an added *s.*

The people's candidate won.

The chain saw was Bill's.

The men's locker room burned.

I borrowed Doris's book.

Have you heard Ray Charles's song?

Do not add an *s* to words that have more than one syllable.

Correct	Aristophanes' death
	for conscience' sake

Do not use an apostrophe to indicate the possessive form of either singular or plural pronouns:

Correct	The book was hers.
	Ours is the best sales record.
	The fault was theirs.

At the end of a plural word that ends in *s,* add an apostrophe only.

Correct	the cows' water supply
	the Jacksons' wine cellar

At the end of a compound noun, add an apostrophe plus *s.*

Correct	my father-in-law's false teeth

At the end of the last word in nouns of joint possession, add an apostrophe plus *s* if both own one item.

Correct	Joe and Sam's lakefront cottage

Add an apostrophe plus *s* to both nouns if each owns specific items.

Correct	Joe's and Sam's passports

Apostrophe Indicating a Contraction. An apostrophe shows that you have omitted one or more letters in a phrase that is usually a combination of a pronoun and a verb.

Correct	I'm	they're
	he's	we're
	you're	who's

Don't confuse *they're* with *their* or *there.*

Faulty	there books
	Their now leaving.
	living their
Correct	their books
	They're now leaving.
	living there

Remember the distinction in this way:

Correct	Their boss knows they're there.

Don't confuse **it's** and **its**. **It's** means "it is." **Its** is the possessive.

Correct	It's watching its reflection in the pond.

Don't confuse **who's** and **whose**. **Who's** means "who is," whereas **whose** indicates the possessive.

Correct	Who's interrupting whose work?

Other contractions are formed from the verb and the negative.

Correct	isn't	can't
	don't	haven't
	won't	wasn't

Apostrophe Indicating the Plural of Numbers, Letters, and Figures.

The 6's on this new typewriter look like smudged G's, the 9's are illegible, and the %'s are unclear.

Quotation Marks

Quotation marks set off exact words borrowed from another speaker or writer. At the end of a quotation the period or comma is placed within quotation marks.

Correct	"Hurry up," he whispered.
	She told me, "I'm depressed."

The colon or semicolon is always placed outside quotation marks:

Correct	Our contract clearly defines "middle-management personnel"; however, it does not state salary range.
	You know what to expect when Honest John offers you a "bargain": a piece of junk.

Sometimes a question mark is used within a quotation that is part of a larger sentence. (Do not follow a question mark with a comma.)

Correct	"Can we stop the flooding?" inquired the supervisor.

When the question mark or exclamation point is part of the quotation, it belongs within quotation marks, replacing the comma or period.

Correct	"Help!" he screamed.
	He asked John, "Can't we agree about anything?"

If, however, the question mark or exclamation point is meant to denote the attitude not of the quotee but of the quoter, it is placed outside the quotation mark.

> **Correct**   Why did he wink and tell me, "It's a big secret"?
>
>   He actually accused me of being an "elitist"!

When quoting a passage of fifty words or longer, indent the entire passage five spaces and single-space between its lines to set it off from the text. Do not enclose the indented passage in quotation marks.

Use quotation marks around titles of articles, book chapters, poems, and unpublished reports.

> **Correct**   The enclosed article, "The Job Market for College Graduates," should provide some helpful insights.

The title of a published work—book, journal, newspaper, brochure, or pamphlet—should be underlined or italicized.

Finally, use quotation marks (with restraint) to indicate your ironic use of a word.

> **Correct**   He is some "friend"!

Ellipses

Three dots in a row (. . .) indicate that you have omitted some material from a quotation. If the omitted words come at the end of the original sentence, a fourth dot indicates the period. Use several dots centered in a line to indicate that a paragraph or more has been left out. Ellipses help you save time and zero in on the important material within a quotation.

> **Correct**   "Three dots. . . . indicate that you have omitted some material. . . . A fourth dot indicates the period. . . . Several dots centered in a line . . . indicate a paragraph or more. . . . Ellipses help you . . . zero in. . . ."

Italics

In typing or longhand writing, indicate italics by *underlining*. On a word processor, use italic print for titles of books, periodicals, films, newspapers, and plays; for the names of ships; for foreign words or scientific names; for emphasizing a word (used sparingly); for indicating the special use of a word.

The Oxford English Dictionary is a handy reference tool.

The *Lusitania* sank rapidly.

She reads *The Boston Globe* often.

My only advice is *caveat emptor*.

Bacillus anthracis is a highly virulent organism.

Do *not* inhale these spores, under any circumstances!

Our contract defines a *full-time employee* as one who works a minimum of thirty-five hours weekly.

Parentheses

Use commas normally to set off parenthetical elements, dashes to give some emphasis to the material that is set off, and parentheses to enclose material that defines or explains the statement that precedes it.

This organism requires an anaerobic (oxygenless) environment.

The cost of manufacturing our Beta II transistors has increased by 10 percent in one year. (See Appendix A for full cost breakdown.)

This new three-colored model (made by Ilco Corporation) is selling well.

Notice that material within parentheses, like all other parenthetical material discussed earlier, can be deleted without harming the logical and grammatical structure of the sentence.

Also, use parentheses to enclose numbers or letters that segment items of information in a series.

The three basic steps in this procedure are (1) . . . , (2) . . . , and (3). . . .

Brackets

Use brackets within a quotation to add material that was not in the original quotation but is needed for clarification. Sometimes a bracketed word will provide an antecedent (or referent) for a pronoun.

"She [Jones] was the outstanding candidate for the job."

Brackets can enclose information from some other location within the context of the quotation.

"It was in early spring [April 2, to be exact] that the tornado hit."

Use brackets to correct a quotation.

> "His report was [full] of mistakes."

Use *sic* ("thus" or "so") when quoting a mistake in spelling, usage, or logic.

> Her secretary's comment was clear: "She don't [*sic*] want any of these."

Dashes

Dashes can be effective indicators of meaning—as long as they are not over-used. Make dashes on your typewriter by placing two hyphens side by side. Parentheses de-emphasize enclosed material; dashes emphasize it.

Used selectively, dashes can provide emphasis but they are not a substitute for all other punctuation. When in doubt, do not use a dash!

Dashes can denote an afterthought.

> Have a good vacation—but don't get sunstroke.

They can enclose an interruption in the middle of a sentence.

> This building's designer—I think it was Wright—was, above all, an artist.
> Our new team—Jones, Smith, and Brown—is already compiling outstanding statistics.

Although they can often be used interchangeably with commas, dashes dramatize a parenthetical statement more than commas do.

> Mary, a true friend, spent hours helping me rehearse for my interview.
> Mary—a true friend—spent hours helping me rehearse for my interview.

Notice the added emphasis in the second version.

Hyphen

Use a hyphen to divide a word at the right-hand margin. Consult your dictionary for the correct syllable breakdown:

> com-puter
> comput-er

Actually, it is best to avoid altogether this practice of dividing words at the ends of lines in a typewritten text.

Use a hyphen to join compound modifiers (two or more words preceding the noun as an adjective), but not compound nouns.

the rough-hewn wood

the well-written report

the all-too-human error

a three-part report

Do not hyphenate these same words if they *follow* the noun.

The wood was rough hewn.

The report is well written.

The error was all too human.

Hyphenate an adverb-participle compound preceding a noun.

the high-flying glider

Do not hyphenate compound modifiers if the adverb ends in **-ly**.

the finely tuned engine

Hyphenate most words that begin with the prefix **self-**. (Check your
dictionary.)

self-reliance

self-discipline

self-actualizing

Hyphenate to avoid ambiguity.

re-creation *(a new creation)*
recreation *(leisure activity)*

Hyphenate words that begin with **ex-** only if **ex-** means "past."

ex-employee

expectant

Hyphenate all fractions, along with ratios that are used as adjectives and
that precede the noun.

a two-thirds majority

In a four-to-one ratio they defeated the proposal.

Do not hyphenate ratios if they do not immediately precede the noun.

The proposal was voted down four to one.

Hyphenate compound numbers from twenty-one through ninety-nine.

Thirty-eight windows were broken.

Hyphenate a series of compound adjectives preceding a noun.

The subjects for the motivation experiment were fourteen-, fifteen-, and sixteen-year-old students.

TRANSITIONS AND OTHER CONNECTORS

Transitions help make your meaning clear, signaling readers that you are in a specific time or place, that you are giving an example, showing a contrast, shifting gears, or concluding your discussion. Here are some common transitions and the relations they indicate:

Addition	I am majoring in naval architecture; **furthermore**, I spent three years crewing on a racing yawl.

moreover	and
in addition	again
also	as well as

Place	Here is the switch that turns on the stage lights. **To the right** is the switch that dims them.

beyond	to the left
over	nearby
under	adjacent to
opposite to	next to
beneath	where

Time	The crew will mow the ball field this morning; **immediately afterward**, we will clean the dugouts.

first	the next day
next	in the meantime
second	in turn
then	subsequently
meanwhile	while
at length	since
later	before
now	after

Comparison	Our reservoir is drying up because of the drought; **similarly**, water supplies in neighboring towns are low.

likewise
in the same way
in comparison

Contrast or alternative	Felix worked hard; **however**, he received poor grades. **Although** Mary worked hard, she was not promoted.

however	but
nevertheless	on the other hand
yet	to the contrary
still	notwithstanding
in contrast	conversely

Results	Jack fooled around; **consequently**, he was fired. **Because** Jack fooled around, he was fired.

thus	thereupon
hence	as a result
therefore	so
accordingly	as a consequence

Example	Competition for part-time jobs is fierce; **for example**, eighty students applied for the clerk's job at Sears.

for instance	namely
to illustrate	specifically

Explanation	She had a terrible semester; **that is**, she flunked four courses.

in other words	in fact
simply stated	put another way

Summary or conclusion	Our credit is destroyed, our bank account is overdrawn, and our debts are piling up; **in short**, we are bankrupt.

in closing	to sum up
to conclude	all in all
to summarize	on the whole
in brief	in retrospect
in summary	in conclusion

Pronouns serve as connectors, because a pronoun refers back to a noun in a preceding clause or sentence.

As the **crew** neared the end of the project, **they** were all willing to work overtime to get the job done.

Low employee morale is damaging our productivity. **This** problem needs immediate attention.

Repetition of key words or phrases is another good connecting device—as long as it is not overdone. Here is another example of effective repetition:

Overuse and drought have depleted our water supply critically. Because of our **depleted supply**, we need to enforce strict **water**-conservation measures.

Here, the repetition also emphasizes a critical problem.

Because this next paragraph lacks transitions, sentences seem choppy and awkward:

Choppy	Technical writing is a difficult but important skill to master. It requires long hours of work and concentration. This time and effort are well spent. Writing is indispensable for success. Good writers derive pride and satisfaction from their effort. A highly disciplined writing course should be part of every student's curriculum.

Here is the same paragraph rewritten to improve coherence:

Revised	Technical writing is a difficult but important skill to master. It requires long hours of work and concentration. This time and effort, **however**, are well spent **because** writing is an indispensable tool for success. **Moreover**, good writers derive pride and satisfaction from their effort. A highly disciplined writing course, **therefore**, should be part of every student's curriculum.

Besides increasing coherence *within* a paragraph, transitions and connectors emphasize relationships *between* paragraphs by linking related groups of ideas. Here are two transitional sentences that could serve as conclusions for some paragraphs or as topic sentences for paragraphs that would follow; or they could stand alone for emphasis as single-sentence paragraphs:

Because the A-12 filter has decreased overall engine wear by 15 percent, it should be included as a standard item in all our new models.

With the camera activated and the watertight cover sealed, the diving bell is ready to be submerged.

Topic headings, like those in this book, are another connecting device. A topic heading is both a link and a separation between related yet distinct groups of ideas.

Sometimes a whole paragraph can serve as a connector between major sections of your report. Assume that you have just completed a section in a report on the advantages of a new oil filter and are now moving to a section on selling the idea to the buying public. Here is a paragraph you might write to link the two sections:

Because the A-12 filter has decreased overall engine wear by 15 percent, it should be included as a standard item in all our new models. However, tooling and installation adjustments will add roughly $100 to the list price of each model. Therefore, we have to explain to customers the filter's long-range advantages. Let's look at ways of explaining these advantages.

Notice that this transitional paragraph *contains* transitional expressions as well.

EFFECTIVE MECHANICS

Correctness in abbreviation, capitalization, use of numbers, and spelling is an important sign of your attention to detail.

Abbreviations

(For correct abbreviations in documentation, see pages 189–202.)

Whenever you abbreviate, consider your audience; never use an abbreviation that might confuse your reader. Abbreviations are often inappropriate in formal writing. When in doubt, write the word out.

Abbreviate some words and titles when they precede or immediately follow a proper name.

Correct   Mr. Jones   Raymond Dumont, Jr.

Dr. Jekyll   Warren Weary, Ph.D.

St. Simeon

Do not, however, write abbreviations such as these:

Faulty   Mary is a Dr.

Pray, and you might become a St.

In general, do not abbreviate military, religious, and political titles.

Correct   Reverend Ormsby

Captain Hook

President Clinton

Abbreviate time designations only when they are used with actual times.

Correct   A.D. 576

400 B.C.

5:15 A.M.

Do not abbreviate these designations when they are used alone.

Faulty   Plato lived sometime in the B.C. period.

She arrived in the A.M.

In formal writing, do not abbreviate days of the week, months, words such as **street** and **road**, or names of disciplines such as **English**. Avoid abbreviating states, such as **Me.** for **Maine**; countries, such as **U.S.** for **United States**; and book parts such as **Chap.** for **Chapter**, **pp.** for **page**, and **fig.** for **figure**.

Use **no.** for **number** only when the actual number is given.

Correct   Check switch No. 3.

Abbreviate a unit of measurement only when it appears often in your report and is written out in full on first use. Use only abbreviations you are sure readers will understand. Abbreviate items in a visual aid only if you need to save space.

Here are common abbreviations for units of measurement:

ac	alternating current	l	liter
amp	ampere	lat	latitude
Å	angstrom	lb	pound
az	azimuth	lin	linear
bbl	barrel	long	longitude
BTU	British Thermal Unit	log	logarithm
C	Celsius	m	meter
cal	calorie	max	maximum
cc	cubic centimeter	mg	milligram
circ	circumference	min	minute
cm	centimeter	ml	milliliter
cps	cycles per second	mm	millimeter
cu ft	cubic foot	mo	month
db	decibel	mph	miles per hour
dc	direct current	oct	octane
dm	decimeter	oz	ounce
doz	dozen	psf	pounds per square foot
dp	dewpoint		
F	Fahrenheit	psf	pounds per square inch
f	farad		
fbm	foot board measure	qt	quart
fl oz	fluid ounce	r	roentgen
FM	frequency modulation	rpm	revolutions per minute
fp	foot pound		
freq	frequency	sec	second
ft	foot	sp gr	specific gravity
g	gram	sq	square
gal	gallon	T	ton
gpm	gallons per minute	temp	temperature
gr	gram	tol	tolerance
hp	horsepower	ts	tensile strength
hr	hour	v	volt
in.	inch	va	volt ampere
iu	international unit	w	watt
j	joule	wk	week
ke	kinetic energy	wl	wavelength
kg	kilogram	yd	yard
km	kilometer	yr	year
kw	kilowatt		
kwh	kilowatt hour		

Here are some common abbreviations for reference in manuscripts:

anon.	anonymous	fig.	figure
app.	appendix	i.e.	that is
b.	born	illus.	illustrated
©	copyright	jour.	journal
c., ca.	about (c. 1988)	l., ll.	line(s)
cf.	compare	ms., mss.	manuscript(s)
ch.	chapter	no.	number
col.	column	p., pp.	page(s)
d.	died	pt., pts.	part(s)
ed.	editor	rev.	revised or review
e.g.	for example	rep.	reprint
esp.	especially	sec.	section
et al.	and others	sic	thus, so (to cite an
etc.	and so on		error in the quo-
ex.	example		tation)
f. or ff.	the following page	trans.	translation
	or pages	vol.	volume

For abbreviations of other words, consult your dictionary. Most dictionaries list abbreviations at the front or back or alphabetically with the word entry.

Capitalization

Capitalize these items: proper nouns, titles of people, books and chapters, languages, days of the week, the months, holidays, names of organizations or groups, races and nationalities, historical events, important documents, and names of structures or vehicles. In titles of books, films, and so on, capitalize first and following words, except articles or prepositions.

A Tale of Two Cities	the Chevrolet Corvette
Protestant	Russian
Wednesday	Labor Day
the *Queen Elizabeth II*	Dupont Chemical Company
the Statue of Liberty	Senator John Pasteur
April	France
Chicago	the War of 1812
the Bill of Rights	the Emancipation Proclamation

Do not capitalize the seasons, names of college classes (**senior**, **junior**), or general groups (**the younger generation**, or **the leisure class**.)
Capitalize adjectives that are derived from proper nouns.

Chaucerian English

Capitalize titles preceding a proper noun but not those following:

State Senator Marsha Smith

Marsha Smith, state senator

Capitalize words such as **street**, **road**, **corporation**, **college** only when they accompany a proper noun.

Bob Jones University

High Street

the Rand Corporation

Capitalize **north**, **south**, **east**, and **west** when they denote specific locations, not when they are simple directions.

the South

the Northwest

Turn east at the next set of lights.

Begin all sentences with capitals.

Use of Numbers

If numbers can be expressed in one or two words, you can write them out or you can use the numerals.

fourteen	14
eighty-one	81
ninety-nine	99

For larger numbers, use numerals.

4,364	2,800,357
543	200 million
3¼	

Use numerals to express decimals, precise technical figures, or any other exact measurements. Numerals are more easily read and better remembered than numbers that are spelled out.

50 kilowatts   15 pounds of pressure
14.3 milligrams   4,000 rpm

Express these in numerals: dates, census figures, addresses, page numbers, exact units of measurement, percentages, ages, times with A.M. or P.M. designations, and monetary and mileage figures.

page 14   1:15 P.M.
18.4 pounds   9 feet
115 miles   12 gallons
the 9-year-old tractor $15
15.1 percent

Do not begin a sentence with a numeral.

Six hundred students applied for the 102 available jobs.

If your figure takes more than two words, revise your word order.

The 102 available jobs brought 780 applicants.

Do not use numerals to express approximate figures, time not designated as A.M. or P.M., or streets named by numbers less than 100.

about seven hundred fifty

four fifteen

108 East Forty-second Street

If one number immediately precedes another, spell out the first and use a numeral for the second:

Please deliver twelve 18-foot rafters.

In contracts and other documents in which precision is vital a number can be stated both in numerals and in words:

The tenant agrees to pay a rental fee of three hundred seventy-five dollars ($375.00) monthly.

Appendix B
Writers and Audiences
on the Job

■   ■   ■

This appendix contains the full (and mostly unedited) text of interviews with four working professionals who write daily. The first two can be classified as part-time writers; the latter two, full-time. The varied reporting tasks of these four respondents typify the range of primary audiences and information needs that writers face on the job:

- Blair Cordasco's audience needs instructions for improving their job performance.
- James North's audience needs to make major investment and marketing decisions.
- Bill Trippe's audience needs to make government policy decisions.
- Pam Herbert's audience needs clear documentation to understand and use her company's mainframe computer.

As the interviews show, each of these writers has secondary audiences as well.

Each respondent was asked eight questions about amount and types of writing, audiences, writing challenges, deadlines, individual writing processes, and advice to students. The collected responses contain much useful advice for anyone whose career will depend in some way on good writing.

FIRST RESPONDENT—WRITING FOR COLLEAGUES

Blair Cordasco is a training specialist for an international bank. Her main job is to develop instructional programs and procedures that help improve employees' performance at all levels.

Question 1: What percentage of your job is spent writing, editing, and dealing with written communication?
Answer: Roughly 75 percent.

Question 2: What types of writing or editing do you do?
Answer: I do a few types:

1. I write memos recommending types of training needed in various divisions. I also write status (periodic) reports on the types and evaluation of training that has been given. Basically, I keep management informed about the activities of the training department.

2. I write materials used for training (lectures, procedures, and manuals). Also, I rewrite technical manuals that accompany high-tech equipment so that nonspecialized managers can understand the operating principles of the equipment, to better supervise the staff using the equipment.

3. I edit my staff's material for clarity, conciseness, fluency, and basic correctness. Reading their material from the viewpoint of the person who has to learn from it, I adjust the level of technicality and detail for the intended audience. Then the edited material goes to the content experts, who review the simplified versions for technical accuracy. Organization and format are *very* important. Word processors and desktop publishing systems make the formatting much easier, and we use them for our training manuals.

Question 3: Who are your audiences, and what are their needs?
Answer: I write for two audiences:

1. Management staff (B.A.'s in finance, economics, et cetera, and M.B.A.'s), who range from college trainees to upper management, with varying degrees of technical and business experience. Senior members of the audience need executive briefings (overviews, summaries), not a lot of detail, with technical material as appropriate. Junior members need training materials either for skills (interviewing, quality-control techniques, et cetera) or for knowledge (overview of international banking policy, or fundamentals of data processing, et cetera).

2. Clerical staff (high school education), such as secretaries and terminal operators, who need to learn specific skills (word processing, key punching, et cetera) they can apply directly to the job. The material has to be at their level of comprehension.

Question 4: What does your audience expect from your documents?
Answer: Upper management asks, "How does it affect the bottom line? How

can I use this information to increase productivity and quality?" Lower management and trainees ask, "What's in it for me? Can it help me do *my* job better?"

I spend much of my time trying to persuade people that the information we offer can help them. Managers have to be persuaded that your recommendations are worthwhile and the best way to go.

Clerical staff want to know, "Will this help me keep my job?" Their lifeline is the information and training we offer. An underlying assumption in training clericals is that the training will increase their self-esteem and sense of belonging: "Will this make me feel like a jerk or help me feel I can accomplish something?" Will the tone be condescending, or the material too technical? Or will the tone be friendly and engaging, and the material at the right level of technicality?

Question 5: What is your biggest writing challenge on the job?
Answer: Aside from the persuasive challenge I just described, the biggest challenge is taking a body of information and packaging it *accurately* in different ways for different audiences.

For instance, a new piece of equipment arrives, and the vendor's documentation is too technical for operators (how to) and managers (overview). We have to take the vendor's information and rewrite (how to) for operators. Then, for managers, we have to explain the operating principles, point out potential problems and solutions, and identify major problem areas. So, from one technical manual, we have to extract instructions for the operators, and explanations and overviews for the managers.

A trainer really is dealing with all kinds of information and all kinds of vehicles, and trying to reduce them to the lowest common denominator— to make the meaning absolutely clear.

Question 6: What about deadlines?
Answer: Deadlines are a way of life. Nothing is more essential than getting information to the right people at the right time.

Question 7: Do you follow a standard and predictable process when writing?
Answer: Deadlines affect how we approach the writing process. With plenty of time, we can afford the luxury of the whole process: careful decisions about audience, purpose, content, organization, and style—and plenty of revisions. At times, we have to take shortcuts.

Each project presents its own problem, so we have no pat way of writing training materials. With limited time, we have limited revisions—so you have to think on your feet.

Say we receive word November 1 that a new check-processing system will be installed and operating by December 1. Operators will begin training on this equipment one week before the system is implemented (last week of November). In this case, we might write backward: the basic "How to" comes first, and the more elaborate documentation (for management) would come whenever we could complete it.

Question 8: What advice do you have for students?
Answer: Make sure that *whatever* you're writing is clear *to you* first, and that the way you've organized it makes sense *to you*. Then take that material and become more objective. Try to understand how your audience thinks: "How can I make this logical to my audience? Will they understand what I want them to understand?"

Organizing is the key. Develop the type of outlining or listing or brainstorming tool that works best for you, but *find* one that works, and use it consistently. Then you'll be comfortable with that general strategy whenever you sit down to write, especially under rigid deadline.

SECOND RESPONDENT—WRITING FOR CLIENTS

James North is a project manager for a market-research firm. Most of his writing is for clients who will make investment decisions based on feasibility and strategy for marketing new products.

Question 1: What percentage of your job is spent writing, editing, and dealing with written communication?
Answer: From 50 to 60 percent.

Question 2: What types of writing and editing do you do?
Answer: Aside from an occasional internal memo, my writing is designed for two purposes: (1) to gather, analyze, and interpret data from surveys, and (2) to report our findings to clients.

For surveys, I have to translate the market-research (or information) needs of clients into precise questions that *cannot* be misinterpreted. The questions have to be so precise and unambiguous that the respondent knows *exactly* what we're asking.

Reports, on the other hand (of findings gathered from questionnaires), have to allow the clients to make marketing decisions based on their understanding of complex data analyses presented in a concise, *nontechnical* way. These people are not in any way experts. They want to know, "What do I do next?" Our reports interpret the findings ("What do they mean?"), and offer recommendations for marketing strategy ("What should I do?").

Question 3: Who are your audiences, and what are their needs?
Answer: Among survey respondents, the range is vast: from scientists mak-

ing a breakthrough in robotics, to homemakers choosing a brand of coffee off a supermarket shelf.

Clients range from an express-mail carrier who's spending $3 million to assess the feasibility of expanding into international express service, to an individual who wants to market a new brand of salad dressing or low-calorie chocolate.

Both respondents and clients need material that is concise, clear, and unambiguous.

Question 4: What does your audience expect from your documents?
Answer: Respondents expect questions they can interpret accurately—that is, in only *one* way.

Clients expect enough information to make basic investment and marketing decisions. Some clients need answers to very specific statistical questions: "How many overnight letters are sent daily from England to the United States? How many potentially could be sent daily—classified by type of industry (pharmaceutical, aerospace, and so on) and size of company?" Other clients have more general questions:

- "How does one break into the salad-dressing market?"
- "What are your recommendations for establishing a new brand of salad dressing?"
- "How can I make any money with my low-calorie chocolate?"
- "I have a million, and want to market this chocolate. What can you do for me?"

For the latter group of clients, we might research the taste, packaging, and price-point (say, $4 a pound for chocolate) that is most popular for that product.

Our audience's expectations range from those of real experts who know exactly what they want—down to the nth decimal point—to somebody who calls with a vague idea and asks, "What can you do for me?"

Question 5: What is your biggest writing challenge on the job?
Answer: My writing has to have *one* interpretation *only.* I have to be certain that respondents are answering *exactly* the question I had in mind, not inventing their own version of the question. Then I have to take these data and translate them into accurate interpretations and recommendations for our clients.

Question 6: What about deadlines?
Answer: The priority is to *meet* the deadline, regardless of the quality of the report. You never have the time you need, and so the people who survive

are those who *are able to write it*—no matter what—with enough quality to satisfy clients' expectations.

The difference between real-world and college writing is that, in the real world, the deadlines are *never* extended—never any excuses for not getting the report done on time.

Whether you keep the client depends on how well you can write under impossible time pressures.

Question 7: Do you follow a standard and predictable process when writing?
Answer: I look at data and move to some kind of prose that gets revised as often as time allows. For long reports, I rely on outlines. The final product is a recommendation, but the process begins with my looking at a computer printout.

Even when we devise survey questions that we're sure are unambiguous to our intended respondents, we pretest them—often finding that the questions have been misinterpreted. Then we write them and test them again, until we get them right.

The basic element in my writing process is checking and rechecking for ambiguous messages, and revising as often as time allows.

Question 8: What advice do you have for students?
Answer: Learn to write as quickly, accurately, and unambiguously as you can in a given amount of time. The issue is to have it there when it's expected—maybe not perfect, but to develop a style so that the document is out and it gets the job done.

Organize your time in such a way that you can always meet the deadline. Prepare to labor under Murphy's Law: there will always be some crisis. The survivors are those who anticipate the crises, and are able to deliver nonetheless.

THIRD RESPONDENT—WRITING FOR GOVERNMENT DECISION MAKERS

Bill Trippe is a communications specialist with a military contract company. He provides all-around communications support for a large group of engineers who perform technical analyses for the government.

Question 1: What percentage of your job is spent writing, editing, and dealing with written communication?
Answer: Writing is probably 30 to 40 percent of my job, with editing tasks taking up another 50 percent, and training and tutoring accounting for the remaining 10 to 20 percent.

Question 2: What types of writing and editing do you do?
Answer: The writing I do falls into four categories:

1. Semitechnical reports from my managers to upper managers and to our military sponsors. These are usually progress reports on our larger projects, and run from 1 to 20 pages.

2. Some sections of highly technical (engineer-to-engineer) reports. The engineers and scientists usually write the "body" of highly technical reports, and I write the abstract, introduction, executive summary, acknowledgments, conclusions, lists of references, and bibliography. I also fill out sections that need more material—add figures and tables, expand explanations for equations and other raw material.

3. Miscellaneous reports and articles for a general audience. These include articles and announcements for the company newspapers, project updates for the board of directors, and short papers about our work, to be used for recruiting new employees.

4. Training materials for engineers. I'm currently writing short essays on grammar, audience analysis, and techniques for oral presentation. I've also put together my own editorial style sheet that I constantly update and expand.

Question 3: Who are your audiences, and what are their needs?
Answer: The typical primary audience is either our military sponsors or our management.

Military. Specifically, the military sponsor is the project officer: a colonel or lieutenant colonel who is commanding officer in charge of procuring a military system. He or she usually has help from other commissioned officers and civilian engineers, who coordinate testing, management and technical reviews, quality assurance, and other specific areas of the program. The project officer is the decision maker—the primary reader—and the supporting staff are the secondary readers.

Their needs? Usually, they've asked our company to perform a specific technical analysis. Our formal documents to them are a technical report, regular progress reports, and related briefings. They want a thorough feasibility analysis that relates the state of the art and our company's expertise to their specific program. If they want to build a radar to operate in extreme cold, we'll tell them if it can be done, how reliable the radar would be, and how much it would probably cost.

Their uses for the papers? As a basis for making decisions, or—once the decision has been made—as a means of persuading their superiors (right on up through the Pentagon and Congress) to go along with their recommendations.

Company Management.   The progress reports that I ghostwrite for my bosses are usually written for what I would call "middle managers"—my bosses are the low-level managers and their superiors (the readers) are just above them and just below vice presidents. They are all ex-engineers-turned-managers, and so the technical content can be anywhere from general to complex. They are looking at how the project team is handling potential problem and risk areas—difficult technical tasks, areas where cost overruns are possible, and any problem with scheduling. I write weekly reports about every project, and other reports as necessary.

Their needs? They are a fairly easy audience to write for, mainly because they can read at almost any level. They are, however, a critical audience—almost to the point of hostility. They are tough on their underlings probably because their superiors are then tough on them, and the stakes get higher up there. But, as I said above, they are looking for highlights about management concerns: cost, schedule, and technical risk. They are also more concerned with the quality of the writing and presentation because they are conscious of company image and how the customer and the public perceive us.

Their use? They use all the information given them as a means for making sure the individual projects and tasks are running smoothly. They put together a picture of their area that they can then present to their superiors and to the customers.

For the other writing I do, the audience is everyone from the readers of the company newspaper to the board of directors: general readers, I guess. Everything in this category is really written for information purposes, for entertainment, or as a bit of fluff.

Question 4: What does your audience expect from your documents?
Answer: The military audience expects a technically complete report with a rigorous organization, a good balance of detail, strong graphics and tables, and a clear and readable writing style. Because not all the military people are engineers, certain sections of the report (the introduction, executive summary, and conclusions) have to be semitechnical. On the other hand, to satisfy the various interests of the specialists in the project office, many areas have to be covered in great detail, but without bogging down the basic paper. For this, we resort to appendixes—lots of them. Occasionally we will publish completely separate reports: an executive summary, the analysis, and supporting appendixes; this type of report usually goes over very well.

I've covered the expectations of our management audience in my response to question 3. But I should add that these readers really expect the writing to be tailored to their needs. When they ask for a one-page memo, they expect a one-page memo. The same is true for briefings; they ask for and get a three-minute presentation.

Question 5: What is your biggest writing challenge on the job?
Answer: All my writing tasks are a big challenge because they are highly visible and fall under the tightest deadlines.

The greatest pressure comes from the writing tasks because they are my product alone, whereas the editing I do is the author's product I've helped along. If the author's paper is unsatisfactory, I don't hear about it (unless there's an editorial problem, but that hasn't happened). If my reports are late or contain a mistake, however, then my bosses come directly to me. Therefore I watch what I write very closely, and rewrite and proofread everything. I also work very closely with the typists and artists, to see that everything is exactly as I want it.

In a corporation, you are invisible unless you do something spectacular or make a mistake. Because the spectacular rarely happens, management tends not to notice you until you make a mistake.

Question 6: What about deadlines?
Answer: I rarely miss a deadline.

I know this isn't my own rule, but I use it anyway: when I need to estimate how much time it will take me to do something, I make a reasonable estimate and then double it. Then when I know something is due, I use my doubled estimate and go to work on it, leaving everything else aside. This technique usually works, unless the computer goes down and I lose an afternoon's writing to that great electronic void, or a secretary is out sick, or everyone I need to see is in a meeting.

The important thing is juggling jobs and deadlines. I work for six projects and really have eight bosses. If everyone is screaming for work, then I go to everyone, explain my schedule, and see if they can come up with a new deadline. If conflicts persist, then I explain the problem to the big boss, and she tells me what to work on first. My negotiations are usually enough; I've had to involve the big boss only once or twice.

I've brought work home, stayed late, gotten extra help from secretaries and co-op students, and done jobs faster than I've wanted to. But by reasonably estimating how long a job will take and by being assertive about how to handle conflicting deadlines, I haven't had to do any of these other things to the point of discomfort.

Question 7: Do you follow a standard and predictable process when writing?
Answer: The process I follow depends on the length of the piece.

For short pieces (say, less than three pages) I outline in my head, draft, and revise. On my first draft of a short piece, I spend 40 to 50 percent of the time on the first one or two paragraphs, and crank out the rest quickly. Then I revise two or three times, and tinker with the mechanics and format right up until printing. I use a word processor for everything I write and edit, and so revising is easy: I'm never afraid to change something because of the extra work mechanical typing would create. I usually don't bother to make a printout until I'm pretty close to a final product, and so it's easy for me to revise something short several times a day, tinker with it a half-dozen times, and then make a printout and tinker some more once I see it on paper.

For longer pieces, I try to follow a rigorous pattern:

a. Interview knowledgeable people first for ideas and inspiration
b. Write a paragraph-long purpose statement and sentence outline
c. Review the outline and thesis with the boss of the project
d. Write a first draft, paying particular attention to the introduction
e. Have the draft reviewed for technical accuracy
f. Revise for accuracy and begin revising for editorial quality
g. Revise (two to five revisions, depending on deadline)
h. Correct the mechanics and format

Question 8: What advice do you have for students?
Answer: Whew! I could say a million things here.

The following paragraphs are for nonengineers interested in technical writing as a career.

First of all, technical writing is as tough a writing job as newspaper writing, advertising writing, creative writing, and any other job where you work primarily with words. And so to enter this field you have to be fully competent as a writer: you need to have an inquisitive mind and critical abilities, have eagerness to interview and do research, be a wide and thoughtful reader, have a college background that includes many writing courses beyond freshman English, have experience writing against deadlines, have at least a strong interest in and an ability to learn about technical areas (courses in math, science, and computers would give you a great advantage), have the social ability to deal with many kinds of people of all ages and backgrounds, and have the instincts and composure to deal with corporate life.

Get practical experience before graduation. I combined my undergraduate writing program with assorted part-time writing and publication jobs. I was on the staffs of the college newspaper, literary magazine, yearbook, and radio station. I wrote for a local daily newspaper and did other freelance writing for newspapers and magazines. I packed in as much practical experience as possible—and learned a great deal about writing, editing, production, graphic design, and photography. Despite the pressure you might feel to learn as much about the technical areas as possible, keep in mind that technical writers and editors are expected to be publications specialists, not engineers.

I also had a strong academic program. I had an excellent writing program in college, minored in philosophy, and took an assortment of interesting and challenging courses. In this job, I have to think on my feet and learn complex things quickly. But I've never felt at a loss—and I've gained the respect of my peers and bosses—by being able to tackle a subject and get things down on paper. My liberal arts background—with its emphasis on reading, and writing—is an asset. On the other hand, you need some math and science knowledge and ability. My only college science courses were in biology and chemistry, but my high school program included physics and trigonometry. In interviews for technical writing jobs, I had to convince interviewers that I had an aptitude for technical subjects. I would recommend getting a few technical courses under your belt, and you won't need anything else after that.

Another point: Learn what the established industry procedures are for writing, editing, production, art, and printing. Specifically:

- Learn how to use the editorial style books. The *U.S. Government Printing Office Style Manual* and the *Chicago Manual of Style* are the two most widely used.

- Learn everything you can about production; a little book by the International Paper Company, *Pocket Pal*, is stuffed with information about production and printing. A part-time job on the school newspaper will get you experience in basic paste-up. Learn typing, word processing, and typesetting; these may be your biggest practical skills on the job. If you know what can be done and how long it takes to do it, then you will be able to reduce your production problems by at least half. Learn about how artwork is prepared, because illustrations are as much as 50 percent of a technical document; I work closely with two illustrators. Learn also what makes a good graphic and what doesn't. Learn something about how maps are made.

- Learn photography (including darkroom work—not just snapshots). I work with photographs all the time, and though I don't have to take pictures myself, I assign photographers to take pictures, work with them when they are taking pictures, choose pictures for papers and for audiovisual presentations, and work them into page layouts, presentations, and three-dimensional displays.

- Learn printing. Read what you can about it in *Pocket Pal*, and then look for a large printer in your area who offers tours and try to arrange one for a group from your school. If you know printing, then you will know a great deal about production—what looks good and what doesn't, what works and what doesn't. I deal with the print shop every day. I often have to rush work through there against the shop's normal schedule; if I weren't reasonable in my demands on these people, they'd have my head.

And a word or two for engineers. According to most surveys I've read, you'll spend about 25 percent of your time on the job writing, and another 25 percent giving formal and informal oral presentations on your work. After your technical knowledge, your most important skill will be your ability to communicate. Your career will soar or die based on your ability to communicate your technical knowledge to other people (especially your bosses); I see this happening all the time.

FOURTH RESPONDENT—WRITING FOR COMPUTER USERS

Pamela Herbert is a technical writer for a nationwide distributor. She writes user documentation (manuals for using the company's mainframe computer in various applications in various network locations), and upgrades existing manuals.

Question 1: What percentage of your time is spent writing, editing, and dealing with written communication?
Answer: The actual physical process of writing constitutes maybe 30 percent of my work time. I spend more time coordinating research (I have to catch the field-support people, who are my information sources, when they are in town), collecting information, organizing notes, planning a manual's general structure, preparing visuals to go with the text, and keeping track of which reviewer has which draft. Because I rely on others' feedback, I circulate materials often. And so I write memos fairly often, about three to five times per week.

Question 2: What types of writing and editing do you do?
Answer: My writing consists of

- user documentation: procedure manuals written for the Hewlett-Packard 3000, a mainframe computer. (Our plants use computer applications such as Accounts Payable, Accounts Receivable, Inventory Control, and Vending. The manuals for these applications are called "control procedures." They are written according to a very structured format.)

- supplements or updates to existing documents. (I'm currently writing a user guide for a "report-writer" program to be used in our plants. The guide provides examples specific to our business, so that users can have models. Also, it defines terms and summarizes important information from the original manual for the software.)
- three to five memos per week, requesting information or reporting on my work.

My editing consists of work on

- revisions of articles written for our company newsletter.
- revisions of procedures written by programmers to be used by computer operators. These have less detail and less supplementary information than the user documentation. They are "skeleton" procedures that simply need rewriting for clarity.

Question 3: Who are your audiences, and what are their needs?
Answer: My audience varies according to the type of writing involved. User documentation is called "control procedures"; the procedures are written both for upper management (so that they can understand the system in general) and for data-entry clerks (so that they can use the system). Therefore, control procedures are written as *tasks* required to perform a function with the computer.

System users in our branch offices are not, for the most part, data-processing people. They are clerical workers using the computer to do the tasks they once did manually; they see the computer as a "black box," and so they need very specific, nontechnical instructions. Managers and other administrators simply can read the task overviews and outlines for a picture of the whole process.

When researching a topic, I try to anticipate my audience's needs, and I ask the technical source person (usually a programmer or systems analyst) specific questions keyed to my audience's needs: Who performs the task? What materials are required? What does the task accomplish? What can go wrong? Otherwise, I would waste time soaking up like a sponge any and all information the source person feels like rattling off, whether it's important or not.

For memos, I try to keep the style consistent regardless of who will be reading it. Anything worth writing as a memo (instead of a scribbled note or phone call) is always in my plainest English—even though my boss initially balked at my writing "here it is . . ." instead of "Enclosed please find. . . ." I have a hard time with the business-ese that so many people use. If they ask me to proofread memos or letters, I can't resist striking all the heretofore's and please-be-advised-that-pursuant-to-your-request's. People dislike surrendering those, though, so I'm rarely asked to proofread those things anymore!

Question 4: What does your audience expect from your documents?

Answer: As I mentioned earlier, my manuals have to be specific and straightforward.

Also, because I'm a member of the Quality Assurance group, one function of my job is to detect potential problems with the computer system: inconsistencies or "user-unfriendliness" that could result in a service call from a branch office to our systems support staff here. I am expected to document a system as it is; then, while reviewing my document, the department managers decide whether something in the system should be changed.

Also, I am expected to submit written reports of any Quality Assurance issues I encounter while researching the applications. Readers expect a crystal-clear description of the problem. Considered a "naive user," I am a fairly good test of how understandable a procedure will be to average data-entry persons. And so I am expected to clarify any procedure that strikes me as unclear.

Question 5: What is your biggest writing challenge on the job?

Answer: The biggest challenge in my job is staying on top of a project. It is too easy to procrastinate, put off sifting through those notes, sitting down with that programmer, or making those phone calls to Jacksonville. My boss doesn't watch over to make sure I'm working, but if a project is unusually delayed, he wants to know why.

The biggest *writing* challenge is getting my reviewers to *review!* They hang on to a draft for weeks, and only after I hound them do they drag the thing out and review it. Then they hound *me* to finish the thing and get it into the field. I have to be diplomatic, of course, but it is frustrating to work hard and meet a deadline and then have the draft stagnating on someone's desk. All I can do is document the delays, so that I'm not held responsible for the lengthy production time on a document.

Question 6: What about deadlines?

Answer: Although I construct project timetables with deadlines, I always end up revising the charts and extending the deadlines. I know pretty much how long it will take to research, organize, and draft a manual. What I can't predict are reviewers who drop my work when they are suddenly called out of town, "emergency" projects that cut into my time, source people who are on the road for weeks, and downtime on our system here. Sometimes I spend a week at a branch office myself to do research, rather than wait for a technical person to become available at the home office. At any rate, I try to meet a project's deadline, but I do not lose sleep over it if I can't do it.

Question 7: Do you follow a standard and predictable process when writing?
Answer: The process involved in producing a typical manual is

a. Once the project is assigned, I set up research time with the technical person.
b. I compile notes, collect report samples, and roughly sketch any data-entry screens I'll have to document.
c. I analyze the notes, organizing information into tasks, procedure steps, knowledge topics, nice-to-knows, rules, cautions, and so on.
d. I write a draft (longhand) and enter it into my word processor.
e. I assemble a complete draft, which is sent around on its in-house review. Major problems are ironed out, and I revise the first draft.
f. The revised draft is mailed to a user who is familiar enough with the system to review the document competently.
g. I usually assemble "dummy data" to run the procedure on the system myself, as a final usability test of what I've written.
h. After "cosmetic" editing, the document is issued.

Deadlines do affect the process. If I'm really strapped for time, I'll scrap the field review (step f). I usually make sure the draft is as accurate as possible before it goes out on its first review in-house. If time gets really tight, though, I won't bother to double-check every issue with the first draft. I'll rap it out quickly and let any errors be caught in the review process. I try not to do this often; shortcuts never save much time, and the glaring gaps in the text make me look like a sloppy writer.

Question 8: What advice do you have for students?
Answer: DON'T let yourself slide into the jargon and business-ese trap. Hold fast to the way you learned to write—correctly.

DO be flexible, though, when it comes to other people's writing. You can't preach English grammar to adults if they don't want to hear it.

DO contribute your ideas. Many times I've found that the only reason something hadn't been done in a better way was that no one had bothered to suggest it.

DO keep lists. They help you to keep tabs on which document is at what stage, who is reviewing what, and so on. It's much easier to concentrate on writing when you have everything written down neatly and don't have to commit it all to memory.

DO try to get hands-on experience with what you're writing about, if possible. If you're writing a procedure, perform it yourself. If you're describing the parts of a device, get your hands on one. Knowing your subject thoroughly can make the writing 100 percent easier.

DO keep up to date with the technical writing field. Subscribe to documentation newsletters (I get lots of info from the Society for Technical Communication). Try to keep learning more about different documentation developments and techniques. You'll be more useful in your present job, and more marketable if you decide to look for a new one.

Appendix C
The Reporting Process for the Course Project: A Document Sequence Culminating in the Final Report

■ ■ ■

Professionals in the workplace engage in one common and continual activity; they struggle with big decisions like these:

- *Should we promote Jones to general manager, or should we bring in someone from outside the company?*
- *Why are we losing customers, and what can we do about it?*
- *How can we decrease work-related injuries?*
- *Should we encourage foreign investment in our corporation?*
- *Will this employee-monitoring program ultimately increase productivity or alienate our workers?*

Research projects are routinely undertaken to answer questions like those above—to provide decision makers the information they need.

A course research project, like any in the workplace, is designed to fill a specific information need: to answer a question, to solve a problem, to recommend a course of action. And so students of technical communication often spend much of the semester preparing a research project.

Virtually any major project in the workplace requires that all vital information at various stages (the plan, the action taken, the results) be recorded *in writing*. The proposal/progress report/final report sequence keeps readers informed at each stage of a project. This appendix presents a unified and sequential view of the documents that record project information.

A SAMPLE RESEARCH PROJECT[1]

Mike Cabral, communications major, works part-time as assistant to the production manager for *Megacrunch*, a computer magazine specializing in small-business applications. (*Production* is the transforming of manuscripts into a published form.) Mike's writing instructor assigns a course project and encourages students to select topics from the workplace, when possible. Mike asks *Megacrunch* production manager Marcia White to suggest a research topic that might be useful to the magazine.

White outlines a problem she thinks needs careful attention: Now six years old, *Megacrunch* has enjoyed steady growth in sales volume and advertising revenue—until recently. In the past year alone, *Megacrunch* has lost $150,000 in subscriptions and one advertising account worth $60,000. White knows that most of these losses are caused by increasing competition (three competing magazines have emerged in 18 months). In response to these pressures, White and the executive staff have been exploring ways of reinvigorating the magazine—through added coverage and "hot" features, more appealing layouts and page design, and creative marketing. But White is concerned about another problem that seems partially responsible for the fall in revenues: too many errors are appearing in recent issues.

When *Megacrunch* first hit the shelves, the occasional error in grammar or accuracy seemed unimportant. But as the magazine grew in volume and complexity, the errors increased. Errors in recent issues include misspellings, inaccurate technical details, unintelligible sentences and paragraphs, and scrambled source code (in sample programs).

White asks Mike if he's interested in researching the error problem and looking into quality-control measures. She feels that his three years' experience with the magazine qualifies him for the task. Mike accepts the assignment.

White cautions Mike that this topic is politically sensitive, especially to the editorial staff and to the investors. She wants to be sure that Mike's investigation doesn't merely turn up a lot of dirty laundry. Above all, White wants to preserve the confidence of investors—not to mention the morale of the editorial staff, who do a good job in a tough environment, plagued by impossible deadlines and constant pressure. White knows that offending people—even unintentionally—can be disastrous.

She therefore insists that all project documents express a supportive rather than critical point of view: "What could we be doing better?" instead of "What are we doing wrong?" Before agreeing to release the information

---

[1]In the interest of privacy, the names of all people, publications, companies, and products in Mike's documents have been changed.

from company files (complaint letters, notes from irate phone calls, and so on) White asks Mike to submit a proposal, in which the intent of this project is made absolutely clear.

THE PROJECT DOCUMENTS

Three types of documents lend shape and sequence to the research project: **the proposal,** which spells out the plan; **the progress report,** which keeps track of the investigation; and **the final report,** which analyzes the findings. This appendix shows how these documents function together in Mike Cabral's reporting process.

The Proposal Stage

Proposals offer plans for meeting needs. A proposal's primary readers are those who will decide whether to approve, fund, or otherwise support the project. Readers of a research proposal usually begin with questions like these:

- *What, exactly, do you intend to find out?*
- *Why is the question worth answering, or the problem worth solving?*
- *What benefits can we expect from this project?*

Once readers agree the project is worthwhile, they will want to know all about the plan:

- *How, exactly, do you plan to do it?*
- *Is the plan realistic?*
- *Is the plan acceptable?*

Besides these questions, readers may have others: *How much will it cost? How long will it take? What makes you qualified to do it?* and so on. See Chapter 22 for more discussion and examples.

Mike Cabral knows his proposal will have only Marcia White (and possibly some executive board members) as primary readers. The secondary reader is Mike's writing instructor, who must approve the topic as well. And at some point, Mike's documents could find their way to his co-workers.

White already knows the background and she needs no persuading that the project is worthwhile, but she does expect a realistic plan before she will approve the project. (For his instructor, Mike attaches a short appendix [not shown here] outlining the background and his qualifications.) Also, Mike concentrates on his emphasis: he wants the proposal to be positive

Rangeley Publishing Company

TO: Marcia White, Production Manager   September 26, 19xx

FROM: Mike Cabral, Production Assistant *MC*

SUBJECT: Proposal for Studying Ways of Improving Quality Control at *Megacrunch*
          Magazine

Introduction

Tells what the problem is, and what it means to the company

The growing number of grammatical, informational, and technical errors in each
monthly issue of *Megacrunch* is raising complaints from authors, advertisers, and
readers. Beyond compromising the magazine's reputation for accurate and dependable information, these errors—almost all of which seem avoidable—endanger our
subscription and advertising revenues.

Summary of the Problem

Further defines the problem and its effects

More and more authors are complaining of errors in published versions of their articles. Software developers assert that errors in reviews and misinformation about
products have damaged reputations and sales. For instance, Osco Scientific, Inc.
claims to have lost $150,000 in software sales because of an erroneous review in
Megacrunch.

    Although we continue to receive a good deal of "fan mail," we also receive let-
ters speculating about whether *Megacrunch* has lost its edge as a leading resource
for small-business users.

Proposed Study

Describes the project and tells who will carry it out

I propose to examine the errors that most frequently recur in our publication,
to analyze their causes, and to search for ways of improving the quality of the
magazine.

Methods and Sources

Tells how the writer plans to carry out the project

In addition to close examination of recent *Megacrunch* issues and competing magazines, my primary data sources will include correspondence and other feedback now
on file from authors, developers, and readers. I also plan telephone interviews with
some of the above sources. In addition, interviews with our editorial staff should
yield valuable insights and suggestions. As secondary material, books and articles
on editing and writing can provide sources of theory and technique.

FIGURE C.1 A Proposal for a Research Project

Encourages reader support by telling *why* the project will be beneficial

Conclusion

We should not allow clearly avoidable errors to eclipse the hard work that has made *Megacrunch* the leading Cosmo resource for small-business users. I hope that my research project will help resolve many such errors. With your approval, I will begin immediately.

FIGURE C.1 A Proposal for a Research Project *Continued*

rather than critical, so as not to offend anyone. He therefore focuses on achieving *greater accuracy* rather than *fewer errors*. So that his instructor can approve the project, Mike submits the proposal in Figure C.1 by the semester's fourth week.

The Progress-Report Stage

The progress report keeps readers up-to-date on the project's activities, new developments, accomplishments or setbacks, and timetable. Depending on the size and length of the particular project, the number of progress reports will vary. (Mike's course project will require only one.) Readers approach any progress report with two big questions:

- *Is the project moving ahead according to plan and schedule?*
- *If not, why not?*

Rangeley Publishing Company

TO: Marcia White, Production Manager   November 6, 19xx

FROM: Mike Cabral, Production Assistant *MC*

SUBJECT: Report of Progress on My Research Project: A Study of Ways for Improving Quality Control at *Megacrunch* Magazine

Work Completed

My topic was approved on September 28, and I immediately began both primary and secondary research. I have since reviewed file letters from contributors and readers, along with notes from phone conversations with various clients and from interviews with *Megacrunch*'s editing staff. I have also surveyed the types and frequency of errors in recent issues. Recent books and articles on writing and editing have rounded out my study. The project has moved ahead without complications. With my research virtually completed, I have begun to interpret the findings.

Preliminary Interpretation of Findings

From my primary research and my own editing experience, I am developing a focused idea of where some of the most avoidable problems lie and how they might be solved. My secondary sources offer support for the solutions I expect to recommend, and they suggest further ideas for implementing the recommendations. With a realistic and efficient plan, I think we can go a long way toward improving our accuracy.

Work Remaining

So far, the project is on schedule. I plan to complete the interpretation of all findings by the week of November 29, and then to organize, draft, and revise my final report in time for the December 14 submission deadline.

FIGURE C.2   A Progress Report for a Research Project

Readers may have a host of subordinate questions as well. See pages 503–506 for more discussion and examples.

Mike Cabral designs his progress report in Figure C.2 for his boss *and* his instructor, and turns it in by the semester's tenth week. Each of these readers will want to know what Mike has accomplished so far.

The Final-Report Stage

The final report presents the results of the research project: findings, interpretations, and recommendations. Readers will expect this document to answer questions like these:

- *What did you find?*
- *What does it all mean?*
- *What should we do?*

Depending on the topic and situation, of course, readers will have specific questions as well. See Chapter 23 for discussion and examples.

During his research, Mike Cabral discovered problems over and above the published errors he had been assigned to investigate. For instance, after looking at competing magazines he decided that *Megacrunch* needed improved page design, along with a higher quality stock (the paper the magazine is printed on). He also concluded that a monthly section on business applications would help. But despite their usefulness, none of these findings or ideas was part of Mike's *original* assignment. White expected him to answer these questions, specifically:

- *Which errors recur most frequently in our publication?*
- *Where are these errors coming from?*
- *What can we do to prevent them?*

Mike therefore decides to focus exclusively on the error problem. (He might later discuss those other issues with White—if the opportunity arises. But if *this* report were to include material that exceeds the assignment *and* the reader's expectations, Mike could end up appearing arrogant or presumptuous.)

Mike tries to give White only what she requested. He analyzes the problem and the causes, and then recommends a solution. Mike adapts the general outline on page 557 to shape the three major sections of his report: *introduction, findings, and conclusions/recommendations.* For the reader's convenience and orientation, he includes the report supplements discussed in Chapter 16: *front matter* (title page, transmittal letter, table of contents, and informative abstract) and *end matter,* as needed (a numbered list of references [as discussed on pages 201–202] and appendixes [not shown here]).

After several revisions, Mike submits copies of the report in Figure C.3 to his boss and to his writing instructor.

Virtually any long
report has a title
page

A forecasting title

The primary
reader's name,
title, and
organization

Author's name

Submission date

Quality-Control Recommendations
for *Megacrunch* Magazine

Prepared for

Marcia S. White
Production Manager
Rangeley Publications

by

Michael T. Cabral

December 14, 19xx

FIGURE C.3 The Final Report for a Research Project

82 Stephens Road
Boca Grande, FL 08754
December 14, 19xx

Marcia S. White, Production Manager and Vice President
Rangeley Publications, Inc.
167 Dolphin Ave.
Englewood, FL 08567

Dear Ms. White:

Here is my report recommending quality-control measures for *Megacrunch*
magazine. The report briefly discusses the history of our quality-control problem,
identifies the types of errors we are up against, analyzes possible causes, and
recommends three realistic solutions.

My research confirmed exactly what you had feared. The problem is big and deeply
rooted: our authors have legitimate complaints; our readers justifiably want
information they can put to work; and developers and advertisers have the right to
demand fair and complete representation. As a result of client dissatisfaction,
competing magazines are gaining readers and authors at our expense.

To have an immediate effect on our quality-control problem, we should act now.
Because of our limited budget, I have tried to recommend low-cost, high-return
solutions. If you have other solutions in mind, I would be happy to research them
for projected effectiveness and feasibility.

Sincerely,

Michael T. Cabral

Michael T. Cabral
Production Assistant
Rangeley Publications

FIGURE C.3   The Final Report for a Research Project   *Continued*

CONTENTS

Table of contents (reports with numerous visuals also have a table of figures)

Front matter (items that precede the report)

Heads and subheads from the report itself

All heads in the table of contents follow the exact phrasing of those in the report text

The various typefaces and indentations reflect the respective rank of various heads in the report

Each head listed in the table of contents is assigned a page number

End matter (items that follow the report)

FIGURE C.3 The Final Report for a Research Project *Continued*

INFORMATIVE ABSTRACT

An investigation of the quality-control problem at *Megacrunch* magazine identifies not only the types of errors and their causes, but also recommends a plan.

Megacrunch suffers from the following avoidable errors:

- *Grammatical errors* are most frequent: misspellings, fragmented and jumbled sentences, misplaced punctuation, and so on.

- *Informational errors:* incorrect prices, products attributed to wrong companies, mismarked visuals, and so on.

- *Technical errors* are less frequent, but the most dangerous: garbled source code, mismarked diagrams, misused technical terms, and so on.

- *Distortions of the author's original meaning:* introduced by editors who attempt to improve clarity and style.

The above errors seem to have the following causes:

- *Poor initial submissions from contributors* ignore basic rules of grammar, clarity, and organization.

- *Lack of structure in the editing cycle* allows for unrestrained and often excessive editing at all stages.

- *Lack of diversity in the editorial staff* leaves language specialists responsible for catching technical and informational errors.

- *Lack of communication with authors and advertisers* leaves the primary sources out of the production process.

On the basis of my findings, I offer three recommendations for improving quality control during the production process:

- *Expanded author's guide* that includes guidelines for effective use of active voice, visuals, direct address, audience analysis, and so on.

- *Five-stage editing cycle* that specifies everyone's duties at each stage. The cycle would require two additional staff members: a technical editor and a fact checker/typist for editorial changes.

- *More communication with contributors* by exchanging galley proofs and increasing our use of the electronic network.

For busy readers, the informative abstract summarizes the report's essential message (findings, conclusions, recommendations). This is the one part of a long report read by most readers

The summary stands alone in meaning—a kind of minireport written for the general reader

Busy readers need to know quickly what is important. A summary gives them enough information to decide whether they should read the whole report, parts of it, or none of it

FIGURE C.3 The Final Report for a Research Project *Continued*

INTRODUCTION

The reputation of *Megacrunch* magazine is jeopardized by grammatical, technical, and other errors appearing in each issue.

Megacrunch has begun to lose some long-time readers, advertisers, and authors. Although many readers continue to praise the usefulness of our information, complaints about errors are increasing and subscriptions are falling. Advertisers and authors increasingly point to articles or layouts in which excessive editing has been introduced, and some have taken their business and articles to competing magazines. One disgruntled reader sums up our problem by asking that we devote "more effort to publishing a magazine without the kinds of elementary errors that distract readers from the content" (1). This kind of complaint is typical of the sample letters in Appendix A.

Granted, complaints are inevitable—as can be seen in a quick review of "Letters to the Editor" in virtually any publication. But if *Megacrunch* is to withstand the competition and uphold its reputation as the leading resource for Cosmo applications in small business, we must minimize such complaints.

This report identifies the major errors that recur in our magazine, and investigates their causes. My data is compiled from interviews with our editorial staff, a review of complaint letters from authors and readers, and a spot-check for errors in the magazine itself. Books and articles on writing and editing provide theory and technique. The report concludes by recommending a three-part solution to our error problem.

The section tells what the report is about, why it was written, and how much it covers

An overview of the problem and its effects on the magazine's revenues

Readers are referred to appendixes for details that would interrupt the report flow

Request for action

Purpose and scope of report; overview of research methods and data sources

Because his primary audience knows the background, Mike keeps the introduction brief

FIGURE C.3 The Final Report for a Research Project *Continued*

This section tells what the writer found and what it means

The first subsection analyzes the problem; the second will examine causes

Introduction and justification for analysis of the problem

A lead-in to the visual

A visual that illustrates parts of the problem

To further segment the report, each major section (INTRODUCTION, FINDINGS AND CONCLUSIONS, RECOMMENDA-TIONS) begins on its own separate page

Discussion of the visual, and lead-in to the analysis

Overview of the next subsection, so that readers know what to expect

One part of the problem defined, with examples and effects of the problem

Citing authorities clarifies and supports the writer's position

2

FINDINGS AND CONCLUSIONS

ELEMENTS OF THE PROBLEM

Errors in *Megacrunch* are limited to no single category. For example, some errors are tied to technical slip-ups, while others result from editors changing the author's intended meaning. My spot-check of *Megacrunch* 8.10, our most recent issue, revealed errors of the types listed in Table 1.

TABLE 1 Sample Errors

Spot-check of *Megacrunch* 8.10		
Error type	As Published	Corrected Version
mechanical	varity of software	variety of software
technical	Dos	DOS
informational	Deluxe Panel	DeluxePanel
grammatical	This will help. . . .	Editing will help. . . .
grammatical	. . . everyone helps for of a program's release date approaches. . . .	. . . everyone helps as a program's release date approaches. . . .
technical	ram	RAM
informational	cosmo	Cosmo
mechanical	We're back, now we will	We're back; now we will

My random analysis of only six pages identified errors in four categories: grammatical/mechanical, technical, informational, and distortions of intended meaning.

Errors in Grammar and Mechanics

Basic correctness is a "given"—and a problem—for any publication. Sentence fragments, confused punctuation, and poor spelling cause readers to "question the professionalism or diligence of both authors and editor" (2, p. 39)—an assertion borne out by the sampling of complaints in Appendix A.

A recent survey of college and workplace writing (3, p. 168) found that the average writer suffers from the following basic problems:

FIGURE C.3 The Final Report for a Research Project *Continued*

- Three-to-four words are misspelled in a memo-length piece.
- One of every ten sentences is a run-on or an "attachable sentence fragment."
- Every fifth possessive is incorrectly formed.

Given these findings, we should not be surprised to receive imperfect manuscripts. However, we must eliminate the slips of the so-called average writer before final copy goes to press.

Errors in Information Accuracy and Access

Beyond basic errors, we have published some inaccurate information. For instance, we sometimes attribute products to the wrong companies or we list incorrect prices. Inaccuracies of this kind infuriate readers, product developers, and suppliers alike. And a retraction printed in the magazine's subsequent issue has little impact once the damage has been done.

Besides inaccurate information, *Megacrunch* too often presents inaccessible information. Mismarked visuals, misplaced headings, and misnumbered page references make the magazine hard to follow and use selectively.

Technical Errors

Technical errors seem one of our biggest problems. While some readers might raise a proverbial eyebrow over grammatical errors or skim over informational errors, technical errors are more frustrating and incapacitating. On a page of text, a misplaced comma or a missing bracket can be irritating, but in a program listing, these same errors can render the program useless. Even worse, a misnamed or misnumbered pin or socket in a hardware diagram might cause users to inadvertently destroy their data or damage their hardware.

Some technical slips in *Megacrunch* have veered close to disaster. Consider, for example, the flawed diagram in Figure 1, from our 7.12 issue.

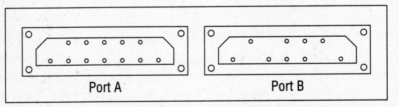

To connect the external drive, plug Cable Y into Port A

FIGURE 1. Partial View of the Port Panel on the AXL 100

FIGURE C.3 The Final Report for a Research Project *Continued*

Our published diagram instructed users to plug a 9-pin external-drive cable into Port A, a 12-pin modem port; the correct connection was to have been made to Port B. Ron Catabia, author of the article and respected tech wizard, explained the flaw in our reproduction of his diagram: "Had any users followed the instructions as printed, the read/write head on their external drive could have suffered permanent damage" (4). Our lengthy correction printed one month later was in no way a sufficient response to an error of this importance. Nor could our belated correction placate an enraged and discredited author.

Such errors do little to encourage reader's perception of *Megacrunch* as the serious user's resource for the latest technical information.

Distortions of the Author's Original Meaning

Authors routinely complain that, in our efforts to increase clarity and readability, we distort their original, intended meaning. After reading the edited version of her article, one author insisted that "too often, edits actually changed what I had said to something I hadn't said—sometimes to the point of altering the facts" (5).

Editorial liberties inevitably alienate authors. Overzealous editors who set out to shorten a sentence or fine-tune a clause—while knowing nothing about the program being discussed—can distort the author's meaning. As a recent study confirms, "When reviewers [editors] criticize in areas outside their expertise, their misguided reviews are seen as an intrusion" (6, p. 37).

Following is an excerpt that typifies the distortions in recent issues of *Megacrunch*. Here, a seemingly minor editorial change (from "but" to "even") radically changes the meaning.

As submitted: A user can complete the Filibond program without ever having typed but a single command.

As published: A user can complete the Filibond program without ever having typed even a single command.

As the irate author later pointed out, "My intent was to indicate that a single command must be typed during the program run" (7).

This type of wholesale editing (of which more examples are shown in Appendix B) is a disservice to all parties: author, reader, and magazine.

CAUSES OF OUR EDITORIAL INACCURACY

Before devising a plan for dealing with our editing difficulties, we have to answer questions such as these:

Discussion of the visual

Effects of the error

Interpretation— what it all means for the magazine

Another part of the problem defined

Example

Interpretation

Citing an authority to support the interpretation

Another example

Effect of the error

Second subsection

Justification for an analysis of causes

FIGURE C.3 The Final Report for a Research Project *Continued*

• Where are these errors coming from?
• Can they be prevented?

Scope of this
subsection, so that
readers know what
to expect

Interviews with our editing staff along with analysis of our editing practices and review of letters on file uncovered the following causes: (1) poor initial submissions from contributors, (2) lack of structure in our editing cycle, (3) lack of diversity in our editing staff, and (4) lack of communication with authors and advertisers.

Poor Initial Submission from Contributors

Some contributors submit poorly written manuscripts. And so we edit heavily whenever "a submission otherwise deserves flatout rejection," as one editor argues. Our editors claim that printing poor writing would be more damaging than the occasional editing excesses that now occur. Although editors can improve clarity and readability without in-depth knowledge of the subject, we often misinterpret the author. Clearer writing guidelines for authors would result in manuscripts needing less editing to begin with. Our single-page author's guide is inadequate.

Lack of Structure in Our Editing Cycle

In our current editing cycle, the most thorough editors see an article repeatedly, as often as time allows. Various editors are free to edit heavily at all stages. And these editors are entirely responsible for judgments about grammatical, informational, and technical accuracy.

Although "having your best give their best" throughout the cycle seems a good idea, this approach leads to inconsistent editing and/or overediting. Some editors do a light editing job, choosing to preserve the original writing. Others prefer to "overhaul" the original. With light-versus-heavy editing styles entering the cycle randomly, errors slip by. As one editor noted, "Sometimes an article doesn't get a tough edit until the third or fourth reading. At that point, we have no time to review these last-minute changes (8)."

Any article heavily edited and rewritten in the final stages stands a chance of containing typographical and mechanical errors, some questionable sentence structures, inadvertent technical changes, and other problems that result from a "rough-and-tumble edit."

Although some articles are edited inconsistently, others are overedited. Our editors tend to be vigilant in pursuit of clarity, conciseness, and tone. Unfortunately, they seem less vigilant about technical accuracy.

Margin notes (left column):

First cause defined

Findings

Conclusion

Second cause defined

Findings

Interpretation— what it means

Findings

Interpretation

FIGURE C.3 The Final Report for a Research Project   *Continued*

Instead of full-scale editing at all stages, we need a cycle that makes a manuscript progress from inadequate (or adequate) to excellent, through different levels of editorial attention. For example, a first edit should be thorough, but a final proofreading should be merely a fine-combing for typographical and mechanical errors.

Lack of Diversity in Our Editing Staff
The variety of errors suggests that our present staff alone cannot spot all problems. Strong writing backgrounds have not prepared our editors to recognize a jumbled line of programming code or a misquoted price. To snag all errors, we must hire technical specialists. We need both a technical editor and a fact checker, to pick up where current editors leave off.

Lack of Communication with Authors and Advertisers
Some of our editing troubles emerge from a gap between the meaning intended by contributors and the interpretation by editors. In the present system, contributors submit manuscripts without seeing any editorial changes until the published version appears. Along with an expanded author's guide, regular communication throughout the editing process (and perhaps the writing process as well) would involve contributors in developing the published piece, and thus make authors more responsible for their work.

FIGURE C.3 The Final Report for a Research Project *Continued*

RECOMMENDATIONS

To eliminate published inaccuracies, I recommend: (1) an expanded author's guide, (2) a five-stage editing cycle, and (3) improved communication with contributors.

EXPANDED AUTHOR'S GUIDE

The obvious way to limit editing changes would be to accept only near-perfect submissions. But as a technical resource we cannot afford to reject poorly written articles that are nonetheless technically valuable.

To reduce editing required on submissions, I recommend we expand our author's guide to include topics like these: audience analysis, use of direct address and active voice, principles of outling and formatting, and use of visuals.

FIVE-STAGE EDITING CYCLE

In place of haphazard editing, I propose a progressive, five-stage cycle: Stages one through three would refine grammar, clarity, and readability. Two additional staff members, a fact checker and a technical editor, would check facts and technical accuracy in the final two stages. Figure 2 outlines responsibilities at each stage.

IMPROVED COMMUNICATION WITH CONTRIBUTORS

The following measures would reduce errors caused by misunderstandings between contributors and editors.

Author-Client Verification of Galley Proofs

Two weeks before our deadline, we could send authors pre-publication galley proofs [which show the text as it will appear in published form]. Authors could check for technical errors or changes in meaning, and return proofs within five days.

Regular Use of Our Electronic-Mail Network

At any time during production, authors and editors could communicate through CompuServe by leaving questions and messages in one another's electronic mailboxes. (Virtually all our regular authors subscribe to CompuServe.) In addition, all parties could log onto CompuServe at one or more scheduled times daily to discuss the manuscript. The E-mail alternative is cheaper than the telephone, eliminates "telephone tag," and could serve as a "hot line" for authors while they prepare a manuscript for submission.

Side annotations (left margin):

- This section tells what should be done
- Scope of this section, so that readers know what to expect
- Lead-in to first recommendation
- The recommendation
- Lead-in to second recommendation
- Lead-in to third, and final, recommendation
- First part of final recommendation
- Second part of final recommendation

FIGURE C.3 The Final Report for a Research Project *Continued*

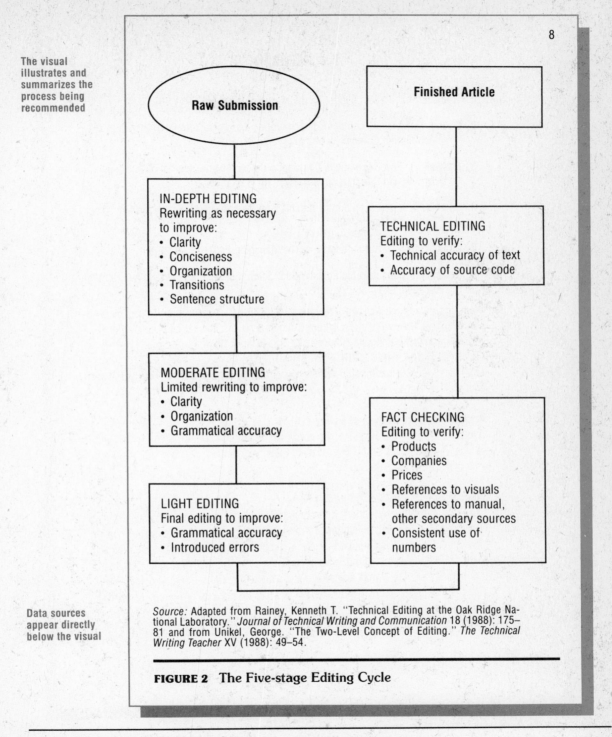

Raw Submission

Finished Article

IN-DEPTH EDITING
Rewriting as necessary to improve:
• Clarity
• Conciseness
• Organization
• Transitions
• Sentence structure

TECHNICAL EDITING
Editing to verify:
• Technical accuracy of text
• Accuracy of source code

MODERATE EDITING
Limited rewriting to improve:
• Clarity
• Organization
• Grammatical accuracy

FACT CHECKING
Editing to verify:
• Products
• Companies
• Prices
• References to visuals
• References to manual, other secondary sources
• Consistent use of numbers

LIGHT EDITING
Final editing to improve:
• Grammatical accuracy
• Introduced errors

Source: Adapted from Rainey, Kenneth T. "Technical Editing at the Oak Ridge National Laboratory." *Journal of Technical Writing and Communication* 18 (1988): 175–81 and from Unikel, George. "The Two-Level Concept of Editing." *The Technical Writing Teacher* XV (1988): 49–54.

FIGURE 2 The Five-stage Editing Cycle

FIGURE C.3 The Final Report for a Research Project *Continued*

REFERENCES

1. Grendel, M.L. Reader's letter to the Managing Editor. 14 May 1993.

2. Chang, F. Revise: A computer-based writing assistant. Journal of Technical Writing and Communication 17(1): 25–41; 1987.

3. Haswell, R. Toward competent writing in the workplace. Journal of Technical Writing and Communication 18(2): 161–72; 1988.

4. Catabia, R. Notes from author Catabia's phone conversation with the Managing Editor. 10 March 1992.

5. Dimmersdale, O. Author's letter to the Managing Editor. 6 July 1993.

6. Barker, T. Feedback in hightech writing. Journal of Technical Writing and Communication 18(1): 35–51; 1988.

7. Klause, M. Author's letter to the Managing Editor. 11 May 1993.

8. *Megacrunch* editing staff. Interviews. 12–14 April 1993.

The list of references clearly identifies each source cited in the report

This writer documents all sources according to the format given on pages 201–202

Specific names omitted from this citation, to protect in-house sources and to allow employees to speak candidly, without fear of reprisal.

FIGURE C.3 The Final Report for a Research Project *Continued*

Works Cited

CHAPTER 1

Barnum, Carol, and Robert Fisher. "Engineering Technologists as Writers: Results of a Survey." *Technical Communication* 31.2 (1984): 9–11.

Florman, Samuel, quoting Donald A. Rikard. "Toward Liberal Learning for Engineers." *Technology Review* Mar. 1986: 18–25.

International Data Corporation. "White Paper." *Fortune* 10 Dec. 1984.

Richards, Thomas O., and Ralph A. Richards. "Technical Writing." Paper presented at the University of Michigan, 11 July 1941. Published by the Society for Technical Communication.

Spruell, Geraldine. "Teaching People Who Already Learned How to Write, to Write." *Training and Development Journal* October 1986: 32–35.

Wickens, Christopher D. *Engineering Psychology and Human Performance.* 2nd ed. New York: Harper, 1992.

CHAPTER 2

Dumont, R. Armand. "Writing, Research, and Computing." *Critical Thinking: An SMU Dialogue,* 1988.

Elbow, Peter. *Writing without Teachers.* New York: Oxford, 1973.

Franke, Earnest A. "The Value of the Retrievable Technical Memorandum System to the Engineering Company." *IEEE Transactions on Professional Communication* 32.1 (March 1989): 12–16.

Grice, Roger A. "Document Development in Industry." *Technical Writing: Theory and Practice.* Ed. Bertie E. Fearing and W. Keats Sparrow. New York: Modern Language Assn., 1989. 27–32.

Hauser, Gerald. *Introduction to Rhetorical Theory.* New York: Harper, 1986.

Selzer, Jack. "Composing Processes for Technical Discourse." *Technical Writing: Theory and Practice*. Ed. Bertie E. Fearing and W. Keats Sparrow. New York: Modern Language Assn., 1989. 43–50.

CHAPTER 4

Cross, Mary. "Aristotle and Business Writing: Why We Need to Teach Persuasion." *The Bulletin of the Association for Business Communication* 54.1 (March 1991): 3–6.

Gilsdorf, Jeanette W. "Executives' and Academics' Perception of the Need for Instruction in Written Persuasion." *The Journal of Business Communication* 23.4 (Fall 1986): 55–68.

———. "Write Me Your Best Case for. . . . *The Bulletin of the Association for Business Communication* 54.1 (March 1991): 7–12.

Goodall, H. Lloyd, Jr., and Christopher L. Waagen. *The Persuasive Presentation*. New York: Harper, 1986.

Harcourt, Jules. "Teaching the Legal Aspects of Business Communication." *The Bulletin of the Association for Business Communication* 53.3 (1990): 63–64.

Hauser, Gerald A. *Introduction to Rhetorical Theory*. New York: Harper, 1986.

Hays, Robert. "Political Realities in Reader/Situation Analysis." *Technical Communication* 31.1 (1984): 16–20.

Kelman, Herbert C. "Compliance, Identification, and Internalization: Three Processes of Attitude Change." *Journal of Conflict Resolution* 2 (1958): 51–60.

Kipnis, David, and Stuart Schmidt. "The Language of Persuasion." *Psychology Today* (April 1985): 40–46. Rpt. in Raymond S. Ross. *Understanding Persuasion*. 3rd ed. Englewood Cliffs: Prentice, 1990: 101–02.

Littlejohn, Stephen W. *Theories of Human Communication*. 2nd ed. Belmont, California: Wadsworth, 1983.

Littlejohn, Stephen W., and David M, Jabusch. *Persuasive Transactions*. Glenview, Illinois: Scott, 1987.

Rokeach, Milton. *The Nature of Human Values*. New York: Free, 1973.

Ross, Raymond S. *Understanding Persuasion*. 3rd ed. Englewood Cliffs: Prentice, 1990.

Rottenberg, Annette T. *Elements of Argument*. 3rd ed. New York: St. Martin's, 1991.

Sherif, Muzafer, et al. *Attitude and Attitude Change: The Social Judgment-Involvement Approach*. Philadelphia: Saunders, 1965.

Stonecipher, Harry. *Editorial and Persuasive Writing*. New York: Hastings House, 1979.

Varner, Iris I., and Carson H. Varner. "Legal Issues in Business Communications." *ABCA Bulletin* (September 1983): 31–40.

CHAPTER 5

Begley, Sharon. "Is Science Censored?" *Newsweek* 14 September 1992.

Brody, Herb. "Great Expectations: Why Technology Predictions Sometimes Go Awry." *Technology Review* July 1991: 38–44.

Brownell, Judi, and Michael Fitzgerald. "Teaching Ethics in Business Communication: The Effective/Ethical Balancing Scale." *The Bulletin of the Association for Business Communication* 55.3 (1992): 15–18.

Bryan, John. "Down the Slippery Slope: Ethics and the Technical Writer as Marketer." *Technical Communication Quarterly* 1.1 (1992): 73–88.

Burghardt, M. David. *Introduction to the Engineering Profession.* New York: Harper, 1991.

Christians, Clifford G., et al. *Media Ethics: Cases and Moral Reasoning.* 2nd ed. White Plains, NY: Longman, 1978.

Clark, Gregory. "Ethics in Technical Communication: A Rhetorical Perspective." *IEEE Transactions on Professional Communication* 30.3 (1987): 190–95.

Dumbrowski, Paul M. "Challenger and the Social Contingency of Meaning: Two Lessons for the Technical Communication Classroom." *Technical Communication Quarterly* 1.3 (1992): 73–86.

Finkelstein, Leo, Jr. "The Social Implications of Computer Technology for the Technical Writer." *Technical Communication* 38.4 (1991): 466–73.

Girill, T. R. "Technical Communication and Ethics." *Technical Communication* 34.3 (1987): 178–79.

Golen, Steven, et al. "How to Teach Ethics in a Basic Business Communication Class." *The Journal of Business Communication* 22.1 (1985): 75–84.

Gouran, Dennis S., et al. "A Critical Analysis of Factors Related to Decisional Processes Involved in the Challenger Disaster." *Central States Speech Journal* 37.3 (1986): 119–35.

Harcourt, Jules. "Teaching the Legal Aspects of Business Communication." *The Bulletin of the Association for Business Communication* 53.3 (1990): 63–64.

Hauser, Gerald A. *Introduction to Rhetorical Theory.* New York: Harper, 1986.

Janis, Irving L. *Victims of Groupthink: A Psychological Study of Foreign Policy Decisions and Fiascos.* Boston: Houghton, 1972.

Johannesen, Richard L. *Ethics in Human Communication.* 2nd ed. Prospect Heights, Illinois: Waveland, 1983.

Lavin, Michael R. *Business Information: How to Find It, How To Use It*. 2nd ed. Phoenix, AZ: Oryx, 1992.

Lewis, Philip L. and N. L. Reinsch, "The Ethics of Business Communicaiton." *Proceedings of the 1981 American Business Communication Conference*. Champaign, IL. In *Technical Communications and Ethics*. Ed. John R. Brockmann and Fern Rook. Washington: Society for Technical Communication, 1989: 29–44.

Littlejohn, Stephen W., and David M. Jabusch. *Persuasive Transactions*. Glenview, IL: Scott, 1987.

Mokhiber, Russell. "Crime in the Suites." *Greenpeace* 14.5 (1989): 14–16.

"On Line." *Chronicle of Higher Education* 21 Sept. 1992, sec. A: 1.

Ornatowski, Cezar M. "Between Efficiency and Politics: Rhetoric and Ethics in Technical Writing." *Technical Communication Quarterly* 1.1 (1992): 91–103.

Pace, Roger C. "Technical Communication, Group Differentiation, and the Decision to Launch the Space Shuttle Challenger." *Journal of Technical Writing and Communication* 18.3 (1988): 207–20.

Porter, James E. "Truth in Technical Advertising: A Case Study." *IEEE Transactions on Professional Communication* 30.3 (1987): 182–89.

Presidential Commission, *Report to the President on the Space Shuttle Challenger Accident*. Vol. I. Washington: GPO, 1986.

Rowland, Robert C. "The Relationship between the Public and the Technical Spheres of Argument: A Case Study of the Challenger Seven Disaster." *Central States Speech Journal* 37.3 (1986): 136–46.

Rubens, Philip M, "Reinventing the Wheel?: Ethics for Technical Communicators." *Journal of Technical Writing and Communication* 11.4 (1981): 329–39.

Ruggiero, Vincent R. *The Art of Thinking*. 3rd ed. New York: Harper, 1991.

Unger, Stephen H. Controlling Technology: *Ethics and the Responsible Engineer*. New York: Holt, 1982.

Walter, Charles, and Thomas F. Marsteller. "Liability for the Dissemination of Defective Information." *IEEE Transactions on Professional Communication* 30.3 (1987): 164–67.

Wicclair, Mark R., and David K. Farkas. "Ethical Reasoning in Technical Communication: A Practical Framework." *Technical Communication* 31.2 (1984): 15–19.

Wickens, Christopher D. *Engineering Psychology and Human Performance*. 2nd ed. New York: Harper, 1992.

Winsor, D. A. "Communication Failures Contributing to the Challenger Accident: An Example for Technical Communicators." *IEEE Transactions on Professional Communication* 31.3 (1988): 101–07.

Yoos, George. "A Revision of the Concept of Ethical Appeal. *Philosophy and Rhetoric* 12.4 (Winter 1979): 41–58.

CHAPTER 6

Alred, Gerald J., Walter E. Oliu, and Charles T. Brusaw. *The Professional Writer: A Guide for Advanced Technical Writing.* New York: St. Martin's, 1992.

Beamer, Linda. "Learning Intercultural Communication Competence." *The Journal of Business Communication* 29.3 (1992): 285–303.

Beard, John D., and Jone Rymer. "The Contexts of Collaborative Writing." *The Bulletin of the Association for Business Communication* 53.2 (1990): 1–3.

Barfield, Woodrow, Mark Haselkorn, and Catherine Weatbrook. "Information Retrieval with a Printed User's Manual and with Online Hypercard Help." *Technical Communication* 37.1 (1990): 22–27.

Bernstein, Mark. "Deeply Intertwingled Hypertext: The Navigation Problem Reconsidered." *Technical Communication* 38.1 (1991): 41–47.

Bogert, Judith, and David Butt. "Opportunities Lost, Challenges Met: Understanding and Applying Group Dynamics in Writing Projects." *The Bulletin of the Association for Business Communication* 53.2 (1990): 51–53.

Brittan, David. "Being There: The Promise of Multimedia Communications." *Technology Review* May/June 1992: 43–50.

Burnett, Rebecca E. "Substantive Conflict in a Cooperative Context: A Way to Improve the Collaborative Planning of Workplace Documents." *Technical Communication* 38.4 (1991): 532–39.

Caswell-Coward, Nancy. "Cross-cultural Communication: Is It Greek to You?" *Technical Communication* 39.2 (1992): 264–66.

Debs, Mary Beth. "Collaborative Writing in Industry." *Technical Writing: Theory and Practice.* Ed. Bertie E. Fearing and W. Keats Sparrow. New York: Modern Language Assn., 1989. 33–42.

———. "Recent Research on Collaborative Writing in Industry." *Technical Communication* 38.4 (1991): 476–85.

Easton, Annette, et al. "Supporting Group Writing with Computer Software." *The Bulletin of the Association for Business Communication* 53.2 (1990): 34–37.

Finkelstein, Leo, Jr. "The Social Implications of Computer Technology for the Technical Writer." *Technical Communication* 38.4 (1991): 466–73.

Halal, William E. "The Information Technology Revolution." *The Futurist* July/Aug. 1992: 10–15.

Hein, Robert G. "Culture and Communication." *Technical Communication* 38.1 (1991): 125–26.

Hill-Duin, Ann. "Terms and Tools: A Theory and Research-Based Approach to Collaborative Writing." *The Bulletin of the Association for Business Communication* 53.2 (June 1990): 45–50.

Horton, William. "Is Hypertext the Best Way to Document Your Product?" *Technical Communication* 38.1 (1991): 20–30.

Kleimann, Susan D. "The Complexity of Workplace Review." *Technical Communication* 38.4 (1991): 520–26.

Lee, Jaesub, and Fredric M. Jablin. "A Cross-cultural Investigation of Exit Voice, Loyalty, and Neglect as Responses to Dissatisfying Job Conditions." *The Journal of Business Communication* 29.3 (1992): 203–28.

Leki, Ilona. "The Technical Editor and the Non-native Speaker of English." *Technical Communication* 37.2 (1990): 148–52.

Mackin, John. "Surmounting the Barrier between Japanese and English Technical Documents." *Technical Communication* 36.4 (1989): 346–51.

Martin, Jeanette S., and Lillian H. Chaney. "Determination of Content for a Collegiate Course in Intercultural Business Communication by Three Delphi Panels." *The Journal of Business Communication* 29.3 (1992): 267–83.

Morgan, Meg. "Patterns of Composing:" Connections between Classroom and Workplace Collaborations." *Technical Communication* 38.4 (1991): 540–42.

Morse, June. "Hypertext—What Can We Expect?" *STC Intercom* [Newsletter of the Society for Technical Communication] Feb. 1992: 6–7.

Nelson, Sandra J., and Douglas C. Smith. "Maximizing Cohesion and Minimizing Conflict in Collaborative Writing Groups." *The Bulletin of the Association for Business Communication* 53.2 (1990): 59–62.

Nickels-Shirk, "'Hyper' Rhetoric: Reflections on Teaching Hypertext." *The Technical Writing Teacher* 18.3 (1991): 189–200.

Rubens, Philip. "Reading and Employing Technical Information in Hypertext." *Technical Communication* 38.1 (1991): 36–40.

Scott, James C., and Diana J. Green. "British Perspectives on Organizing Bad-News Letters: Organizational Patterns Used by Major U.K. Companies. *The Bulletin of the Association for Business Communication* 55.1 (1992): 17–19.

Sturges, David L. "Internationalizing the Business Communication Curriculum." *The Bulletin of the Association for Business Communication* 55.1 (1992): 30–39.

Weiss, Julian M. "Adding Vision to Telecommuting." *The Futurist* May/June 1992: 16–18.

Weymouth, L. C. "Establishing Quality Standards and Trade Regulations for Technical Writing in World Trade." *Technical Communication* 37.2 (1990): 143–47.

Wilson, David L. "Huge Computer Network Quickens Pace of Academic Exchange and Collaboration." *Chronicle of Higher Education* 30 Oct. 1992, sec. A: 1+.

CHAPTER 8

Adams, Gerald R., and Jay D. Schvaneveldt. *Understanding Research Methods.* New York: Longman, 1985.

Burghardt, M. David. *Introduction to the Engineering Profession*. New York: Harper, 1991.

Garfield, Eugene. "What Scientific Journals Can Tell Us about Scientific Journals." *IEEE Transactions on Professional Communication* 16.4 (1973): 200–02.

Hayakawa, S. I. *Language in Thought and Action*. 3rd ed. New York: Harcourt, 1972.

Lavin, Michael R. *Business Information: How to Find It, How to Use It*. 2nd ed. Phoenix: Oryx, 1992.

McDonald, Kim A. "Some Physicists Criticize Research Purporting to Show Links between Low-level Electromagnetic Fields and Cancer." *Chronicle of Higher Education* 38 May 1991, sec. A: 5 +.

"On Line." *Chronicle of Higher Education* 14 Oct. 1992, sec. A: 1.

Pugliano, Fiore. Unpublished review of *Technical Writing*. 5th ed.

Schenk, Margaret T., and James K. Webster. *Engineering Information Resources*. New York: Decker, 1984.

Velotta, Christopher. "How to Design and Implement a Questionnaire." *Technical Communication* 38.3 (1991): 387–92.

Watkins, Beverly T. "Many Campuses Start Building Tomorrow's Electronic Library." *Chronicle of Higher Education* 2 Sep. 1992, sec. A: 1 +.

CHAPTER 9

Adams, Gerald R., and Jay D. Schvaneveldt. *Understanding Research Methods*. New York: Longman, 1985.

Fineman, Howard. "The Power of Talk." *Newsweek* 8 Feb. 1993: 24–28.

Gibaldi, Joseph, and Walter S. Achtert. *MLA Handbook for Writers of Research Papers*. 3rd ed. New York: Modern Language Assn., 1988.

Huff, Darrell. *How to Lie with Statistics*. New York: Norton, 1954.

Lavin, Michael R. *Business Information: How to Find It, How to Use It*. 2nd ed. Phoenix: Oryx, 1992.

Weinstein, Edith K. Unpublished review of *Technical Writing*. 5th ed.

Ruggiero, Vincent R. *The Art of Thinking*. 3rd ed. New York: Harper, 1991.

CHAPTER 10

Vaughan, David K. "Abstracts and Summaries: Some Clarifying Distinctions." *The Technical Writing Teacher* 18.2 (1991): 132–41.

CHAPTER 11

Felker, Daniel B., et al. *Guidelines for Document Designers*. Washington: American Institutes for Research, 1981.

Halpern, Jean W. "An Electronic Odyssey." *Writing in Nonacademic Settings*. Ed. Dixie Goswami and Lee Odell. New York: Guilford, 1985. 157–201.

Redish, Janice C., et al. "Making Information Accessible to Readers." *Writing in Nonacademic Settings*. Ed. Dixie Goswami and Lee Odell. New York: Guilford, 1985. 129–53.

Roundy, N., and D. Mair, "The Composing Process of Technical Writers: A Preliminary Study." *Journal of Advanced Composition* 3 (1982): 89–101.

CHAPTER 12

Consumer Product Safety Commission. *Fact Sheet No. 65*. Washington: GPO, 1979.

Kidder, Tracy. *The Soul of a New Machine*. Boston: Little, 1981.

U.S. Air Force Academy. *Executive Writing Course*. Washington: GPO, 1981.

CHAPTER 13

Bailey, Edward P. *Writing Clearly: A Contemporary Approach*. Columbus, OH: Merrill, 1984.

Cross, Mary. "Aristotle and Business Writing: Why We Need to Teach Persuasion." *The Bulletin of the Association for Business Communication* 54.1 (March 1991): 3–6.

Felker, Daniel B., et al. *Guidelines for Document Designers*. Washington: American Institutes for Research, 1981.

Kremers, Marshall. "Teaching Ethical Thinking in a Technical Writing Course." *IEEE Transactions on Professional Communication* 32.2 (1989): 58–61.

MacKenzie, Nancy. Unpublished review of *Technical Writing*. 5th ed.

Ornatowski, Cezar M. "Between Efficiency and Politics: Rhetoric and Ethics in Technical Writing." *Technical Communication Quarterly* 1.1 (1992): 91–103.

Pugliano, Fiore. Unpublished review of *Technical Writing*. 5th ed.

Williams, Joseph. *Style*. Glenview, IL: Scott, 1981.

Zinsser, William. *On Writing Well*. New York: Harper, 1980.

CHAPTER 14

The Aldus Guide to Basic Design. Aldus Corporation, 1988.

Callahan, Sean. "Eye Tech." *Forbes ASAP* June 7, 1993: 57+.

Cochran, Jeffrey K., et al. "Guidelines for Evaluating Graphical Designs." *Technical Communication* 36.1 (1989): 25–32.

Dragga, Sam, and Gwendolyn Gong. *Editing: The Design of Rhetoric.* Amityville, NY: Baywood, 1989.

Felker, Daniel B., et al. *Guidelines for Document Designers.* Washington: American Institutes for Research, 1981.

Girill, T. R. "Technical Communication and Art." *Technical Communication* 31.2 (1984): 35.

Hartley, James. *Designing Instructional Text.* 2nd ed. London: Kogan Page, 1985.

Journet, Debra. Unpublished review of *Technical Writing.* 3rd ed.

Lambert, Steve. *Presentation Graphics on the Apple® Macintosh.* Bellevue, WA: Microsoft, 1984.

Murch, Gerald R. "Using Color Effectively: Designing to Human Specifications." *Technical Communication* 32.4 (1985): 14–20.

Schmeupe, Ken D. "Upgrading Your Business Graphics." *Popular Computing* Dec. 1985: 51–56.

"Using Icons As Communication." *Simply Stated* [Newsletter of the Document Design Center, American Institutes for Research] 75 (Sept./Oct. 1987): 1+.

Van Pelt, William. Unpublished review of *Technical Writing.* 3rd ed.

Wickens, Christopher D. *Engineering Psychology and Human Performance.* 2nd ed. New York: Harper, 1992.

Williams, Robert I. "Playing with Format, Style, and Reader Assumptions." *Technical Communication* 30.3 (1983): 11–13.

CHAPTER 15

The Aldus Guide to Basic Design. Aldus Corporation, 1988.

Benson, Phillipa J. "Visual Design Considerations in Technical Publications." *Technical Communication* 32.4 (1985): 35–39.

Felker, Daniel B., et al. *Guidelines for Document Designers.* Washington: American Institutes for Research, 1981.

Hartley, James. *Designing Instructional Text.* 2nd ed. London: Kogan Page, 1985.

Kirsh, Lawrence. "Take It from the Top." *MacWorld* Apr. 1986: 112–15.

Pinelli, Thomas E., et al. "A Survey of Typography, Graphic Design, and Physical Media in Technical Reports." *Technical Communication* 33.2 (1986): 75–80.

Redish, Janice C., et al. "Making Information Accessible to Readers." *Writing in Nonacademic Settings*. Ed. Lee Odell and Dixie Goswami. New York: Guilford, 1985.

White, Jan. *Visual Design for the Electronic Age*. New York: Watson-Guptill, 1988.

Wight, Eleanor. "How Creativity Turns Facts into Usable Information." *Technical Communication* 32.1 (1985): 9–12.

Williams, Robert I. "Playing with Format, Style, and Reader Assumptions." *Technical Communication* 30.3 (1983): 11–13.

CHAPTER 17

"Earthquake Hazard Analysis for Nuclear Power Plants." *Energy and Technology Review* June 1984: 8.

Gartaganis, Arthur. "Lasers." *Occupational Outlook Quarterly* Winter 1984: 22–26.

Horton, William. "Is Hypertext the Best Way to Document Your Product?" *Technical Communication* 38.1 (1991): 20–30.

Stanton, Mike. "Fiber Optics." *Occupational Outlook Quarterly* Winter 1984: 27–30.

CHAPTER 18

Coletta, W. John. "The Ideologically Biased Use of Language in Scientific and Technical Writing." *Technical Communication Quarterly* 1.1 (1992): 59–70.

Glidden, H. K. *Reports, Technical Writing, and Specifications*. New York: McGraw, 1964.

MacKenzie, Nancy. Unpublished review of *Technical Writing*. 5th ed.

Riney, Larry A. *Technical Writing for Industry*. Englewood Cliffs: Prentice, 1989.

CHAPTER 19

Bedford, Marilyn S., and F. Cole Stearns. "The Technical Writer's Responsibility for Safety." *IEEE Transactions on Professional Communication* 30.3 (1987): 127–32.

Boiarsky, Carolyn. "Using Usability Testing to Teach Reader Response." *Technical Communication* 39.1 (1992): 100–02.

Clement, David E. "Human Factors, Instructions and Warnings, and Product Liability." *IEEE Transactions on Professional Communication* 30.3 (1987): 149–56.

Girill, T. R. "Technical Communication and Law." *Technical Communication* 32.3 (1985): 37.

Haynes, Kathleen J. M., and Linda K. Robertson. "An Application of Usability Criteria in the Classroom." *The Technical Writing Teacher* 18.3 (1991): 236–42.

Meyer, Benjamin D. "The ABCs of New-Look Publications." *Technical Communication* 33.1 (1986): 16–20.

Mirel, Barbara, Susan Feinberg, and Leif Allmendinger. "Designing Manuals for Active Learning Styles." *Technical Communication* 38.1 (1991): 75–87.

Redish, Janice C., and David A. Schell. "Writing and Testing Instructions for Usability." *Technical Writing: Theory and Practice.* Ed. Bertie E. Fearing and W. Keats Sparrow. New York: Modern Language Assn., 1989. 61–71.

Ruhs, Michael A. "Usability Testing: A Definition Analyzed." *Boston Broadside* [Newsletter of the Society for Technical Communication] May/June 1992: 8 + .

Spyridakis, Jan H., and Michael J. Wenger. "Writing for Human Performance: Relating Reading Research to Document Design." *Technical Communication* 39.2 (1992): 202–15.

Van Pelt, William. Unpublished review of *Technical Writing.* 3rd ed.

Walter, Charles, and Thomas F. Marsteller. "Liability for the Dissemination of Defective Information." *IEEE Transactions on Professional Communication* 30.3 (1987): 164–67.

Weiss, Edmond H. *How to Write a Usable User Manual.* Philadelphia: ISI, 1985.

Wickens, Christopher D. *Engineering Psychology and Human Performance.* 2nd ed. New York: Harper, 1992.

CHAPTER 24

Goodall, H. Lloyd, Jr., and Christopher L. Waagen. *The Persuasive Presentation.* New York: Harper, 1986.

Index

Abbreviations, 462, 468, 480, 626–629
Abstracts. *See also* Summaries
  bound, 149–150, 151, 157
  descriptive, 215, 216, 217
  from database retrieval services, 153, 154, 155, 156, 157
  of government publications, 151
  informative, 24, 215, 216, 217, 374, 376, 540, 659
  placement in reports, 217
Abstract words, 287
Absolutism, as an ethical fallacy, 72. *See also* Ethical relativism
Access points, for library research, 138, 139, 140
Acronyms, 282–283, 389. *See also* Jargon
Action verbs, 255, 267
  for instructions, 256, 432, 435, 436, 444
  in a job application letter, 487
  in a resume, 481
Active voice, 256, 257, 258, 259, 608
  for conversational tone, 295
  for instructions, 434
  for job-application letters, 487
Adaptation problem faced by writers, 10, 14, 88–89, 111, 114
Agreement
  pronoun-referent, 298, 601–602
  subject-verb, 600–601
Almanacs, as research tools, 144
Ambiguity, avoidance of
  in instructions, 434
  in pronoun reference, 602
  in research findings, 182
  in survey questions, 165
  in visuals, 338
  in writing style, 255, 280, 286, 293, 294
Ampersand, 201
Analogy, 288–289, 392

Analysis of parts, in definition, 391
Analysis of samples, in research, 170
Analytical reports
  audience-and-use profile for, 568–569
  audience questions about, 551–553
  definition of, 550
  elements of, 553–557
  outline for, 557–558
  purpose of, 551
  purpose statement for, 553, 554
  reasoning in (*see* Evidence)
  recommendations in, 555, 556
  revision checklist for, 579–580
  supplements to, 567–568
  typical analytical problems, 551–553
Annotated bibliography, 217
Antecedent, of a pronoun, 601, 602, 620
APA documentation style, 189, 196–197, 200, 201, 568, 570–578
Apostrophe, 616–618
Appeals, in persuasion, 47, 48–49, 51, 52, 53, 54, 55
Appendix, 23–24, 169, 377–379, 398, 516, 524, 554. *See also* Supplements to documents
Application letters, for employment, 477, 483, 486–490
Appositive, 614
Asterisks, uses for, 358, 437
Attention line, in letters, 466, 475
Audience (readers)
  definition of, 18
  for a highly technical message, 20–21, 23
  needs, 5, 25, 26, 27, 28
  for a non-technical message, 22–23, 24
  political realities, 35–37, 49
  primary versus secondary, 23–24
  for a semitechnical message, 21–22
Audience analysis, human factors in, 439–440

Comma, 253, 595
Comma splice, 596–597
Communication
  cost of, 5
  effect of, 4, 5
Communication abuses, types of, 78–79
Compact disk, information systems on,
    152–153. *See also* Electronic
    information services
Comparison/contrast
  as analysis, 551, 553, 579
  block structure in, 248
  in definitions, 392–393
  outline sequence, 229
  in paragraph development, 247–248
  point-by-point structure in, 248
  versus analogy, 288
Complaint letters. *See* Claim letters
Compliance, as a response to persuasion,
    37, 38
Complimentary close, in a letter, 465
Compound modifiers, 622
Computer documentation, 428, 438, 634,
    643, 644. *See also* Online
    documentation
Computer graphics, 93, 329–336
Computerized formatting. *See* Automated
    formatting
Computerized information retrieval. *See*
    Electronic information services
Computer's effects on communication, 6,
    7, 14, 88–96
Computer's limitations, for writers, 6
Conceptual knowledge, versus declara-
    tive and procedural knowledge, 438,
    439
Conciseness
  in oral reports, 590–591
  in a paraphrase, 179
  in sentences, 263–271
  in summaries, 208, 209
  in word choice, 286
Conclusions drawn from research,
    assessing the legitimacy of, 137, 165,
    181, 182, 184, 185, 187, 555, 557,
    566
Concrete and specific language, 286–288
Conference proceedings, indexes to, 149
Conjunctive (connecting) adverb, 596,
    610
Connecting with readers, in persuasion,
    38–41, 49–51. *See also* Persuasion
    problem faced by writers
Contingencies, statement of, in proposals,
    525, 529, 530

Constraints, as a measurement of
    usability, 439, 440
Constraints in a persuasive situation, 42–
    46
Contractions, 293–294
Contradictions in research findings, 137,
    555, 557
Conversational tone. *See* Informal tone
Coordinating conjunctions, 597, 611
Coordination in sentences, 275, 597–598,
    599
Copyright laws, 77, 188
Copyrighted software, legal protection for,
    77
Correction symbols for grammar, usage,
    and mechanics, 593
Correspondence. *See* Letters; Short reports
Cover letter, 165, 166–167. *See also* Letter
    of transmittal
Covers, for reports, 368
Critical thinking. *See also* Writing process,
    the
  for ethical decisions, 71–74
  in making recommendations, 501, 555,
    556, 557
  in preparing analytical reports, 550
  in the research process, 130–137, 175,
    180, 187
  role of visuals in, 303
  in the summary process, 209
  in technical writing, 15–17
  in the writing process, 111
Critical tone, avoiding a, 295
Cultural differences, among global
    audiences, 103–105, 107
Cultural expectations among readers,
    differences in, 104–105
Cultural fluency, 106
Cultural preferences, in page design, 357
Cultural stereotypes in audience analysis,
    limitations of, 106, 107
Culture, definition of, 103–104

Dangling modifiers, 254
Dashes, 358
Data. *See also* Evidence
  accuracy of, 181
  adequacy of, 180, 181
  choosing reliable sources of, 180, 181
  gathering of (*see* Research)
  interpretation of, 135–136, 181, 187,
    232
  recording of, (*see* Research)
Database retrieval services. *See* Electronic
    information services

682 INDEX

through visuals, 303, 331, 337, 339
through word arrangement, 255, 256
Employment reference, letters of, 481–
482
Enclosure notation, in a letter, 467
Encyclopedias, as research tools, 143
End matter, 368
Endnotes, 189, 190, 368, 379
End punctuation, 609
Ethical considerations
  in arranging employment interviews,
    491
  in citing and documenting sources, 178,
    188, 668
  in citing statistics, 187
  in composing survey questions, 166
  in definitions, 383
  in formatting, 355
  in global communication, 108
  in intercultural documents, 109
  in making recommendations, 556
  in memoranda, 499, 508
  in organizing a document, 377, 472
  in paraphrasing, 178
  in product and mechanism descriptions,
    411
  in proposals, 525, 526
  in a resume, 480
  in reporting on research, 560
  in specifications, 407
  in visuals, 304, 311, 317, 329, 338, 341
  in writing style, 257, 279, 293, 298
Ethical constraints for communicators, 45
Ethical decisions and choices, 74–75
Ethical dilemmas, 74
Ethical employees, 82
Ethical fallacies, 71–72
Ethical judgment, criteria for, 72–74, 75
Ethical relativism, 71–72. *See also*
  Absolutism
Ethical stand, taking an, 80–82
Ethical versus legal standards, 76–78
Ethics checklist for communicators, 82–83
Ethics guidelines for communicators, 79–
  80
Ethics problem faced by writers, 10, 13,
  14, 62, 63
Etymology, for definitions, 388–389
Euphemism, 283–284, 389
Evaluation of proposals, criteria for, 516,
  523
Evaluation of research findings, 134–135,
  180–187, 555
Evidence. *See also* Data
  evaluation of, 135, 181

in persuasion, 47–49
reasoning from, 181, 183, 184
Examples, to enhance communication
  in definitions, 393–394, 395, 397
  in instructions, 430, 432
  in persuasion, 47–48
Exclamation point, 609
Executive summaries, 216
Expanded definition, 385, 387–398
Experiment, as a form of primary
  research, 170
Expert knowledge, interviews as sources
  of, 158
Expert opinion, limitations of, 132–133,
  158, 181
Expert references, in support of a claim,
  188
Expert testimony, in persuasion, 48
Exploded diagram, 325–326

Factual databases, 153. *See also* Electronic
  information services
Fallacies
  in causal reasoning, 183–84
  in statistical reasoning, 185–187
  in ethical reasoning, 71–72
FAX networks, 92
Feasibility
  analysis of, 552, 553, 570–578, 638
  of a proposal, 529, 534, 539, 545
Faulty causal reasoning, 183–184
Flow chart for the writing process, 113
Flow charts, 113, 136, 321, 322
Fluency, of sentences, 251, 274–277
Footnotes, use of, 189, 190. *See also*
  Documentation
Formal reports. *See* Analytical reports
Formal tone, 292, 293, 295
Formal topic outline, 225, 226
Format
  audience considerations for designing,
    364
  automated formatting, 350
  definition of, 4, 346
  differing cultural expectations about,
    105
  guidelines, 350–363
  revision checklist, 364–365
Formats for various documents
  for instructions, 437
  for letters, 467–468
  for memos, 498, 499
  for proposals, 523–524
Fragment, of a sentence, 592, 594–596
Fraudulent advertising, laws against, 77

Correction Symbols

Symbol	Meaning	Page	Symbol	Meaning	Page
ab	abbreviation	626	– – /	dashes	621
agr p	pronoun/referent agreement	601	. . . /	ellipses	619
			! /	exclamation point	609
agr sv	subject/verb agreement	600	– /	hyphen	621
amb	ambiguity	251	*ital*	italics	619
appr	inappropriate diction	292	( ) /	parentheses	620
bias	biased tone	295	. /	period	609
ca	pronoun case	602	? /	question mark	609
cap	capitalization	629	" / "	quotation	618
chop	choppy sentences	274	; /	semicolon	609
cl	clutter word	270	*qual*	needless qualifier	271
coh	paragraph coherence	243	*red*	redundancy	265
cont	contraction	617	*rep*	needless repetition	265
coord	coordination	597	*ref*	faulty reference	602
cs	comma splice	596	*ro*	run-on sentence	597
dgl	dangling modifier	604	*seq*	sequence of development in a paragraph	244
euph	euphemism	283	*sexist*	sexist usage	297
exact	inexact word	285	*shift*	sentence shift	607
frag	sentence fragment	592	*st mod*	stacked modifiers	254
gen	generalization	284	*str*	paragraph structure	238
jarg	needless jargon	281	*sub*	subordination	598
len	paragraph length	248	*th op*	"th" sentence openers	266
lev	level of technicality	20	*trans*	transition	623
mng	meaning unclear	251	*trite*	triteness	283
mod	misplaced modifier	604	*un*	paragraph unity	242
noun ad	noun addiction	268	*v*	voice	256
om	omitted word	253	*var*	sentence variety	276
over	overstatement	284	*w*	wordiness	263
par	parallelism	606	*wo*	word order	604
pct	punctuation	608	*ww*	wrong word	285
ap/	apostrophe	616	#	numbers	630
[ ]/	brackets	620	¶	begin new paragraph	238
: /	colon	610	*ts*	topic sentence	239
, /	comma	611			